Student's Solutions Manual

Jeffery A. Cole
Anoka-Ramsey Community College

EIGHTH EDITION

SWOKOWSKI

COLE

ALGEBRA AND TRIGONOMETRY
With Analytic Geometry

PWS Publishing Company
Boston

710 BOOK STORE
Top CASH For BOOKS Anytime

"YOURS — FOR LOWER COSTS
OF HIGHER EDUCATION"

PWS PUBLISHING COMPANY

20 Park Plaza, Boston, MA 02116-4324

International Thomson Publishing
The trademark ITP is used under license.

PWS Publishing Company is a division of Wadsworth, Inc.

ISBN 0-534-93193-6

Printer: Malloy Lithographing

Printed in the United States of America

3 4 5 6 7 8 9 10 - - 98 97 96 95 94

PREFACE

This *Student's Solutions Manual* contains selected solutions and strategies for solving typical exercises in the text, *Algebra and Trigonometry with Analytic Geometry, Eighth Edition*, by Earl W. Swokowski and Jeffery A. Cole. No particular problem pattern has been followed in this manual, but the emphasis has been placed on the solutions of the applied "word" problems. I have tried to illustrate enough solutions so that the student will be able to obtain an understanding of all types of problems in each section.

A significant number of today's students are involved in various outside activities, and find it difficult, if not impossible, to attend all class sessions. This manual should help meet the needs of these students. In addition, it is my hope that this manual's solutions will enhance the understanding of all readers of the material and provide insights to solving other exercises.

I would appreciate any feedback concerning errors, solution correctness, solution style, or manual style. These and any other comments may be sent directly to me at the address below or in care of the publisher.

I would like to thank: Editor Tim Anderson, of PWS–KENT Publishing Company, for his guidance and continued support; Gary Rockswold, of Mankato State University, for supplying solutions for many of the new applied problems and calculator exercises; Thomas Vanden Eynden, of Thomas More College, for doing a detailed accuracy check of the exercises; George Morris, of Scientific Illustrators, for creating the mathematically precise art package; and Sally Lifland, Stacy Hinck, and Quica Ostrander, of Lifland et al. Bookmakers, for assembling the final manuscript. I dedicate this book to my children, Becky and Brad.

Jeffery A. Cole

Anoka-Ramsey Community College

11200 Mississippi Blvd. NW

Coon Rapids, MN 55433

Table of Contents

To the Student

This manual is a text supplement and should be read along *with* the text. Read all exercise solutions in this manual since explanations of concepts are given and then appear in subsequent solutions. All concepts necessary to solve a particular problem are not reviewed for every exercise. If you are having difficulty with a previously covered concept, look back to the section where it was covered for more complete help. The writing style I have used in this manual reflects the way I explain concepts to my own students. It is not as mathematically precise as that of the text, including phrases such as "goes down" or "touches and turns around." My students have told me that these terms help them understand difficult concepts with ease.

The most common complaint about solutions manuals that I receive from my students is that there are not enough exercise solutions in them. I believe there is a sufficient number of solutions in this manual, with about one-third of the exercises solved in every section—most of them are odd-numbered exercises.

Lengthier explanations and more steps are given for the more difficult problems. Additional information that my students have found helpful is included—see page 22. The guidelines given in the text are followed for some solutions—see page 87.

In the review sections, the solutions are somewhat abbreviated since more detailed solutions were given in previous sections. However, this is not true for the word problems in these sections since they are unique. In easier groups of exercises, representative solutions are shown. Occasionally, alternate solutions are also given.

All figures are new for this edition. They have been plotted using computer software, offering a high degree of precision. The calculator graphs are from the TI-81 screen, and any specific instructions are for the TI-81. When possible, each piece of art was made with the same scale to show a realistic and consistent graph.

This manual was done using **EXP**: *The Scientific Word Processor*. I have used a variety of display formats for the mathematical equations, including centering, vertical alignment, and flushing text to the right. I hope that these make reading and comprehending the material easier for you.

Notations

The following notations are used in the manual.

Note:	{ Notes to the student pertaining to hints on solutions, common mistakes, or conventions to follow. }
{ }	{ comments to the reader are in braces }
LS	{ Left Side of an equation }
RS	{ Right Side of an equation }
$\approx$	{ approximately equal to }
$\Rightarrow$	{ implies, next equation, logically follows }
$\Leftrightarrow$	{ if and only if, is equivalent to }
$\bullet$	{ bullet, used to separate problem statement from solution or explanation }
$\star$	{ used to identify the answer to the problem }
§	{ *section* references }
$\forall$	{ For all, i.e., $\forall x$ means "for all x". }
$\mathbb{R} - \{a\}$	{ The set of all real numbers except a. }
$\therefore$	{ therefore }
QI–QIV	{ quadrants I, II, III, IV }

Chapter 1: Fundamental Concepts of Algebra

1 (a) Since x and y have opposite signs, the product xy is negative.

   (b) Since $x^2 > 0$ and $y > 0$, $x^2 y > 0$.

   (c) Since $x < 0$ { x is negative } and $y > 0$ { y is positive }, $\frac{x}{y}$ is negative.

   Thus, $\frac{x}{y} + x$ is the sum of two negatives, which is *negative*.

   (d) Since $y > 0$ and $x < 0$, $y - x > 0$.

3 (a) Since -7 is to the left of -4 on a coordinate line, $-7\boxed{<}-4$.

   (b) Using a calculator, we see that $\frac{\pi}{2} \approx 1.5708$. Hence, $\frac{\pi}{2}\boxed{>}1.57$.

   (c) $\sqrt{225}\boxed{=}15$ *Note:* $\sqrt{225} \neq \pm 15$

Note: An informal definition of absolute value that may be helpful is

$$|\, something \,| = \begin{cases} itself & \text{if } itself \text{ is positive or zero} \\ -(itself) & \text{if } itself \text{ is negative} \end{cases}$$

9 (a) $|-3-2| = |-5| = -(-5) \; \{\text{since } -5 < 0\} = 5$

   (b) $|-5| - |2| = -(-5) - 2 = 5 - 2 = 3$

   (c) $|7| + |-4| = 7 + [-(-4)] = 7 + 4 = 11$

13 (a) Since $(4 - \pi)$ is positive, $|4 - \pi| = 4 - \pi$.

   (b) Since $(\pi - 4)$ is negative, $|\pi - 4| = -(\pi - 4) = 4 - \pi$.

   (c) Since $(\sqrt{2} - 1.5)$ is negative, $|\sqrt{2} - 1.5| = -(\sqrt{2} - 1.5) = 1.5 - \sqrt{2}$.

17 (a) $d(A, B) = |1 - (-9)| = |10| = 10$ (b) $d(B, C) = |10 - 1| = |9| = 9$

   (c) $d(C, B) = d(B, C) = 9$ (d) $d(A, C) = |10 - (-9)| = |19| = 19$

Note: Exer. 19–24: Since $|a| = |-a|$, the answers could have a different form.

   For example, $|-3 - x| \geq 8$ is equivalent to $|x + 3| \geq 8$.

21 $d(A, B) = |-3 - x| \Rightarrow |-3 - x| \geq 8$

25 Pick an arbitrary value for x that is less than -3, say -5.

   Since $3 + (-5) = -2$ is negative, we conclude that if $x < -3$, then $3 + x$ is negative.

   Hence, $|3 + x| = -(3 + x) = -x - 3$.

29 If $a < b$, then $a - b < 0$, and $|a - b| = -(a - b) = b - a$.

31 Since $x^2 + 4 > 0$ for every x, $|x^2 + 4| = x^2 + 4$.

33 LS $= \frac{ab + ac}{a} = \frac{ab}{a} + \frac{ac}{a} = b + c \boxed{\neq} $ RS $(b + ac)$.

37 LS $= (a \div b) \div c = \frac{a}{b} \cdot \frac{1}{c} = \frac{a}{bc}$. RS $= a \div (b \div c) = a \div \frac{b}{c} = a \cdot \frac{c}{b} = \frac{ac}{b}$. LS$\boxed{\neq}$RS

39 LS $= \frac{a - b}{b - a} = \frac{-(b - a)}{b - a} = -1\boxed{=}$RS.

45 (a) Since the decimal point is 5 places to the right of the first nonzero digit,
$$427{,}000 = 4.27 \times 10^5.$$

(b) Since the decimal point is 8 places to the left of the first nonzero digit,
$$0.000\,000\,098 = 9.8 \times 10^{-8}.$$

47 (a) Moving the decimal point 4 places to the left, we have $85{,}200 = 8.52 \times 10^4$.

(b) Moving the decimal point 6 places to the right, we have $0.000\,005\,5 = 5.5 \times 10^{-6}$.

51 It is helpful to write the units of any fraction, and then "cancel" those units to determine the units of the final answer.

$$\frac{186{,}000 \text{ miles}}{\text{second}} \cdot \frac{60 \text{ seconds}}{1 \text{ minute}} \cdot \frac{60 \text{ minutes}}{1 \text{ hour}} \cdot \frac{24 \text{ hours}}{1 \text{ day}} \cdot \frac{365 \text{ days}}{1 \text{ year}} \cdot 1 \text{ year} \approx 5.87 \times 10^{12} \text{ mi}$$

53 $\dfrac{\dfrac{1.01 \text{ grams}}{\text{mole}}}{\dfrac{6.02 \times 10^{23} \text{ atoms}}{\text{mole}}} \cdot 1 \text{ atom} = \dfrac{1.01 \text{ grams}}{6.02 \times 10^{23}} \approx 0.1678 \times 10^{-23} \text{ g} = 1.678 \times 10^{-24} \text{ g}$

57 (a) $\text{IQ} = \dfrac{\text{mental age (MA)}}{\text{chronological age (CA)}} \times 100 = \dfrac{15}{12} \times 100 = 125.$

(b) $\text{CA} = 15$ and $\text{IQ} = 140 \Rightarrow 140 = \text{MA}/15 \times 100 \Rightarrow \text{MA} = 21.$

1.2 Exercises

1 $\left(-\frac{2}{3}\right)^4 = \left(-\frac{2}{3}\right) \cdot \left(-\frac{2}{3}\right) \cdot \left(-\frac{2}{3}\right) \cdot \left(-\frac{2}{3}\right) = \frac{16}{81}$

Note: Do not confuse $(-x)^4$ and $-x^4$ since $(-x)^4 = x^4$ and $-x^4$ is the negative of x^4.

5 $-2^4 + 3^{-1} = -16 + \frac{1}{3} = -\frac{48}{3} + \frac{1}{3} = \frac{-47}{3}$

7 $16^{-3/4} = 1/16^{3/4} = 1/(\sqrt[4]{16})^3 = 1/2^3 = \frac{1}{8}$

9 $(-0.008)^{2/3} = (\sqrt[3]{-0.008})^2 = (-0.2)^2 = 0.04 = \frac{4}{100} = \frac{1}{25}$

13 A common mistake is to write $x^3 x^2 = x^6$, and another is to write $(x^2)^3 = x^5$.

The following solution illustrates the proper use of the exponent rules.

$$\frac{(2x^3)(3x^2)}{(x^2)^3} = \frac{(2 \cdot 3)x^{3+2}}{x^{2 \cdot 3}} = \frac{6x^5}{x^6} = 6x^{5-6} = 6x^{-1} = \frac{6}{x}$$

19 $(3u^7 v^3)(4u^4 v^{-5}) = 12u^{7+4} v^{3+(-5)} = 12u^{11} v^{-2} = \frac{12u^{11}}{v^2}$

23 $(\frac{1}{3}x^4 y^{-3})^{-2} = (\frac{1}{3})^{-2}(x^4)^{-2}(y^{-3})^{-2} = (\frac{3}{1})^2 x^{-8} y^6 = 3^2 x^{-8} y^6 = \frac{9y^6}{x^8}$

31 $\left(\dfrac{3x^5 y^4}{x^0 y^{-3}}\right)^2$ {remember that $x^0 = 1$} $= \dfrac{9x^{10} y^8}{y^{-6}} = 9x^{10} y^{14}$

35 $(3x^{5/6})(8x^{2/3}) = 24x^{(5/6)+(4/6)} = 24x^{9/6} = 24x^{3/2}$

41 $\left(\dfrac{-8x^3}{y^{-6}}\right)^{2/3} = \dfrac{(-8)^{2/3}(x^3)^{2/3}}{(y^{-6})^{2/3}} = \dfrac{(\sqrt[3]{-8})^2\, x^{(3)(2/3)}}{y^{(-6)(2/3)}} = \dfrac{(-2)^2 x^2}{y^{-4}} = \dfrac{4x^2}{y^{-4}} = 4x^2 y^4$

45 $\dfrac{(x^6 y^3)^{-1/3}}{(x^4 y^2)^{-1/2}} = \dfrac{(x^6)^{-1/3}(y^3)^{-1/3}}{(x^4)^{-1/2}(y^2)^{-1/2}} = \dfrac{x^{-2} y^{-1}}{x^{-2} y^{-1}} = 1$

49 $\sqrt[3]{(a+b)^2} = [(a+b)^2]^{1/3} = (a+b)^{2/3}$

53 (a) $4x^{3/2} = 4x^1 x^{1/2} = 4x\sqrt{x}$

     (b) $(4x)^{3/2} = (4x)^1 (4x)^{1/2} = (4x)^1\, 4^{1/2} x^{1/2} = 4x \cdot 2 \cdot x^{1/2} = 8x\sqrt{x}$

59 $\sqrt[5]{-64} = \sqrt[5]{-32}\,\sqrt[5]{2} = \sqrt[5]{(-2)^5}\,\sqrt[5]{2} = -2\sqrt[5]{2}$

63 $\sqrt{9x^{-4}y^6} = (9x^{-4}y^6)^{1/2} = 9^{1/2}(x^{-4})^{1/2}(y^6)^{1/2} = 3x^{-2}y^3 = \dfrac{3y^3}{x^2}$

Note: For exercises similar to those in 67–74, pick a multiplier that will make

     all of the exponents of the terms in the denominator a multiple of the index.

67 The index is 2. Choose the multiplier to be $\sqrt{2y}$ so that the denominator contains

only terms with even exponents. $\sqrt{\dfrac{3x}{2y^3}} = \sqrt{\dfrac{3x}{2y^3}} \cdot \dfrac{\sqrt{2y}}{\sqrt{2y}} = \sqrt{\dfrac{6xy}{4y^4}} = \dfrac{\sqrt{6xy}}{2y^2}$, or $\dfrac{1}{2y^2}\sqrt{6xy}$

69 The index is 3. Choose the multiplier to be $\sqrt[3]{3x^2}$ so that the denominator contains

only terms with exponents that are multiples of 3.

$$\sqrt[3]{\dfrac{2x^4y^4}{9x}} = \sqrt[3]{\dfrac{2x^4y^4}{9x}} \cdot \dfrac{\sqrt[3]{3x^2}}{\sqrt[3]{3x^2}} = \dfrac{\sqrt[3]{6x^6y^4}}{\sqrt[3]{27x^3}} = \dfrac{\sqrt[3]{x^6y^3}\,\sqrt[3]{6y}}{3x} = \dfrac{x^2 y\,\sqrt[3]{6y}}{3x} = \dfrac{xy}{3}\sqrt[3]{6y}$$

71 The index is 4. Choose the multiplier to be $\sqrt[4]{3x^2}$ so that the denominator contains

only terms with exponents that are multiples of 4.

$$\sqrt[4]{\dfrac{5x^8y^3}{27x^2}} = \sqrt[4]{\dfrac{5x^8y^3}{27x^2}} \cdot \dfrac{\sqrt[4]{3x^2}}{\sqrt[4]{3x^2}} = \dfrac{\sqrt[4]{15x^{10}y^3}}{\sqrt[4]{81x^4}} = \dfrac{\sqrt[4]{x^8}\,\sqrt[4]{15x^2y^3}}{3x} = \dfrac{x^2\,\sqrt[4]{15x^2y^3}}{3x}$$

$$= \dfrac{x}{3}\sqrt[4]{15x^2y^3}$$

75 $\sqrt[4]{(3x^5y^{-2})^4} = 3x^5y^{-2} = \dfrac{3x^5}{y^2}$

77 $\sqrt[5]{\dfrac{8x^3}{y^4}}\,\sqrt[5]{\dfrac{4x^4}{y^2}} = \sqrt[5]{\dfrac{8x^3}{y^4}}\,\sqrt[5]{\dfrac{4x^4}{y^2}} \cdot \dfrac{\sqrt[5]{y^4}}{\sqrt[5]{y^4}} = \dfrac{\sqrt[5]{32x^5}\,\sqrt[5]{x^2y^4}}{y^2} = \dfrac{2x}{y^2}\sqrt[5]{x^2y^4}$

79 $\sqrt[3]{3t^4v^2}\,\sqrt[3]{-9t^{-1}v^4} = \sqrt[3]{-27t^3v^6} = -3tv^2$

81 $\sqrt{x^6y^4} = \sqrt{(x^3)^2(y^2)^2} = \sqrt{(x^3)^2}\,\sqrt{(y^2)^2} = |x^3|\,|y^2| = |x^3|\,y^2$ since y^2 is always

     nonnegative. *Note:* $|x^3|$ could be written as $x^2\,|x|$.

83   $\sqrt[4]{x^5(y-1)^6} = \sqrt[4]{x^4(y-1)^4}\,\sqrt[4]{x(y-1)^2} = |\,x(y-1)\,|\,\sqrt[4]{x(y-1)^2}.$

Note: The answer could be written as $x\,|\,(y-1)\,|\,\sqrt[4]{x(y-1)^2}$ since x must be

nonnegative to make the original expression defined.

85   $(a^r)^2 = a^{2r}\boxed{\neq}a^{(r^2)}$ since $2r \neq r^2$ for every r.

93   $W = 230$ kg $\Rightarrow L = 0.46\sqrt[3]{W} = 0.46\sqrt[3]{230} \approx 2.82$ m

95   $b = 75$ and $w = 180 \Rightarrow W = \dfrac{w}{\sqrt[3]{b-35}} = \dfrac{180}{\sqrt[3]{75-35}} \approx 52.6.$

$b = 120$ and $w = 250 \Rightarrow W = \dfrac{w}{\sqrt[3]{b-35}} = \dfrac{250}{\sqrt[3]{120-35}} \approx 56.9.$

It is interesting to note that the 75-kg lifter can lift 2.4 times his/her body weight

and the 120-kg lifter can lift approximately 2.08 times his/her body weight, but

the formula ranks the 120-kg lifter as the superior lifter.

1.3 Exercises

5   $(2x+5)(3x-7) = (2x)(3x) + (2x)(-7) + (5)(3x) + (5)(-7)$

$= 6x^2 - 14x + 15x - 35$

$= 6x^2 + x - 35$

9   $(2u+3)(u-4) + 4u(u-2) = (2u^2 - 5u - 12) + (4u^2 - 8u) = 6u^2 - 13u - 12$

15   $(x+1)(2x^2-2)(x^3+5) = 2\,[(x+1)(x^2-1)](x^3+5)$

$= 2(x^3 + x^2 - x - 1)(x^3 + 5)$

$= 2(x^6 + x^5 - x^4 + 4x^3 + 5x^2 - 5x - 5)$

$= 2x^6 + 2x^5 - 2x^4 + 8x^3 + 10x^2 - 10x - 10$

19   $\dfrac{3u^3v^4 - 2u^5v^2 + (u^2v^2)^2}{u^3v^2} = \dfrac{3u^3v^4}{u^3v^2} - \dfrac{2u^5v^2}{u^3v^2} + \dfrac{u^4v^4}{u^3v^2} = 3v^2 - 2u^2 + uv^2$

21   We recognize this product as the difference of two squares.

$(2x+3y)(2x-3y) = (2x)^2 - (3y)^2 = 4x^2 - 9y^2$

25   $(x^2+9)(x^2-4) = x^4 - 4x^2 + 9x^2 - 36 = x^4 + 5x^2 - 36$

31   We could expand $(x+2)^2$ and $(x-2)^2$ and then multiply the resulting expressions,

but the following solution is simpler.

$(x+2)^2(x-2)^2 = [(x+2)(x-2)]^2$

$= (x^2 - 4)^2$

$= (x^2)^2 - 2(x^2)(4) + (4)^2$

$= x^4 - 8x^2 + 16$

35 $(x^{1/3} - y^{1/3})(x^{2/3} + x^{1/3}y^{1/3} + y^{2/3})$

$$= x^{1/3}(x^{2/3} + x^{1/3}y^{1/3} + y^{2/3}) - y^{1/3}(x^{2/3} + x^{1/3}y^{1/3} + y^{2/3})$$

$$= x + x^{2/3}y^{1/3} + x^{1/3}y^{2/3} - x^{2/3}y^{1/3} - x^{1/3}y^{2/3} - y = x - y$$

This exercise illustrates how the difference of *any* two terms can be factored as the difference of cubes. Another example of this concept is

$$x - 5 = (\sqrt[3]{x} - \sqrt[3]{5})(\sqrt[3]{x^2} + \sqrt[3]{5x} + \sqrt[3]{25}).$$

39 $(2x + 3y)^3 = (2x)^3 + 3(2x)^2(3y) + 3(2x)(3y)^2 + (3y)^3$

$$= 8x^3 + 3(4x^2)(3y) + 3(2x)(9y^2) + 27y^3$$

$$= 8x^3 + 36x^2y + 54xy^2 + 27y^3$$

Note: Exer. 41–44: Treat these as "the sum of the squares plus twice the product of all possible pairs of terms", that is, $(x + y + z)^2 = x^2 + y^2 + z^2 + 2xy + 2xz + 2yz$.

43 $(2x + y - 3z)^2 = (2x)^2 + (y)^2 + (-3z)^2 + 2(2x)(y) + 2(2x)(-3z) + 2(y)(-3z)$

$$= 4x^2 + y^2 + 9z^2 + 4xy - 12xz - 6yz$$

47 Always factor out the greatest common factor first. $3a^2b^2 - 6a^2b = 3a^2b(b - 2)$

51 $15x^3y^5 - 25x^4y^2 + 10x^6y^4 = 5x^3y^2(3y^3 - 5x + 2x^3y^2)$

53 We recognize this as a trinomial that may be able to be factored into the product of two binomials. $8x^2 - 53x - 21 = (8x + 3)(x - 7)$

55 The factors for $x^2 + 3x + 4$ would have to be of the form $(x + \_)$ and $(x + \_)$.

The factors of 4 are 1 & 4 and 2 & 2, but their sums are 5 and 4, respectively.

Thus, $x^2 + 3x + 4$ is *irreducible*.

61 $4x^2 - 20x + 25 = (2x - 5)(2x - 5) = (2x - 5)^2$

65 $45x^2 + 38xy + 8y^2 = (5x + 2y)(9x + 4y)$

69 $z^4 - 64w^2 = (z^2)^2 - (8w)^2 = (z^2 + 8w)(z^2 - 8w)$

71 $x^4 - 4x^2 = x^2(x^2 - 4) = x^2(x^2 - 2^2) = x^2(x + 2)(x - 2)$

73 $x^2 + 25$ is irreducible. *Note:* A common mistake is to confuse

the *sum* of two squares with the *difference* of two squares.

75 $75x^2 - 48y^2 = 3(25x^2 - 16y^2) = 3[(5x)^2 - (4y)^2] = 3(5x + 4y)(5x - 4y)$

79 We recognize this expression as the difference of two cubes.

$$64x^3 - y^6 = (4x)^3 - (y^2)^3$$

$$= (4x - y^2)[(4x)^2 + (4x)(y^2) + (y^2)^2]$$

$$= (4x - y^2)(16x^2 + 4xy^2 + y^4)$$

81 We recognize this expression as the sum of two cubes.

$$343x^3 + y^9 = (7x)^3 + (y^3)^3$$
$$= (7x + y^3)\,[(7x)^2 - (7x)(y^3) + (y^3)^2]$$
$$= (7x + y^3)(49x^2 - 7xy^3 + y^6)$$

87 Since there are more than 3 terms, we will try to factor by grouping first.

$$x^4 + 2x^3 - x - 2 = x^3(x + 2) - 1(x + 2) = (x^3 - 1)(x + 2) = (x - 1)(x + 2)(x^2 + x + 1)$$

91 We could treat this expression as the difference of two squares or the difference of two cubes. Factoring as the difference of two squares and then as the sum and difference of two cubes leads to the following:

$$a^6 - b^6 = (a^3)^2 - (b^3)^2$$
$$= (a^3 + b^3)(a^3 - b^3)$$
$$= (a + b)(a - b)(a^2 - ab + b^2)(a^2 + ab + b^2)$$

95 We might first try to factor this expression by grouping since it has more than 3 terms, but this would prove to be unsuccessful. Instead, we will group the terms containing y and the constant term together, and then proceed as in Example 10 part (c).

$$y^2 - x^2 + 8y + 16 = (y^2 + 8y + 16) - x^2$$
$$= (y + 4)^2 - (x)^2$$
$$= (y + 4 + x)(y + 4 - x)$$

97 We should first note that one of the two variable terms, y^6, is the square of the other, y^3. Thus, we may treat this expression as a simple trinomial that can be factored into the product of two binomials.

$$y^6 + 7y^3 - 8 = (y^3 + 8)(y^3 - 1) = (y + 2)(y^2 - 2y + 4)(y - 1)(y^2 + y + 1)$$

101 The dimensions of I are (x) and $(x - y)$. The area of I is $(x - y)x$, and the area of II is $(x - y)y$. The area $A = x^2 - y^2 = (x - y)x + (x - y)y = \underline{(x - y)(x + y)}$.

1.4 Exercises

3 $\dfrac{5}{24} - \dfrac{3}{20} = \dfrac{5}{2^3 \cdot 3} - \dfrac{3}{2^2 \cdot 5} = \dfrac{5 \cdot 5 - 3(2 \cdot 3)}{2^3 \cdot 3 \cdot 5} = \dfrac{25 - 18}{2^3 \cdot 3 \cdot 5} = \dfrac{7}{2^3 \cdot 3 \cdot 5} = \dfrac{7}{120}$

7 $\dfrac{y^2 - 25}{y^3 - 125} = \dfrac{(y + 5)(y - 5)}{(y - 5)(y^2 + 5y + 25)} = \dfrac{y + 5}{y^2 + 5y + 25}$

9 $\dfrac{12 + r - r^2}{r^3 + 3r^2} = \dfrac{(3 + r)(4 - r)}{r^2(r + 3)} = \dfrac{4 - r}{r^2}$

13 $\dfrac{5a^2 + 12a + 4}{a^4 - 16} \div \dfrac{25a^2 + 20a + 4}{a^2 - 2a} = \dfrac{(5a + 2)(a + 2)}{(a^2 + 4)(a + 2)(a - 2)} \cdot \dfrac{a(a - 2)}{(5a + 2)(5a + 2)} =$

$$\dfrac{a}{(a^2 + 4)(5a + 2)}$$

$\boxed{15}$ $\dfrac{6}{x^2-4}-\dfrac{3x}{x^2-4}=\dfrac{6-3x}{x^2-4}=\dfrac{3(2-x)}{(x+2)(x-2)}=\dfrac{-3}{x+2}.$　　　Since $2-x=-(x-2)$, we

canceled the two factors, $2-x$ and $x-2$, and replaced them with -1. In general,

you may do this whenever you encounter factors of the form $a-b$ and $b-a$ in the

numerator and the denominator, respectively, of a fractional expression.

$\boxed{19}$ $\dfrac{2}{x}+\dfrac{3x+1}{x^2}-\dfrac{x-2}{x^3}=\dfrac{2x^2+(3x+1)x-x+2}{x^3}=\dfrac{5x^2+2}{x^3}$

$\boxed{21}$ $\dfrac{3t}{t+2}+\dfrac{5t}{t-2}-\dfrac{40}{t^2-4}=\dfrac{3t}{t+2}+\dfrac{5t}{t-2}-\dfrac{40}{(t+2)(t-2)}$

$$=\dfrac{3t(t-2)}{(t+2)(t-2)}+\dfrac{5t(t+2)}{(t+2)(t-2)}-\dfrac{40}{(t+2)(t-2)}$$

$$=\dfrac{3t^2-6t+5t^2+10t-40}{(t+2)(t-2)}$$

$$=\dfrac{8t^2+4t-40}{(t+2)(t-2)}$$

$$=\dfrac{4(2t+5)(t-2)}{(t+2)(t-2)}=\dfrac{4(2t+5)}{t+2}$$

$\boxed{25}$ $\dfrac{2x}{x+2}-\dfrac{8}{x^2+2x}+\dfrac{3}{x}=\dfrac{2x(x)-8+3(x+2)}{x(x+2)}=\dfrac{2x^2+3x-2}{x(x+2)}=\dfrac{(2x-1)(x+2)}{x(x+2)}=\dfrac{2x-1}{x}$

$\boxed{27}$ $\dfrac{p^4+3p^3-8p-24}{p^3-2p^2-9p+18}=\dfrac{p^3(p+3)-8(p+3)}{p^2(p-2)-9(p-2)}=\dfrac{(p^3-8)(p+3)}{(p^2-9)(p-2)}=$

$$\dfrac{(p-2)(p^2+2p+4)(p+3)}{(p+3)(p-3)(p-2)}=\dfrac{p^2+2p+4}{p-3}$$

$\boxed{31}$ $\dfrac{2x+1}{x^2+4x+4}-\dfrac{6x}{x^2-4}+\dfrac{3}{x-2}=\dfrac{(2x+1)(x-2)-6x(x+2)+3(x^2+4x+4)}{(x+2)^2(x-2)}=$

$$\dfrac{-x^2-3x+10}{(x+2)^2(x-2)}=-\dfrac{x^2+3x-10}{(x+2)^2(x-2)}=-\dfrac{(x+5)(x-2)}{(x+2)^2(x-2)}=-\dfrac{x+5}{(x+2)^2}$$

$\boxed{33}$ The lcd of the entire expression is ab.

Thus, we will multiply both the numerator and denominator by ab.

$$\dfrac{\dfrac{b}{a}-\dfrac{a}{b}}{\dfrac{1}{a}-\dfrac{1}{b}}=\dfrac{\left(\dfrac{b}{a}-\dfrac{a}{b}\right)\cdot ab}{\left(\dfrac{1}{a}-\dfrac{1}{b}\right)\cdot ab}=\dfrac{b^2-a^2}{b-a}=\dfrac{(b+a)(b-a)}{b-a}=a+b$$

$\boxed{35}$ $\dfrac{\dfrac{x}{y^2}-\dfrac{y}{x^2}}{\dfrac{1}{y^2}-\dfrac{1}{x^2}}=\dfrac{\left(\dfrac{x}{y^2}-\dfrac{y}{x^2}\right)\cdot x^2y^2}{\left(\dfrac{1}{y^2}-\dfrac{1}{x^2}\right)\cdot x^2y^2}=\dfrac{x^3-y^3}{x^2-y^2}=\dfrac{(x-y)(x^2+xy+y^2)}{(x+y)(x-y)}=\dfrac{x^2+xy+y^2}{x+y}$

$\boxed{37}$ $\dfrac{\dfrac{5}{x+1}+\dfrac{2x}{x+3}}{\dfrac{x}{x+1}+\dfrac{7}{x+3}}=\dfrac{\dfrac{5(x+3)+2x(x+1)}{(x+1)(x+3)}}{\dfrac{x(x+3)+7(x+1)}{(x+1)(x+3)}}=\dfrac{5x+15+2x^2+2x}{x^2+3x+7x+7}=\dfrac{2x^2+7x+15}{x^2+10x+7}$

$\boxed{41}$ $\dfrac{\dfrac{1}{(x+h)^3}-\dfrac{1}{x^3}}{h}=\dfrac{\dfrac{x^3-(x+h)^3}{(x+h)^3x^3}}{h}=\dfrac{x^3-(x+h)^3}{hx^3(x+h)^3}=$

$\dfrac{[x-(x+h)][x^2+x(x+h)+(x+h)^2]}{hx^3(x+h)^3}=\dfrac{-h(3x^2+3xh+h^2)}{hx^3(x+h)^3}=-\dfrac{3x^2+3xh+h^2}{x^3(x+h)^3}$

$\boxed{43}$ $\dfrac{\dfrac{4}{3x+3h-1}-\dfrac{4}{3x-1}}{h}=\dfrac{\dfrac{4(3x-1)-4(3x+3h-1)}{(3x+3h-1)(3x-1)}}{h}=\dfrac{12x-4-12x-12h+4}{h(3x+3h-1)(3x-1)}=$

$\dfrac{-12h}{h(3x+3h-1)(3x-1)}=\dfrac{-12}{(3x+3h-1)(3x-1)}$

$\boxed{47}$ $\dfrac{81x^2-16y^2}{3\sqrt{x}-2\sqrt{y}}=\dfrac{81x^2-16y^2}{3\sqrt{x}-2\sqrt{y}}\cdot\dfrac{3\sqrt{x}+2\sqrt{y}}{3\sqrt{x}+2\sqrt{y}}$

$=\dfrac{(9x+4y)(9x-4y)(3\sqrt{x}+2\sqrt{y})}{9x-4y}=(9x+4y)(3\sqrt{x}+2\sqrt{y})$

$\boxed{49}$ $\dfrac{1}{\sqrt[3]{a}-\sqrt[3]{b}}=\dfrac{1}{\sqrt[3]{a}-\sqrt[3]{b}}\cdot\dfrac{\sqrt[3]{a^2}+\sqrt[3]{ab}+\sqrt[3]{b^2}}{\sqrt[3]{a^2}+\sqrt[3]{ab}+\sqrt[3]{b^2}}=\dfrac{\sqrt[3]{a^2}+\sqrt[3]{ab}+\sqrt[3]{b^2}}{a-b}$

$\boxed{53}$ $\dfrac{\sqrt{2(x+h)+1}-\sqrt{2x+1}}{h}=\dfrac{\sqrt{2(x+h)+1}-\sqrt{2x+1}}{h}\cdot\dfrac{\sqrt{2(x+h)+1}+\sqrt{2x+1}}{\sqrt{2(x+h)+1}+\sqrt{2x+1}}$

$=\dfrac{(2x+2h+1)-(2x+1)}{h(\sqrt{2(x+h)+1}+\sqrt{2x+1})}=\dfrac{2}{\sqrt{2(x+h)+1}+\sqrt{2x+1}}$

$\boxed{55}$ $\dfrac{\sqrt{1-x-h}-\sqrt{1-x}}{h}=\dfrac{\sqrt{1-x-h}-\sqrt{1-x}}{h}\cdot\dfrac{\sqrt{1-x-h}+\sqrt{1-x}}{\sqrt{1-x-h}+\sqrt{1-x}}$

$=\dfrac{(1-x-h)-(1-x)}{h(\sqrt{1-x-h}+\sqrt{1-x})}=\dfrac{-1}{\sqrt{1-x-h}+\sqrt{1-x}}$

$\boxed{59}$ $\dfrac{(x^2+2)^2}{x^5}=\dfrac{x^4+4x^2+4}{x^5}=\dfrac{x^4}{x^5}+\dfrac{4x^2}{x^5}+\dfrac{4}{x^5}=x^{-1}+4x^{-3}+4x^{-5}$

Note: Exercises 61–76 are worked using the factoring concept given as the third method

of simplification in Example 9.

$\boxed{61}$ The *smallest* exponent that appears on the factor x is -3.

$x^{-3}+x^2$ {factor out x^{-3}} $=x^{-3}(1+x^{2-(-3)})=x^{-3}(1+x^5)=\dfrac{1+x^5}{x^3}$

$\boxed{67}$ The smallest exponent that appears on the factor $(x^2 - 4)$ is $-\frac{1}{2}$ and the smallest exponent that appears on the factor $(2x + 1)$ is 2. Thus, we will factor out $(x^2 - 4)^{-1/2}(2x + 1)^2$.

$(x^2 - 4)^{1/2}(3)(2x + 1)^2(2) + (2x + 1)^3(\frac{1}{2})(x^2 - 4)^{-1/2}(2x)$

$\qquad = (x^2 - 4)^{-1/2}(2x + 1)^2[6(x^2 - 4) + x(2x + 1)]$

If you are unsure of this factoring, it is easy to visually check at this stage by merely multiplying the expression—that is, we mentally add the exponents on the factor $(x^2 - 4)$, $-\frac{1}{2}$ and 1, and we get $\frac{1}{2}$, which is the exponent we started with.

Proceeding, we may simplify as follows:

$$= \frac{(2x + 1)^2(8x^2 + x - 24)}{(x^2 - 4)^{1/2}}$$

$\boxed{69}$ $(3x + 1)^6(\frac{1}{2})(2x - 5)^{-1/2}(2) + (2x - 5)^{1/2}(6)(3x + 1)^5(3) =$

$\qquad (3x + 1)^5(2x - 5)^{-1/2}[(3x + 1) + 18(2x - 5)] = \dfrac{(3x + 1)^5(39x - 89)}{(2x - 5)^{1/2}}$

$\boxed{71}$ $\dfrac{(6x + 1)^3(27x^2 + 2) - (9x^3 + 2x)(3)(6x + 1)^2(6)}{(6x + 1)^6} =$

$\qquad \dfrac{(6x + 1)^2[(6x + 1)(27x^2 + 2) - 18(9x^3 + 2x)]}{(6x + 1)^6} = \dfrac{27x^2 - 24x + 2}{(6x + 1)^4}$

$\boxed{75}$ $\dfrac{(4x^2 + 9)^{1/2}(2) - (2x + 3)(\frac{1}{2})(4x^2 + 9)^{-1/2}(8x)}{[(4x^2 + 9)^{1/2}]^2} =$

$\qquad \dfrac{(4x^2 + 9)^{-1/2}[2(4x^2 + 9) - 4x(2x + 3)]}{(4x^2 + 9)} = \dfrac{18 - 12x}{(4x^2 + 9)^{3/2}} = \dfrac{6(3 - 2x)}{(4x^2 + 9)^{3/2}}$

Chapter 1 Review Exercises

$\boxed{4}$ (c) $|3^{-1} - 2^{-1}| = |\frac{1}{3} - \frac{1}{2}| = |\frac{2}{6} - \frac{3}{6}| = |-\frac{1}{6}| = -(-\frac{1}{6}) = \frac{1}{6}$

$\boxed{6}$ (a) $(x + y)^2 = x^2 + 2xy + y^2 \boxed{\neq} x^2 + y^2$ for every nonzero x and nonzero y.

(b) $\dfrac{1}{\sqrt{x + y}} = \dfrac{1}{\sqrt{x}} + \dfrac{1}{\sqrt{y}}$ is not true if $x = y = 1$ { we need only find one set of values of the variables for which the expression is false }.

(c) $\dfrac{1}{\sqrt{c} - \sqrt{d}} = \dfrac{1}{\sqrt{c} - \sqrt{d}} \cdot \dfrac{\sqrt{c} + \sqrt{d}}{\sqrt{c} + \sqrt{d}} \boxed{=} \dfrac{\sqrt{c} + \sqrt{d}}{c - d}$

$\boxed{10}$ If $2 < x < 3$, then $x - 2 > 0$ { $x - 2$ is positive } and $x - 3 < 0$ { $x - 3$ is negative }. Thus, $(x - 2)(x - 3) < 0$ { positive times negative is negative }, and since the absolute value of an expression that is negative is the negative of the expression,

$$|(x - 2)(x - 3)| = -(x - 2)(x - 3), \text{ or, equivalently, } (2 - x)(x - 3).$$

$\boxed{11}$ $-3^2 + 2^0 + 27^{-2/3} = -9 + 1 + \dfrac{1}{(\sqrt[3]{27})^2} = -8 + \dfrac{1}{3^2} = -\dfrac{72}{9} + \dfrac{1}{9} = \dfrac{-71}{9}$

$\boxed{15}$ $\dfrac{(3x^2 y^{-3})^{-2}}{x^{-5} y} = \dfrac{3^{-2} x^{-4} y^6}{x^{-5} y} = \dfrac{x^5 y^5}{3^2 x^4} = \dfrac{xy^5}{9}$

$\boxed{18}$ $c^{-4/3} c^{3/2} c^{1/6} = c^{-8/6} c^{9/6} c^{1/6} = c^{(-8+9+1)/6} = c^{2/6} = c^{1/3}$

$\boxed{21}$ $\left[(a^{2/3} b^{-2})^3 \right]^{-1} = (a^2 b^{-6})^{-1} = a^{-2} b^6 = \dfrac{b^6}{a^2}$

$\boxed{24}$ Do not expand $(u+v)^3$ since it can be combined with $(u+v)^{-2}$.

$$(u+v)^3 (u+v)^{-2} = (u+v)^1 = u+v$$

$\boxed{29}$ Since $\sqrt[3]{4} = \sqrt[3]{2^2}$, we need to multiply the numerator and the denominator by $\sqrt[3]{2}$ to

obtain a cube in the radicand of the denominator. $\dfrac{1}{\sqrt[3]{4}} = \dfrac{1}{\sqrt[3]{4}} \cdot \dfrac{\sqrt[3]{2}}{\sqrt[3]{2}} = \dfrac{\sqrt[3]{2}}{\sqrt[3]{8}} = \dfrac{1}{2} \sqrt[3]{2}$

$\boxed{32}$ $\sqrt[4]{(-4a^3 b^2 c)^2} = \sqrt[4]{16 a^6 b^4 c^2} = \sqrt[4]{2^4 a^4 b^4} \sqrt[4]{a^2 c^2} = 2ab \sqrt[4]{(ac)^2} = 2ab \sqrt{ac}$

$\boxed{33}$ $\dfrac{1}{\sqrt{t}} \left(\dfrac{1}{\sqrt{t}} - 1 \right) = \dfrac{1}{\sqrt{t}} \left(\dfrac{1}{\sqrt{t}} - \dfrac{\sqrt{t}}{\sqrt{t}} \right) = \dfrac{1}{\sqrt{t}} \left(\dfrac{1 - \sqrt{t}}{\sqrt{t}} \right) = \dfrac{1 - \sqrt{t}}{t}$

$\boxed{37}$ $\sqrt[3]{\dfrac{1}{2\pi^2}} = \dfrac{1}{\sqrt[3]{2\pi^2}} \cdot \dfrac{\sqrt[3]{4\pi}}{\sqrt[3]{4\pi}} = \dfrac{\sqrt[3]{4\pi}}{\sqrt[3]{8\pi^3}} = \dfrac{\sqrt[3]{4\pi}}{2\pi}$, or $\dfrac{1}{2\pi} \sqrt[3]{4\pi}$

$\boxed{40}$ $\dfrac{1}{\sqrt{a} + \sqrt{a-2}} = \dfrac{1}{\sqrt{a} + \sqrt{a-2}} \cdot \dfrac{\sqrt{a} - \sqrt{a-2}}{\sqrt{a} - \sqrt{a-2}} = \dfrac{\sqrt{a} - \sqrt{a-2}}{a - (a-2)} = \dfrac{\sqrt{a} - \sqrt{a-2}}{2}$

$\boxed{47}$ $(3y^3 - 2y^2 + y + 4)(y^2 - 3)\ = (3y^3 - 2y^2 + y + 4)y^2 + (3y^3 - 2y^2 + y + 4)(-3)$

$\qquad\qquad = (3y^5 - 2y^4 + y^3 + 4y^2) + (-9y^3 + 6y^2 - 3y - 12)$

$\qquad\qquad = 3y^5 - 2y^4 - 8y^3 + 10y^2 - 3y - 12$

$\boxed{50}$ $\dfrac{9p^4 q^3 - 6p^2 q^4 + 5p^3 q^2}{3p^2 q^2} = \dfrac{9p^4 q^3}{3p^2 q^2} - \dfrac{6p^2 q^4}{3p^2 q^2} + \dfrac{5p^3 q^2}{3p^2 q^2} = 3p^2 q - 2q^2 + \dfrac{5}{3}p$

$\boxed{54}$ $\qquad\qquad (a^3 - a^2)^2 = (a^3)^2 - 2(a^3)(a^2) + (a^2)^2 = a^6 - 2a^5 + a^4$

Alternatively, we could factor out a^2 first.

$$(a^3 - a^2)^2 = \left[a^2(a-1) \right]^2 = (a^2)^2 (a-1)^2 = a^4 (a^2 - 2a + 1) = a^6 - 2a^5 + a^4$$

$\boxed{57}$ $(3x+2y)^2 (3x-2y)^2 = \left[(3x+2y)(3x-2y) \right]^2 = (9x^2 - 4y^2)^2 = 81x^4 - 72x^2 y^2 + 16y^4$

$\boxed{64}$ $2c^3 - 12c^2 + 3c - 18 = 2c^2(c-6) + 3(c-6) = (2c^2 + 3)(c-6)$

$\boxed{67}$ $p^8 - q^8\ = (p^4)^2 - (q^4)^2$

$\qquad = (p^4 + q^4)(p^4 - q^4)$

$\qquad = (p^4 + q^4)(p^2 + q^2)(p^2 - q^2)$

$\qquad = (p^4 + q^4)(p^2 + q^2)(p+q)(p-q)$

$\boxed{69}$ $w^6 + 1 = (w^2)^3 + (1)^3 = (w^2 + 1)(w^4 - w^2 + 1)$, which cannot be factored any further.

$\boxed{72}$ $x^2 - 49y^2 - 14x + 49 = (x^2 - 14x + 49) - 49y^2$

$$= (x - 7)^2 - (7y)^2$$

$$= (x - 7 + 7y)(x - 7 - 7y)$$

$\boxed{76}$ $\dfrac{r^3 - t^3}{r^2 - t^2} = \dfrac{(r - t)(r^2 + rt + t^2)}{(r + t)(r - t)} = \dfrac{r^2 + rt + t^2}{r + t}$

$\boxed{80}$ $\dfrac{x + x^{-2}}{1 + x^{-2}} = \dfrac{x + \dfrac{1}{x^2}}{1 + \dfrac{1}{x^2}} = \dfrac{\left(x + \dfrac{1}{x^2}\right) \cdot x^2}{\left(1 + \dfrac{1}{x^2}\right) \cdot x^2} = \dfrac{x^3 + 1}{x^2 + 1}$. We could factor the numerator,

but since it doesn't lead to a reduction of the fraction, we leave it in this form.

$\boxed{81}$ $\dfrac{1}{x} - \dfrac{2}{x^2 + x} - \dfrac{3}{x + 3} = \dfrac{1(x + 1)(x + 3) - 2(x + 3) - 3x(x + 1)}{x(x + 1)(x + 3)}$

$$= \dfrac{x^2 + 4x + 3 - 2x - 6 - 3x^2 - 3x}{x(x + 1)(x + 3)} = \dfrac{-2x^2 - x - 3}{x(x + 1)(x + 3)}$$

$\boxed{83}$ $\dfrac{x + 2 - \dfrac{3}{x + 4}}{\dfrac{x}{x + 4} + \dfrac{1}{x + 4}} = \dfrac{\dfrac{(x + 2)(x + 4) - 3}{x + 4}}{\dfrac{x + 1}{x + 4}} = \dfrac{x^2 + 6x + 5}{x + 1} = \dfrac{(x + 1)(x + 5)}{x + 1} = x + 5$

$\boxed{85}$ $(x^2 + 1)^{3/2}(4)(x + 5)^3 + (x + 5)^4(\tfrac{3}{2})(x^2 + 1)^{1/2}(2x)$

$$= (x^2 + 1)^{1/2}(x + 5)^3[4(x^2 + 1) + 3x(x + 5)]$$

$$= (x^2 + 1)^{1/2}(x + 5)^3(7x^2 + 15x + 4)$$

$\boxed{86}$ $\dfrac{(4 - x^2)(\tfrac{1}{3})(6x + 1)^{-2/3}(6) - (6x + 1)^{1/3}(-2x)}{(4 - x^2)^2} = \dfrac{(6x + 1)^{-2/3}[2(4 - x^2) + 2x(6x + 1)]}{(4 - x^2)^2}$

$$= \dfrac{10x^2 + 2x + 8}{(6x + 1)^{2/3}(4 - x^2)^2}$$

$$= \dfrac{2(5x^2 + x + 4)}{(6x + 1)^{2/3}(4 - x^2)^2}$$

$\boxed{87}$ $N = 5$ and $x_1 = \tfrac{5}{2} \Rightarrow x_2 = \tfrac{1}{2}\left(x_1 + \dfrac{N}{x_1}\right) = \tfrac{1}{2}\left(2.5 + \dfrac{5}{2.5}\right) = 2.25 \Rightarrow$

$x_3 = \tfrac{1}{2}\left(2.25 + \dfrac{5}{2.25}\right) \approx 2.236111 \Rightarrow x_4 \approx 2.236068 \Rightarrow x_5 \approx 2.236068.$

Thus, $\sqrt{5} \approx 2.236068.$

$\boxed{89}$ $(5.5 \text{ liters})\left(10^6 \dfrac{\text{mm}^3}{\text{liter}}\right)\left(5 \times 10^6 \dfrac{\text{cells}}{\text{mm}^3}\right) = 2.75 \times 10^{13}$ red blood cells $\{27.5 \text{ trillion}\}$

$\boxed{91}$ $h = 86$ cm and $w = 13$ kg $\Rightarrow$

$$S = (0.007184)w^{0.425}h^{0.725} = (0.007184)(13)^{0.425}(86)^{0.725} \approx 0.54 \text{ m}^2.$$

Chapter 2: Equations and Inequalities

[5] $4(2y + 5) = 3(5y - 2)$ given

       $8y + 20 = 15y - 6$ multiply terms

       $26 = 7y$ get constants on one side, variables on the other

       $y = \frac{26}{7}$ solve for y

[7] $\frac{1}{5}x + 2 = 3 - \frac{2}{7}x$ given

     $(\frac{1}{5}x + 2) \cdot 35 = (3 - \frac{2}{7}x) \cdot 35$ multiply by the lcd, 35

       $7x + 70 = 105 - 10x$ simplify

       $17x = 35$ simplify

       $x = \frac{35}{17}$ solve for x

[9] Decimal values can be thought of as equivalent rationals, i.e., $0.3 = \frac{3}{10}$. We then multiply all terms by the lcd, 10 in this case, just as we multiplied by 35 in Exercise 7. A common mistake is to multiply 10 times a term such as $0.3(x + 1)$ and get $3(10x + 10)$. However, this is just like multiplying 10 times ab, which is $10ab$. Thus, $10[0.3(x + 1)] = 10(0.3)(x + 1) = 3(x + 1)$. With that in mind, we proceed with the problem: $[0.3(3 + 2x) + 1.2x = 3.2] \cdot 10 \Rightarrow 3(3 + 2x) + 12x = 32 \Rightarrow$

$$9 + 6x + 12x = 32 \Rightarrow 18x = 23 \Rightarrow x = \frac{23}{18}$$

Note: You may have solved equations such as those in Exercise 13 using a process called *cross-multiplication* in the past. This method is sufficient for problems of the form $\frac{P}{Q} = \frac{R}{S}$, but the guidelines for solving an equation containing rational expressions (page 58 in the text) apply to rational equations of a more complex form.

[13] The lcd is $4(4x + 1)$ and we need to remember that x cannot equal $-\frac{1}{4}$.

$$\left[\frac{13 + 2x}{4x + 1} = \frac{3}{4}\right] \cdot 4(4x + 1) \Rightarrow 4(13 + 2x) = 3(4x + 1) \Rightarrow 52 + 8x = 12x + 3 \Rightarrow 49 = 4x \Rightarrow$$

$$x = \frac{49}{4}. \text{ Since } x \neq -\frac{1}{4}, x = \frac{49}{4} \text{ is a solution.}$$

[17] $(3x - 2)^2 = (x - 5)(9x + 4) \Rightarrow 9x^2 - 12x + 4 = 9x^2 - 41x - 20 \Rightarrow 29x = -24 \Rightarrow$

$$x = -\frac{24}{29}$$

[21] $\left[\frac{3x + 1}{6x - 2} = \frac{2x + 5}{4x - 13}\right] \cdot (6x - 2)(4x - 13) \Rightarrow 12x^2 - 35x - 13 = 12x^2 + 26x - 10 \Rightarrow$

$$-3 = 61x \Rightarrow x = -\frac{3}{61} \text{ \{note that } x \neq \frac{1}{3}, \frac{13}{4}\}$$

$\boxed{25}$ Since $2x - 4 = 2(x - 2)$ and $3x - 6 = 3(x - 2)$, the lcd is $2 \cdot 3 \cdot 5(x - 2) = 30(x - 2)$.

$$\left[\frac{3}{2x - 4} - \frac{5}{3x - 6} = \frac{3}{5}\right] \cdot 30(x - 2) \Rightarrow 3(15) - 5(10) = 3(6)(x - 2) \Rightarrow 18x = 31 \Rightarrow x = \frac{31}{18}$$

$\boxed{27}$ $2 - \dfrac{5}{3x - 7} = 2 \Rightarrow \dfrac{5}{3x - 7} = 0 \Rightarrow$ *no solution* since the numerator is never 0.

$\boxed{29}$ $\dfrac{1}{2x - 1} = \dfrac{4}{8x - 4} \Rightarrow \dfrac{1}{2x - 1} = \dfrac{4}{4(2x - 1)} \Rightarrow \dfrac{1}{2x - 1} = \dfrac{1}{2x - 1}.$ This is an identity,

and the solutions consist of every number in the domains of the given expressions.

Thus, the solutions are all real numbers except $\frac{1}{2}$, which we denote by $\mathbb{R} - \{\frac{1}{2}\}$.

$\boxed{35}$ $\left[\dfrac{9x}{3x - 1} = 2 + \dfrac{3}{3x - 1}\right] \cdot (3x - 1) \Rightarrow 9x = 2(3x - 1) + 3 \Rightarrow 9x = 6x + 1 \Rightarrow 3x = 1 \Rightarrow$

$x = \frac{1}{3}$, which is not in the domain of the given expressions. No solution

$\boxed{37}$ $\left[\dfrac{1}{x + 4} + \dfrac{3}{x - 4} = \dfrac{3x + 8}{x^2 - 16}\right] \cdot (x + 4)(x - 4) \Rightarrow x - 4 + 3(x + 4) = 3x + 8 \Rightarrow$

$4x + 8 = 3x + 8 \Rightarrow x = 0$

$\boxed{41}$ $\left[\dfrac{2}{2x + 1} - \dfrac{3}{2x - 1} = \dfrac{-2x + 7}{4x^2 - 1}\right] \cdot (2x + 1)(2x - 1) \Rightarrow 2(2x - 1) - 3(2x + 1) = -2x + 7 \Rightarrow$

$-2x - 5 = -2x + 7 \Rightarrow -5 = 7$, a contradiction. No solution

$\boxed{43}$ $\left[\dfrac{5}{2x + 3} + \dfrac{4}{2x - 3} = \dfrac{14x + 3}{4x^2 - 9}\right] \cdot (2x + 3)(2x - 3) \Rightarrow 5(2x - 3) + 4(2x + 3) = 14x + 3 \Rightarrow$

$18x - 3 = 14x + 3 \Rightarrow 4x = 6 \Rightarrow x = \frac{3}{2},$

which is not in the domain of the given expressions. No solution

Note: For Exercises 45–50, we must show that LS = RS.

$\boxed{47}$ LS $= \dfrac{x^2 - 9}{x + 3} = \dfrac{(x + 3)(x - 3)}{x + 3} = x - 3 = $ RS, $\forall x$ except $x = -3$.

$\boxed{51}$ Substituting -2 for x in $4x + 1 + 2c = 5c - 3x + 6$ yields $-7 + 2c = 5c + 12 \Rightarrow$

$3c = -19 \Rightarrow c = -\frac{19}{3}.$

$\boxed{53}$ (a) $\dfrac{7x}{x - 5} = \dfrac{42}{x - 5} \Rightarrow 7x = 42 \Rightarrow x = 6,$

and the two equations are equivalent since they have the same solution.

(b) No, 5 is not a solution of the first equation since it is undefined if $x = 5$.

$\boxed{55}$ Substituting $\frac{5}{3}$ for x yields $\frac{5}{3}a + b = 0$, or, equivalently, $b = -\frac{5}{3}a$.

Choose any a and b such that $b = -\frac{5}{3}a$. For example, let $a = 3$ and $b = -5$.

$\boxed{57}$ Going from the second line to the third line, we must remember that division by the

variable expression $x - 2$ is not allowed. $\bigstar\ x + 1 = x + 2$

$\boxed{65}$ $P = 2l + 2w$ { get all terms containing w on one side } $\Rightarrow P - 2l = 2w \Rightarrow w = \dfrac{P - 2l}{2}$

$\boxed{69}$
$$S = \frac{p}{q + p(1-q)} \qquad \text{given}$$

$$Sq + Sp(1-q) = p \qquad \text{eliminate the fraction}$$

$$Sq + Sp - Spq = p \qquad \text{multiply terms}$$

$$Sq - Spq = p - Sp \qquad \text{isolate terms containing } q$$

$$Sq(1-p) = p(1-S) \qquad \text{factor}$$

$$q = \frac{p(1-S)}{S(1-p)} \qquad \text{solve for } q$$

$\boxed{71}$ $\frac{1}{f} = \frac{1}{p} + \frac{1}{q}$ { multiply by the lcd fpq } $\Rightarrow$

$$pq = fq + fp \Rightarrow pq - fq = fp \Rightarrow q(p-f) = fp \Rightarrow q = \frac{fp}{p-f}$$

2.2 Exercises

$\boxed{3}$ Let x denote the gross pay. Gross pay $-$ deductions $=$ Net (take-home) pay $\Rightarrow$

$$x - 0.40x = 492 \Rightarrow 0.60x = 492 \Rightarrow x = 820.$$

$\boxed{7}$ Let x denote the amount invested in the 8% account and $100{,}000 - x$ the amount invested in the 6.4% account. Use the formula $I = Prt$

$$(\text{interest} = \text{principal} \times \text{rate} \times \text{time}) \text{ with } t = 1.$$

Interest$_{\text{first account}}$ + Interest$_{\text{second account}}$ = Interest$_{\text{total}}$ $\Rightarrow$

$$x(0.08) + (100{,}000 - x)(0.064) = 7500 \Rightarrow 0.08x + 6400 - 0.064x = 7500 \Rightarrow$$

$0.016x = 1100 \Rightarrow x = 68{,}750.$ Since only \$50,000 can be insured in the 8% account,

we *cannot* fully insure the money and earn annual interest of \$7,500.

$\boxed{9}$ Let x denote the number of children and $600 - x$ the number of adults.

Receipts$_{\text{children}}$ + Receipts$_{\text{adults}}$ = Receipts$_{\text{total}}$ $\Rightarrow x(2) + (600 - x)(5) = 2400 \Rightarrow$

$$-3x = -600 \Rightarrow x = 200 \text{ children.}$$

$\boxed{13}$ Let x denote the number of grams of British sterling silver and $200 - x$ the number of grams of pure copper. We will compare the amounts of pure copper.

{ The percentages are 7.5%, 100%, and 10% or, equivalently, 0.075, 1, and 0.10. }

Copper$_{\text{British sterling silver}}$ + Copper$_{\text{pure}}$ = Copper$_{\text{alloy}}$ $\Rightarrow$

$$(0.075)x + 1(200 - x) = (0.10)(200) \Rightarrow 180 = 0.925x \Rightarrow$$

$$x = 194.6. \text{ Use 194.6 g of British sterling silver and 5.4 g of copper.}$$

$\boxed{17}$ Let r denote the rate of the snowplow. At 8:30 A.M., the car has traveled 15 miles and the snowplow has been traveling for $2\frac{1}{2} = \frac{5}{2}$ hours.

$$\text{Using the relationship time} \times \text{rate} = \text{distance}, \ \tfrac{5}{2}r = 15 \Rightarrow r = 6 \text{ mi/hr.}$$

[19] (a) Let r denote the rate of the river's current.

The rates of the boat upstream and downstream are $5 - r$ and $5 + r$, respectively.

$\text{Distance}_{\text{upstream}} = \text{Distance}_{\text{downstream}} \Rightarrow (5 - r)\frac{15}{60} = (5 + r)\frac{12}{60} \Rightarrow$

$5(5 - r) = 4(5 + r) \Rightarrow 25 - 5r = 20 + 4r \Rightarrow 5 = 9r \Rightarrow r = \frac{5}{9}$ mi/hr.

(b) The distance upstream is $(5 - \frac{5}{9})\frac{1}{4} = \frac{10}{9}$. The total distance is $2 \cdot \frac{10}{9} = \frac{20}{9}$, or $2\frac{2}{9}$ mi.

[21] Let x denote the distance to the target. We know the total time involved and need a

formula for time. Solving $d = rt$ for t gives us $t = d/r$.

$\text{Time}_{\text{to target}} + \text{Time}_{\text{from target}} = \text{Time}_{\text{total}} \Rightarrow$

$\frac{x}{3300} + \frac{x}{1100} = 1.5$ { multiply by the lcd, 3300 } $\Rightarrow$

$x + 3x = 1.5(3300) \Rightarrow 4x = 4950 \Rightarrow x = 1237.5$ ft.

[25] $\text{Area}_{\text{semicircle}} + \text{Area}_{\text{rectangle}} = \text{Area}_{\text{total}} \Rightarrow \frac{1}{2}\pi r^2 + lw = 24 \Rightarrow$

$\frac{1}{2}\pi(\frac{3}{2})^2 + (h - \frac{3}{2})3 = 24 \Rightarrow (h - \frac{3}{2})3 = 24 - \frac{9\pi}{8} \Rightarrow h - \frac{3}{2} = 8 - \frac{3\pi}{8} \Rightarrow h = \frac{19}{2} - \frac{3\pi}{8} \approx 8.32$ ft.

[27] Let h_1 denote the height of the cylinder. $V = \frac{2}{3}\pi r^3 + \pi r^2 h_1 = 11{,}250\pi$ and $r = 15 \Rightarrow$

$2250\pi + 225\pi h_1 = 11{,}250\pi \Rightarrow h_1 = 40$. The total height is 40 ft $+ 15$ ft $= 55$ ft.

[31] Let x denote the desired time.

Using the rates (in minutes), $\frac{1}{45} + \frac{1}{x} = \frac{1}{20} \Rightarrow 4x + 180 = 9x \Rightarrow x = 36$ min.

[33] Let x denote the number of additional games. The team has won $0.650(100) = 65$

games and will win $0.5x$ more. $\dfrac{\text{games won}}{\text{games played}} = \text{record} \Rightarrow \dfrac{65 + 0.5x}{100 + x} = 0.600 \Rightarrow$

$65 + 0.5x = 60 + 0.6x \Rightarrow 5 = 0.1x \Rightarrow x = 50.$

2.3 Exercises

[3] $15x^2 - 12 = -8x$ { get all terms on one side of the equals sign, zero on the other side }

$\Rightarrow 15x^2 + 8x - 12 = 0$ { factor } $\Rightarrow (5x + 6)(3x - 2) = 0 \Rightarrow x = -\frac{6}{5}, \frac{2}{3}$

[5] A common mistake for this exercise is to write $2x = 27$ or $4x + 15 = 27$.

However, remember that you want to get 0 on one side of the equals sign.

$2x(4x + 15) = 27 \Rightarrow 8x^2 + 30x - 27 = 0 \Rightarrow (2x + 9)(4x - 3) = 0 \Rightarrow x = -\frac{9}{2}, \frac{3}{4}$

[9] $12x^2 + 60x + 75 = 0$ { factor out the gcf, 3 } $\Rightarrow$

$3(4x^2 + 20x + 25) = 0$ { divide by 3 } $\Rightarrow$

$4x^2 + 20x + 25 = 0 \Rightarrow (2x + 5)^2 = 0 \Rightarrow x = -\frac{5}{2}$

[13] We will use the same process for solving rational equations as outlined in §2.1.

$\left[\dfrac{5x}{x - 3} + \dfrac{4}{x + 3} = \dfrac{90}{x^2 - 9}\right] \cdot (x + 3)(x - 3) \Rightarrow 5x(x + 3) + 4(x - 3) = 90 \Rightarrow$

$5x^2 + 19x - 102 = 0 \Rightarrow (5x + 34)(x - 3) = 0 \Rightarrow$

$x = -\frac{34}{5}$ { 3 is not in the domain of the given expressions }

[15] (a) The first equation, $x^2 = 16$, has solutions $x = \pm 4$. The equations are not

equivalent since -4 is not a solution of the second equation, $x = 4$.

(b) First note that $x = \sqrt{9} = 3$.

Thus, the equations are equivalent since they have exactly the same solutions.

[19] $25x^2 = 9 \Rightarrow x^2 = \frac{9}{25} \Rightarrow x = \pm\sqrt{\frac{9}{25}} = \pm\frac{3}{5}$

[23] $4(x+2)^2 = 11 \Rightarrow (x+2)^2 = \frac{11}{4} \Rightarrow x+2 = \pm\sqrt{\frac{11}{4}} \Rightarrow x = -2 \pm \frac{1}{2}\sqrt{11}$

[25] For this exercise, consider the general expression $x^2 + bx + c$.

(a) In general, $d = (\frac{1}{2}b)^2$. In this case, $d = \left[\frac{1}{2}(9)\right]^2 = \frac{81}{4}$.

(b) As in part (a), $d = (\frac{1}{2}b)^2 = \left[\frac{1}{2}(-8)\right]^2 = 16$. *Note:* It is appropriate to use 8 or -8.

(c) In general, $d = 2(\pm\sqrt{c})$ for $c > 0$.

In this case, $c = 36 \Rightarrow \sqrt{c} = 6$, and $d = 2(\pm 6) = \pm 12$.

(d) $c = \frac{49}{4} \Rightarrow \sqrt{c} = \frac{7}{2}$, and $d = 2(\pm\frac{7}{2}) = \pm 7$.

[29]

$4x^2 - 12x - 11 = 0$	given
$x^2 - 3x - \frac{11}{4} = 0$	divide by 4
$x^2 - 3x = \frac{11}{4}$	isolate x^2 and x terms
$x^2 - 3x + \frac{9}{4} = \frac{11}{4} + \frac{9}{4}$	add $(\frac{1}{2} \cdot 3)^2 = \frac{9}{4}$
$(x - \frac{3}{2})^2 = 5$	factor and simplify
$x - \frac{3}{2} = \pm\sqrt{5}$	take the square root
$x = \frac{3}{2} \pm \sqrt{5}$	solve for x

[31] $6x^2 - x = 2 \Rightarrow 6x^2 - x - 2 = 0$.

Use the quadratic formula, $x = \dfrac{-b \pm \sqrt{b^2 - 4ac}}{2a}$, with $a = 6$, $b = -1$, and $c = -2$.

$$x = \frac{-(-1) \pm \sqrt{(-1)^2 - 4(6)(-2)}}{2(6)} = \frac{1 \pm \sqrt{1 + 48}}{12} = \frac{1 \pm 7}{12} = -\frac{1}{2}, \frac{2}{3}$$

[35] $2x^2 - 3x - 4 = 0 \Rightarrow x = \dfrac{-(-3) \pm \sqrt{(-3)^2 - 4(2)(-4)}}{2(2)} = \dfrac{3 \pm \sqrt{9 + 32}}{4} = \frac{3}{4} \pm \frac{1}{4}\sqrt{41}$

Note: A common mistake is to not divide 4 into both terms of the numerator.

[37] $\frac{3}{2}z^2 - 4z - 1 = 0$ {multiply by 2} $\Rightarrow 3z^2 - 8z - 2 = 0 \Rightarrow$

$$z = \frac{8 \pm \sqrt{64 + 24}}{6} = \frac{8 \pm \sqrt{88}}{6} = \frac{2(4 \pm \sqrt{22})}{2 \cdot 3} = \frac{4}{3} \pm \frac{1}{3}\sqrt{22}$$

[41] $4x^2 + 81 = 36x \Rightarrow 4x^2 - 36x + 81 = 0 \Rightarrow x = \dfrac{36 \pm \sqrt{1296 - 1296}}{8} = \dfrac{36}{8} = \dfrac{9}{2}$

[43] $\dfrac{5x}{x^2 + 9} = -1 \Rightarrow 5x = -x^2 - 9 \Rightarrow x^2 + 5x + 9 = 0 \Rightarrow x = \dfrac{-5 \pm \sqrt{25 - 36}}{2}$.

Since the discriminant is negative, there are no real solutions.

$\boxed{45}$ (a) For this exercise, we must recognize the equation as a quadratic in x, that is,

$$(A)x^2 + (B)x + (C) = 0,$$

where A is the coefficient of x^2, B is the coefficient of x, and C is the collection of all terms that do not contain x^2 or x.

$$4x^2 - 4xy + 1 - y^2 = 0 \Rightarrow (4)x^2 + (-4y)x + (1 - y^2) = 0 \Rightarrow$$

$$x = \frac{4y \pm \sqrt{16y^2 - 16(1 - y^2)}}{2(4)} = \frac{4y \pm \sqrt{16[y^2 - (1 - y^2)]}}{2(4)} =$$

$$\frac{4y \pm 4\sqrt{2y^2 - 1}}{2(4)} = \frac{y \pm \sqrt{2y^2 - 1}}{2}$$

(b) Similar to part (a), we must now recognize the equation as a quadratic equation in y. $4x^2 - 4xy + 1 - y^2 = 0 \Rightarrow (-1)y^2 + (-4x)y + (4x^2 + 1) = 0 \Rightarrow$

$$y = \frac{4x \pm \sqrt{16x^2 + 4(4x^2 + 1)}}{-2} = \frac{4x \pm 2\sqrt{8x^2 + 1}}{-2} = -2x \pm \sqrt{8x^2 + 1}$$

$\boxed{49}$ $A = 2\pi r(r + h) \Rightarrow A = 2\pi r^2 + 2\pi rh \Rightarrow (2\pi)r^2 + (2\pi h)r - A = 0$ { a quadratic in r } $\Rightarrow$

$$r = \frac{-(2\pi h) \pm \sqrt{(2\pi h)^2 - 4(2\pi)(-A)}}{2(2\pi)} = \frac{-2\pi h \pm \sqrt{4\pi^2 h^2 + 8\pi A}}{2(2\pi)} =$$

$$\frac{-2\pi h \pm 2\sqrt{\pi^2 h^2 + 2\pi A}}{2(2\pi)} = \frac{-\pi h \pm \sqrt{\pi^2 h^2 + 2\pi A}}{2\pi}.$$

Since $r > 0$, we must use the plus sign, and $r = \dfrac{-\pi h + \sqrt{\pi^2 h^2 + 2\pi A}}{2\pi}$.

$\boxed{53}$ Using $V = \pi r^2 h$ with $V = 3000$ and $h = 20$ gives us:

$$3000 = \pi r^2(20) \Rightarrow r^2 = 150/\pi \Rightarrow r = \sqrt{150/\pi} \approx 6.9 \text{ cm}$$

$\boxed{57}$ (a) $T = 98 \Rightarrow h = 1000(100 - T) + 580(100 - T)^2 = 1000(2) + 580(2)^2 = 4320$ m.

(b) If $x = 100 - T$ and $h = 8840$, then $8840 = 1000x + 580x^2 \Rightarrow$

$$29x^2 + 50x - 442 = 0 \Rightarrow x = \frac{-25 \pm \sqrt{13{,}443}}{29} \approx -4.86, \ 3.14.$$

$x = -4.86 \Rightarrow T = 100 - x = 104.86\,°C$, which is outside the allowable range of T.

$x = 3.14 \Rightarrow T = 100 - x = 96.86\,°C$ for $95 \le T \le 100$.

$\boxed{59}$ Let x denote the width of the walk. The area (including the walk) has dimensions $(26 + x + x)$ by $(30 + x + x)$ or, equivalently, $(26 + 2x)$ by $(30 + 2x)$.

$\text{Area}_{plot} + \text{Area}_{walk} = \text{Area}_{total} \Rightarrow 26 \cdot 30 + 240 = (26 + 2x)(30 + 2x) \Rightarrow$

$26 \cdot 30 + 240 = 26 \cdot 30 + 52x + 60x + 4x^2 \Rightarrow 240 = 4x^2 + 112x \Rightarrow$

$$x^2 + 28x - 60 = 0 \Rightarrow (x + 30)(x - 2) = 0 \Rightarrow x = 2 \text{ ft, since } x \text{ is positive.}$$

[63] Let $d(A, P) = x$ and $d(P, B) = 6 - x$.

$x^2 + (6 - x)^2 = 5^2 \Rightarrow 2x^2 - 12x + 11 = 0 \Rightarrow x = 3 \pm \frac{1}{2}\sqrt{14} \approx 4.9, 1.1$ mi.

There are 4 possible roads since P could be on either side of segment AB.

[65] (a) The distances of the northbound and eastbound planes are $100 + 200t$ and $400t$, respectively. Using the Pythagorean theorem,

$$d = \sqrt{(100 + 200t)^2 + (400t)^2} = \sqrt{100^2(1 + 2t)^2 + 100^2(4t)^2} = 100\sqrt{20t^2 + 4t + 1}.$$

(b) $d = 500 \Rightarrow 500 = 100\sqrt{20t^2 + 4t + 1} \Rightarrow 5 = \sqrt{20t^2 + 4t + 1} \Rightarrow$

$5^2 = 20t^2 + 4t + 1 \Rightarrow 5t^2 + t - 6 = 0 \Rightarrow$

$(5t + 6)(t - 1) = 0 \Rightarrow t = 1$ hour after 2:30 P.M., or 3:30 P.M.

[69] Let x denote the rate of the canoeist in still water. $x - 5$ is the rate upstream and

$x + 5$ is the rate downstream. $\text{Time}_{\text{up}} = \text{Time}_{\text{down}} + \frac{1}{2} \Rightarrow \left\{t = \frac{d}{r}\right\} \frac{1.2}{x - 5} = \frac{1.2}{x + 5} + \frac{1}{2}$

$\Rightarrow 2.4(x + 5) = 2.4(x - 5) + x^2 - 25 \Rightarrow x^2 = 49 \Rightarrow x = 7$ mi/hr.

[71] Let x denote the number of pairs ordered. The price per pair is the discount subtracted from \$40. Since the discount is \$0.04 times the number ordered, x, the cost per pair is $40 - 0.04x$. Cost $= (\# \text{ of pairs})(\text{cost per pair}) \Rightarrow$

$8400 = x(40 - 0.04x) \Rightarrow \frac{1}{25}x^2 - 40x + 8400 = 0$ { multiply by 25 } $\Rightarrow$

$x^2 - 1000x + 210,000 = 0 \Rightarrow (x - 300)(x - 700) = 0 \Rightarrow x = 300$ for $0 \le x \le 600$.

[73] The total surface area is the sum of the surface area of the cylinder and that of the top and bottom. $S = 2\pi r h + 2\pi r^2 \Rightarrow 10\pi = 8\pi r + 2\pi r^2$ { divide by 2π } $\Rightarrow$

$r^2 + 4r - 5 = 0 \Rightarrow (r + 5)(r - 1) = 0 \Rightarrow r = 1$, and the diameter is 2 ft.

[75] V is 95% of $V_0 \Rightarrow V = 0.95V_0 \Rightarrow \frac{V}{V_0} = 0.95$.

$0.95 = 0.8197 + 0.007752t + 0.0000281t^2 \Rightarrow 0.281t^2 + 77.52t - 1303 = 0 \Rightarrow$

$t = \dfrac{-77.52 \pm \sqrt{(77.52)^2 - 4(0.281)(-1303)}}{2(0.281)} \approx -291.76, 15.89.$

Thus, the volume of the fireball will be 95% of the maximum volume

approximately 15.89 seconds after the explosion.

$\boxed{77}$ (a) $x = \dfrac{-4{,}500{,}000 \pm \sqrt{4{,}500{,}000^2 - 4(1)(-0.96)}}{2} \approx 0$ and $-4{,}500{,}000$

(b) $x = \dfrac{-b \pm \sqrt{b^2 - 4ac}}{2a} \cdot \dfrac{-b \mp \sqrt{b^2 - 4ac}}{-b \mp \sqrt{b^2 - 4ac}} = \dfrac{b^2 - (b^2 - 4ac)}{2a(-b \mp \sqrt{b^2 - 4ac})} =$

$\dfrac{4ac}{2a(-b \mp \sqrt{b^2 - 4ac})} = \dfrac{2c}{-b \mp \sqrt{b^2 - 4ac}}$. The root near zero was obtained in part

(a) using the plus sign, In the second formula, it corresponds to the minus sign.

$$x = \dfrac{2(-0.96)}{-4{,}500{,}000 - \sqrt{4{,}500{,}000^2 - 4(1)(-0.96)}} \approx 2.13 \times 10^{-7}$$

2.4 Exercises

$\boxed{5}$ $(3 + 5i)(2 - 7i) = (3 + 5i)2 + (3 + 5i)(-7i) = 6 + 10i - 21i - 35i^2$ { group the real

parts and the imaginary parts } $= (6 - 35i^2) + (10 - 21)i = (6 + 35) - 11i = 41 - 11i$

$\boxed{11}$ $i(3 + 4i)^2 = i\big[(9 - 16) + 2(3)(4i)\big] = i(-7 + 24i) = -24 - 7i$

$\boxed{17}$ Since $i^k = 1$ if k is a multiple of 4, we will write i^{73} as $i^{72}i^1$,

knowing that i^{72} will reduce to 1. $i^{73} = i^{72}i = (i^4)^{18}i = 1^{18}i = i$

$\boxed{21}$ Multiply by the conjugate of the denominator to eliminate all i's in the denominator.

The new denominator is the sum of the squares of the coefficients—in this case, 6^2

and 2^2. $\dfrac{1 - 7i}{6 - 2i} \cdot \dfrac{6 + 2i}{6 + 2i} = \dfrac{(6 + 14) + (2 - 42)i}{36 - (-4)} = \dfrac{20 - 40i}{40} = \dfrac{1}{2} - i$

$\boxed{25}$ Multiplying the denominator by i will eliminate the i's in the denominator.

$$\dfrac{4 - 2i}{-5i} = \dfrac{4 - 2i}{-5i} \cdot \dfrac{i}{i} = \dfrac{4i - 2i^2}{-5i^2} = \dfrac{2 + 4i}{5} = \dfrac{2}{5} + \dfrac{4}{5}i$$

$\boxed{27}$ $(2 + 5i)^3 = (2)^3 + 3(2)^2(5i) + 3(2)(5i)^2 + (5i)^3 = (8 + 150i^2) + (60i + 125i^3) =$

$(8 - 150) + (60 - 125)i = -142 - 65i$

$\boxed{29}$ A common mistake is to multiply $\sqrt{-4}\sqrt{-16}$ and obtain $\sqrt{64}$, or 8.

The correct procedure is $\sqrt{-4}\sqrt{-16} = \sqrt{4}\,i \cdot \sqrt{16}\,i = (2i)(4i) = 8i^2 = -8$.

$(2 - \sqrt{-4})(3 - \sqrt{-16}) = (2 - 2i)(3 - 4i) = (6 - 8) + (-6i - 8i) = -2 - 14i$

$\boxed{33}$ $\dfrac{\sqrt{-36}\sqrt{-49}}{\sqrt{-16}} = \dfrac{(6i)(7i)}{4i} \cdot \dfrac{-i}{-i} = \dfrac{(-42)(-i)}{-4i^2} = \dfrac{42i}{4} = \dfrac{21}{2}i$

$\boxed{37}$ We need to equate the real parts and the imaginary parts.

$(3x + 2y) - y^3i = 9 - 27i \Rightarrow y^3 = 27$ { $y = 3$ } and $3x + 2y = 9 \Rightarrow x = 1,\ y = 3$

$\boxed{41}$ $x^2 + 4x + 13 = 0 \Rightarrow x = \dfrac{-4 \pm \sqrt{16 - 52}}{2} = \dfrac{-4 \pm 6i}{2} = -2 \pm 3i$

$\boxed{47}$ $x^3 + 125 = 0 \Rightarrow (x+5)(x^2 - 5x + 25) = 0 \Rightarrow$

$$x = -5 \text{ or } x = \frac{5 \pm \sqrt{25-100}}{2} = \frac{5 \pm 5\sqrt{3}\,i}{2}. \quad \text{The three solutions are } -5, \frac{5}{2} \pm \frac{5}{2}\sqrt{3}\,i.$$

$\boxed{49}$ $x^4 = 256 \Rightarrow x^4 - 256 = 0 \Rightarrow (x^2 - 16)(x^2 + 16) = 0 \Rightarrow x = \pm 4, \ \pm 4i$

$\boxed{51}$ $4x^4 + 25x^2 + 36 = 0 \Rightarrow (x^2 + 4)(4x^2 + 9) = 0 \Rightarrow x = \pm 2i, \ \pm\frac{3}{2}i$

$\boxed{53}$ $x^3 + 3x^2 + 4x = 0 \Rightarrow x(x^2 + 3x + 4) = 0 \Rightarrow x = 0, \ -\frac{3}{2} \pm \frac{1}{2}\sqrt{7}\,i$

$\boxed{55}$ If $w = c + di$, then

$$\overline{z+w} = \overline{(a+bi) + (c+di)} \qquad \text{definition of } z \text{ and } w$$

$$= \overline{(a+c) + (b+d)i} \qquad \text{write in complex number form}$$

$$= (a+c) - (b+d)i \qquad \text{definition of conjugate}$$

$$= (a - bi) + (c - di) \qquad \text{rearrange terms}$$

$$= \bar{z} + \bar{w}. \qquad \text{definition of conjugates of } z \text{ and } w$$

In the step described by "rearrange terms",

we are really looking ahead to the terms we want to obtain, $\bar{z}$ and $\bar{w}$.

$\boxed{59}$ If $\bar{z} = z$, then $a - bi = a + bi$ and hence $-bi = bi$, or $2bi = 0$.

Thus, $b = 0$ and $z = a$ is real. Conversely, if z is real, then $b = 0$ and hence

$$\bar{z} = \overline{a + 0i} = a - 0i = a + 0i = z.$$

2.5 Exercises

$\boxed{3}$ $|3x - 2| + 3 = 7 \Rightarrow |3x - 2| = 4 \Rightarrow 3x - 2 = 4 \text{ or } 3x - 2 = -4 \Rightarrow$

$$3x = 6 \text{ or } 3x = -2 \Rightarrow x = 2 \text{ or } x = -\tfrac{2}{3}$$

$\boxed{5}$ $3|x+1| - 2 = -11 \Rightarrow 3|x+1| = -9 \Rightarrow |x+1| = -3.$

Since the absolute value of an expression is nonnegative, $|x+1| = -3$ has no solution.

$\boxed{7}$ $9x^3 - 18x^2 - 4x + 8 = 0 \Rightarrow 9x^2(x-2) - 4(x-2) = 0 \Rightarrow (9x^2 - 4)(x-2) = 0 \Rightarrow$

$$x = \pm\tfrac{2}{3}, 2$$

$\boxed{9}$ $4x^4 + 10x^3 = 6x^2 + 15x \Rightarrow x(4x^3 + 10x^2 - 6x - 15) = 0 \Rightarrow$

$$x\big[2x^2(2x+5) - 3(2x+5)\big] = 0 \Rightarrow x(2x^2 - 3)(2x+5) = 0 \Rightarrow x = 0, \ \pm\tfrac{1}{2}\sqrt{6}, \ -\tfrac{5}{2}$$

Note: The comment on page 94 about raising both sides to a reciprocal power addresses what is often a difficult concept for many students. The following illustration may help in understanding this concept. Note that if m is even, we have to use the $\pm$ symbol.

Problem	Solution
$x^{1/2} = 4$	$(x^{1/2})^{2/1} = 4^{2/1} \Rightarrow x = 16$
$x^{-1/2} = 5$	$(x^{-1/2})^{-2/1} = 5^{-2/1} \Rightarrow x = \frac{1}{25}$
$x^{3/4} = 8$	$(x^{3/4})^{4/3} = 8^{4/3} \Rightarrow x = 16$
$x^{4/3} = 16$	$(x^{4/3})^{3/4} = 16^{3/4} \Rightarrow x = \pm 8$

This principle is used in many exercises, especially Exercises 51 and 52.

$\boxed{11}$ $y^{3/2} = 5y \Rightarrow y^{3/2} - 5y = 0 \Rightarrow y(y^{1/2} - 5) = 0 \Rightarrow y = 0$ or $y^{1/2} = 5$.

$$y^{1/2} = 5 \Rightarrow (y^{1/2})^2 = 5^2 \Rightarrow y = 25. \quad y = 0, \, 25$$

Note: The following guidelines may be helpful when solving radical equations.

Guidelines for Solving a Radical Equation

---

(1) Isolate the radical. If we cannot get the radical isolated on one side of the equals sign because there is more than one radical, then we will split up the radical terms as evenly as possible on each side of the equals sign. For example, if there are 2 radicals, we put one on each side; if there are 3 radicals, we put 2 on one side and 1 on the the other.

(2) Raise both sides to the same power as the root index. *Note:* Remember here that

$$\boxed{(a + b\sqrt{n})^2 = a^2 + 2ab\sqrt{n} + b^2 n} \quad \text{and that it is } not \quad a^2 + b^2 n.$$

(3) If your equation contains no radicals, proceed to part (4). If there are still radicals in the equation, go back to part (1).

(4) Solve the resulting equation.

(5) Check the answers found in part (4) in the original equation to determine the valid solutions. *Note:* You may check the solutions in any equivalent equation of the original equation, i.e., an equation which occurs prior to raising both sides to a power. Also, extraneous solutions are introduced when raising both sides to an even power. Hence, all solutions *must* be checked in this case. Checking solutions when raising each side to an odd power is up to the individual instructor.

$\boxed{15}$ $2 + \sqrt[3]{1 - 5t} = 0 \Rightarrow \sqrt[3]{1 - 5t} = -2 \Rightarrow (\sqrt[3]{1 - 5t})^3 = (-2)^3 \Rightarrow 1 - 5t = -8 \Rightarrow t = \frac{9}{5}$

$\boxed{19}$ $\sqrt{7-x} = x - 5$ {square both sides} $\Rightarrow$ $7 - x = x^2 - 10x + 25$ {set equal to zero} $\Rightarrow$

$$x^2 - 9x + 18 = 0 \text{ \{factor\}} \Rightarrow (x-3)(x-6) = 0 \Rightarrow x = 3, 6.$$

Check $x = 6$: LS $= \sqrt{7-6} = 1$; RS $= 6 - 5 = 1$.

Since both sides have the same value, $x = 6$ is a valid solution.

Check $x = 3$: LS $= \sqrt{7-3} = 2$; RS $= 3 - 5 = -2$.

Since both sides do not have the same value, $x = 3$ is an extraneous solution.

$\boxed{21}$ $3\sqrt{2x-3} + 2\sqrt{7-x} = 11$ $\qquad\qquad$ given

$\qquad$ $3\sqrt{2x-3} = 11 - 2\sqrt{7-x}$ $\qquad$ split radicals evenly

$\qquad$ $9(2x-3) = 121 - 44\sqrt{7-x} + 4(7-x)$ square both sides

$\qquad$ $44\sqrt{7-x} = -22x + 176$ $\qquad$ isolate radical again and simplify

$\qquad$ $2\sqrt{7-x} = 8 - x$ $\qquad\qquad$ divide by gcf, 22

$\qquad$ $4(7-x) = 64 - 16x + x^2$ $\qquad$ square both sides

$\qquad$ $x^2 - 12x + 36 = 0$ $\qquad\qquad$ collect terms on one side

$\qquad$ $(x-6)^2 = 0$ $\qquad\qquad$ factor

$\qquad$ $x - 6 = 0$ $\qquad\qquad$ take the square root

$\qquad$ $x = 6$ $\qquad\qquad$ solve for x

Check $x = 6$: LS $= 3(3) + 2(1) = 11 =$ RS $\Rightarrow x = 6$ is the solution.

$\boxed{25}$ $x + \sqrt{5x+19} = -1 \Rightarrow \sqrt{5x+19} = -x - 1 \Rightarrow 5x + 19 = x^2 + 2x + 1 \Rightarrow$

$$x^2 - 3x - 18 = 0 \Rightarrow (x-6)(x+3) = 0 \Rightarrow x = -3, 6.$$

Check $x = -3$: LS $= -3 + 2 = -1 =$ RS $\Rightarrow x = -3$ is a solution.

Check $x = 6$: LS $= 6 + 7 = 13 \neq$ RS $\Rightarrow x = 6$ is an extraneous solution.

$\boxed{27}$ $\sqrt{7-2x} - \sqrt{5+x} = \sqrt{4+3x}$ {square both sides} $\Rightarrow$

$(7-2x) - 2\sqrt{(7-2x)(5+x)} + (5+x) = 4 + 3x$ {isolate the radical} $\Rightarrow$

$-4x + 8 = 2\sqrt{-2x^2 - 3x + 35}$ {divide by 2} $\Rightarrow$

$-2x + 4 = \sqrt{-2x^2 - 3x + 35}$ {square both sides} $\Rightarrow$

$4x^2 - 16x + 16 = -2x^2 - 3x + 35$ {simplify} $\Rightarrow$

$$6x^2 - 13x - 19 = 0 \Rightarrow (x+1)(6x-19) = 0 \Rightarrow x = -1, \tfrac{19}{6}.$$

Check $x = -1$: LS $= 3 - 2 = 1 =$ RS $\Rightarrow x = -1$ is a solution.

Check $x = \frac{19}{6}$: LS $= \sqrt{\frac{2}{3}} - \sqrt{\frac{49}{6}}$ {note that this is negative} $\neq \sqrt{\frac{27}{2}} =$ RS $\Rightarrow$

$$x = \tfrac{19}{6} \text{ is an extraneous solution.}$$

$\boxed{31}$ $\sqrt{2\sqrt{x+1}} = \sqrt{3x-5} \Rightarrow 2\sqrt{x+1} = 3x - 5 \Rightarrow 4(x+1) = 9x^2 - 30x + 25 \Rightarrow$

$$9x^2 - 34x + 21 = 0 \Rightarrow (x-3)(9x-7) = 0 \Rightarrow x = 3, \tfrac{7}{9}.$$

Check $x = 3$: LS $= 2 =$ RS $\Rightarrow x = 3$ is a solution.

Check $x = \frac{7}{9}$: LS $= \sqrt{2 \cdot \frac{4}{3}} = \sqrt{\frac{8}{3}} \neq \sqrt{-\frac{8}{3}} =$ RS $\Rightarrow x = \frac{7}{9}$ is an extraneous solution.

$\boxed{35}$ $x^4 - 25x^2 + 144 = 0 \Rightarrow (x^2 - 9)(x^2 - 16) = 0 \Rightarrow x = \pm 3, \ \pm 4$

$\boxed{37}$ $5y^4 - 7y^2 + 1 = 0 \Rightarrow y^2 = \dfrac{7 \pm \sqrt{29}}{10} \cdot \dfrac{10}{10} = \dfrac{70 \pm 10\sqrt{29}}{100} \Rightarrow y = \pm \dfrac{1}{10}\sqrt{70 \pm 10\sqrt{29}}$

Alternatively, let $u = y^2$ and solve $5u^2 - 7u + 1 = 0$.

$\boxed{39}$ $36x^{-4} - 13x^{-2} + 1 = 0 \Rightarrow (4x^{-2} - 1)(9x^{-2} - 1) = 0 \Rightarrow x^{-2} = \frac{1}{4}, \frac{1}{9} \Rightarrow x^2 = 4, \ 9 \Rightarrow$

$$x = \pm 2, \ \pm 3$$

Alternatively, let $u = x^{-2}$ and solve $36u^2 - 13u + 1 = 0$.

$\boxed{41}$ $3x^{2/3} + 4x^{1/3} - 4 = 0 \Rightarrow (3x^{1/3} - 2)(x^{1/3} + 2) = 0 \Rightarrow \sqrt[3]{x} = \frac{2}{3}, \ -2 \Rightarrow x = \frac{8}{27}, \ -8$

Alternatively, let $u = x^{1/3}$ and solve $3u^2 + 4u - 4 = 0$.

$\boxed{45}$ $\left(\dfrac{t}{t+1}\right)^2 - \dfrac{2t}{t+1} - 8 = 0 \Rightarrow \left(\dfrac{t}{t+1} - 4\right)\left(\dfrac{t}{t+1} + 2\right) = 0 \Rightarrow \dfrac{t}{t+1} = 4, \ -2 \Rightarrow$

$$t = 4t + 4, \ -2t - 2 \Rightarrow t = -\tfrac{4}{3}, \ -\tfrac{2}{3}$$

Alternatively, let $u = \dfrac{t}{t+1}$ and solve $u^2 - 2u - 8 = 0$.

$\boxed{47}$ For this exercise, we must note that the variable terms are both cubed, and then combine them into one term.

$$27x^3 = (x+5)^3 \Rightarrow \left(\dfrac{x+5}{x}\right)^3 = 27 \Rightarrow \dfrac{x+5}{x} = 3 \Rightarrow x + 5 = 3x \Rightarrow x = \tfrac{5}{2}$$

$\boxed{49}$ The least common multiple of 3 and 4 is 12—so by raising both sides to the 12th power we will eliminate the radicals. $\sqrt[3]{x} = 2\sqrt[4]{x} \Rightarrow (\sqrt[3]{x})^{12} = (2\sqrt[4]{x})^{12} \Rightarrow$

$$x^4 = 2^{12}x^3 \Rightarrow x^4 - 4096x^3 = 0 \Rightarrow x^3(x - 4096) = 0 \Rightarrow x = 0, \ 4096.$$

Check $x = 0$: $\text{LS} = 0 = \text{RS} \Rightarrow x = 0$ is a solution.

Check $x = 4096 = 2^{12}$: $\text{LS} = \sqrt[3]{2^{12}} = 2^4$; $\text{RS} = 2\sqrt[4]{2^{12}} = 2 \cdot 2^3 = 2^4 \Rightarrow$

$$x = 4096 \text{ is a solution.}$$

$\boxed{51}$ See the note before the solution for Exercise 11.

(a) $x^{5/3} = 32 \Rightarrow (x^{5/3})^{3/5} = (32)^{3/5} \Rightarrow x = (\sqrt[5]{32})^3 = 2^3 = 8$

(b) $x^{4/3} = 16 \Rightarrow (x^{4/3})^{3/4} = \pm(16)^{3/4} \Rightarrow x = \pm(\sqrt[4]{16})^3 = \pm 2^3 = \pm 8$

(c) $x^{2/3} = -36 \Rightarrow (x^{2/3})^{3/2} = \pm(-36)^{3/2} \Rightarrow x = \pm(\sqrt{-36})^3,$

which are not real numbers. No real solutions

(d) $x^{3/4} = 125 \Rightarrow (x^{3/4})^{4/3} = (125)^{4/3} \Rightarrow x = (\sqrt[3]{125})^4 = 5^4 = 625$

(e) $x^{3/2} = -27 \Rightarrow (x^{3/2})^{2/3} = (-27)^{2/3} \Rightarrow x = (\sqrt[3]{-27})^2 = (-3)^2 = 9,$

which is an extraneous solution. No real solutions

$\boxed{55}$ $S = \pi r\sqrt{r^2 + h^2} \Rightarrow \dfrac{S}{\pi r} = \sqrt{r^2 + h^2} \Rightarrow \dfrac{S^2}{\pi^2 r^2} = h^2 + r^2 \Rightarrow \dfrac{S^2}{\pi^2 r^2} - r^2 = h^2 \Rightarrow$

$$h^2 = \dfrac{1}{\pi^2 r^2}(S^2 - \pi^2 r^4) \Rightarrow h = \pm\dfrac{1}{\pi r}\sqrt{S^2 - \pi^2 r^4} \Rightarrow h = \dfrac{1}{\pi r}\sqrt{S^2 - \pi^2 r^4} \text{ since } h > 0$$

59 $k = 10^5$ and $c = \frac{1}{2} \Rightarrow Q = kP^{-c} = 10^5 P^{-1/2} \Rightarrow Q = 10^5/\sqrt{P} \Rightarrow$

$$\sqrt{P} = \frac{10^5}{Q} \Rightarrow P = \left(\frac{10^5}{Q}\right)^2 = \left(\frac{100,000}{5000}\right)^2 = (20)^2 = 400 \text{ cents, or, } \$4.00.$$

63 $y = 60\% \Rightarrow \dfrac{x^3}{x^3 + (1-x)^3} = \dfrac{3}{5} \Rightarrow 5x^3 = 3x^3 + 3(1-x)^3 \Rightarrow 2x^3 = 3(1-x)^3 \Rightarrow$

$$\left(\frac{x}{1-x}\right)^3 = \frac{3}{2} \Rightarrow \frac{x}{1-x} = \sqrt[3]{1.5} \Rightarrow x = \sqrt[3]{1.5} - \sqrt[3]{1.5}\,x \Rightarrow x + \sqrt[3]{1.5}\,x = \sqrt[3]{1.5} \Rightarrow$$

$$(1 + \sqrt[3]{1.5})x = \sqrt[3]{1.5} \Rightarrow x = \frac{\sqrt[3]{1.5}}{1 + \sqrt[3]{1.5}} \approx 0.534, \text{ or } 53.4\%.$$

65 $\text{Cost}_{\text{underwater}} + \text{Cost}_{\text{overland}} = \text{Cost}_{\text{total}} \Rightarrow$

$7500 \cdot (\text{underwater miles}) + 6000 \cdot (\text{overland miles}) = 35,000 \Rightarrow$

$7500\sqrt{x^2 + 1} + 6000(5 - x) = 35,000 \Rightarrow 15\sqrt{x^2 + 1} = 12x + 10 \Rightarrow$

$225(x^2 + 1) = 144x^2 + 240x + 100 \Rightarrow 81x^2 - 240x + 125 = 0 \Rightarrow$

$x = \dfrac{240 \pm \sqrt{17,100}}{162} = \dfrac{40 \pm 5\sqrt{19}}{27} \approx 2.2887, \ 0.6743 \text{ mi. There are two possible routes.}$

67 $x_1 = 2$ and $x_2 = \frac{1}{3}\sqrt[3]{x_1} + 2 \Rightarrow x_2 \approx 2.419974 \Rightarrow x_3 \approx 2.447523 \Rightarrow x_4 \approx 2.449215 \Rightarrow$

$$x_5 \approx 2.449319 \Rightarrow x_6 \approx 2.449325. \text{ The root is approximately } 2.4493.$$

2.6 Exercises

Note: The bracket symbols "[" and "]", are used with $\leq$ or $\geq$ to denote that the end
point of the interval is part of the solution. Parentheses, "(" and ")" are used
with $<$ or $>$ and denote that the end point is *not* part of the solution.

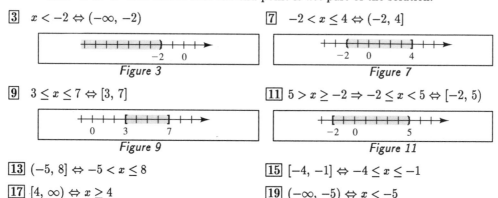

3 $x < -2 \Leftrightarrow (-\infty, -2)$

Figure 3

7 $-2 < x \leq 4 \Leftrightarrow (-2, 4]$

Figure 7

9 $3 \leq x \leq 7 \Leftrightarrow [3, 7]$

Figure 9

11 $5 > x \geq -2 \Rightarrow -2 \leq x < 5 \Leftrightarrow [-2, 5)$

Figure 11

13 $(-5, 8] \Leftrightarrow -5 < x \leq 8$

15 $[-4, -1] \Leftrightarrow -4 \leq x \leq -1$

17 $[4, \infty) \Leftrightarrow x \geq 4$

19 $(-\infty, -5) \Leftrightarrow x < -5$

23 $-2 - 3x \geq 2 \Rightarrow -3x \geq 4$ { Remember to change the direction of the inequality when
multiplying or dividing by a negative value. } $\Rightarrow x \leq -\frac{4}{3} \Leftrightarrow (-\infty, -\frac{4}{3}]$

27 $\left[9 + \frac{1}{3}x \geq 4 - \frac{1}{2}x\right] \cdot 6$ { multiply by the lcd, 6 } $\Rightarrow 54 + 2x \geq 24 - 3x \Rightarrow 5x \geq -30 \Rightarrow$

$$x \geq -6 \Leftrightarrow [-6, \infty)$$

$\boxed{29}$ $-3 < 2x - 5 < 7$ { add 5 to all three expressions } $\Rightarrow$

$\qquad$ $2 < 2x < 12$ { divide all three expressions by 2 } $\Rightarrow$ $1 < x < 6 \Leftrightarrow (1, 6)$

$\boxed{33}$ $4 > \dfrac{2 - 3x}{7} \geq -2$ { multiply all three expressions by 7 } $\Rightarrow$ $28 > 2 - 3x \geq -14 \Rightarrow$

$\qquad$ $26 > -3x \geq -16$ { divide by -3 and change directions of *both* inequality signs } $\Rightarrow$

$\qquad\qquad$ $-\dfrac{26}{3} < x \leq \dfrac{16}{3} \Leftrightarrow (-\dfrac{26}{3}, \dfrac{16}{3}]$

$\boxed{37}$ $(2x - 3)(4x + 5) \leq (8x + 1)(x - 7) \Rightarrow 8x^2 - 2x - 15 \leq 8x^2 - 55x - 7 \Rightarrow 53x \leq 8 \Rightarrow$

$\qquad\qquad$ $x \leq \dfrac{8}{53} \Leftrightarrow (-\infty, \dfrac{8}{53}]$

$\boxed{41}$ By the law of signs, a quotient is positive if the sign of the numerator and the sign of

$\qquad$ the denominator are the same. Since the numerator is positive, $\dfrac{4}{3x + 2} > 0 \Rightarrow$

$\qquad$ $3x + 2 > 0 \Rightarrow x > -\dfrac{2}{3} \Leftrightarrow (-\dfrac{2}{3}, \infty)$. The expression is never equal to 0 since the

$\qquad$ numerator is never 0. Thus, the solution of $\dfrac{4}{3x + 2} \geq 0$ is $(-\dfrac{2}{3}, \infty)$.

$\boxed{43}$ $\dfrac{-2}{4 - 3x} > 0 \Rightarrow 4 - 3x < 0$ { denominator must also be negative } $\Rightarrow x > \dfrac{4}{3} \Leftrightarrow (\dfrac{4}{3}, \infty)$

$\boxed{45}$ $(1 - x)^2 > 0 \; \forall x$ except 1. Thus, $\dfrac{2}{(1 - x)^2} > 0$ has solution $\mathbb{R} - \{1\}$.

$\boxed{51}$ $|x + 3| < 0.01 \Rightarrow -0.01 < x + 3 < 0.01 \Rightarrow -3.01 < x < -2.99 \Leftrightarrow (-3.01, -2.99)$

$\boxed{53}$ $|x + 2| + 0.1 \geq 0.2$ { isolate the absolute value expression } $\Rightarrow |x + 2| \geq 0.1 \Rightarrow$

$\qquad$ $x + 2 \geq 0.1$ or $x + 2 \leq -0.1 \Rightarrow x \geq -1.9$ or $x \leq -2.1 \Leftrightarrow (-\infty, -2.1] \cup [-1.9, \infty)$

$\boxed{57}$ $-\dfrac{1}{3}|6 - 5x| + 2 \geq 1 \Rightarrow -\dfrac{1}{3}|6 - 5x| \geq -1 \Rightarrow |6 - 5x| \leq 3 \Rightarrow -3 \leq 6 - 5x \leq 3 \Rightarrow$

$\qquad\qquad$ $-9 \leq -5x \leq -3 \Rightarrow \dfrac{9}{5} \geq x \geq \dfrac{3}{5} \Leftrightarrow [\dfrac{3}{5}, \dfrac{9}{5}]$

$\boxed{59}$ Since $|7x + 2| \geq 0 \; \forall x$, $|7x + 2| > -2$ has solution $(-\infty, \infty)$.

$\boxed{61}$ $|3x - 9| > 0 \; \forall x$ except when $3x - 9 = 0$, or $x = 3$. The solution is $(-\infty, 3) \cup (3, \infty)$.

$\boxed{63}$ $\left|\dfrac{2 - 3x}{5}\right| \geq 2 \Rightarrow \dfrac{|2 - 3x|}{|5|} \geq 2 \Rightarrow |2 - 3x| \geq 10 \Rightarrow 2 - 3x \geq 10$ or $2 - 3x \leq -10 \Rightarrow$

$\qquad\qquad$ $-3x \geq 8$ or $-3x \leq -12 \Rightarrow x \leq -\dfrac{8}{3}$ or $x \geq 4 \Leftrightarrow (-\infty, -\dfrac{8}{3}] \cup [4, \infty)$

$\boxed{65}$ Since $|5 - 2x| \geq 0 \; \forall x$, we can multiply the inequality by $|5 - 2x|$ without

$\qquad$ changing the direction of the inequality sign. We must exclude $x = \dfrac{5}{2}$ from the

$\qquad$ solution since it makes the original inequality undefined.

$\qquad$ $\dfrac{3}{|5 - 2x|} < 2 \Rightarrow |5 - 2x| > \dfrac{3}{2} \Rightarrow 5 - 2x > \dfrac{3}{2}$ or $5 - 2x < -\dfrac{3}{2} \Rightarrow$

$\qquad$ $-2x > -\dfrac{7}{2}$ or $-2x < -\dfrac{13}{2} \Rightarrow x < \dfrac{7}{4}$ or $x > \dfrac{13}{4}$ { $\dfrac{5}{2}$ doesn't fall in this region } $\Leftrightarrow$

$\qquad\qquad$ $(-\infty, \dfrac{7}{4}) \cup (\dfrac{13}{4}, \infty)$

$\boxed{67}$ From the definition of absolute value, $|x - 2| =$ either $x - 2$ or $-(x - 2)$.

Thus, $1 < |x - 2| < 4 \Rightarrow 1 < x - 2 < 4$ or $1 < -(x - 2) < 4 \Rightarrow$

$1 < x - 2 < 4$ or $-1 > x - 2 > -4 \Rightarrow 3 < x < 6$ or $1 > x > -2 \Leftrightarrow (-2, 1) \cup (3, 6)$.

An alternative method is to rewrite the inequality as $|x - 2| > 1$ *and* $|x - 2| < 4$.

Solving independently gives us

$x - 2 > 1$ or $x - 2 < -1 \Rightarrow x > 3$ or $x < 1$ *and* $-4 < x - 2 < 4 \Rightarrow -2 < x < 6$.

Taking the *intersection* of these intervals gives $(-2, 1) \cup (3, 6)$.

$\boxed{69}$ (a) $|x + 5| = 3 \Rightarrow x + 5 = 3$ or $x + 5 = -3 \Rightarrow x = -2$ or $x = -8$.

(b) $|x + 5| < 3$ has solutions between the values found in part (a), that is, $(-8, -2)$.

(c) The solutions of $|x + 5| > 3$ are the portions of the real line that are not in

parts (a) and (b), that is, $(-\infty, -8) \cup (-2, \infty)$.

$\boxed{73}$ The difference of two temperatures T_1 and T_2 can be represented by $T_1 - T_2$.

Since there is no indication as to whether T_1 is larger than T_2, or vice versa,

we will use $|T_1 - T_2|$. $5 < |T_1 - T_2| < 10$

$\boxed{77}$ Since $V = 110$, $R = \frac{110}{I}$, or equivalently, $I = \frac{110}{R}$. If the current is not to exceed 10,

we want to solve the inequality $I \le 10$. $I \le 10 \Rightarrow \frac{110}{R} \le 10 \Rightarrow 110 \le 10R$ $\{ R > 0$, so

we may multiply by R without changing the direction of the inequality $\} \Rightarrow R \ge 11$

$\boxed{79}$ We want to know what condition will assure us that an object's image is at least 3

times as large as the object, or, equivalently, when $M \ge 3$.

$M \ge 3$ $\{ f = 6 \} \Rightarrow \frac{6}{6 - p} \ge 3 \Rightarrow 6 \ge 18 - 3p$ $\{$ since $6 - p > 0$, we can multiply by

$6 - p$ and not change the direction of the inequality $\} \Rightarrow 3p \ge 12 \Rightarrow$

$p \ge 4$, but $p < 6$ since $p < f$. Thus, $4 \le p < 6$.

$\boxed{81}$ Let x denote the number of years before A becomes more economical than B.

The costs are the initial costs plus the yearly costs times the number of years.

$\text{Cost}_A < \text{Cost}_B \Rightarrow 50{,}000 + 4000x < 40{,}000 + 5500x \Rightarrow 10{,}000 < 1500x \Rightarrow$

$x > \frac{20}{3}$, or $6\frac{2}{3}$ yr.

2.7 Exercises

Note: Many solutions for exercises involving inequalities contain a sign diagram. You may want to read Example 3 again if you have trouble interpreting the sign diagrams.

1 $(3x + 1)(5 - 10x) > 0$ has solutions in the interval $(-\frac{1}{3}, \frac{1}{2})$. See *Diagram 1* for details concerning the signs of the individual factors and the resulting sign.

Resulting sign:	$\ominus$	$\oplus$	$\ominus$
Sign of $5 - 10x$:	$+$	$+$	$-$
Sign of $3x + 1$:	$-$	$+$	$+$
x values:	$-1/3$		$1/2$

Diagram 1

5 $x^2 - x - 6 < 0 \Rightarrow (x - 3)(x + 2) < 0$; $(-2, 3)$

Resulting sign:	$\oplus$	$\ominus$	$\oplus$
Sign of $x - 3$:	$-$	$-$	$+$
Sign of $x + 2$:	$-$	$+$	$+$
x values:	-2		3

Diagram 5

7 $x^2 - 2x - 5 > 3 \Rightarrow x^2 - 2x - 8 > 0 \Rightarrow (x - 4)(x + 2) > 0$; $(-\infty, -2) \cup (4, \infty)$

Resulting sign:	$\oplus$	$\ominus$	$\oplus$
Sign of $x - 4$:	$-$	$-$	$+$
Sign of $x + 2$:	$-$	$+$	$+$
x values:	-2		4

Diagram 7

11 $6x - 8 > x^2 \Rightarrow x^2 - 6x + 8 < 0 \Rightarrow (x - 2)(x - 4) < 0$; $(2, 4)$

Resulting sign:	$\oplus$	$\ominus$	$\oplus$
Sign of $x - 4$:	$-$	$-$	$+$
Sign of $x - 2$:	$-$	$+$	$+$
x values:	2		4

Diagram 11

Note: Solving $x^2 < $ (or $>$) a^2 for $a > 0$ may be solved using factoring, that is, $x^2 - a^2 < 0 \Rightarrow (x + a)(x - a) < 0 \Rightarrow -a < x < a$; or by taking the square root of each side, that is, $\sqrt{x^2} < \sqrt{a^2} \Rightarrow |x| < a \Rightarrow -a < x < a$.

13   *Note:* The most common mistake is not remembering that $\sqrt{x^2} = |x|$.

$$x^2 < 16 \Rightarrow |x| < 4 \Rightarrow -4 < x < 4 \Leftrightarrow (-4, 4).$$

17 $16x^2 \geq 9x \Rightarrow x(16x - 9) \geq 0$; $(-\infty, 0] \cup [\frac{9}{16}, \infty)$

Resulting sign:	$\oplus$	$\ominus$	$\oplus$
Sign of $16x - 9$:	$-$	$-$	$+$
Sign of x:	$-$	$+$	$+$
x values:	0		$9/16$

Diagram 17

$\boxed{19}$ $x^4 + 5x^2 \geq 36 \Rightarrow x^4 + 5x^2 - 36 \geq 0 \Rightarrow (x^2 + 9)(x^2 - 4) \geq 0 \Rightarrow x^2 - 4 \geq 0 \; \{x^2 + 9 > 0\}$

$\Rightarrow x^2 \geq 4 \Rightarrow |x| \geq 2 \Rightarrow x \geq 2 \text{ or } x \leq -2 \Leftrightarrow (-\infty, -2] \cup [2, \infty)$

$\boxed{21}$ $x^3 + 2x^2 - 4x - 8 \geq 0 \Rightarrow x^2(x + 2) - 4(x + 2) \geq 0 \Rightarrow (x^2 - 4)(x + 2) \geq 0 \Rightarrow$

$(x - 2)(x + 2)^2 \geq 0$. The expression $(x + 2)^2$ is positive except when $x = -2$.

The sign is determined by the sign of $x - 2$, which is positive if $x > 2$. Since $x = \pm 2$

make the expression zero, the solution is $x \geq 2$ or $x = -2 \Leftrightarrow \{-2\} \cup [2, \infty)$.

$\boxed{23}$ $\dfrac{x^2(x + 2)}{(x + 2)(x + 1)} \leq 0 \Rightarrow \dfrac{x^2}{x + 1} \leq 0$ { we will exclude $x = -2$ since it makes the original

expression undefined} $\Rightarrow \dfrac{1}{x + 1} \leq 0$ { we can divide by x^2 since $x^2 \geq 0$ and we will

include $x = 0$ since it makes x^2 equal to zero and we want all solutions less than *or*

equal to zero} $\Rightarrow x + 1 < 0$ { the fraction cannot equal zero and $x + 1$ must be

negative so that the fraction is negative} $\Rightarrow x < -1; \; (-\infty, -2) \cup (-2, -1) \cup \{0\}$

$\boxed{25}$ $\dfrac{x^2 - x}{x^2 + 2x} \leq 0 \Rightarrow \dfrac{x(x - 1)}{x(x + 2)} \leq 0 \Rightarrow \dfrac{x - 1}{x + 2} \leq 0$ { we will exclude $x = 0$ from the solution };

$(-2, 0) \cup (0, 1]$

Resulting sign:	$\oplus$	$\ominus$	$\oplus$
Sign of $x - 1$:	$-$	$-$	$+$
Sign of $x + 2$:	$-$	$+$	$+$
x values:		-2	1

Diagram 25

$\boxed{29}$ $\dfrac{-3x}{x^2 - 9} > 0 \Rightarrow \dfrac{x}{(x + 3)(x - 3)} < 0$ { divide by -3}; $(-\infty, -3) \cup (0, 3)$

Resulting sign:	$\ominus$	$\oplus$	$\ominus$	$\oplus$
Sign of $x - 3$:	$-$	$-$	$-$	$+$
Sign of x:	$-$	$-$	$+$	$+$
Sign of $x + 3$:	$-$	$+$	$+$	$+$
x values:		-3	0	3

Diagram 29

$\boxed{31}$ $\dfrac{x + 1}{2x - 3} > 2 \Rightarrow \dfrac{x + 1 - 2(2x - 3)}{2x - 3} > 0 \Rightarrow \dfrac{-3x + 7}{2x - 3} > 0$.

From *Diagram 31*, the solution is $\left(\frac{3}{2}, \frac{7}{3}\right)$. Note that you should *not* multiply by the

factor $2x - 3$ as we did with rational *equations* because $2x - 3$ may be positive or

negative, and multiplying by it would require solving two inequalities. This method

of solution tends to be more difficult than the sign diagram method.

Resulting sign:	$\ominus$	$\oplus$	$\ominus$
Sign of $-3x + 7$:	$+$	$+$	$-$
Sign of $2x - 3$:	$-$	$+$	$+$
x values:		$3/2$	$7/3$

Diagram 31

33 $\dfrac{1}{x-2} \geq \dfrac{3}{x+1} \Rightarrow \dfrac{1(x+1)-3(x-2)}{(x-2)(x+1)} \geq 0 \Rightarrow \dfrac{-2x+7}{(x-2)(x+1)} \geq 0;\ (-\infty,\,-1) \cup (2,\,\tfrac{7}{2}]$

Resulting sign:	$\oplus$	$\ominus$	$\oplus$	$\ominus$
Sign of $-2x+7$:	$+$	$+$	$+$	$-$
Sign of $x-2$:	$-$	$-$	$+$	$+$
Sign of $x+1$:	$-$	$+$	$+$	$+$
x values:		-1	2	$7/2$

Diagram 33

37 $\dfrac{x}{3x-5} \leq \dfrac{2}{x-1} \Rightarrow \dfrac{x(x-1)-2(3x-5)}{(3x-5)(x-1)} \leq 0 \Rightarrow \dfrac{(x-2)(x-5)}{(3x-5)(x-1)} \leq 0;\ (1,\,\tfrac{5}{3}) \cup [2,\,5]$

Res. sign:	$\oplus$	$\ominus$	$\oplus$	$\ominus$	$\oplus$
$x-5$:	$-$	$-$	$-$	$-$	$+$
$x-2$:	$-$	$-$	$-$	$+$	$+$
$3x-5$:	$-$	$-$	$+$	$+$	$+$
$x-1$:	$-$	$+$	$+$	$+$	$+$
x values:		1	$5/3$	2	5

Diagram 37

39 $x^3 > x \Rightarrow x^3 - x > 0 \Rightarrow x(x^2-1) > 0 \Rightarrow x(x+1)(x-1) > 0;\ (-1,\,0) \cup (1,\,\infty)$

Resulting sign:	$\ominus$	$\oplus$	$\ominus$	$\oplus$
Sign of $x-1$:	$-$	$-$	$-$	$+$
Sign of x:	$-$	$-$	$+$	$+$
Sign of $x+1$:	$-$	$+$	$+$	$+$
x values:		-1	0	1

Diagram 39

41 $v \geq k \Rightarrow t^3 - 3t^2 - 4t + 20 \geq 8 \Rightarrow t^3 - 3t^2 - 4t + 12 \geq 0 \Rightarrow t^2(t-3) - 4(t-3) \geq 0 \Rightarrow$

$(t^2-4)(t-3) \geq 0 \Rightarrow (t+2)(t-2)(t-3) \geq 0.$ For $[0,\,5]$, we have $[0,\,2] \cup [3,\,5]$.

Resulting sign:	$\ominus$	$\oplus$	$\ominus$	$\oplus$
Sign of $t-3$:	$-$	$-$	$-$	$+$
Sign of $t-2$:	$-$	$-$	$+$	$+$
Sign of $t+2$:	$-$	$+$	$+$	$+$
t values:		-2	2	3

Diagram 41

45 $d < 75 \Rightarrow v + \tfrac{1}{20}v^2 < 75 \ \{\text{multiply by } 20\} \Rightarrow 20v + v^2 < 1500 \Rightarrow$

$v^2 + 20v - 1500 < 0 \Rightarrow (v+50)(v-30) < 0 \ \{\text{use a sign diagram}\} \Rightarrow$

$$-50 < v < 30 \Rightarrow 0 \leq v < 30 \ \{\text{since } v \geq 0\}$$

47 $R > S \Rightarrow \dfrac{4500\,S}{S+500} > S \Rightarrow \dfrac{S(S-4000)}{S+500} < 0 \ \{\text{use a sign diagram}\} \Rightarrow$

$$S < -500 \text{ or } 0 < S < 4000 \Rightarrow 0 < S < 4000 \ \{\text{since } S > 0\}$$

49 $W < 5 \Rightarrow 125\left(\dfrac{6400}{6400+x}\right)^2 < 5 \Rightarrow \left(\dfrac{6400}{6400+x}\right)^2 < \left(\dfrac{1}{5}\right)^2 \ \{\text{take the square root}\} \Rightarrow$

$$\dfrac{6400}{6400+x} < \dfrac{1}{5}\left\{\text{since } \dfrac{6400}{6400+x} > 0\right\} \Rightarrow 32{,}000 < x + 6400 \Rightarrow x > 25{,}600 \text{ km.}$$

Chapter 2 Review Exercises

$\boxed{3}$ $\left[\dfrac{2}{x+5} - \dfrac{3}{2x+1} = \dfrac{5}{6x+3}\right] \cdot 3(x+5)(2x+1) \Rightarrow 6(2x+1) - 9(x+5) = 5(x+5) \Rightarrow$

$$3x - 39 = 5x + 25 \Rightarrow -2x = 64 \Rightarrow x = -32$$

$\boxed{5}$ $\text{LS} = \dfrac{1}{\sqrt{x}} - 2 = \dfrac{1 - 2\sqrt{x}}{\sqrt{x}} = \text{RS}$, an identity.

The given equation is true for every $x > 0$.

$\boxed{9}$ $(x-2)(x+1) = 3 \Rightarrow x^2 - x - 2 = 3 \Rightarrow x^2 - x - 5 = 0 \Rightarrow x = \dfrac{1 \pm \sqrt{1 + 20}}{2} = \dfrac{1}{2} \pm \dfrac{1}{2}\sqrt{21}$

$\boxed{11}$ $x^{2/3} - 2x^{1/3} - 15 = 0 \Rightarrow (x^{1/3} + 3)(x^{1/3} - 5) = 0 \Rightarrow \sqrt[3]{x} = -3,\, 5 \Rightarrow x = -27,\, 125$

$\boxed{15}$ $6x^4 + 29x^2 + 28 = 0 \Rightarrow (2x^2 + 7)(3x^2 + 4) = 0 \Rightarrow x^2 = -\dfrac{7}{2},\, -\dfrac{4}{3} \Rightarrow$

$$x = \pm\dfrac{1}{2}\sqrt{14}\,i,\ \pm\dfrac{2}{3}\sqrt{3}\,i$$

$\boxed{19}$ $\left[\dfrac{1}{x} + 6 = \dfrac{5}{\sqrt{x}}\right] \cdot x \Rightarrow 1 + 6x = 5\sqrt{x} \Rightarrow$

$6x - 5\sqrt{x} + 1 = 0$ { factoring or substituting would be appropriate } $\Rightarrow$

$$(2\sqrt{x} - 1)(3\sqrt{x} - 1) = 0 \Rightarrow \sqrt{x} = \dfrac{1}{2},\, \dfrac{1}{3} \Rightarrow x = \dfrac{1}{4},\, \dfrac{1}{9}$$

Check $x = \dfrac{1}{4}$: $\text{LS} = 4 + 6 = 10$; $\text{RS} = 5/\dfrac{1}{2} = 10 \Rightarrow x = \dfrac{1}{4}$ is a solution.

Check $x = \dfrac{1}{9}$: $\text{LS} = 9 + 6 = 15$; $\text{RS} = 5/\dfrac{1}{3} = 15 \Rightarrow x = \dfrac{1}{9}$ is a solution.

$\boxed{23}$ $\sqrt{3x+1} - \sqrt{x+4} = 1 \Rightarrow \sqrt{3x+1} = 1 + \sqrt{x+4} \Rightarrow 3x + 1 = 1 + 2\sqrt{x+4} + x + 4 \Rightarrow$

$2\sqrt{x+4} = 2x - 4 \Rightarrow \sqrt{x+4} = x - 2 \Rightarrow (\sqrt{x+4})^2 = (x-2)^2 \Rightarrow$

$$x + 4 = x^2 - 4x + 4 \Rightarrow x^2 - 5x = 0 \Rightarrow x(x-5) = 0 \Rightarrow x = 0,\, 5.$$

Check $x = 0$: $\text{LS} = 1 - 2 = -1 \neq \text{RS} \Rightarrow x = 0$ is an extraneous solution.

Check $x = 5$: $\text{LS} = 4 - 3 = 1 = \text{RS} \Rightarrow x = 5$ is a solution.

$\boxed{27}$ The expression $(x-3)^2$ is never less than 0, but it is equal to 0 when $x = 3$.

Thus, $(x-3)^2 \le 0$ has solution $x = 3$.

$\boxed{31}$ $\dfrac{6}{10x+3} < 0 \Rightarrow 10x + 3 < 0$ { since $6 > 0$ } $\Rightarrow x < -\dfrac{3}{10} \Leftrightarrow (-\infty,\, -\dfrac{3}{10})$

$\boxed{35}$ $|16 - 3x| \ge 5 \Rightarrow 16 - 3x \ge 5$ or $16 - 3x \le -5 \Rightarrow -3x \ge -11$ or $-3x \le -21 \Rightarrow$

$$x \le \dfrac{11}{3} \text{ or } x \ge 7 \Leftrightarrow (-\infty,\, \tfrac{11}{3}] \cup [7,\, \infty)$$

$\boxed{39}$ $\dfrac{x^2(3-x)}{x+2} \le 0 \Rightarrow \dfrac{3-x}{x+2} \le 0$ { include 0 }; $(-\infty,\, -2) \cup \{\, 0\, \} \cup [3,\, \infty)$

Resulting sign:	$\ominus$	$\oplus$	$\ominus$
Sign of $3 - x$:	$+$	$+$	$-$
Sign of $x + 2$:	$-$	$+$	$+$
x values:		-2 $\qquad$ 3	

Diagram 39

41 $\dfrac{3}{2x+3} < \dfrac{1}{x-2} \Rightarrow \dfrac{3(x-2)-1(2x+3)}{(2x+3)(x-2)} < 0 \Rightarrow \dfrac{x-9}{(2x+3)(x-2)} < 0;\ (-\infty,\ -\tfrac{3}{2}) \cup (2,\ 9)$

Resulting sign:	⊖	⊕	⊖	⊕
Sign of $x-9$:	−	−	−	+
Sign of $x-2$:	−	−	+	+
Sign of $2x+3$:	−	+	+	+
x values:		−3/2	2	9

Diagram 41

43 $x^3 > x^2 \Rightarrow x^2(x-1) > 0 \ \{x^2 \ge 0\} \Rightarrow x-1 > 0 \Rightarrow x > 1 \Leftrightarrow (1,\ \infty)$

48 $V = \tfrac{1}{3}\pi h(r^2 + R^2 + rR) \Rightarrow r^2 + Rr + R^2 - \dfrac{3V}{\pi h} = 0 \Rightarrow$

$(\pi h)r^2 + (\pi hR)r + (\pi hR^2 - 3V) = 0 \Rightarrow$

$r = \dfrac{-\pi hR \pm \sqrt{\pi^2 h^2 R^2 - 4\pi h(\pi hR^2 - 3V)}}{2\pi h} = \dfrac{-\pi hR \pm \sqrt{12\pi hV - 3\pi^2 h^2 R^2}}{2\pi h}.$

Since $r > 0$, we must use the plus sign, and $r = \dfrac{-\pi hR + \sqrt{12\pi hV - 3\pi^2 h^2 R^2}}{2\pi h}.$

52 $\dfrac{1}{9 - \sqrt{-4}} = \dfrac{1}{9 - 2i} = \dfrac{1}{9 - 2i} \cdot \dfrac{9 + 2i}{9 + 2i} = \dfrac{9 + 2i}{81 + 4} = \dfrac{9}{85} + \dfrac{2}{85}i$

56 Let P denote the principal that will be invested, and r the yield rate of the stock fund. Income$_{\text{stocks}}$ − 28% federal tax − 7% state tax = Income$_{\text{bonds}}$ ⇒

$(Pr) - 0.28(Pr) - 0.07(Pr) = 0.07186P$ { divide by P } ⇒

$1r - 0.28r - 0.07r = 0.07186 \Rightarrow 0.65r = 0.07186 \Rightarrow r \approx 0.11055,$ or, 11.055%.

59 Let x denote the number of grams of 95% ethyl alcohol solution used, $400 - x$ the number of grams of water. $95(x) + 0(400 - x) = 75(400)$ { all in % } ⇒

$95x = 75(400) \Rightarrow x = \dfrac{6000}{19} \approx 315.8.$ Use 315.8 g of ethyl alcohol and 84.2 g of water.

64 Let $50 + r$ denote the rate the automobile, that is, r is the rate over 50 mi/hr. The automobile must travel $40 + 20 = 60$ ft more than the truck (traveling at 50 mi/hr) in 5 seconds. Since 1 mi/hr $= \dfrac{5280}{3600} = \dfrac{22}{15}$ ft/sec, the automobile's rate in *excess* of 50 mi/hr is $\dfrac{22}{15}r$. Thus, $d = rt \Rightarrow 60 = (\tfrac{22}{15}r)(5) \Rightarrow r = \dfrac{90}{11}.$

The rate is $50 + \dfrac{90}{11} = \dfrac{640}{11} \approx 58.2$ mi/hr.

65 Let x denote the number of hours needed to fill an empty bin.

Using the hourly rates, $\left[\dfrac{1}{2} - \dfrac{1}{5} = \dfrac{1}{x}\right] \cdot 10x \Rightarrow 5x - 2x = 10 \Rightarrow 3x = 10 \Rightarrow x = \dfrac{10}{3}$ hr.

Since the bin was half-full at the start, $\dfrac{1}{2}x = \dfrac{1}{2} \cdot \dfrac{10}{3} = \dfrac{5}{3}$ hr, or, 1 hr 40 min.

$\boxed{69}$ (a) The eastbound car has distance $20t$ and the southbound car has distance

$$(-2+50t). \quad d^2 = (20t)^2 + (-2+50t)^2 \Rightarrow d = \sqrt{2900t^2 - 200t + 4}$$

(b) $104 = \sqrt{2900t^2 - 200t + 4} \Rightarrow 2900t^2 - 200t - 10{,}812 = 0 \Rightarrow$

$$725t^2 - 50t - 2703 = 0 \Rightarrow t = \frac{50 \pm \sqrt{7{,}841{,}200}}{1450} \{t > 0\} = \frac{5 + 2\sqrt{19{,}603}}{145} \approx 1.97,$$

or approximately 11:58 A.M.

$\boxed{71}$ Let x denote the length of one side of an end.

(a) $V = lwh \Rightarrow 48 = 6 \cdot x \cdot x \Rightarrow x^2 = 8 \Rightarrow x = 2\sqrt{2}$ ft

(b) $S = lw + 2wh + 2lh \Rightarrow 44 = 6x + 2(x^2) + 2(6x) \Rightarrow 44 = 2x^2 + 18x \Rightarrow$

$$x^2 + 9x - 22 = 0 \Rightarrow (x + 11)(x - 2) = 0 \Rightarrow x = 2 \text{ ft}$$

$\boxed{73}$ Let x denote the width of the tiled area, $2x$ the length.

The bathing area has measurements $x - 2$ and $2x - 2$. For the bathing area,

width $\cdot$ length = area $\Rightarrow (x - 2)(2x - 2) = 40 \Rightarrow 2x^2 - 6x + 4 = 40 \Rightarrow$

$x^2 - 3x + 2 = 20 \Rightarrow x^2 - 3x - 18 = 0 \Rightarrow (x - 6)(x + 3) = 0 \Rightarrow x = 6 \{x > 0\}.$

The tiled area is 12 ft by 6 ft and the bathing area is 10 ft by 4 ft.

$\boxed{76}$ Let x denote the amount of yearly business (in dollars).

$\text{Pay}_B > \text{Pay}_A \Rightarrow \$20{,}000 + 0.10x > \$25{,}000 + 0.05x \Rightarrow 0.05x > \$5000 \Rightarrow x > \$100{,}000$

$\boxed{78}$ $T = 2\pi\sqrt{\dfrac{l}{980}} \Rightarrow l = \dfrac{980\,T^2}{4\pi^2}. \quad 98 \le l \le 100 \Rightarrow 98 \le \dfrac{980\,T^2}{4\pi^2} \le 100 \Rightarrow$

$\dfrac{2\pi^2}{5} \le T^2 \le \dfrac{20\pi^2}{49} \Rightarrow \dfrac{10\pi^2}{25} \le T^2 \le \dfrac{20\pi^2}{49} \Rightarrow \dfrac{\pi}{5}\sqrt{10} \le T \le \dfrac{2\pi}{7}\sqrt{5} \{T \ge 0\},$

or, approximately, $1.987 \le T \le 2.007$ sec.

$\boxed{82}$ Let x denote the number of \$10 increases in rent. Then the

number of occupied apartments is $180 - 5x$ and the rent per apartment is $300 + 10x$.

Total income = (# of occupied apartments)(rent per apartment) =

$(180 - 5x)(300 + 10x) = -50x^2 + 300x + 54{,}000. \quad$ Income $\ge 54{,}400 \Rightarrow$

$-50x^2 + 300x + 54{,}000 \ge 54{,}400 \Rightarrow -50x^2 + 300x - 400 \ge 0 \Rightarrow x^2 - 6x + 8 \le 0 \Rightarrow$

$(x - 2)(x - 4) \le 0 \Rightarrow 2 \le x \le 4.$ Hence, the rent charged should be \$320 to \$340.

Chapter 3: Functions and Graphs

$\boxed{7}$ (a) $x = -2$ is the line parallel to the y-axis that intersects the x-axis at $(-2, 0)$.

(b) $y = 3$ is the line parallel to the x-axis that intersects the y-axis at $(0, 3)$.

(c) $x \ge 0$ { x is zero or positive }

is the set of all points to the right of and on the y-axis.

(d) $xy > 0$ { x and y have the same sign, that is, either both are positive or both are

negative } is the set of all points in quadrants I and III.

(e) $y < 0$ { y is negative } is the set of all points below the x-axis.

(f) $x = 0$ is the set of all points on the y-axis.

$\boxed{9}$ (a) $A(4, -3)$, $B(6, 2) \Rightarrow d(A, B) = \sqrt{(6-4)^2 + [2-(-3)]^2} = \sqrt{4+25} = \sqrt{29}$

(b) $M_{AB} = \left(\dfrac{4+6}{2}, \dfrac{-3+2}{2}\right) = (5, -\tfrac{1}{2})$

$\boxed{11}$ (a) $A(-5, 0)$, $B(-2, -2) \Rightarrow d(A, B) = \sqrt{[-2-(-5)]^2 + (-2-0)^2} = \sqrt{9+4} = \sqrt{13}$

(b) $M_{AB} = \left(\dfrac{-5+(-2)}{2}, \dfrac{0+(-2)}{2}\right) = (-\tfrac{7}{2}, -1)$

$\boxed{15}$ We need to show that the sides satisfy the Pythagorean theorem. Finding the distances, we have $d(A, B) = \sqrt{98}$, $d(B, C) = \sqrt{32}$, and $d(A, C) = \sqrt{130}$. Since $d(A, C)$ is the largest of the three values, it must be the hypotenuse, hence, we need to check if $d(A, C)^2 = d(A, B)^2 + d(B, C)^2$. Since $(\sqrt{130})^2 = (\sqrt{98})^2 + (\sqrt{32})^2$, we know that $\triangle ABC$ is a right triangle. The area of a triangle is given by $A = \tfrac{1}{2}(\text{base})(\text{height})$. We can use $d(B, C)$ for the base and $d(A, B)$ for the height. Hence, area $= \tfrac{1}{2}bh = \tfrac{1}{2}(\sqrt{32})(\sqrt{98}) = \tfrac{1}{2}(4\sqrt{2})(7\sqrt{2}) = \tfrac{1}{2}(28)(2) = 28$.

$\boxed{17}$ We need to show that all 4 sides are the same length. Checking, we find that $d(A, B) = d(B, C) = d(C, D) = d(D, A) = \sqrt{29}$. This guarantees that we have a rhombus { a parallelogram with 4 equal sides }. Thus, we also need to show that adjacent sides meet at right angles. This can be done by showing that two adjacent sides and a diagonal form a right triangle. Using $\triangle ABC$, we see that $d(A, C) = \sqrt{58}$ and hence $d(A, C)^2 = d(A, B)^2 + d(B, C)^2$. We conclude that $ABCD$ is a square.

$\boxed{19}$ Let $B = (x, y)$. $A(-3, 8) \Rightarrow M_{AB} = \left(\dfrac{-3+x}{2}, \dfrac{8+y}{2}\right)$. $M_{AB} = C(5, -10) \Rightarrow$

$-3 + x = 2(5)$ and $8 + y = 2(-10) \Rightarrow x = 13$ and $y = -28$. $B = (13, -28)$.

21 The perpendicular bisector of AB is the line that passes through the midpoint of segment AB and intersects segment AB at a right angle. The points on the perpendicular bisector are all equidistant from A and B. Thus, we need to show that $d(A, C) = d(B, C)$. Since each of these is $\sqrt{145}$, we conclude that C is on the perpendicular bisector of AB.

23 We must have $d(A, P) = d(B, P)$. $\sqrt{(x+4)^2 + (y+3)^2} = \sqrt{(x-6)^2 + (y-1)^2} \Rightarrow$
$x^2 + 8x + 16 + y^2 + 6y + 9 = x^2 - 12x + 36 + y^2 - 2y + 1 \Rightarrow$
$$8x + 6y + 25 = -12x - 2y + 37 \Rightarrow 20x + 8y = 12 \Rightarrow 5x + 2y = 3$$

25 Let $O(0, 0)$ represent the origin. Applying the distance formula with O and $P(x, y)$, we have $d(O, P) = 5 \Rightarrow \sqrt{(x-0)^2 + (y-0)^2} = 5 \Rightarrow \sqrt{x^2 + y^2} = 5$.

This is a circle of radius 5 with center at the origin.

27 Let $Q(0, y)$ be an arbitrary point on the y-axis.

Applying the distance formula with Q and $P(5, 3)$, we have $6 = d(P, Q) \Rightarrow$
$$6 = \sqrt{(0-5)^2 + (y-3)^2} \Rightarrow 36 = 25 + y^2 - 6y + 9 \Rightarrow$$
$$y^2 - 6y - 2 = 0 \Rightarrow y = 3 \pm \sqrt{11}. \text{ The points are } (0, 3 + \sqrt{11}) \text{ and } (0, 3 - \sqrt{11}).$$

29 $5 = \sqrt{(2a-1)^2 + (a-3)^2} \Rightarrow 25 = 4a^2 - 4a + 1 + a^2 - 6a + 9 \Rightarrow$
$5a^2 - 10a - 15 = 0 \Rightarrow a^2 - 2a - 3 = 0 \Rightarrow (a-3)(a+1) = 0 \Rightarrow a = 3, -1$. Since the
y-coordinate is negative in the third quadrant, $a = -1$, and $(2a, a) = (-2, -1)$.

31 With $P(a, 3)$ and $Q(5, 2a)$, $d(P, Q) > \sqrt{26} \Rightarrow \sqrt{(5-a)^2 + (2a-3)^2} > \sqrt{26} \Rightarrow$
$25 - 10a + a^2 + 4a^2 - 12a + 9 > 26 \Rightarrow 5a^2 - 22a + 8 > 0 \Rightarrow (5a-2)(a-4) > 0$.

Using *Diagram 31*, we see that $a < \frac{2}{5}$ or $a > 4$ will assure us that $d(P, Q) > \sqrt{26}$.

Resulting sign:	$\oplus$	$\ominus$	$\oplus$
Sign of $5a - 2$:	$-$	$+$	$+$
Sign of $a - 4$:	$-$	$-$	$+$
a values:		2/5	4

Diagram 31

33 Let M be the midpoint of the hypotenuse. Then $M = (\frac{1}{2}a, \frac{1}{2}b)$.

Show that $d(A, M) = d(B, M) = d(O, M) = \frac{1}{2}\sqrt{a^2 + b^2}$.

3.2 Exercises

1 As in Example 1, we expect the graph to be a line.

Creating a table of values similar to those in the text, we have:

x	-2	-1	0	1	2
y	-7	-5	-3	-1	1

By plotting these points and connecting them, we obtain *Figure 1*.

To find the x-intercept, let $y = 0$ in $y = 2x - 3$, and solve for x. (1.5, 0)

To find the y-intercept, let $x = 0$ in $y = 2x - 3$, and solve for y. (0, −3)

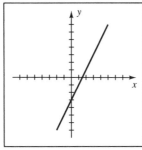

Figure 1

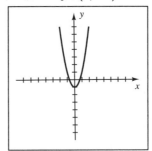

Figure 7

⑦ $y = 2x^2 - 1$ • x-intercepts: $(\pm \frac{1}{2}\sqrt{2}, 0)$ y-intercept: $(0, -1)$

Since we can substitute $-x$ for x in the equation and obtain an equivalent equation, we know the graph is symmetric with respect to the y-axis. We will make use of this fact when constructing our table. As in Example 2, we obtain a parabola.

x	± 2	$\pm \frac{3}{2}$	± 1	$\pm \frac{1}{2}$	0
y	7	$\frac{7}{2}$	1	$-\frac{1}{2}$	−1

⑪ $x = -y^2 + 3$ • x-intercept: (3, 0) y-intercepts: $(0, \pm\sqrt{3})$

Since we can substitute $-y$ for y in the equation and obtain an equivalent equation, we know the graph is symmetric with respect to the x-axis. We will make use of this fact when constructing our table. As in Example 4, we obtain a parabola.

x	−13	−6	−1	2	3
y	± 4	± 3	± 2	± 1	0

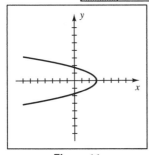

Figure 11

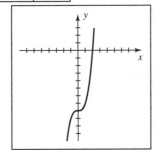

Figure 15

⑮ $y = x^3 - 8$ • x-intercept: (2, 0) y-intercept: (0, −8)

x	−2	−1	0	1	2
y	−16	−9	−8	−7	0

19 $y = \sqrt{x} - 4$ • x-intercept: $(16, 0)$ y-intercept: $(0, -4)$

x	0	1	4	9	16
y	-4	-3	-2	-1	0

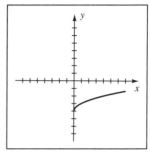

Figure 19

21 You may be able to do this exercise mentally. For example (using Exercise 1 with $y = 2x - 3$), we see that substituting $-x$ for x gives us $y = -2x - 3$; substituting $-y$ for y gives us $-y = 2x - 3$ or, equivalently, $y = -2x + 3$; and substituting $-x$ for x and $-y$ for y gives us $-y = -2x - 3$ or, equivalently, $y = 2x + 3$. None of the resulting equations are equivalent to the original equation, so there is no symmetry with respect to the y-axis, x-axis, or the origin.

(a) The graphs of the equations in Exercises 5 and 7 are symmetric with respect to the y-axis.

(b) The graphs of the equations in Exercises 9 and 11 are symmetric with respect to the x-axis.

(c) The graph of the equation in Exercise 13 is symmetric with respect to the origin.

25 $(x + 3)^2 + (y - 2)^2 = 9$ is a circle of radius $r = \sqrt{9} = 3$ with center $C(-3, 2)$.

See *Figure 25*.

29 $4x^2 + 4y^2 = 25 \Rightarrow x^2 + y^2 = \frac{25}{4}$ is a circle of radius $r = \sqrt{\frac{25}{4}} = \frac{5}{2}$ with center $C(0, 0)$.

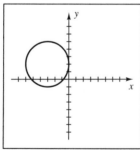
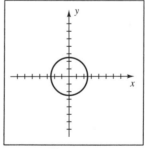
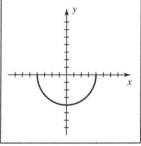

Figure 25   *Figure 29*   *Figure 31*

31 As in Example 8, $y = -\sqrt{16 - x^2}$ is the lower half of the circle $x^2 + y^2 = 16$.

39 An equation of a circle with center $C(-4, 6)$ is $(x + 4)^2 + (y - 6)^2 = r^2$.

Since the circle passes through $P(1, 2)$, we know that $x = 1$ and $y = 2$ is one solution of the general equation. Letting $x = 1$ and $y = 2$ yields $5^2 + (-4)^2 = r^2 \Rightarrow r^2 = 41$.

An equation is $(x + 4)^2 + (y - 6)^2 = 41$.

41 "Tangent to the y-axis" means that the circle will intersect the y-axis at exactly one point. The distance from the center $C(-3,\ 6)$ to this point of tangency is 3 units— this is the length of the radius of the circle. An equation is $(x+3)^2 + (y-6)^2 = 9$.

43 Since the radius is 4 and $C(h,\ k)$ is in QII, $h = -4$ and $k = 4$.

An equation is $(x+4)^2 + (y-4)^2 = 16$.

45 The center of the circle is the midpoint M of $A(4,\ -3)$ and $B(-2,\ 7)$. $M = (1,\ 2)$.

The radius of the circle is $\frac{1}{2} \cdot d(A,\ B) = \frac{1}{2}\sqrt{136} = \sqrt{34}$.

An equation is $(x-1)^2 + (y-2)^2 = 34$.

47 $x^2 + y^2 - 4x + 6y - 36 = 0$ { complete the square on x and y } $\Rightarrow$

$x^2 - 4x + \underline{4} + y^2 + 6y + \underline{9} = 36 + \underline{4} + \underline{9} \Rightarrow$

$(x-2)^2 + (y+3)^2 = 49$.   This is a circle with center $C(2,\ -3)$ and radius $r = 7$.

51 $2x^2 + 2y^2 - 12x + 4y - 15 = 0$ { add 15 to both sides and divide by 2 } $\Rightarrow$

$x^2 + y^2 - 6x + 2y = \frac{15}{2}$ { complete the square on x and y } $\Rightarrow$

$x^2 - 6x + \underline{9} + y^2 + 2y + \underline{1} = \frac{15}{2} + \underline{9} + \underline{1} \Rightarrow$

$(x-3)^2 + (y+1)^2 = \frac{35}{2}$.   This is a circle with center $C(3,\ -1)$ and radius $r = \frac{1}{2}\sqrt{70}$.

55 $x^2 + y^2 - 2x - 8y + 19 = 0 \Rightarrow x^2 - 2x + \underline{1} + y^2 - 8y + \underline{16} = -19 + \underline{1} + \underline{16} \Rightarrow$

$(x-1)^2 + (y-4)^2 = -2$. This is not a circle since r^2 cannot equal -2.

59 To obtain equations for the upper and lower halves, we solve the given equation for y in terms of x. $(x-2)^2 + (y+1)^2 = 49 \Rightarrow (y+1)^2 = 49 - (x-2)^2 \Rightarrow$

$y + 1 = \pm\sqrt{49 - (x-2)^2} \Rightarrow y = -1 \pm\sqrt{49 - (x-2)^2}$.

The upper half is $y = -1 + \sqrt{49 - (x-2)^2}$ and

the lower half is $y = -1 - \sqrt{49 - (x-2)^2}$.

To obtain equations for the right and left halves, we solve for x in terms of y.

$(x-2)^2 + (y+1)^2 = 49 \Rightarrow (x-2)^2 = 49 - (y+1)^2 \Rightarrow$

$x - 2 = \pm\sqrt{49 - (y+1)^2} \Rightarrow x = 2 \pm\sqrt{49 - (y+1)^2}$. The right half is

$x = 2 + \sqrt{49 - (y+1)^2}$ and the left half is $x = 2 - \sqrt{49 - (y+1)^2}$.

61 We need to determine if the distance from P to C is *less than r, greater than r,* or *equal to r* and hence, P will be *inside* the circle, *outside* the circle, or *on* the circle, respectively.

(a) $P(2,\ 3)$, $C(4,\ 6) \Rightarrow d(P,\ C) = \sqrt{4+9} = \sqrt{13} < r \ \{r = 4\} \Rightarrow P$ is *inside* C.

(b) $P(4,\ 2)$, $C(1,\ -2) \Rightarrow d(P,\ C) = \sqrt{9+16} = 5 = r \ \{r = 5\} \Rightarrow P$ is *on* C.

(c) $P(-3,\ 5)$, $C(2,\ 1) \Rightarrow d(P,\ C) = \sqrt{25+16} = \sqrt{41} > r \ \{r = 6\} \Rightarrow P$ is *outside* C.

63 (a) To find the x-intercepts, let $y = 0$ and solve the resulting equation for x.

$$x^2 - 4x + 4 = 0 \Rightarrow (x-2)^2 = 0 \Rightarrow x = 2.$$

(b) To find the y-intercepts, let $x = 0$ and solve the resulting equation for y.

$$y^2 - 6y + 4 = 0 \Rightarrow y = \frac{6 \pm \sqrt{36-16}}{2} = 3 \pm \sqrt{5}.$$

65 $x^2 + y^2 + 4x - 6y + 4 = 0 \Leftrightarrow (x+2)^2 + (y-3)^2 = 9$. This is a circle with center $C(-2,\ 3)$ and radius 3. The circle we want has the same center, $C(-2,\ 3)$, and radius that is equal to the distance from C to $P(2,\ 6)$.

$$d(P,\ C) = \sqrt{16+9} = 5 \text{ and an equation is } (x+2)^2 + (y-3)^2 = 25.$$

67 Assuming that the x- and y-values of each point of intersection are integers, and that the tics each represent one unit, we see that the intersection points are $(-3,\ -5)$ and $(2,\ 0)$. The viewing rectangle (VR) is $[-15,\ 15]$ by $[-10,\ 10]$.

$$Y_1 < Y_2 \text{ on } [-15,\ -3) \cup (2,\ 15].$$

71 Assign $x^3 - \frac{9}{10}x^2 - \frac{43}{25}x + \frac{24}{25}$ to Y_1. After trying a standard viewing rectangle, we see that the x-intercepts are near the origin and we choose the viewing rectangle $[-6,\ 6]$ by $[-4,\ 4]$. This is simply one choice, not necessarily the best choice. For most ⓒ exercises, we have selected viewing rectangles that are in a $3:2$ proportion (horizontal : vertical) to maintain a true proportion. From the graph, there are three x-intercepts. Use a zoom-in feature to determine that they are approximately -1.2, 0.5, and 1.6.

$[-6,\ 6]$ by $[-4,\ 4]$ $[-6,\ 6]$ by $[-4,\ 4]$

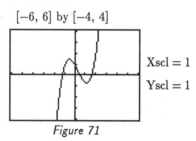   Xscl $= 1$ 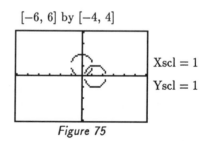   Xscl $= 1$

Yscl $= 1$ Yscl $= 1$

Figure 71   *Figure 75*

75 Depending on the type of graphing utility used, you may need to solve for y first.

$$x^2 + (y-1)^2 = 1 \Rightarrow y = 1 \pm \sqrt{1-x^2}; \qquad (x-\tfrac{5}{4})^2 + y^2 = 1 \Rightarrow y = \pm \sqrt{1 - (x-\tfrac{5}{4})^2}.$$

Make the assignments $Y_1 = \sqrt{1-x^2}$, $Y_2 = 1 + Y_1$, $Y_3 = 1 - Y_1$, $Y_4 = \sqrt{1 - (x-\tfrac{5}{4})^2}$, and $Y_5 = -Y_4$. If a Y_5 is not available, you will need to use other function assignments or alternate methods. For example, on the TI-81, you can graph Y_5 by using DrawF $-Y_4$. Be sure to "turn off" Y_1 before graphing. From the graph, there are two points of intersection.

$$\text{They are approximately } (0.999,\ 0.968) \text{ and } (0.251,\ 0.032).$$

3.3 Exercises

$\boxed{1}$ $A(-3, 2), B(5, -4) \Rightarrow m_{AB} = \dfrac{(-4) - 2}{5 - (-3)} = \dfrac{-6}{8} = -\dfrac{3}{4}$

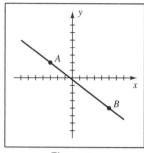

Figure 1

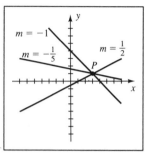

Figure 15

$\boxed{7}$ Show that the slopes of opposite sides are equal.

$A(-3, 1), B(5, 3), C(3, 0), D(-5, -2) \Rightarrow m_{AB} = \frac{1}{4} = m_{DC}$ and $m_{DA} = \frac{3}{2} = m_{CB}$.

$\boxed{11}$ $A(-1, -3)$ is 5 units to the left and 5 units down from $B(4, 2)$. D will have the

same relative position from $C(-7, 5)$, that is, $(-7 - 5, 5 - 5) = (-12, 0)$.

$\boxed{15}$ $P(3, 1)$; $m = \frac{1}{2}, -1, -\frac{1}{5}$ • See *Figure 15*.

$\boxed{19}$ (a) Parallel to the y-axis implies the equation is of the form $x = k$.

The x-value of $A(5, -2)$ is 5, hence $x = 5$ is the equation.

(b) Perpendicular to the y-axis implies the equation is of the form $y = k$.

The y-value of $A(5, -2)$ is -2, hence $y = -2$ is the equation.

$\boxed{23}$ $A(4, 0)$; slope -3 { use the point-slope form of a line } $\Rightarrow$

$$y - 0 = -3(x - 4) \Rightarrow y = -3x + 12 \Rightarrow 3x + y = 12.$$

$\boxed{25}$ $A(4, -5), B(-3, 6) \Rightarrow m_{AB} = -\frac{11}{7}$.

$$y + 5 = -\tfrac{11}{7}(x - 4) \Rightarrow 7(y + 5) = -11(x - 4) \Rightarrow 7y + 35 = -11x + 44 \Rightarrow 11x + 7y = 9.$$

$\boxed{27}$ $5x - 2y = 4 \Leftrightarrow y = \frac{5}{2}x - 2$. Using the same slope, $\frac{5}{2}$, with $A(2, -4)$, gives us

$$y + 4 = \tfrac{5}{2}(x - 2) \Rightarrow 2(y + 4) = 5(x - 2) \Rightarrow 2y + 8 = 5x - 10 \Rightarrow 5x - 2y = 18.$$

$\boxed{29}$ $2x - 5y = 8 \Leftrightarrow y = \frac{2}{5}x - \frac{8}{5}$. Using the negative reciprocal of $\frac{2}{5}$ for the slope,

$$y + 3 = -\tfrac{5}{2}(x - 7) \Rightarrow 2(y + 3) = -5(x - 7) \Rightarrow 2y + 6 = -5x + 35 \Rightarrow 5x + 2y = 29.$$

$\boxed{33}$ $A(5, 2), B(-1, 4) \Rightarrow m = -\frac{1}{3}$.

$$y - 2 = -\tfrac{1}{3}(x - 5) \Rightarrow y = -\tfrac{1}{3}x + \tfrac{5}{3} + 2 \Rightarrow y = -\tfrac{1}{3}x + \tfrac{11}{3}.$$

$\boxed{35}$ We need the line through the midpoint of segment AB that is perpendicular to

segment AB. $A(3, -1), B(-2, 6) \Rightarrow M_{AB} = (\frac{1}{2}, \frac{5}{2})$ and $m_{AB} = -\frac{7}{5}$.

$$y - \tfrac{5}{2} = \tfrac{5}{7}(x - \tfrac{1}{2}) \Rightarrow 7(y - \tfrac{5}{2}) = 5(x - \tfrac{1}{2}) \Rightarrow 7y - \tfrac{35}{2} = 5x - \tfrac{5}{2} \Rightarrow 5x - 7y = -15.$$

39 We can solve the given equation for y to obtain the slope-intercept form, $y = mx + b$.

$$2x = 15 - 3y \Rightarrow 3y = -2x + 15 \Rightarrow y = -\tfrac{2}{3}x + 5;\ m = -\tfrac{2}{3},\ b = 5$$

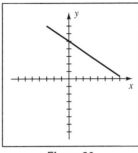

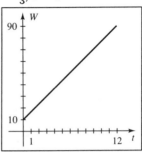

Figure 39 Figure 55

43 (a) An equation of the horizontal line with y-intercept 3 is $y = 3$.

 (b) An equation of the line through the origin with slope $-\tfrac{1}{2}$ is $y = -\tfrac{1}{2}x$.

 (c) An equation of the line with slope $-\tfrac{3}{2}$ and y-intercept 1 is $y = -\tfrac{3}{2}x + 1$.

 (d) An equation of the line through $(3, -2)$ with slope -1 is $y + 2 = -(x - 3)$.

  Alternatively, we have a slope of -1 and a y-intercept of 1, i.e., $y = -x + 1$.

45 Since we want to obtain a "1" on the right side of the equation, we will divide by 6.

$$\left[4x - 2y = 6\right] \cdot \tfrac{1}{6} \Rightarrow \tfrac{4x}{6} - \tfrac{2y}{6} = \tfrac{6}{6} \Rightarrow \tfrac{2x}{3} - \tfrac{y}{3} = 1 \Rightarrow \tfrac{x}{\frac{3}{2}} + \tfrac{y}{-3} = 1$$

The x-intercept is $(\tfrac{3}{2}, 0)$ and the y-intercept is $(0, -3)$.

47 The radius of the circle is the vertical distance from the center of the circle to the line

$y = 5$, that is, $r = 5 - (-2) = 7$. An equation is $(x - 3)^2 + (y + 2)^2 = 49$.

51 (a) $L = 40 \Rightarrow W = 1.70(40) - 42.8 = 25.2$ tons

 (b) Error in $L = \pm 2 \Rightarrow$ Error in $W = 1.70(\pm 2) = \pm 3.4$ tons

53 (a) $y = mx = \dfrac{\text{change in } y \text{ from the beginning of the season}}{\text{change in } x \text{ from the beginning of the season}}(x) = \dfrac{5 - 0}{14 - 0}x = \dfrac{5}{14}x$.

 (b) $x = 162 \Rightarrow y = \tfrac{5}{14}(162) \approx 58$.

55 (a) Using the slope-intercept form, $W = mt + b = mt + 10$.

  $W = 30$ when $t = 3 \Rightarrow 30 = 3m + 10 \Rightarrow m = \tfrac{20}{3}$ and $W = \tfrac{20}{3}t + 10$.

 (b) $t = 6 \Rightarrow W = \tfrac{20}{3}(6) + 10 \Rightarrow W = 50$ lb

 (c) $W = 70 \Rightarrow 70 = \tfrac{20}{3}t + 10 \Rightarrow 60 = \tfrac{20}{3}t \Rightarrow t = 9$ years old

 (d) The graph has end points at $(0, 10)$ and $(12, 90)$. See *Figure 55*.

59 (a) Using the slope-intercept form with $m = 0.032$ and $b = 13.5$,

  we have $T = 0.032t + 13.5$.

 (b) $t = 2000 - 1915 = 85 \Rightarrow T = 0.032(85) + 13.5 = 16.22\,°\text{C}$.

61 (a) Expenses $(E) = (\$1000) + (5\% \text{ of } R) + (\$2600) + (50\% \text{ of } R) \Rightarrow$

$$E = 1000 + 0.05R + 2600 + 0.50R = 0.55R + 3600.$$

(b) Profit $(P) = $ Revenue $(R) - $ Expenses $(E) \Rightarrow$

$$P = R - (0.55R + 3600) = R - 0.55R - 3600 = 0.45R - 3600.$$

(c) *Break even* means P would be 0. $P = 0 \Rightarrow 0 = 0.45R - 3600 \Rightarrow 0.45R = 3600 \Rightarrow$

$$R = 3600(\tfrac{100}{45}) = \$8000/\text{month}$$

63 The targets are on the x-axis $\{$ which is the line $y = 0\}$.

To determine if a target is hit, set $y = 0$ and solve for x.

(a) $y - 2 = -1(x - 1) \Rightarrow x + y = 3$. $y = 0 \Rightarrow x = 3$ and a creature is hit.

(b) $y - \frac{5}{3} = -\frac{4}{9}(x - \frac{3}{2}) \Rightarrow 4x + 9y = 21$. $y = 0 \Rightarrow x = 5.25$ and no creature is hit.

67 From the graph, we can see that the points of intersection are $A(-0.8, -0.6)$,

$B(4.8, -3.4)$, and $C(2, 5)$. The lines intersecting at A are perpendicular since they

have slopes of 2 and $-\frac{1}{2}$. Since $d(A, B) = \sqrt{39.2}$ and $d(A, C) = \sqrt{39.2}$,

the triangle is isosceles. Thus, the polygon is a right isosceles triangle.

$[-15, 15]$ by $[-10, 10]$

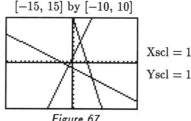

Xscl $= 1$

Yscl $= 1$

Figure 67

69 The slope of AB is $\dfrac{-1.11905 - (-1.3598)}{-0.55 - (-1.3)} = 0.321$. Similarly, the slopes of BC and

CD are also 0.321. Therefore, the points all lie on the same line. Since the common

slope is 0.321, let $a = 0.321$. $y = 0.321x + b \Rightarrow -1.3598 = 0.321(-1.3) + b \Rightarrow$

$b = -0.9425$. Thus, the points are linearly related by the equation

$$y = 0.321x - 0.9425.$$

3.4 Exercises

3 $f(x) = \sqrt{x - 4} - 3x \Rightarrow f(4) = \sqrt{4 - 4} - 3(4) = \sqrt{0} - 12 = 0 - 12 = -12$. Similarly,

$f(8) = -22$ and $f(13) = -36$. Note that $f(a)$, with $a < 4$, would be undefined.

7 (a) $f(x) = x^2 - x + 3 \Rightarrow f(a) = (a)^2 - (a) + 3 = a^2 - a + 3$

(b) $f(-a) = (-a)^2 - (-a) + 3 = a^2 + a + 3$

(c) $-f(a) = -1 \cdot (a^2 - a + 3) = -a^2 + a - 3$

(d) $f(a + h) = (a + h)^2 - (a + h) + 3 = a^2 + 2ah + h^2 - a - h + 3$

(e) $f(a) + f(h) = (a^2 - a + 3) + (h^2 - h + 3) = a^2 + h^2 - a - h + 6$

(f) $\dfrac{f(a+h)-f(a)}{h} = \dfrac{(a^2+2ah+h^2-a-h+3)-(a^2-a+3)}{h} = \dfrac{2ah+h^2-h}{h} =$

$$\dfrac{h(2a+h-1)}{h} = 2a+h-1$$

$\boxed{11}$ (a) $g(x) = \dfrac{2x}{x^2+1} \Rightarrow g\!\left(\dfrac{1}{a}\right) = \dfrac{2(1/a)}{(1/a)^2+1} = \dfrac{2/a}{1/a^2+1}\cdot\dfrac{a^2}{a^2} = \dfrac{2a}{1+a^2} = \dfrac{2a}{a^2+1}$

(b) $\dfrac{1}{g(a)} = \dfrac{1}{\dfrac{2a}{a^2+1}} = \dfrac{a^2+1}{2a}$
    (c) $g(\sqrt{a}) = \dfrac{2\sqrt{a}}{(\sqrt{a})^2+1} = \dfrac{2\sqrt{a}}{a+1}$

(d) $\sqrt{g(a)} = \sqrt{\dfrac{2a}{a^2+1}} = \dfrac{\sqrt{a^2+1}}{\sqrt{a^2+1}} = \dfrac{\sqrt{2a(a^2+1)}}{a^2+1}$, or, equivalently, $\dfrac{\sqrt{2a^3+2a}}{a^2+1}$

$\boxed{13}$ (a) The domain of a function f is the set of all x-values for which the function is defined. In this case, the graph extends from $x = -3$ to $x = 4$. Hence, the domain is $[-3, 4]$.

(b) The range of a function f is the set of all y-values that the function takes on. In this case, the graph includes all values from $y = -2$ to $y = 2$. Hence, the range is $[-2, 2]$.

(c) $f(1)$ is the y-value of f corresponding to $x = 1$. In this case, $f(1) = 0$.

(d) If we were to draw the horizontal line $y = 1$ on the same coordinate plane, it would intersect the graph at $x = -1$, $\tfrac{1}{2}$, and 2. Hence, $f(x) = 1 \Rightarrow x = -1, \tfrac{1}{2}, 2$.

(e) The function is above 1 between $x = -1$ and $x = \tfrac{1}{2}$, and also to the right of $x = 2$. Hence, $f(x) > 1 \Rightarrow x \in (-1, \tfrac{1}{2}) \cup (2, 4]$.

$\boxed{15\text{–}26}$ We need to make sure that the radicand {the expression under the radical sign} is greater than or equal to zero and that the denominator is not equal to zero.

$\boxed{17}$ $f(x) = \sqrt{9-x^2}$ • $9-x^2 \ge 0 \Rightarrow 3 \ge |x| \Rightarrow -3 \le x \le 3$ ★ $[-3, 3]$

$\boxed{19}$ $f(x) = \dfrac{x+1}{x^3-4x}$ • $x^3-4x = 0 \Rightarrow x(x+2)(x-2) = 0$ ★ $\mathbb{R} - \{\pm 2, 0\}$

$\boxed{21}$ $f(x) = \dfrac{\sqrt{2x-3}}{x^2-5x+4}$ • For this function we must have the radicand greater than or equal to 0 *and* the denominator not equal to 0. The radicand is greater than or equal to 0 if $2x-3 \ge 0$, or, equivalently, $x \ge \tfrac{3}{2}$. The denominator is $(x-1)(x-4)$, so $x \ne 1, 4$. The solution is then all real numbers greater than or equal to $\tfrac{3}{2}$, excluding 4. In interval notation, we have $[\tfrac{3}{2}, 4) \cup (4, \infty)$.

$\boxed{25}$ $f(x) = \sqrt{x+2} + \sqrt{2-x}$ • $x+2 \ge 0 \Rightarrow x \ge -2$; $2-x \ge 0 \Rightarrow x \le 2$.

The domain is the intersection of $x \ge -2$ and $x \le 2$, that is, $[-2, 2]$.

29 (a) To sketch the graph of $f(x) = 4 - x^2$, we can make use of the symmetry with respect to the y-axis. See *Figure 29*.

x	± 4	± 3	± 2	± 1	0
y	-12	-5	0	3	4

(b) Since we can substitute any number for x, the domain is all real numbers, that is, $D = \mathbb{R}$. By examining *Figure 29*, we see that the values of y are at most 4. Hence, the range of f is all reals less than or equal to 4, that is, $R = (-\infty, 4]$.

(c) A common mistake is to confuse the function values, the y's, with the input values, the x's. We are not interested in the specific y-values for determining if the function is increasing, decreasing, or constant. We are only interested if the y-values are going up, going down, or staying the same. For the function $f(x) = 4 - x^2$, we say f *is increasing on* $(-\infty, 0]$ since the y-values are getting larger as we move from left to right over the x-values from $-\infty$ to 0. Also, f *is decreasing on* $[0, \infty)$ since the y-values are getting smaller as we move from left to right over the x-values from 0 to ∞. Note that this answer would have been the same if the function was $f(x) = 500 - x^2$, $f(x) = -300 - x^2$, or any function of the form $f(x) = a - x^2$, where a is any real number.

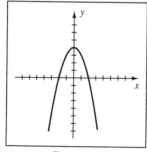

Figure 29

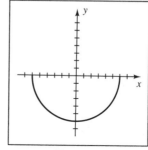

Figure 35

35 (a) We recognize $y = f(x) = -\sqrt{36 - x^2}$ as the lower half of the circle $x^2 + y^2 = 36$.

(b) To find the domain, we solve $36 - x^2 \geq 0$. $36 - x^2 \geq 0 \Rightarrow x^2 \leq 36 \Rightarrow$ $|x| \leq 6 \Rightarrow D = [-6, 6]$. From *Figure 35*, we see that the y-values vary from
$$y = -6 \text{ to } y = 0. \text{ Hence, the range } R \text{ is } [-6, 0].$$

(c) As we move from left to right, $x = -6$ to $x = 0$, the y-values are decreasing. From $x = 0$ to $x = 6$, the y-values increase.
$$\text{Hence, } f \text{ is decreasing on } [-6, 0] \text{ and increasing on } [0, 6].$$

37 As in Example 6, $a = \dfrac{2-1}{3-(-3)} = \dfrac{1}{6}$ and f has the form $f(x) = \frac{1}{6}x + b$.

$f(3) = \frac{1}{6}(3) + b = \frac{1}{2} + b$. But $f(3) = 2$, so $\frac{1}{2} + b = 2 \Rightarrow b = \frac{3}{2}$, and $f(x) = \frac{1}{6}x + \frac{3}{2}$.

Note: For Exercises 39–48, a good question to consider is "Given a particular value of x, can a unique value of y be found?" If the answer is yes, the value of y (general formula) is given. If no, two ordered pairs satisfying the relation having x in the first position are given.

39 $2y = x^2 + 5 \Rightarrow y = \dfrac{x^2 + 5}{2}$, a function

41 $x^2 + y^2 = 4 \Rightarrow y^2 = 4 - x^2 \Rightarrow y = \pm\sqrt{4 - x^2}$, not a function, $(0, \pm 2)$

45 Any ordered pair with x-coordinate 0 satisfies $xy = 0$.

         Two such ordered pairs are $(0, 0)$ and $(0, 1)$. Not a function

51 (a) The formula for the area of a rectangle is $A = lw$ { Area $=$ length $\times$ width }.

$$A = 500 \Rightarrow xy = 500 \Rightarrow y = \frac{500}{x}$$

   (b) We need to determine the number of linear feet (P) first. There are two walls of length y, two walls of length $(x - 3)$, and one wall of length x.

Thus, $P = $ Linear feet of wall $= x + 2(y) + 2(x - 3) = 3x + 2\left(\dfrac{500}{x}\right) - 6.$

The cost C is 100 times P, so $C = 100P = 300x + \dfrac{100{,}000}{x} - 600.$

53 (a) Using $(6, 48)$ and $(7, 50.5)$, we have

$$y - 48 = \frac{50.5 - 48}{7 - 6}(t - 6), \text{ or } y = 2.5t + 33.$$

   (b) The slope represents the
yearly increase in height, 2.5 in./yr.

   (c) $t = 10 \Rightarrow y = 2.5(10) + 33 = 58$ in.

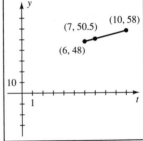

Figure 53

57 (a) CTP forms a right angle, so the Pythagorean theorem may be applied.

$$(CT)^2 + (PT)^2 = (PC)^2 \Rightarrow r^2 + y^2 = (h + r)^2 \Rightarrow r^2 + y^2 = h^2 + 2hr + r^2 \Rightarrow$$
$$y^2 = h^2 + 2hr \ \{y > 0\} \Rightarrow y = \sqrt{h^2 + 2hr}$$

   (b) $y = \sqrt{(200)^2 + 2(4000)(200)} = \sqrt{(200)^2(1 + 40)} = 200\sqrt{41} \approx 1280.6$ mi

59 Form a right triangle with the control booth and the beginning of the runway. Let y denote the distance from the control booth to the beginning of the runway and apply the Pythagorean theorem. $y^2 = 300^2 + 20^2 \Rightarrow y^2 = 90{,}400.$ Now form a right triangle, in a different plane, with sides y and x and hypotenuse d.

Then $d^2 = y^2 + x^2 \Rightarrow d^2 = 90{,}400 + x^2 \Rightarrow d = \sqrt{90{,}400 + x^2}.$

61 (b) The maximum y-value of 0.75 occurs when $x \approx 0.55$ and

the minimum y-value of -0.75 occurs when $x \approx -0.55$.

Therefore, the range of f is approximately $[-0.75, 0.75]$.

(c) f is decreasing on $[-2, -0.55]$ and on $[0.55, 2]$. f is increasing on $[-0.55, 0.55]$.

$[-2, 2]$ by $[-2, 2]$

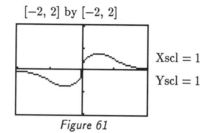

Xscl $= 1$

Yscl $= 1$

Figure 61

65 For each of (a)–(e), an assignment to Y_1, an appropriate viewing rectangle, and the

solution(s) are listed.

(a) $Y_1 = (x\char94 5)\char94(1/3)$, VR: $[-40, 40]$ by $[-40, 40]$, $x = 8$

(b) $Y_1 = (x\char94 4)\char94(1/3)$, VR: $[-20, 20]$ by $[-20, 20]$, $x = \pm 8$

(c) $Y_1 = (x\char94 2)\char94(1/3)$, VR: $[-40, 40]$ by $[-40, 40]$, no real solutions

(d) $Y_1 = (x\char94 3)\char94(1/4)$, VR: $[0, 650]$ by $[0, 650]$, $x = 625$

(e) $Y_1 = (x\char94 3)\char94(1/2)$, VR: $[-30, 30]$ by $[-30, 30]$, no real solutions

3.5 Exercises

3 $f(x) = 3x^4 + 2x^2 - 5 \Rightarrow f(-x) = 3(-x)^4 + 2(-x)^2 - 5 = 3x^4 + 2x^2 - 5 = f(x)$

Since $f(-x) = f(x)$, f is even and its graph is symmetric with respect to the y-axis.

Note that this means if (a, b) is a point on the graph of f,

then the point $(-a, b)$ is also on the graph.

5 $f(x) = 8x^3 - 3x^2 \Rightarrow f(-x) = 8(-x)^3 - 3(-x)^2 = -8x^3 - 3x^2$

$-f(x) = -1 \cdot f(x) = -1(8x^3 - 3x^2) = -8x^3 + 3x^2$

Since $f(-x) \neq f(x)$ and $f(-x) \neq -f(x)$, f is neither even nor odd.

9 $f(x) = \sqrt[3]{x^3 - x} \Rightarrow f(-x) = \sqrt[3]{(-x)^3 - (-x)} = \sqrt[3]{-x^3 + x} = \sqrt[3]{-1(x^3 - x)} =$

$\sqrt[3]{-1}\sqrt[3]{x^3 - x} = -\sqrt[3]{x^3 - x}; \ -f(x) = -1 \cdot f(x) = -1 \cdot \sqrt[3]{x^3 - x} = -\sqrt[3]{x^3 - x}$

Since $f(-x) = -f(x)$, f is odd and its graph is symmetric with respect to the origin.

Note that this means if (a, b) is a point on the graph of f,

then the point $(-a, -b)$ is also on the graph.

15 $f(x) = 2\sqrt{x} + c$, $c = -3, 0, 2$ • The graph of $y^2 = x$ is shown in Figure 11 in the text. The top half of this graph is the graph of the **square root function**, $h(x) = \sqrt{x}$. The second value of c, 0, gives us the graph of $g(x) = 2\sqrt{x}$, which is a vertical stretching of h by a factor of 2. The effect of *adding* -3 and 2 is to vertically shift g down 3 units and up 2 units, respectively.

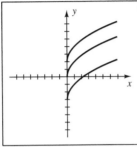

Figure 15

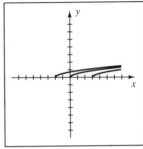

Figure 17

17 $f(x) = \frac{1}{2}\sqrt{x - c}$, $c = -2, 0, 3$ • The graph of $g(x) = \frac{1}{2}\sqrt{x}$ is a vertical compression of the square root function by a factor of $1/(1/2) = 2$. The effect of *subtracting* -2 and 3 from x will be to horizontally shift g left 2 units and right 3 units, respectively. If you forget which way to shift the graph, it is helpful to find the domain of the function. For example, if $h(x) = \sqrt{x - 2}$, then $x - 2$ must be nonnegative. $x - 2 \geq 0 \Rightarrow x \geq 2$, which also indicates a shift of 2 units to the right.

19 $f(x) = c\sqrt{4 - x^2}$, $c = -2, 1, 3$ • For $c = 1$, the graph of $g(x) = \sqrt{4 - x^2}$ is the upper half of the circle $x^2 + y^2 = 4$. For $c = -2$, reflect g through the x-axis and vertically stretch it by a factor of 2. For $c = 3$, vertically stretch g by a factor of 3.

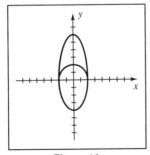

Figure 19

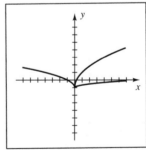

Figure 23

23 $f(x) = \sqrt{cx} - 1$, $c = -1, \frac{1}{9}, 4$ • If $c = 1$, then the graph of $g(x) = \sqrt{x} - 1$ is the graph of the square root function vertically shifted down one unit. For $c = -1$, reflect g through the y-axis. For $c = \frac{1}{9}$, horizontally stretch g by a factor $1/(1/9) = 9$ { x-intercept changes from 1 to 9 }. For $c = 4$, horizontally compress g by a factor 4 { x-intercept changes from 1 to $\frac{1}{4}$ }.

25 (a) $y = f(x+3)$ • shift f left 3 units

(b) $y = f(x-3)$ • shift f right 3 units

(c) $y = f(x)+3$ • shift f up 3 units

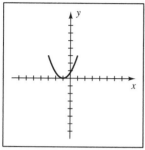
Figure 25(a)

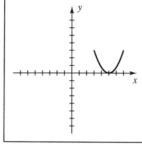
Figure 25(b)

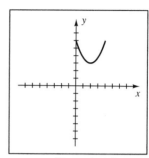
Figure 25(c)

(d) $y = f(x)-3$ • shift f down 3 units

(e) $y = -3f(x)$ •

reflect f through the x-axis and vertically stretch it by a factor of 3

(f) $y = -\frac{1}{3}f(x)$ •

reflect f through the x-axis and vertically compress it by a factor of $1/(1/3) = 3$

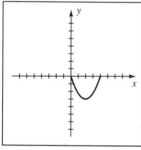
Figure 25(d)

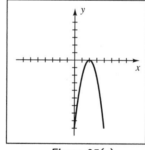
Figure 25(e)

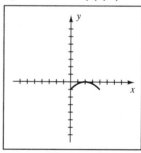
Figure 25(f)

(g) $y = f(-\frac{1}{2}x)$ •

reflect f through the y-axis and horizontally stretch it by a factor of $1/(1/2) = 2$

(h) $y = f(2x)$ • horizontally compress f by a factor of 2

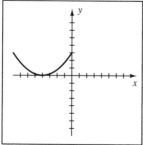
Figure 25(g)

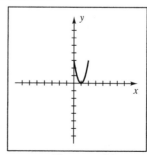
Figure 25(h)

(i) $y = -f(x+2) - 3$ • reflect f about the x-axis, shift it left 2 units and down 3

(j) $y = f(x-2) + 3$ • shift f right 2 units and up 3

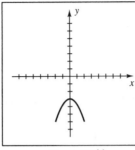

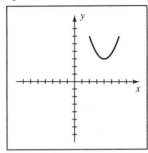

Figure 25(i)   *Figure 25(j)*

27 (a) The minimum point on $y = f(x)$ is $(2, -1)$.

On the graph labeled (a), the minimum point is $(-7, 0)$.

It has been shifted left 9 units and up 1. Hence, $y = f(x+9) + 1$.

(b) f is reflected about the x-axis $\Rightarrow y = -f(x)$

(c) f is reflected about the x-axis and shifted left 7 units and down 1 $\Rightarrow$

$$y = -f(x+7) - 1$$

35 $f(x) = \begin{cases} x+2 & \text{if } x \leq -1 \\ x^3 & \text{if } |x| < 1 \\ -x+3 & \text{if } x \geq 1 \end{cases}$

If $x \leq -1$, we want the graph of $y = x+2$. To determine the end point of this part of the graph, merely substitute $x = -1$ in $y = x+2$, obtaining $y = 1$. If $|x| < 1$, or, equivalently, $-1 < x < 1$, we want the graph of $y = x^3$. We do not include the end points $(-1, -1)$ and $(1, 1)$. If $x \geq 1$, we want the graph of $y = -x+3$ and include its end point $(1, 2)$.

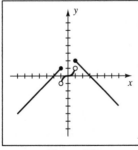

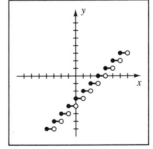

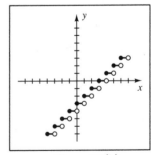

Figure 35   *Figure 37(a)*   *Figure 37(b)*

37 (a) $f(x) = [\![x-3]\!]$ • shift $g(x) = [\![x]\!]$ right 3 units

(b) $f(x) = [\![x]\!] - 3$ • shift g down 3 units, which is the same graph as in part (a).

(c) $f(x) = 2[\![x]\!]$ •   vertically stretch g by a factor of 2

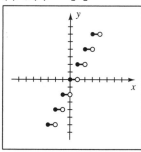
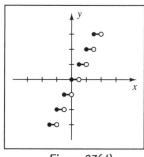
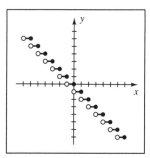

Figure 37(c)   *Figure 37(d)*   *Figure 37(e)*

(d) $f(x) = [\![2x]\!]$ •   horizontally compress g by a factor of 2

Alternatively, we could determine the pattern of "steps" for this function by

finding the values of x that make $f(x)$ change from 0 to 1, then from 1 to 2, etc.

If $2x = 0$, then $x = 0$, and if $2x = 1$, then $x = \frac{1}{2}$.

Thus, the function will equal 0 from $x = 0$ to $x = \frac{1}{2}$ and then jump to 1 at $x = \frac{1}{2}$.

If $2x = 2$, then $x = 1$. The pattern is established: each step will be $\frac{1}{2}$ unit long.

(e) $f(x) = [\![-x]\!]$ •   reflect g through the y-axis

39 A question you can ask to help determine if a relationship is a function is "If x is a

particular value, can I find a unique y-value?" In this case, if x was 16, then

$16 = y^2 \Rightarrow y = \pm 4$. Since we cannot find a unique y-value, this is not a function.

Graphically, {see Figure 11 in §3.2 in the text} given any x-value greater than 0,

there are two points on the graph and a vertical line intersects the graph in more

than one point.

41 We need to examine $(fg)(-x)$. If we obtain $(fg)(x)$, then fg is an even function,

whereas if we obtain $-(fg)(x)$, then fg is an odd function.

$(fg)(-x)$

$= f(-x)g(-x)$ { we may substitute $-x$ into f and $-x$ into g and then multiply }

$= -f(x)g(x)$ { f is odd, so $f(-x) = -f(x)$; g is even, so $g(-x) = g(x)$ }

$= -(fg)(x)$ { substituting x into f and x into g and then multiplying is the

same as substituting x into the function f times g }

Since $(fg)(-x) = -(fg)(x)$, fg is an odd function.

43 $y = |9 - x^2|$ •   First sketch $y = 9 - x^2$,

then reflect the portions of the graph below the x-axis through the x-axis.

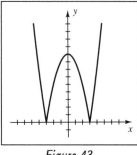

Figure 43

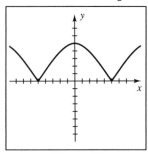

Figure 47

47 Reflect each portion of the graph that is below the x-axis through the x-axis.

49 If $x \leq 20{,}000$, then $T(x) = 0.15x$. If $x > 20{,}000$, then the tax is 15% of the first 20,000, which is 3000, plus 20% of the amount over 20,000, which is $(x - 20{,}000)$. We may summarize and simplify as follows:

$$T(x) = \begin{cases} 0.15x & \text{if } x \leq 20{,}000 \\ 3000 + 0.20(x - 20{,}000) & \text{if } x > 20{,}000 \end{cases} = \begin{cases} 0.15x & \text{if } x \leq 20{,}000 \\ 0.20x - 1000 & \text{if } x > 20{,}000 \end{cases}$$

51 The author receives $1.20 on the first 10,000 copies,

$1.50 on the next 5000, and $1.80 on each additional copy.

$$R(x) = \begin{cases} 1.20x & \text{if } 0 \leq x \leq 10{,}000 \\ 12{,}000 + 1.50(x - 10{,}000) & \text{if } 10{,}000 < x \leq 15{,}000 \\ 19{,}500 + 1.80(x - 15{,}000) & \text{if } x > 15{,}000 \end{cases}$$

$$= \begin{cases} 1.20x & \text{if } 0 \leq x \leq 10{,}000 \\ 1.50x - 3000 & \text{if } 10{,}000 < x \leq 15{,}000 \\ 1.80x - 7500 & \text{if } x > 15{,}000 \end{cases}$$

55 Assign $\text{ABS}(1.2x^2 - 10.8)$ to Y_1 and $1.36x + 4.08$ to Y_2. The standard viewing rectangle $[-15, 15]$ by $[-10, 10]$ shows intersection points at approximately 1.87 and 4.13. The solution is $(-\infty, -3) \cup (-3, 1.87) \cup (4.13, \infty)$.

3.6 Exercises

1   We can use the standard equation of a parabola with a vertical axis.

$$V(-3, 1) \Rightarrow y = a[x - (-3)]^2 + 1 \Rightarrow y = a(x + 3)^2 + 1.$$

$\boxed{7}$ Examples 3 and 4 in the text illustrate a method for expressing $f(x)$ in the desired form. The approach shown here is slightly different.

$f(x) = 2x^2 - 12x + 22$ { divide the equation by the coefficient of x^2, i.e., 2 } $\Rightarrow$

$\frac{1}{2}f(x) = x^2 - 6x + 11$ { complete the square } $\Rightarrow$

$\frac{1}{2}f(x) = x^2 - 6x + \underline{9} + 11 - \underline{9}$ { factor and combine constants } $\Rightarrow$

$\frac{1}{2}f(x) = (x-3)^2 + 2$ { multiply by 2 to isolate $f(x)$ } $\Rightarrow \underline{f(x) = 2(x-3)^2 + 4}$

$\boxed{11}$ $f(x) = -\frac{3}{4}x^2 + 9x - 34$ { divide by $-\frac{3}{4}$ and proceed as in Exercise 7 } $\Rightarrow$

$-\frac{4}{3}f(x) = x^2 - 12x + \frac{136}{3} = x^2 - 12x + \underline{36} + \frac{136}{3} - \underline{36} = (x-6)^2 + \frac{28}{3} \Rightarrow$

$$f(x) = -\tfrac{3}{4}(x-6)^2 - 7$$

$\boxed{15}$ (a) $f(x) = -12x^2 + 11x + 15$ •

The x-coordinate of the vertex is given by $x = -\dfrac{b}{2a} = -\dfrac{11}{2(-12)} = \dfrac{11}{24}$.

The y-coordinate of the vertex is then $f(\frac{11}{24}) = \frac{841}{48} \approx 17.52$.

This is a maximum since $a < 0$.

(b) $-12x^2 + 11x + 15 = 0 \Rightarrow x = \dfrac{-11 \pm \sqrt{121 + 720}}{-24} = \dfrac{-11 \pm 29}{-24} = -\dfrac{3}{4}, \dfrac{5}{3}$

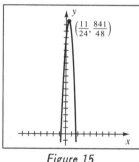

Figure 15

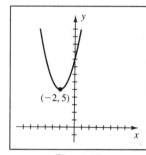

Figure 19

$\boxed{19}$ (a) $f(x) = x^2 + 4x + 9 \Rightarrow -\dfrac{b}{2a} = -\dfrac{4}{2(1)} = -2$. $f(-2) = 5$ is a minimum since $a > 0$.

(b) $x^2 + 4x + 9 = 0 \Rightarrow x = \dfrac{-4 \pm \sqrt{16 - 36}}{2} = -2 \pm \sqrt{5}\,i$.

The imaginary part indicates that there are *no* x-intercepts.

$\boxed{23}$ $V(4, -1) \Rightarrow y = a(x-4)^2 - 1$. $x = 0$, $y = 1 \Rightarrow 1 = a(0-4)^2 - 1 \Rightarrow 2 = 16a \Rightarrow a = \frac{1}{8}$.

Hence, $y = \frac{1}{8}(x-4)^2 - 1$.

$\boxed{27}$ $V(0, -2) \Rightarrow (h, k) = (0, -2)$. $x = 3$, $y = 25 \Rightarrow 25 = a(3-0)^2 - 2 \Rightarrow$

$27 = 9a \Rightarrow a = 3$. Hence, $y = 3(x-0)^2 - 2$, or $y = 3x^2 - 2$.

$\boxed{29}$ $V(3, 5) \Rightarrow y = a(x-3)^2 + 5$ (*). If the x-intercept is 0, then the point $(0, 0)$ is on the parabola. Substituting $x = 0$ and $y = 0$ into (*) gives us $0 = a(0-3)^2 + 5 \Rightarrow$

$-5 = 9a \Rightarrow a = -\frac{5}{9}$. Hence, $y = -\frac{5}{9}(x-3)^2 + 5$.

31 Since the x-intercepts are -3 and 5, the x-coordinate of the vertex of the parabola is

1 { the average of the x-intercept values }. Since the highest point has y-coordinate 4,

the vertex is $(1, 4)$. $V(1, 4) \Rightarrow y = a(x-1)^2 + 4$. We can use the point $(-3, 0)$ since

there is an x-intercept of -3. $x = -3$, $y = 0 \Rightarrow 0 = a(-3-1)^2 + 4 \Rightarrow -4 = 16a \Rightarrow$

$a = -\frac{1}{4}$. Hence, $y = -\frac{1}{4}(x-1)^2 + 4$.

Note: We may find the vertex of a parabola in the following problems using either:

      (1) the complete the square method,

      (2) the formula method, or

      (3) the fact that the vertex lies halfway between the x-intercepts.

33 The vertex is located at $h = \frac{-b}{2a} = \frac{-2.867}{2(-0.058)} \approx 24.72$ km. Since $a < 0$,

this will produce a maximum value.

35 Since the x-intercepts of $y = cx(21 - x)$ are 0 and 21, the maximum will occur

halfway between them, that is, when the infant weighs 10.5 lb.

37 (a) $s(t) = -16t^2 + 144t + 100$ will be a maximum when $t = \frac{-b}{2a} = \frac{-144}{2(-16)} = \frac{9}{2}$.

$$s\left(\tfrac{9}{2}\right) = -16\left(\tfrac{9}{2}\right)^2 + 144\left(\tfrac{9}{2}\right) + 100 = -324 + 648 + 100 = 424 \text{ ft.}$$

(b) When $t = 0$, $s(t) = 100$ ft, which is the height of the building.

39 Let x and $40 - x$ denote the numbers, and their product P is $x(40 - x)$.

P has zeros at 0 and 40 and is a maximum (since $a < 0$) when $x = \frac{0 + 40}{2} = 20$.

The product will be a maximum when both numbers are 20.

41 (a) The 1000 ft of fence is made up of 3 sides of length x and 4 sides of length y.

To express y as a function of x, we need to solve $3x + 4y = 1000$ for y.

$$3x + 4y = 1000 \Rightarrow 4y = 1000 - 3x \Rightarrow y = 250 - \tfrac{3}{4}x.$$

(b) Using the value of y from part (a), $A = xy = x(250 - \tfrac{3}{4}x) = -\tfrac{3}{4}x^2 + 250x$.

(c) A will be a maximum when $x = \frac{-b}{2a} = \frac{-250}{2(-3/4)} = \frac{500}{3} = 166\frac{2}{3}$ ft. Using part (a)

to find the corresponding value of y, $y = 250 - \tfrac{3}{4}(\tfrac{500}{3}) = 250 - 125 = 125$ ft.

45 (a) Since the vertex is at $(0, 10)$, an equation for the parabola is $y = ax^2 + 10$.

The points $(200, 90)$ and $(-200, 90)$ are on the parabola.

Substituting $(200, 90)$ for (x, y) yields $90 = a(200)^2 + 10 \Rightarrow a = \frac{80}{40,000} = \frac{1}{500}$.

Hence, $y = \frac{1}{500}x^2 + 10$.

(b) The cables are spaced 40 ft apart. Using $y = \frac{1}{500}x^2 + 10$ with

$x = 40, 80, 120,$ and 160, we get $y = \frac{66}{5}, \frac{114}{5}, \frac{194}{5},$ and $\frac{306}{5}$, respectively.

There is one cable of length 10 ft and 2 cables of each of the other lengths.

Thus, the total length is $10 + 2(\frac{66}{5} + \frac{114}{5} + \frac{194}{5} + \frac{306}{5}) = 282$ ft.

47 An equation describing the doorway is $y = ax^2 + 9$. Since the doorway is 6 feet wide

at the base, $x = 3$ when $y = 0 \Rightarrow 0 = 9a + 9 \Rightarrow a = -1$. Thus, the equation is

$y = -x^2 + 9$. To fit an 8 foot high box through the doorway, we must find x when

$\qquad y = 8.$ $y = 8 \Rightarrow 8 = -x^2 + 9 \Rightarrow x = \pm 1$. Hence, the box can only be 2 feet wide.

49 Let x denote the number of pairs of shoes that are ordered.

$$A(x) = \begin{cases} 40x & \text{if } x < 50 \\ (40 - 0.04x)x & \text{if } 50 \le x \le 600 \end{cases}$$

The maximum value of the first part of A is $(\$40)(49) = \$1960$. For the second part

of A, $A = -0.04x^2 + 40x$ has a maximum when $x = \dfrac{-b}{2a} = \dfrac{-40}{2(-0.04)} = 500$ pairs.

$\qquad A(500) = 10,000 > 1960$, so $x = 500$ produces a maximum for both parts of A.

51 (a) Let y denote the number of $1 decreases in the monthly charge.

$\qquad\begin{aligned} R(y) &= (\text{\# of customers})(\text{monthly charge per customer}) \\ &= (5000 + 500y)(20 - y) \\ &= 500(10 + y)(20 - y) \end{aligned}$

Now let x denote the monthly charge, which is $20 - y$.

R becomes $500[10 + (20 - x)](x) = 500x(30 - x)$.

(b) R has x-intercepts at 0 and 30, and must have its

vertex halfway between them at $x = 15$.

Note that this gives us $y = 5$, and we have

7500 customers for a revenue of $112,500.

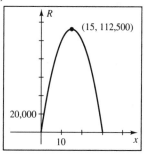

Figure 51

3.7 Exercises

---

3 (a) $(f + g)(x) = f(x) + g(x) = (x^2 + 2) + (2x^2 - 1) = 3x^2 + 1$;

$(f - g)(x) = f(x) - g(x) = (x^2 + 2) - (2x^2 - 1) = 3 - x^2$;

$(fg)(x) = f(x) \cdot g(x) = (x^2 + 2) \cdot (2x^2 - 1) = 2x^4 + 3x^2 - 2$;

$\left(\dfrac{f}{g}\right)(x) = \dfrac{f(x)}{g(x)} = \dfrac{x^2 + 2}{2x^2 - 1}$

(b) The domain of $f + g$, $f - g$, and fg is the set of all real numbers, $\mathbb{R}$.

(c) The domain of f/g is the same as in (b), except we must exclude the zeros of g.

$\qquad$ Hence, the domain of f/g is all real numbers except $\pm\frac{1}{2}\sqrt{2}$.

7 (a) $(f + g)(x) = f(x) + g(x) = \dfrac{2x}{x - 4} + \dfrac{x}{x + 5} = \dfrac{2x(x + 5) + x(x - 4)}{(x - 4)(x + 5)} = \dfrac{3x^2 + 6x}{(x - 4)(x + 5)}$;

$(f - g)(x) = f(x) - g(x) = \dfrac{2x}{x - 4} - \dfrac{x}{x + 5} = \dfrac{2x(x + 5) - x(x - 4)}{(x - 4)(x + 5)} = \dfrac{x^2 + 14x}{(x - 4)(x + 5)}$;

$(fg)(x) = f(x) \cdot g(x) = \dfrac{2x}{x - 4} \cdot \dfrac{x}{x + 5} = \dfrac{2x^2}{(x - 4)(x + 5)}$;   (cont.)

$$\left(\frac{f}{g}\right)(x) = \frac{f(x)}{g(x)} = \frac{2x/(x-4)}{x/(x+5)} = \frac{2(x+5)}{x-4}$$

(b) The domain of f is $\mathbb{R} - \{4\}$ and the domain of g is $\mathbb{R} - \{-5\}$. The intersection of these two domains, $\mathbb{R} - \{-5, 4\}$, is the domain of the three functions.

(c) To determine the domain of the quotient f/g, we also exclude any values that make the denominator g equal to zero. Hence, we exclude $x = 0$ and the domain of the quotient is all real numbers except -5, 0, and 4, that is, $\mathbb{R} - \{-5, 0, 4\}$.

9 (a) $(f \circ g)(x) = f(g(x)) = f(-x^2) = 2(-x^2) - 1 = -2x^2 - 1$

(b) $(g \circ f)(x) = g(f(x)) = g(2x - 1) = -(2x - 1)^2 = -(4x^2 - 4x + 1) = -4x^2 + 4x - 1$

(c) $(f \circ f)(x) = f(f(x)) = f(2x - 1) = 2(2x - 1) - 1 = (4x - 2) - 1 = 4x - 3$

(d) $(g \circ g)(x) = g(g(x)) = g(-x^2) = -(-x^2)^2 = -(x^4) = -x^4$

Note: Let $h(x) = (f \circ g)(x) = f(g(x))$ and $k(x) = (g \circ f)(x) = g(f(x))$.

$h(-2)$ and $k(3)$ could be worked two ways, as in Example 3(c) in the text.

11 (a) $h(x) = f(3x + 7) = 2(3x + 7) - 5 = 6x + 9$

(b) $k(x) = g(2x - 5) = 3(2x - 5) + 7 = 6x - 8$

(c) Using the result from part (a), $h(-2) = 6(-2) + 9 = -12 + 9 = -3$.

(d) Using the result from part (b), $k(3) = 6(3) - 8 = 18 - 8 = 10$.

15 (a) $h(x) = f(2x - 1) = 2(2x - 1)^2 + 3(2x - 1) - 4 = 8x^2 - 2x - 5$

(b) $k(x) = g(2x^2 + 3x - 4) = 2(2x^2 + 3x - 4) - 1 = 4x^2 + 6x - 9$

(c) $h(-2) = 8(-2)^2 - 2(-2) - 5 = 32 + 4 - 5 = 31$

(d) $k(3) = 4(3)^2 + 6(3) - 9 = 36 + 18 - 9 = 45$

19 (a) $h(x) = f(-7) = |-7| = 7$ (b) $k(x) = g(|x|) = -7$

(c) $h(-2) = 7$ since $h(\text{any value}) = 7$ (d) $k(3) = -7$ since $k(\text{any value}) = -7$

21 (a) $h(x) = f(\sqrt{x+2}) = (\sqrt{x+2})^2 - 3(\sqrt{x+2}) = x + 2 - 3\sqrt{x+2}$. The domain of $f \circ g$ is the set of all x in the domain of g, $x \geq -2$, such that $g(x)$ is in the domain of f. Since the domain of f is $\mathbb{R}$, any value of $g(x)$ is in its domain.

Thus, the domain is all x such that $x \geq -2$.

(b) $k(x) = g(x^2 - 3x) = \sqrt{(x^2 - 3x) + 2} = \sqrt{x^2 - 3x + 2}$. The domain of $g \circ f$ is the set of all x in the domain of f, $\mathbb{R}$, such that $f(x)$ is in the domain of g. Since the domain of g is $x \geq -2$, we must solve $f(x) \geq -2$. $x^2 - 3x \geq -2 \Rightarrow$ $x^2 - 3x + 2 \geq 0 \Rightarrow (x - 1)(x - 2) \geq 0 \Rightarrow x \in (-\infty, 1] \cup [2, \infty)$ { use a sign diagram }. Thus, the domain is all x such that $x \in (-\infty, 1] \cup [2, \infty)$.

23 (a) $h(x) = f(\sqrt{3x}) = (\sqrt{3x})^2 - 4 = 3x - 4$.

Domain of $g = [0, \infty)$. Domain of $f = \mathbb{R}$. Since $g(x)$ is always in the domain of
f, the domain of $f \circ g$ is the same as the domain of g, $[0, \infty)$.

(b) $k(x) = g(x^2 - 4) = \sqrt{3(x^2 - 4)} = \sqrt{3x^2 - 12}$.

Domain of $f = \mathbb{R}$. Domain of $g = [0, \infty)$.
$$f(x) \geq 0 \Rightarrow x^2 - 4 \geq 0 \Rightarrow x^2 \geq 4 \Rightarrow |x| \geq 2 \Rightarrow x \in (-\infty, -2] \cup [2, \infty).$$

25 (a) $h(x) = f(\sqrt{x+5}) = \sqrt{\sqrt{x+5} - 2}$. Domain of $g = [-5, \infty)$. Domain of
$f = [2, \infty)$. $g(x) \geq 2 \Rightarrow \sqrt{x+5} \geq 2 \Rightarrow x + 5 \geq 4 \Rightarrow x \geq -1$ or $x \in [-1, \infty)$.

(b) $k(x) = g(\sqrt{x-2}) = \sqrt{\sqrt{x-2} + 5}$. Domain of $f = [2, \infty)$.

Domain of $g = [-5, \infty)$. $f(x) \geq -5 \Rightarrow \sqrt{x-2} \geq -5$. This is always true since
the result of a square root is nonnegative. The domain is $[2, \infty)$.

27 (a) $h(x) = f(\sqrt{x^2 - 16}) = \sqrt{3 - \sqrt{x^2 - 16}}$. Domain of $g = (-\infty, -4] \cup [4, \infty)$.

Domain of $f = (-\infty, 3]$. $g(x) \leq 3 \Rightarrow \sqrt{x^2 - 16} \leq 3 \Rightarrow x^2 - 16 \leq 9 \Rightarrow x^2 \leq 25 \Rightarrow$
$x \in [-5, 5]$. But $|x| \geq 4$ from the domain of g.

Hence, the domain of $f \circ g$ is $[-5, -4] \cup [4, 5]$.

(b) $k(x) = g(\sqrt{3-x}) = \sqrt{(\sqrt{3-x})^2 - 16} = \sqrt{3 - x - 16} = \sqrt{-x - 13}$.

Domain of $f = (-\infty, 3]$. Domain of $g = (-\infty, -4] \cup [4, \infty)$.
$$f(x) \geq 4 \ \{ f(x) \text{ cannot be less than } 0 \} \Rightarrow \sqrt{3-x} \geq 4 \Rightarrow 3 - x \geq 16 \Rightarrow x \leq -13.$$

29 (a) $h(x) = f\left(\dfrac{2x-5}{3}\right) = \dfrac{3\left(\dfrac{2x-5}{3}\right) + 5}{2} = \dfrac{2x - 5 + 5}{2} = \dfrac{2x}{2} = x$.

Domain of $g = \mathbb{R}$. Domain of $f = \mathbb{R}$. All values of $g(x)$ are in the domain of f.

Hence, the domain of $f \circ g$ is $\mathbb{R}$.

(b) $k(x) = g\left(\dfrac{3x+5}{2}\right) = \dfrac{2\left(\dfrac{3x+5}{2}\right) - 5}{3} = \dfrac{3x + 5 - 5}{3} = \dfrac{3x}{3} = x$.

Domain of $f = \mathbb{R}$. Domain of $g = \mathbb{R}$. All values of $f(x)$ are in the domain of g.

Hence, the domain of $g \circ f$ is $\mathbb{R}$.

31 (a) $h(x) = f\left(\dfrac{1}{x^3}\right) = \left(\dfrac{1}{x^3}\right)^2 = \dfrac{1}{x^6}$. Domain of $g = \mathbb{R} - \{0\}$. Domain of $f = \mathbb{R}$.

All values of $g(x)$ are in the domain of f. Hence, the domain of $f \circ g$ is $\mathbb{R} - \{0\}$.

(b) $k(x) = g(x^2) = \dfrac{1}{(x^2)^3} = \dfrac{1}{x^6}$. Domain of $f = \mathbb{R}$. Domain of $g = \mathbb{R} - \{0\}$.

All values of $f(x)$ are in the domain of g except for 0.

Since f is 0 when x is 0, the domain of $f \circ g$ is $\mathbb{R} - \{0\}$.

33 (a) $h(x) = f\left(\dfrac{x-3}{x-4}\right) = \dfrac{\dfrac{x-3}{x-4} - 1}{\dfrac{x-3}{x-4} - 2} \cdot \dfrac{x-4}{x-4} = \dfrac{x-3-1(x-4)}{x-3-2(x-4)} = \dfrac{1}{5-x}.$

Domain of $g = \mathbb{R} - \{4\}$. Domain of $f = \mathbb{R} - \{2\}$.

$g(x) \neq 2 \Rightarrow \dfrac{x-3}{x-4} \neq 2 \Rightarrow x - 3 \neq 2x - 8 \Rightarrow x \neq 5$. The domain is $\mathbb{R} - \{4, 5\}$.

(b) $k(x) = g\left(\dfrac{x-1}{x-2}\right) = \dfrac{\dfrac{x-1}{x-2} - 3}{\dfrac{x-1}{x-2} - 4} \cdot \dfrac{x-2}{x-2} = \dfrac{x-1-3(x-2)}{x-1-4(x-2)} = \dfrac{-2x+5}{-3x+7}.$

Domain of $f = \mathbb{R} - \{2\}$. Domain of $g = \mathbb{R} - \{4\}$.

$f(x) \neq 4 \Rightarrow \dfrac{x-1}{x-2} \neq 4 \Rightarrow x - 1 \neq 4x - 8 \Rightarrow x \neq \frac{7}{3}$. The domain is $\mathbb{R} - \{2, \frac{7}{3}\}$.

35 $(f \circ g)(x) = f(g(x)) = f(x+3) = (x+3)^2 - 2.$

$(f \circ g)(x) = 0 \Rightarrow (x+3)^2 - 2 = 0 \Rightarrow (x+3)^2 = 2 \Rightarrow x + 3 = \pm\sqrt{2} \Rightarrow x = -3 \pm \sqrt{2}$

37 (a) $(f \circ g)(6) = f(g(6)) = f(8) = 5$ (b) $(g \circ f)(6) = g(f(6)) = g(7) = 6$

(c) $(f \circ f)(6) = f(f(6)) = f(7) = 6$ (d) $(g \circ g)(6) = g(g(6)) = g(8) = 5$

39 $(D \circ R)(x) = D(R(x)) = D(20x) =$

$$\sqrt{400 + (20x)^2} = \sqrt{400 + 400x^2} = \sqrt{400(1 + x^2)} = 20\sqrt{x^2 + 1}$$

43 $V = \frac{1}{3}\pi r^2 h = \frac{1}{3}\pi r^3 \Rightarrow r = \sqrt[3]{\frac{3V}{\pi}}$. $V = 243\pi t \Rightarrow r = \sqrt[3]{729t} = 9\sqrt[3]{t}$ ft.

45 Let l denote the length of the rope. At $t = 0$, $l = 20$.

At time t, $l = 20 + 5t$, *not* just $5t$. We have a right triangle with sides 20, h, and l.

$$h^2 + 20^2 = l^2 \Rightarrow h = \sqrt{(20 + 5t)^2 - 20^2} = \sqrt{25t^2 + 200t} = \sqrt{25(t^2 + 8t)} = 5\sqrt{t^2 + 8t}.$$

47 From Exercise 59 of Section 3.4, $d = \sqrt{90{,}400 + x^2}$. The distance x of the plane from the control tower is 500 feet plus 150 feet per second, that is, $x = 500 + 150t$.

$$\text{Thus, } d = \sqrt{90{,}400 + (500 + 150t)^2} = 10\sqrt{225t^2 + 1500t + 3404}.$$

49 $y = (x^2 + 3x)^{1/3}$ •   Suppose you were to find the value of y if x was equal to 3. Using a calculator, you might compute the value of $x^2 + 3x$ first, and then raise that result to the $\frac{1}{3}$ power. Thus, we would choose $y = u^{1/3}$ and $u = x^2 + 3x$.

55 For $y = \dfrac{\sqrt{x+4} - 2}{\sqrt{x+4} + 2}$, there is not a "simple" choice for y as in previous exercises.

One choice for u is $u = x + 4$. Then y would be $\dfrac{\sqrt{u} - 2}{\sqrt{u} + 2}$.

Another choice for u is $u = \sqrt{x+4}$. Then y would be $\dfrac{u-2}{u+2}$.

$\boxed{57}$ $(f \circ g)(x) = f(g(x)) = f(x^3 + 1) = \sqrt{x^3 + 1} - 1$. We will multiply this expression by

$\dfrac{\sqrt{x^3 + 1} + 1}{\sqrt{x^3 + 1} + 1}$, treating it as though it was one factor of a difference of two squares.

$$\left(\sqrt{x^3 + 1} - 1\right) \times \frac{\sqrt{x^3 + 1} + 1}{\sqrt{x^3 + 1} + 1} = \frac{(\sqrt{x^3 + 1})^2 - 1^2}{\sqrt{x^3 + 1} + 1} = \frac{x^3}{\sqrt{x^3 + 1} + 1}$$

$$\text{Thus, } (f \circ g)(0.0001) = f(g(10^{-4})) \approx \frac{(10^{-4})^3}{2} = 5 \times 10^{-13}.$$

3.8 Exercises

Note: To help determine if you should try to prove the function is one-to-one or look for
a counterexample to show that it is not one-to-one, consider the question: "If y
was a particular value, could I find a unique x?" If the answer is yes, try to prove
the function is one-to-one. Also, consider the Horizontal Line Test listed near the
top of page 205 in the text.

$\boxed{1}$ If y was a particular value, say 5, we would have $5 = 3x - 7$. Trying to solve for x
would yield $12 = 3x \Rightarrow x = 4$. Since we could find a unique x, we will try to prove
that the function is one-to-one. **Proof** Suppose that $f(a) = f(b)$ for some numbers a
and b in the domain. This gives us $3a - 7 = 3b - 7 \Rightarrow 3a = 3b \Rightarrow a = b$. Since
$f(a) = f(b)$ implies that $a = b$, we conclude that f is one-to-one.

$\boxed{3}$ If y was a particular value, say 7, we would have $7 = x^2 - 9$. Trying to solve for x
would yield $16 = x^2 \Rightarrow x = \pm 4$. Since we could not find a *unique* x, we will show
that two different numbers have the same function value. Using the information
already obtained, $f(4) = 7$ and $f(-4) = 7$, but $4 \ne -4$ and hence, f is *not* one-to-one.

$\boxed{5}$ Suppose $f(a) = f(b)$ with $a, b \ge 0$. $\sqrt{a} = \sqrt{b} \Rightarrow (\sqrt{a})^2 = (\sqrt{b})^2 \Rightarrow a = b$.

f is one-to-one.

$\boxed{9}$ For $f(x) = \sqrt{4 - x^2}$, $f(-1) = \sqrt{3} = f(1)$. f is *not* one-to-one.

Note: For Exercises 13–16, we need to show that $f(g(x)) = x = g(f(x))$.

$\boxed{15}$ $f(g(x)) = -(\sqrt{3-x})^2 + 3 = -(3-x) + 3 = x$.

$\quad g(f(x)) = \sqrt{3 - (-x^2 + 3)} = \sqrt{x^2} = |x| = x$ {since $x \geq 0$}.

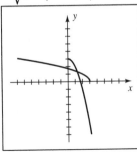

Figure 15

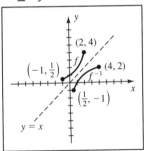

Figure 35

$\boxed{19}$ $f(x) = \dfrac{1}{3x-2} \Rightarrow 3xy - 2y = 1 \Rightarrow 3xy = 2y + 1 \Rightarrow x = \dfrac{2y+1}{3y} \Rightarrow f^{-1}(x) = \dfrac{2x+1}{3x}$

$\boxed{21}$ $f(x) = \dfrac{3x+2}{2x-5} \Rightarrow 2xy - 5y = 3x + 2 \Rightarrow 2xy - 3x = 5y + 2 \Rightarrow$

$$x(2y - 3) = 5y + 2 \Rightarrow x = \frac{5y+2}{2y-3} \Rightarrow f^{-1}(x) = \frac{5x+2}{2x-3}$$

$\boxed{23}$ $f(x) = 2 - 3x^2,\ x \leq 0 \Rightarrow y + 3x^2 = 2 \Rightarrow$

$$x^2 = \frac{2-y}{3} \Rightarrow x = \pm\sqrt{\frac{2-y}{3}} \text{ \{choose minus since } x \leq 0 \} \Rightarrow f^{-1}(x) = -\sqrt{\frac{2-x}{3}}$$

$\boxed{27}$ $f(x) = \sqrt{3-x} \Rightarrow y^2 = 3 - x \Rightarrow$

$$x = 3 - y^2 \text{ \{ Since } y \geq 0 \text{ for } f,\ x \geq 0 \text{ for } f^{-1}. \} \Rightarrow f^{-1}(x) = 3 - x^2,\ x \geq 0$$

$\boxed{29}$ $f(x) = \sqrt[3]{x} + 1 \Rightarrow y - 1 = \sqrt[3]{x} \Rightarrow x = (y-1)^3 \Rightarrow f^{-1}(x) = (x-1)^3$

$\boxed{35}$ Remember that the domain of f is the range of f^{-1} and

$\qquad\qquad$ that the range of f is the domain of f^{-1}. See *Figure 35*.

$\quad$ (b) $D = [-1,\ 2];\ R = [\frac{1}{2},\ 4]$ $\qquad\qquad$ (c) $D_1 = R = [\frac{1}{2},\ 4];\ R_1 = D = [-1,\ 2]$

$\boxed{39}$ (a) $f(x) = -x + b \Rightarrow y = -x + b \Rightarrow x = -y + b$, or $f^{-1}(x) = -x + b$.

$\quad$ (b) $f(x) = \dfrac{ax+b}{cx-a}$ for $c \neq 0 \Rightarrow y = \dfrac{ax+b}{cx-a} \Rightarrow cyx - ya = ax + b \Rightarrow$

$$cyx - ax = ay + b \Rightarrow x(cy - a) = ay + b \Rightarrow x = \frac{ay+b}{cy-a}, \text{ or } f^{-1}(x) = \frac{ax+b}{cx-a}.$$

$\quad$ (c) The graph of f is symmetric about the line $y = x$. Thus, $f(x) = f^{-1}(x)$.

$\boxed{41}$ From a graph of f, we see that f is always increasing. Thus, f is one-to-one.

$\boxed{45}$ $D = 0.0833x^2 - 0.4996x + 3.5491 \Rightarrow 0.0833x^2 - 0.4996x + (3.5491 - D) = 0 \Rightarrow$

$$x = \frac{0.4996 \pm \sqrt{(-0.4996)^2 - 4(0.0833)(3.5491 - D)}}{2(0.0833)} \text{ \{quadratic formula\}. From the}$$

graph of D with VR $[-20,\ 20]$ by $[0,\ 40]$, we see that $3 \leq x \leq 15$ corresponds to the

right half of the parabola. Hence, we choose the plus sign in the preceding equation.

3.9 Exercises

5 y is directly proportional to the square of x and inversely proportional to the cube of

$z \Rightarrow y = k\dfrac{x^2}{z^3}$. Substituting $x = 5$, $z = 3$, and $y = 25$ gives us $25 = \dfrac{k(25)}{27}$.

Solving for the constant of proportionality, k, we have $k = 27$.

7 z is directly proportional to the product of the square of x and the cube of

$y \Rightarrow z = kx^2y^3$. Substituting $x = 7$, $y = -2$, and $z = 16$ gives us $16 = k(49)(-8)$.

Solving for the constant of proportionality, k, we have $k = -\dfrac{2}{49}$.

11 y is directly proportional to the square root of x and inversely proportional to the

cube of $z \Rightarrow y = k\dfrac{\sqrt{x}}{z^3}$. Substituting $x = 9$, $z = 2$, and $y = 5$ gives us $5 = \dfrac{k(3)}{8}$.

Solving for the constant of proportionality, k, we have $k = \dfrac{40}{3}$.

15 (a) $R = k\dfrac{l}{d^2} = \dfrac{kl}{d^2}$ (b) $25 = \dfrac{k(100)}{(0.01)^2} \Rightarrow k = \dfrac{1}{40,000}$

(c) $R = \dfrac{50}{(40,000)(0.015)^2} = \dfrac{50}{9}$ ohms

19 (a) $T = kd^{3/2}$ (b) $365 = k(93)^{3/2} \Rightarrow k = \dfrac{365}{(93)^{3/2}}$

(c) $T = \dfrac{365}{(93)^{3/2}} \cdot (67)^{3/2} \approx 223.2$ days

23 (a) $W = kh^3$ (b) $200 = k(6)^3 \Rightarrow k = \dfrac{25}{27}$

(c) 5 feet 6 inches is $\dfrac{11}{2}$ feet. Hence, $W = \dfrac{25}{27}(\dfrac{11}{2})^3 \approx 154.1$ lb, or 154 lb.

25 (a) $F = kPr^4 \Rightarrow P = \dfrac{F}{kr^4}$ under normal conditions.

(b) "Normal flow rates triple" means that we will use $3F$ for the flow rate F.

"Radius increases by 10%" means that we will use $(r + 10\%r) = 1.1r$ for the

radius. Hence, $3F = kP(1.1r)^4 \Rightarrow P = \dfrac{3F}{(1.1)^4kr^4} \approx 2.05\left(\dfrac{F}{kr^4}\right)$,

or about 2.05 times as hard as normal.

27 $C = \dfrac{kDE}{Vt} \Rightarrow D = \left(\dfrac{Ct}{k}\right)\dfrac{V}{E}$, where $\dfrac{Ct}{k}$ is constant. If V is twice its original value and

E is 0.8 of its original value (reduced by 20%), then D becomes $D_1 = \left(\dfrac{Ct}{k}\right)\dfrac{2V}{0.8E}$.

Comparing D_1 to D, we have $\dfrac{D_1}{D} = \dfrac{\left(\dfrac{Ct}{k}\right)\dfrac{2V}{0.8E}}{\left(\dfrac{Ct}{k}\right)\dfrac{V}{E}} = \dfrac{2}{0.8} = 2.5 = 250\%$ of its original

value. Thus, D increases by 250%.

29 We need to examine y/x for the given set of data points.

$$\frac{y}{x} = \frac{0.72}{0.6} = \frac{1.44}{1.2} = \frac{5.04}{4.2} = \frac{8.52}{7.1} = \frac{11.16}{9.3} = 1.2 \Rightarrow y = 1.2x.$$

Thus, y varies directly as x with constant of variation $k = 1.2$.

Chapter 3 Review Exercises

3 (a) $P(-5, 9)$, $Q(-8, -7) \Rightarrow$

$$d(P, Q) = \sqrt{[-8 - (-5)]^2 + (-7 - 9)^2} = \sqrt{9 + 256} = \sqrt{265}.$$

(b) $P(-5, 9)$, $Q(-8, -7) \Rightarrow M_{PQ} = \left(\frac{-5 + (-8)}{2}, \frac{9 + (-7)}{2}\right) = (-\frac{13}{2}, 1)$.

(c) Let $R = (x, y)$. $Q = M_{PR} \Rightarrow$

$$(-8, -7) = \left(\frac{-5 + x}{2}, \frac{9 + y}{2}\right) \Rightarrow -8 = \frac{-5 + x}{2} \text{ and } -7 = \frac{9 + y}{2} \Rightarrow$$

$-5 + x = -16$ and $9 + y = -14 \Rightarrow x = -11$ and $y = -23 \Rightarrow R = (-11, -23)$.

5 With $P(a, 1)$ and $Q(-2, a)$, $d(P, Q) < 3 \Rightarrow \sqrt{(-2 - a)^2 + (a - 1)^2} < 3 \Rightarrow$
$4 + 4a + a^2 + a^2 - 2a + 1 < 9 \Rightarrow 2a^2 + 2a - 4 < 0 \Rightarrow a^2 + a - 2 < 0 \Rightarrow$
$(a + 2)(a - 1) < 0$. Using *Diagram 5*,

we see that $-2 < a < 1$ will assure us that $d(P, Q) < 3$.

Resulting sign:	$\oplus$	$\ominus$	$\oplus$
Sign of $a + 2$:	$-$	$+$	$+$
Sign of $a - 1$:	$-$	$-$	$+$
a values:	-2		1

Diagram 5

7 The center of the circle is the midpoint of $A(8, 10)$ and $B(-2, -14)$.

$$M_{AB} = \left(\frac{8 + (-2)}{2}, \frac{10 + (-14)}{2}\right) = (3, -2). \text{ The radius of the circle is}$$

$$\frac{1}{2} \cdot d(A, B) = \frac{1}{2}\sqrt{(-2 - 8)^2 + (-14 - 10)^2} = \frac{1}{2}\sqrt{100 + 576} = \frac{1}{2} \cdot 26 = 13.$$

An equation is $(x - 3)^2 + (y + 2)^2 = 13^2 = 169$.

11 (a) $6x + 2y + 5 = 0 \Leftrightarrow y = -3x - \frac{5}{2}$. Using the same slope, -3, with $A(\frac{1}{2}, -\frac{1}{3})$,

we have $y + \frac{1}{3} = -3(x - \frac{1}{2}) \Rightarrow 6y + 2 = -18x + 9 \Rightarrow 18x + 6y = 7$.

(b) Using the negative reciprocal of -3 for the slope,

$$y + \frac{1}{3} = \frac{1}{3}(x - \frac{1}{2}) \Rightarrow 6y + 2 = 2x - 1 \Rightarrow 2x - 6y = 3.$$

13 The radius of the circle is the distance from the line $x = 4$ to the x-value of the

center $C(-5, -1)$; $r = 4 - (-5) = 9$. An equation is $(x + 5)^2 + (y + 1)^2 = 81$.

16 $A(-1, 2)$ and $B(3, -4) \Rightarrow M_{AB} = (1, -1)$ and $m_{AB} = -\frac{3}{2}$. We want the

equation of the line through $(1, -1)$ with slope $\frac{2}{3}$ { the negative reciprocal of $-\frac{3}{2}$ }.

$$y + 1 = \tfrac{2}{3}(x - 1) \Rightarrow 3y + 3 = 2x - 2 \Rightarrow 2x - 3y = 5.$$

22 $\dfrac{f(a+h) - f(a)}{h} = \dfrac{\frac{1}{a+h+2} - \frac{1}{a+2}}{h} = \dfrac{\frac{(a+2) - (a+h+2)}{(a+h+2)(a+2)}}{h} = \dfrac{-h}{(a+h+2)(a+2)h} =$

$$-\frac{1}{(a+h+2)(a+2)}$$

23 $f(x) = ax + b$ is the desired form. $a = \text{slope} = \dfrac{7-2}{3-1} = \dfrac{5}{2}$. $f(x) = \tfrac{5}{2}x + b \Rightarrow$

$f(1) = \tfrac{5}{2} + b$, but $f(1) = 2$, so $\tfrac{5}{2} + b = 2$, and $b = -\tfrac{1}{2}$. Thus, $f(x) = \tfrac{5}{2}x - \tfrac{1}{2}$.

27 $2y + 5x - 8 = 0 \Leftrightarrow y = -\tfrac{5}{2}x + 4$, a line with slope $-\tfrac{5}{2}$ and y-intercept 4;

x-intercept: $(1.6, 0)$

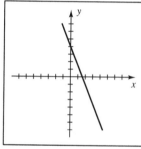

Figure 27

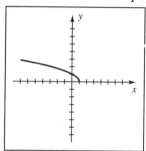

Figure 31

31 $y = \sqrt{1 - x}$ • The radicand must be nonnegative for the radical to be defined.

$1 - x \geq 0 \Rightarrow 1 \geq x$ or $x \leq 1$. The domain is $(-\infty, 1]$ and the range is $[0, \infty)$.

x-intercept: $(1, 0)$, y-intercept: $(0, 1)$

35 $x^2 + y^2 - 8x = 0 \Leftrightarrow x^2 - 8x + \underline{16} + y^2 = \underline{16} \Leftrightarrow (x-4)^2 + y^2 = 16$;

$C(4, 0)$, $r = \sqrt{16} = 4$; x-intercepts: $(0, 0)$ and $(8, 0)$, y-intercept: $(0, 0)$

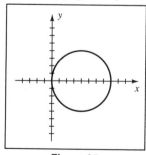

Figure 35

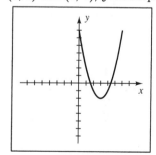

Figure 37

37 $y = (x-3)^2 - 2$ has vertex $(3, -2)$; x-intercepts: $(3 \pm \sqrt{2}, 0)$, y-intercept: $(0, 7)$

$\boxed{41}$ (a) The graph of $f(x) = |x+3|$ can be thought of as

the graph of $g(x) = |x|$ shifted left 3 units.

(b) The function is defined for all x, so the domain D is the set of all real numbers.

The range is the set of all nonnegative numbers, that is, $R = [0, \infty)$.

(c) The function f is decreasing on $(-\infty, -3]$ and is increasing on $[-3, \infty)$.

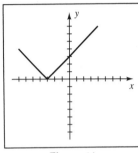

Figure 41

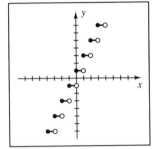

Figure 48

$\boxed{48}$ (a) The "2" in front of $[\![x]\!]$ has the effect of doubling all the y-values of $g(x) = [\![x]\!]$.

The "+1" has the effect of vertically shifting the graph of $h(x) = 2[\![x]\!]$ up 1 unit.

(b) The function is defined for all x, so the domain D is the set of all real numbers.

The range of $g(x) = [\![x]\!]$ is the set of integers, that is, $\{\ldots, -2, -1, 0, 1, 2, \ldots\}$.

The range of $h(x) = 2[\![x]\!]$ is the set of even integers since we are doubling the

values of g—that is, $\{\ldots, -4, -2, 0, 2, 4, \ldots\}$. Since $f(x) = 1 + h(x)$, the range

R of f is $\{\ldots, -3, -1, 1, 3, \ldots\}$.

(c) The function f is constant on intervals such as $[0, 1)$, $[1, 2)$, and $[2, 3)$. In

general, f is constant on $[n, n+1)$, where n is any integer.

$\boxed{50}$ (a) $y = f(x-2)$ •   shift f right 2 units

(b) $y = f(x) - 2$ •   shift f down 2 units

(c) $y = f(-x)$ •   reflect f through the y-axis

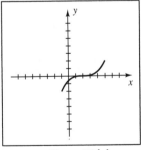

Figure 50(a)

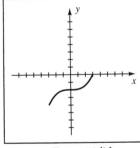

Figure 50(b)

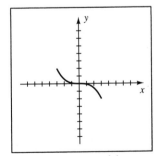

Figure 50(c)

(d) $y = f(2x)$ •   horizontally compress f by a factor of 2

(e) $y = f(\frac{1}{2}x)$ •   horizontally stretch f by a factor of $1/(1/2) = 2$

(f) $y = f^{-1}(x)$ •   reflect f through the line $y = x$

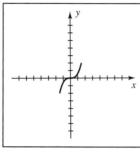

Figure 50(d)

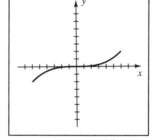

Figure 50(e)

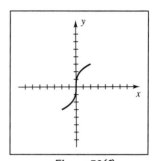

Figure 50(f)

[51] Using the intercept form of a line with x-intercept 5 and y-intercept -2, an equation

is $\frac{x}{5} + \frac{y}{-2} = 1$. Multiplying by 10 gives us $\left[\frac{x}{5} + \frac{y}{-2} = 1\right] \cdot 10$, or, equivalently,

$$2x - 5y = 10.$$

[54] The graph could be made by taking the graph of $y = |x|$, reflecting it through the

x-axis $\{y = -|x|\}$, then shifting that graph to the right 2 units $\{y = -|x-2|\}$,

and then shifting that graph down 1 unit, resulting in the graph of the equation

$y = -|x-2| - 1$.

[55] $f(x) = 5x^2 + 30x + 49 \Rightarrow -\frac{b}{2a} = -\frac{30}{2(5)} = -3$. $f(-3) = 4$ is a minimum since $a > 0$.

[59] The domain of $f(x) = \sqrt{4 - x^2}$ is $[-2, 2]$. The domain of $g(x) = \sqrt{x}$ is $[0, \infty)$.

(a) The domain of fg is the intersection of those two domains, $[0, 2]$.

(b) The domain of f/g is the same as that of fg,

excluding any values that make g equal to 0. Thus, the domain of f/g is $(0, 2]$.

[63] (a) $h(x) = f(\sqrt{x - 3}) = \sqrt{25 - (\sqrt{x - 3})^2} = \sqrt{25 - (x - 3)} = \sqrt{28 - x}$.

Domain of $g = [3, \infty)$. Domain of $f = [-5, 5]$.

$g(x) \le 5 \{g(x)$ cannot be less than 0 $\} \Rightarrow \sqrt{x - 3} \le 5 \Rightarrow$

$$x - 3 \le 25 \Rightarrow x \le 28. \quad [3, \infty) \cap (-\infty, 28] = [3, 28]$$

(b) $k(x) = g(\sqrt{25 - x^2}) = \sqrt{\sqrt{25 - x^2} - 3}$.

Domain of $f = [-5, 5]$. Domain of $g = [3, \infty)$.

$$f(x) \ge 3 \Rightarrow \sqrt{25 - x^2} \ge 3 \Rightarrow 25 - x^2 \ge 9 \Rightarrow x^2 \le 16 \Rightarrow x \in [-4, 4].$$

68 $f(x) = 9 - 2x^2$, $x \le 0 \Rightarrow$

$y + 2x^2 = 9 \Rightarrow x^2 = \dfrac{9 - y}{2} \Rightarrow$

$x = \pm \sqrt{\dfrac{9 - y}{2}}$ {choose minus since $x \le 0$} $\Rightarrow$

$f^{-1}(x) = -\sqrt{\dfrac{9 - x}{2}}$

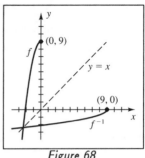

Figure 68

71 (a) $V = at + b$ is the desired form. $V = 89{,}000$ when $t = 0 \Rightarrow V = at + 89{,}000$.

$V = 125{,}000$ when $t = 6 \Rightarrow 125{,}000 = 6a + 89{,}000 \Rightarrow a = \dfrac{36{,}000}{6} = 6000$ and hence,

$$V = 6000t + 89{,}000.$$

(b) $V = 103{,}000 \Rightarrow 103{,}000 = 6000t + 89{,}000 \Rightarrow t = \frac{7}{3}$, or $2\frac{1}{3}$.

73 (a) $C_1(x) = \left(1.25 \, \dfrac{\text{dollars}}{\text{gallon}} \right) \div \left(20 \, \dfrac{\text{miles}}{\text{gallon}} \right) \cdot x \text{ miles} = \dfrac{1.25}{20} x = 0.0625x$, or $\frac{1}{16}x$.

(b) After the tune-up, the gasoline mileage will be 10% more than 20 mi/gal, that is,

22 mi/gal. $C_2(x) = \frac{1.25}{22}x + 50 = \frac{5}{88}x + 50 \approx 0.0568x + 50$.

(c) $C_2 < C_1 \Rightarrow \left[\frac{5}{88}x + 50 < \frac{1}{16}x \right] \cdot 16(11) \Rightarrow 10x + 8800 < 11x \Rightarrow x > 8800$ miles.

76 (a) $V = (10 \text{ ft}^3 \text{ per minute})(t \text{ minutes}) = 10t$

(b) The height and length of the bottom triangular region are in the proportion 6–60, or 1–10, and the length is 10 times the height. When $0 \le h \le 6$, the volume is

$V = (\text{cross sectional area})(\text{pool width}) = \frac{1}{2}bh(40) = \frac{1}{2}(10h)(h)(40) = 200h^2 \text{ ft}^3$.

When $6 < h \le 9$, the triangular region is full and

$$V = 200(6)^2 + (h - 6)(80)(40) = 7200 + 3200(h - 6).$$

(c) $10t = 200h^2 \Rightarrow h = \sqrt{t/20}$;

$$0 \le h \le 6 \Rightarrow 0 \le \sqrt{t/20} \le 6 \Rightarrow 0 \le t/20 \le 36 \Rightarrow 0 \le t \le 720.$$

$10t = 7200 + 3200(h - 6) \Rightarrow h - 6 = \dfrac{t - 720}{320} \Rightarrow h = 6 + \dfrac{t - 720}{320}$; $6 < h \le 9 \Rightarrow$

$$6 < 6 + \dfrac{t - 720}{320} \le 9 \Rightarrow 0 < \dfrac{t - 720}{320} \le 3 \Rightarrow 0 < t - 720 \le 960 \Rightarrow 720 < t \le 1680.$$

77 (a) Using similar triangles, $\dfrac{r}{x} = \dfrac{2}{4} \Rightarrow r = \frac{1}{2}x$.

(b) $\text{Volume}_{\text{cone}} + \text{Volume}_{\text{cup}} = \text{Volume}_{\text{total}} \Rightarrow \frac{1}{3}\pi r^2 h + \pi r^2 h = 5 \Rightarrow$

$$\tfrac{1}{3}\pi\left(\tfrac{1}{2}x\right)^2(x) + \pi(2)^2(y) = 5 \Rightarrow 5 - \tfrac{\pi}{12}x^3 = 4\pi y \Rightarrow y = \tfrac{5}{4\pi} - \tfrac{1}{48}x^3$$

$\boxed{78}$ (a) $\frac{y}{b} = \frac{y+h}{a} \Rightarrow ay = by + bh \Rightarrow y(a-b) = bh \Rightarrow y = \frac{bh}{a-b}$

(b) $V = \frac{1}{3}\pi a^2(y+h) - \frac{1}{3}\pi b^2 y = \frac{\pi}{3}[(a^2-b^2)y + a^2h] =$

$$\frac{\pi}{3}\Big[(a^2-b^2)\frac{bh}{a-b} + a^2h\Big] = \frac{\pi}{3}h\big[(a+b)b + a^2\big] = \frac{\pi}{3}h(a^2+ab+b^2)$$

(c) $a=6$, $b=3$, $V=600 \Rightarrow \frac{\pi}{3}h(6^2 + 6\cdot 3 + 3^2) = 600 \Rightarrow h = \frac{1800}{63\pi} = \frac{200}{7\pi} \approx 9.1$ ft

$\boxed{80}$ Let r denote the radius of the semicircles and x the length of the rectangle.

Perimeter = half-mile $\Rightarrow 2x + 2\pi r = \frac{1}{2} \Rightarrow x = -\pi r + \frac{1}{4}$.

$A = 2rx = 2r(-\pi r + \frac{1}{4}) = -2\pi r^2 + \frac{1}{2}r$. The maximum value of A occurs when

$$r = -\frac{b}{2a} = -\frac{1/2}{2(-2\pi)} = \frac{1}{8\pi} \text{ mi.} \quad x = -\pi\Big(\frac{1}{8\pi}\Big) + \frac{1}{4} = \frac{1}{8} \text{ mi.}$$

$\boxed{82}$ (a) Solving $-0.016x^2 + 1.6x = \frac{1}{5}x$ for x represents the intersection between the

parabola and the line. $-0.08x^2 + 8x = x \Rightarrow 7x - \frac{8}{100}x^2 = 0 \Rightarrow x(7 - \frac{8}{100}x) = 0 \Rightarrow$

$x = 0, \frac{175}{2}$. The rocket lands at $(\frac{175}{2}, \frac{35}{2}) = (87.5, 17.5)$.

(b) The *difference d* between the parabola and the line is to be maximized here.

$d = (-0.016x^2 + 1.6x) - (\frac{1}{5}x) = -0.016x^2 + 1.4x$. d obtains a maximum when

$x = -\frac{b}{2a} = -\frac{1.4}{2(-0.016)} = 43.75$. The maximum height of the rocket

above the ground is $d = -0.016(43.75)^2 + 1.4(43.75) = 30.625$ units.

Chapter 4: Polynomial and Rational Functions

$\boxed{3}$ $f(x) = ax^3 + 2$ • The "+2" will shift each graph up 2 units.

(a) The effect of the "2" for a is to vertically stretch $g(x) = x^3 + 2$ by a factor of 2

and make it appear "steeper."

(b) The effect of the "$\frac{1}{3}$" for a is to vertically compress $g(x) = x^3$ by a factor of

$1/(1/3) = 3$, making it appear "flatter." The "$-$" reflects the graph of

$h(x) = \frac{1}{3}x^3$ about the x-axis.

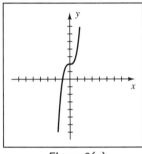

Figure 3(a)

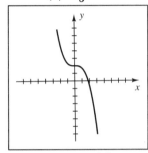

Figure 3(b)

$\boxed{5}$ We need to show that there is a sign change between $f(3)$ and $f(4)$ for $f(x) = x^3 - 4x^2 + 3x - 2$.

$$f(3) = 27 - 36 + 9 - 2 = -2 \quad \text{and} \quad f(4) = 64 - 64 + 12 - 2 = 10.$$

Since $f(3) = -2 < 0$ and $f(4) = 10 > 0$, the intermediate value theorem for polynomial functions assures us that f takes on every value between -2 and 10 in the interval $[3, 4]$, namely, 0.

$\boxed{11}$ $f(x) = \frac{1}{4}x^3 - 2 = \frac{1}{4}(x^3 - 8) = \frac{1}{4}(x - 2)(x^2 + 2x + 4)$. The general shape of the graph is that of $f(x) = x^3$. To find the x-intercepts, set $y\ \{f(x)\}$ equal to 0. This means either $x - 2 = 0$ or $x^2 + 2x + 4 = 0$. $x - 2 = 0 \Rightarrow x = 2$ and $x^2 + 2x + 4 = 0 \Rightarrow$ $x = -1 \pm \sqrt{3}\,i$. The imaginary solutions mean that we have no x-intercepts from the factor $x^2 + 2x + 4$ and the only x-intercept is 2. To find the y-intercept, set x equal to 0. $x = 0 \Rightarrow y = -2$. $f(x) > 0$ if $x > 2$, $f(x) < 0$ if $x < 2$

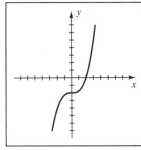

Figure 11   *Figure 17*

$\boxed{17}$ $f(x) = -x^3 + 3x^2 + 10x = -x(x^2 - 3x - 10) = -x(x + 2)(x - 5)$;

$f(x) > 0$ if $x < -2$ or $0 < x < 5$, $f(x) < 0$ if $-2 < x < 0$ or $x > 5$

$\boxed{19}$ $f(x) = \frac{1}{6}(x + 2)(x - 3)(x - 4)$ •

The y-intercept is $f(0) = \frac{1}{6}(0 + 2)(0 - 3)(0 - 4) = \frac{1}{6}(2)(-3)(-4) = 4$. To determine the intervals on which $f(x) > 0$ and $f(x) < 0$, you may want to refer back to the concept of a sign diagram. Below is a sign diagram for this function. Note how the sign of each region correlates to the sign of $f(x)$.

Resulting sign:	$\ominus$	$\oplus$	$\ominus$	$\oplus$
Sign of $x - 4$:	$-$	$-$	$-$	$+$
Sign of $x - 3$:	$-$	$-$	$+$	$+$
Sign of $x + 2$:	$-$	$+$	$+$	$+$
x values:		-2	3	4

Diagram 19

$f(x) > 0$ if $-2 < x < 3$ or $x > 4$, $f(x) < 0$ if $x < -2$ or $3 < x < 4$

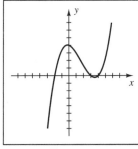

Figure 19   *Figure 21*

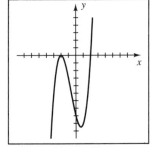

$\boxed{21}$ $f(x) = x^3 + 2x^2 - 4x - 8 = x^2(x + 2) - 4(x + 2) = (x + 2)^2(x - 2)$;

$f(x) > 0$ if $x > 2$, $f(x) < 0$ if $x < -2$ or $|x| < 2$

$\boxed{23}$ $f(x) = x^4 - 6x^2 + 8 = (x^2 - 2)(x + 2)(x - 2)$ •

The general shape of the graph is that of $g(x) = x^4$.

$$f(x) > 0 \text{ if } |x| > 2 \text{ or } |x| < \sqrt{2},\ f(x) < 0 \text{ if } \sqrt{2} < |x| < 2$$

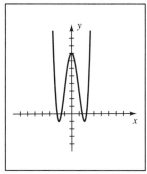

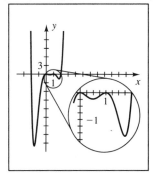

Figure 23   Figure 25

$\boxed{25}$ $f(x) = x^2(x + 2)(x - 1)^2(x - 2)$ •

If an intercept value has a corresponding *linear* factor, the graph will go through that intercept—that is, there will be a sign change from positive to negative or negative to positive in the function. Thus, at $x = -2$ and $x = 2$ the graph of the function "goes through" the point. If an intercept value has a corresponding quadratic factor (as in the case of 0 and x^2; and 1 and $(x - 1)^2$), then the function will not change sign. Hence, at $x = 0$ and $x = 1$ the graph of the function "touches the point and turns around."

The above discussion can be generalized as follows:

(1) If an intercept value has a corresponding factor raised to an <u>odd</u> power,

then the function <u>will</u> change sign at that point.

(2) If an intercept value has a corresponding factor raised to an <u>even</u> power,

then the function <u>will not</u> change sign at that point.

$f(x) > 0$ if $|x| > 2$, $f(x) < 0$ if $|x| < 2$, $x \neq 0$, $x \neq 1$

$\boxed{29}$ If a graph contains the point $(-1, 4)$, then $f(-1) = 4$.

For this function, $f(x) = 3x^3 - kx^2 + x - 5k$,

$$f(-1) = 3(-1)^3 - k(-1)^2 + (-1) - 5k = -3 - k - 1 - 5k = -4 - 6k.$$

Thus, $-4 - 6k$ must equal 4. Solving for k yields $-4 - 6k = 4 \Rightarrow k = -\frac{4}{3}$.

33 (a) $V(x) = lwh = (30 - x - x)(20 - x - x)x = x(20 - 2x)(30 - 2x) =$

$$4x(10 - x)(15 - x) = 4x(x - 10)(x - 15).$$

(b) $V(x) > 0$ on $(0, 10)$ and $(15, \infty)$. Allowable values for x are in $(0, 10)$.

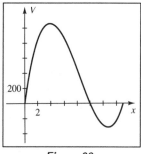

Figure 33

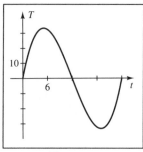

Figure 35

35 We will work part (b) first.

(b) T has the general shape of a cubic with zeros at 0, 12, and 24.

Its sign pattern is negative, positive, negative, positive.

(a) $T = \frac{1}{20}t(t - 12)(t - 24) = 0 \Rightarrow t = 0, 12, 24$. $T > 0$ for

$0 < t < 12$ {6 A.M. to 6 P.M.}; $T < 0$ for $12 < t < 24$ {6 P.M. to 6 A.M.}.

(c) 12 noon corresponds to $t = 6$, $T(6) = 32.4 > 32°F$ and $T(7) = 29.75 < 32°F$

37 (a) $N(t) = -t^4 + 21t^2 + 100$

$= -(t^4 - 21t^2 - 100)$

$= -(t^2 - 25)(t^2 + 4)$

$= -(t + 5)(t - 5)(t^2 + 4)$

If $t > 0$, then $N(t) > 0$ for $0 < t < 5$.

(b) The population becomes extinct when $N = 0$.

This occurs after 5 years.

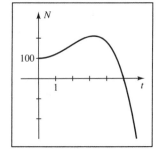

Figure 37

41 From the graph, f has three zeros. They are approximately -1.88, 0.35, and 1.53.

[−4.5, 4.5] by [−3, 3]

[−4.5, 4.5] by [−3, 3]

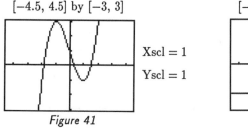

Xscl = 1

Yscl = 1

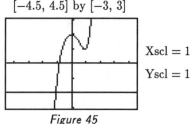

Xscl = 1

Yscl = 1

Figure 41

Figure 45

45 If $f(x) = x^5 - 2x^2 + 2$ and $k = -2$, then $f(x) > k$ on $(-1.10, \infty)$.

4.2 Exercises

Note: Refer to the illustration on page 234. We will walk through the steps to obtain the quotient and remainder for this example.

(1) Determine the quotient of the first terms of the dividend $(x^4 - 16)$ and the divisor $(x^2 + 3x + 1)$. $\dfrac{x^4}{x^2} = x^2$—this is the first term of the resulting quotient.

(2) Multiply the result in step (1), x^2, by the divisor, $x^2 + 3x + 1$. This product is $x^4 + 3x^3 + x^2$.

(3) Subtract the result in step (2) from the dividend. This result is $-3x^3 - x^2$ (there is no need to worry about the lower degree terms here, namely, -16). A common mistake is to forget to subtract <u>all</u> of the terms and obtain (in this case) $3x^3 + x^2$.

(4) Repeat steps (1) through (3) for the remaining dividend until the degree of the remainder is less than the degree of the divisor.

1. $f(x) = 2x^4 - x^3 - 3x^2 + 7x - 12; \quad p(x) = x^2 - 3$ •

$$
\begin{array}{r}
2x^2 - x + 3 \\
x^2 - 3 \overline{\smash{\big)}\, 2x^4 - x^3 - 3x^2 + 7x - 12} \\
2x^4 - 6x^2 \\
\hline
-x^3 + 3x^2 \\
-x^3 + 3x \\
\hline
3x^2 + 4x \\
3x^2 - 9 \\
\hline
4x - 3
\end{array}
$$

★ $2x^2 - x + 3$; $4x - 3$

9. We wish to find $f(2)$ by using the remainder theorem. Thus, we need to divide $f(x) = 3x^3 - x^2 + 5x - 4$ by $x - 2$ using either long division or synthetic division. Using synthetic division, we have

$$
\begin{array}{r|rrrr}
2 & 3 & -1 & 5 & -4 \\
  &   & 6 & 10 & 30 \\
\hline
  & 3 & 5 & 15 & 26
\end{array}
$$

The last number in the third row, 26, is our remainder. Hence, $f(2) = 26$.

[11] We divide by $x + 3$ or $x - (-3)$. Note that we include a 0 for the missing x^3 term in $f(x) = x^4 - 6x^2 + 4x - 8$.

$$
\begin{array}{r|rrrrr}
-3 & 1 & 0 & -6 & 4 & -8 \\
   &   & -3 & 9 & -9 & 15 \\
\hline
   & 1 & -3 & 3 & -5 & 7
\end{array}
$$

Hence, $f(-3) = 7$.

[13] To show that $x + 3$ is a factor of $f(x) = x^3 + x^2 - 2x + 12$, we must show that $f(-3) = 0$. $f(-3) = -27 + 9 + 6 + 12 = 0$, and, hence, $x + 3$ is a factor of $f(x)$.

[17] f has degree 3 with zeros $-2, 0, 5 \Rightarrow$

$$f(x) = a\big[x - (-2)\big](x - 0)(x - 5) \ \{\text{let } a = 1\}$$
$$= x(x + 2)(x - 5) = x(x^2 - 3x - 10) = x^3 - 3x^2 - 10x$$

[19] f has degree 4 with zeros $-2, \pm 1, 4 \Rightarrow$

$$f(x) = a(x + 2)(x + 1)(x - 1)(x - 4) \ \{\text{let } a = 1\}$$
$$= (x^2 - 1)(x^2 - 2x - 8) = x^4 - 2x^3 - 9x^2 + 2x + 8$$

[21]
$$
\begin{array}{r|rrrr}
2 & 2 & -3 & 4 & -5 \\
  &   & 4 & 2 & 12 \\
\hline
  & 2 & 1 & 6 & 7
\end{array}
$$

The synthetic division indicates that the quotient is $2x^2 + x + 6$ and the remainder is 7.

[23]
$$
\begin{array}{r|rrrr}
-3 & 1 & 0 & -8 & -5 \\
   &   & -3 & 9 & -3 \\
\hline
   & 1 & -3 & 1 & -8
\end{array}
$$

The synthetic division indicates that the quotient is $x^2 - 3x + 1$ and the remainder is -8.

[29]
$$
\begin{array}{r|rrrr}
3 & 2 & 3 & -4 & 4 \\
  &   & 6 & 27 & 69 \\
\hline
  & 2 & 9 & 23 & 73
\end{array}
$$

The synthetic division indicates that $f(3) = 73$.

[31]
$$
\begin{array}{r|rrrr}
-0.2 & 0.3 & 0 & 0.04 & -0.034 \\
     &   & -0.06 & 0.012 & -0.0104 \\
\hline
     & 0.3 & -0.06 & 0.052 & -0.0444
\end{array}
$$

The remainder is -0.0444 and the remainder theorem indicates that this is $f(-0.2)$.

[33]
$$
\begin{array}{r|rrr}
2 + \sqrt{3} & 1 & 3 & -5 \\
             &   & 2 + \sqrt{3} & 13 + 7\sqrt{3} \\
\hline
             & 1 & 5 + \sqrt{3} & 8 + 7\sqrt{3}
\end{array}
$$

The remainder is $8 + 7\sqrt{3}$ and the remainder theorem indicates that this is the value of $f(2 + \sqrt{3})$.

35 -2 is a zero of $f(x)$ if we can show that $f(-2) = 0$.

$$
\begin{array}{r|rrrrr}
-2 & 3 & 8 & -2 & -10 & 4 \\
   &   & -6 & -4 & 12 & -4 \\
\hline
   & 3 & 2 & -6 & 2 & 0
\end{array}
$$

Hence, $f(-2) = 0$ and -2 is a zero of $f(x)$.

37
$$
\begin{array}{r|rrrr}
\frac{1}{2} & 4 & -6 & 8 & -3 \\
   &   & 2 & -2 & 3 \\
\hline
   & 4 & -4 & 6 & 0
\end{array}
$$

Hence, $f(\frac{1}{2}) = 0$ and $\frac{1}{2}$ is a zero of $f(x)$.

39 $f(-2) = k(-2)^3 + (-2)^2 + k^2(-2) + 3k^2 + 11$

$\qquad = -8k + 4 - 2k^2 + 3k^2 + 11 = k^2 - 8k + 15.$

The remainder, $k^2 - 8k + 15$, must be zero if $f(x)$ is to be divisible by $x + 2$.

$$k^2 - 8k + 15 = 0 \Rightarrow (k - 3)(k - 5) = 0 \Rightarrow k = 3, 5.$$

41 $x - c$ will not be a factor of $f(x) = 3x^4 + x^2 + 5$ for any real number c if the remainder of dividing $f(x)$ by $x - c$ is not zero for any real number c. The remainder is $f(c)$, or $3c^4 + c^2 + 5$. Since $3c^4 + c^2 + 5$ is greater than or equal to 5 for any real number c, it is never zero and hence, $x - c$ is never a factor of $f(x)$.

45 If $f(x) = x^n - y^n$ and n is even, then $f(-y) = (-y)^n - (y)^n = y^n - y^n = 0.$

Hence, $x + y$ is a factor of $x^n - y^n$.

47 (a) The radius of the resulting cylinder is x and the height is y, that is, $6 - x$.

$$V = \pi r^2 h = \pi x^2 (6 - x)$$

(b) The volume of the cylinder of radius 1 and altitude 5 is $\pi(1)^2 5 = 5\pi$. To determine another value of x which would result in the same volume, we need to solve the equation $5\pi = \pi x^2 (6 - x)$. $5\pi = \pi x^2 (6 - x) \Rightarrow 5 = 6x^2 - x^3 \Rightarrow x^3 - 6x^2 + 5 = 0$. We know that $x = 1$ is a solution to this equation. Synthetically dividing, we obtain

$$
\begin{array}{r|rrrr}
1 & 1 & -6 & 0 & 5 \\
   &   & 1 & -5 & -5 \\
\hline
   & 1 & -5 & -5 & 0
\end{array}
$$

The quotient is $x^2 - 5x - 5$. Using the quadratic formula, we see that the solutions are $x = \dfrac{5 \pm \sqrt{45}}{2}$. Since x must be positive in the first quadrant, $\dfrac{5 + \sqrt{45}}{2} \approx 5.85$ would be an allowable value of x. If $x = \frac{1}{2}(5 + \sqrt{45})$, then

$y = 6 - x = 6 - \frac{1}{2}(5 + \sqrt{45}) = \frac{12}{2} - \frac{5}{2} - \frac{1}{2}\sqrt{45} = \frac{1}{2}(7 - \sqrt{45}).$

The point P is $P(x, y) = \left(\frac{1}{2}(5 + \sqrt{45}), \frac{1}{2}(7 - \sqrt{45}) \right) \approx (5.85, 0.15).$

[49] (a) $A = lw = (2x)(y) = 2x(4 - x^2) = 8x - 2x^3$

(b) This solution is similar to the solution in Exercise 47(b).

$$A = 6 \Rightarrow 6 = 8x - 2x^3 \Rightarrow x^3 - 4x + 3 = 0 \Rightarrow (x - 1)(x^2 + x - 3) = 0 \Rightarrow$$

$$x = 1, \frac{-1 \pm \sqrt{13}}{2}. \quad \frac{\sqrt{13} - 1}{2} \text{ would be an allowable value of } x.$$

The base would then be $\sqrt{13} - 1 \approx 2.61$.

[51] $f(x) = 4x^4 - 5x^2 + 7x - 3 = (((4x)x - 5)x + 7)x - 3.$ Start with 4 times -2, or -8.

Multiply by -2 to yield 16. Subtract 5 $\{11\}$, multiply by -2 $\{-22\}$, add 7 $\{-15\}$,

multiply by -2 $\{30\}$, and subtract 3 $\{27\}$. Hence, $f(-2) = 27$.

[53] Similar to Example 7, store 0.0325 in a memory location and use the nested form for

$$f, f(x) = ((((2x + 3)x - 6)x - 4)x + 1)x - 9. \text{ The result is } f(0.0325) \approx -8.9719.$$

[57] $f(1.6) = -2k^4 + 2.56k^3 + 3.2k + 4.096.$ Graph $y = -2k^4 + 2.56k^3 + 3.2k + 4.096$ (that

is, $y = -2x^4 + 2.56x^3 + 3.2x + 4.096$). From the graph, we see that $y = 0$ when

$k \approx -0.75, 1.96.$ Thus, if k assumes either of these values, $f(1.6) = 0$ and f will be

divisible by $x - 1.6$ by the factor theorem.

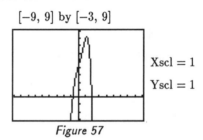

[−9, 9] by [−3, 9]

Xscl = 1

Yscl = 1

Figure 57

4.3 Exercises

[1] The polynomial is of the form $f(x) = a(x - (-1))(x - 2)(x - 3).$

$f(-2) = a(-1)(-4)(-5) = 80 \Rightarrow -20a = 80 \Rightarrow a = -4.$

Hence, the polynomial is $-4(x + 1)(x - 2)(x - 3)$, or $-4x^3 + 16x^2 - 4x - 24$.

[3] The polynomial is of the form $f(x) = a(x - (-4))(x - 3)(x - 0).$

$f(2) = a(6)(-1)(2) = -36 \Rightarrow -12a = -36 \Rightarrow a = 3.$

Hence, the polynomial is $3(x + 4)(x - 3)(x)$, or $3x^3 + 3x^2 - 36x$.

[5] The polynomial is of the form $f(x) = a(x - (-2i))(x - 2i)(x - 3).$

$f(1) = a(1 + 2i)(1 - 2i)(-2) = 20 \Rightarrow -10a = 20 \Rightarrow a = -2.$

Hence, the polynomial is $-2(x + 2i)(x - 2i)(x - 3) = -2(x^2 + 4)(x - 3),$

or $-2x^3 + 6x^2 - 8x + 24$.

$\boxed{7}$ $f(x) = a(x+4)^2(x-3)^2 = (x^2 + x - 12)^2 \{a = 1\} = x^4 + 2x^3 - 23x^2 - 24x + 144$

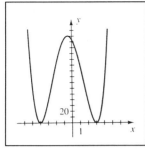

Figure 7

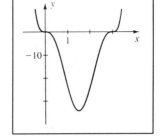

Figure 9

$\boxed{9}$ $f(x) = a(x)^3(x-3)^3$ so $f(2) = a(8)(-1) = -8a$. But $f(2) = -24$, so $-8a = -24$, or,

$$a = 3. \quad f(x) = 3(x)^3(x^3 - 9x^2 + 27x - 27) = 3x^6 - 27x^5 + 81x^4 - 81x^3.$$

$\boxed{11}$ The graph has x-intercepts at -1, $\frac{3}{2}$, 3, and $f(0) = \frac{7}{2}$.

$$f(x) = a(x+1)(x - \tfrac{3}{2})(x-3); \; f(0) = a(1)(-\tfrac{3}{2})(-3) = \tfrac{7}{2} \Rightarrow \tfrac{9}{2}a = \tfrac{7}{2} \Rightarrow a = \tfrac{7}{9}.$$

$$f(x) = \tfrac{7}{9}(x+1)(x - \tfrac{3}{2})(x-3)$$

$\boxed{13}$ 3 is a zero of multiplicity one, 1 is a zero of multiplicity two, and $f(0) = 3$.

$$f(x) = a(x-1)^2(x-3); \; f(0) = a(1)(-3) = 3 \Rightarrow a = -1. \quad f(x) = -1(x-1)^2(x-3).$$

$\boxed{17}$ $f(x) = 4x^5 + 12x^4 + 9x^3 = x^3(4x^2 + 12x + 9) = x^3(2x+3)^2$

$$\bigstar \; -\tfrac{3}{2} \text{ (multiplicity 2); } 0 \text{ (multiplicity 3)}$$

$\boxed{19}$ $f(x) = (x^2 + x - 12)^3(x^2 - 9)^2 = \left[(x+4)(x-3)\right]^3\left[(x+3)(x-3)\right]^2$

$$= (x+4)^3(x-3)^3(x+3)^2(x-3)^2 = (x+4)^3(x+3)^2(x-3)^5$$

$$\bigstar \; -4 \text{ (multiplicity 3); } -3 \text{ (multiplicity 2); } 3 \text{ (multiplicity 5)}$$

$\boxed{23}$ Using synthetic division,

$$
\begin{array}{r|rrrrr}
-3 & 1 & 7 & 13 & -3 & -18 \\
   &   & -3 & -12 & -3 & 18 \\
\hline
   & 1 & 4 & 1 & -6 & 0 \\
\end{array}
$$

$$
\begin{array}{r|rrrr}
-3 & 1 & 4 & 1 & -6 \\
   &   & -3 & -3 & 6 \\
\hline
   & 1 & 1 & -2 & 0 \\
\end{array}
$$

The remaining polynomial, $x^2 + x - 2$, can be factored as $(x+2)(x-1)$.

$$\text{Hence, } f(x) = (x+3)^2(x+2)(x-1).$$

$\boxed{25}$ *Note:* If the sum of the coefficients of the polynomial is 0, then 1 is a zero of the polynomial. Synthetically dividing 1 five times, we obtain the remaining polynomial,

$$x + 1. \text{ Hence, } f(x) = x^6 - 4x^5 + 5x^4 - 5x^2 + 4x - 1 = (x-1)^5(x+1).$$

Note: For the following exercises, let $f(x)$ denote the polynomial, P, the number of sign changes in $f(x)$, and N, the number of sign changes in $f(-x)$. The types of possible solutions are listed in the order positive, negative, nonreal complex.

27 $f(x) = 4x^3 - 6x^2 + x - 3$. The sign pattern for coefficients is $+, -, +, -$. Since there are 3 sign changes in $f(x)$, P = 3. $f(-x) = -4x^3 - 6x^2 - x - 3$. *Note:* Simply negate the coefficients of the odd-powered terms of $f(x)$ to find $f(-x)$. The sign pattern for coefficients is $-, -, -, -$. Since there are no sign changes in $f(-x)$, N = 0. If there are 3 positive solutions, then there are no negative or nonreal complex. If there is 1 positive solution, there are no negative and 2 nonreal complex solutions.

31 $f(x) = 3x^4 + 2x^3 - 4x + 2$ and P = 2.

$f(-x) = 3x^4 - 2x^3 + 4x + 2$ and N = 2.   ★ 2, 2, 0; 2, 0, 2; 0, 2, 2; 0, 0, 4

35 Synthetically dividing 5 into the polynomial yields a bottom row consisting of all nonnegative numbers (see below). Synthetically dividing -2 into the polynomial yields a bottom row that alternates in sign (see below).

5	1	-4	-5	7
		5	5	0
	1	1	0	7

-2	1	-4	-5	7
		-2	12	-14
	1	-6	7	-7

These results indicate that the upper bound is 5 and the lower bound is -2.

37 Synthetically dividing 2 into the polynomial yields a bottom row consisting of all nonnegative numbers (see below). Synthetically dividing -2 into the polynomial yields a bottom row that alternates in sign (see below).

2	1	-1	-2	3	6
		2	2	0	6
	1	1	0	3	12

-2	1	-1	-2	3	6
		-2	6	-8	10
	1	-3	4	-5	16

These results indicate that the upper bound is 2 and the lower bound is -2.

41 The zeros are 0, 5, 19, 24, and $f(12) = 10$. $f(t) = a(t)(t-5)(t-19)(t-24)$;

$f(12) = a(12)(7)(-7)(-12) = 10 \Rightarrow 7056a = 10 \Rightarrow a = \frac{5}{3528}$.

$$f(t) = \frac{5}{3528}t(t-5)(t-19)(t-24)$$

43 The graph of f does not cross the x-axis at a zero of even multiplicity, but does cross the x-axis at a zero of odd multiplicity. The higher the multiplicity of a zero, the more horizontal the graph of f is near that zero.

[−3, 3] by [−2, 2]

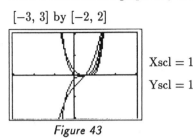

Xscl = 1

Yscl = 1

Figure 43

[−3, 3] by [−3, 1]

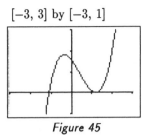

Xscl = 1

Yscl = 1

Figure 45

45 From the graph of f there are two zeros. They are -1.2 and 1.1. The zero at -1.2 has even multiplicity and the zero at 1.1 has odd multiplicity. Since f has degree 3, the zero at -1.2 must have multiplicity 2 and the zero at 1.1 has multiplicity 1.

47 Since the zeros are 2, 5.2, and 10.1, the polynomial must have the form
$f(x) = a(x-2)(x-5.2)(x-10.1)$. Now, $f(1.1) = a(-33.21) = -49.815 \Rightarrow a = 1.5$.
Let $f(x) = 1.5(x-2)(x-5.2)(x-10.1)$. Since f has been completely determined, we must check the remaining data points. $f(3.5) = 25.245$ and $f(6.4) = -29.304$.

Thus, a third-degree polynomial does fit the data points.

49 From the graph of $A(t) = -\frac{1}{2400}t^3 + \frac{1}{20}t^2 + \frac{7}{6}t + 340$, we see that $A = 400$ when $t \approx 27.1$. Thus, the carbon dioxide concentration will be 400 in $1980 + 27.1 = 2007.1$, or, during the year 2007.

[0, 60] by [0, 600]

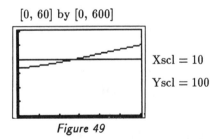

Xscl = 10

Yscl = 100

Figure 49

4.4 Exercises

Note: It is helpful to remember that if $a + bi$ is a zero, then $\left[x^2 - 2ax + (a^2 + b^2)\right]$
is the associated quadratic factor for the following exercises.

1 Since $3 + 2i$ is a root, so is $3 - 2i$. As in the preceding note, $a = 3$ and $b = -2$, and hence, $-2a = -6$ and $a^2 + b^2 = 13$. Thus, the polynomial is of the form
$\left[x - (3 + 2i)\right]\left[x - (3 - 2i)\right] = x^2 - 6x + 13$.

[3] Since 2 is a zero, $x - 2$ is a factor. The associated quadratic factor for $-2 \pm 5i$ is
$x^2 - 2(-2)x + (-2)^2 + (5)^2$, or, equivalently, $x^2 + 4x + 29$. Hence, the polynomial is
of the form $(x - 2)(x^2 + 4x + 29)$.

[5] Since -1 and 0 are zeros, $x + 1$ and x are factors. The associated quadratic factor for
$3 \pm i$ is $x^2 - 2(3)x + (3)^2 + (1)^2$, or, equivalently, $x^2 - 6x + 10$. Hence, the polynomial
is of the form $(x)(x + 1)(x^2 - 6x + 10)$.

[9] If $-2i$ is a zero, then $2i$ is a zero. Thus, $(x - 2i)(x - (-2i)) = x^2 - 4i^2 = x^2 + 4$ is the
associated factor. Hence, the polynomial is of the form $(x)(x^2 + 4)(x^2 - 2x + 2)$.

[11] The constant term, 6, has positive integer divisors 1, 2, 3, 6. The leading coefficient,
1, has integer divisors ± 1. By the theorem on rational zeros of a polynomial, any
rational root will be of the form

$$\frac{\text{positive divisors of } 6}{\text{divisors of } 1} = \frac{1,\, 2,\, 3,\, 6}{\pm 1} \longrightarrow \pm 1,\ \pm 2,\ \pm 3,\ \pm 6.$$

Since ± 1 are always potential solutions, they are as good as any other values to
begin with. If the coefficients of the polynomial ($x^3 + 3x^2 - 4x + 6$ in this exercise)
sum to 0, then 1 is a zero. In this case, the sum is $1 + 3 + (-4) + 6 = 6 \neq 0$. Thus, 1
is eliminated and we will try -1. If $f(x)$ is the polynomial, then -1 will be a zero if
the coefficients of $f(-x)$ sum to 0. In this case, $f(-x) = -x^3 + 3x^2 - 10x + 6$ and the
sum of the coefficients is $(-1) + 3 + (-10) + 6 = -2 \neq 0$. Thus, -1 is not a zero. We
now use synthetic division to show that ± 2, ± 3, and ± 6 are not zeros of $f(x)$.

[15] The constant term, -8, has positive integer divisors 1, 2, 4, 8. The leading coefficient
has integer divisors ± 1. Hence, any rational root will be of the form

$$\frac{\text{positive divisors of } -8}{\text{divisors of } 1} = \frac{1,\, 2,\, 4,\, 8}{\pm 1} \longrightarrow \pm 1,\ \pm 2,\ \pm 4,\ \pm 8.$$

We now need to try to find one solution of the equation. Once we have one solution,
the resulting polynomial will be a quadratic and then we can use the quadratic
formula to find the other 2 solutions. As in the solution to Exercise 11, we will try
± 1 first. After determining that 1 is not a solution, we try -1. In this case,
$f(-x) = -x^3 - x^2 + 10x - 8$ and   the   sum   of   the   coefficients   is
$(-1) + (-1) + 10 + (-8) = 0$. Thus, -1 is a zero and we will use synthetic division to
find the remaining polynomial.

$$
\begin{array}{r|rrrr}
-1 & 1 & -1 & -10 & -8 \\
   &   & -1 &   2 & 8 \\
\hline
   & 1 & -2 & -8 & 0
\end{array}
$$

The remaining polynomial is $(x^2 - 2x - 8) = (x - 4)(x + 2)$.

Hence, the solutions are -2, -1, and 4.

17 As previously noted, we first check ± 1. Neither of these values are solutions so we list the possible rational roots. The values listed below are obtained by taking quotients of each number in the numerator with ± 1, and then each number in the numerator with ± 2, while discarding any repeat choices.

$$\frac{1,\ 2,\ 3,\ 5,\ 6,\ 10,\ 15,\ 30}{\pm 1,\ \pm 2} \longrightarrow$$
$$\pm 1,\ \pm 2,\ \pm 3,\ \pm 5,\ \pm 6,\ \pm 10,\ \pm 15,\ \pm 30,\ \pm \tfrac{1}{2},\ \pm \tfrac{3}{2},\ \pm \tfrac{5}{2},\ \pm \tfrac{15}{2}$$

Trying 2 we obtain

$$
\begin{array}{r|rrrr}
2 & 2 & -3 & -17 & 30 \\
  &   & 4 & 2   & -30 \\
\hline
  & 2 & 1 & -15 & 0
\end{array}
$$

The remaining polynomial is $(2x^2 + x - 15) = (2x - 5)(x + 3)$.

Hence, the solutions are -3, 2, and $\tfrac{5}{2}$.

19 $\dfrac{1,\ 2,\ 4,\ 7,\ 8,\ 14,\ 28,\ 56}{\pm 1} \longrightarrow \pm 1,\ \pm 2,\ \pm 4,\ \pm 7,\ \pm 8,\ \pm 14,\ \pm 28,\ \pm 56$. Again, ± 1

are not solutions. Trying 4 and then -7 (using a slightly different format) we obtain

$$
\begin{array}{r|rrrrr}
4 & 1 & 3 & -30 & -6 & 56 \\
  &   & 4 & 28 & -8 & -56 \\
\hline
-7 & 1 & 7 & -2 & -14 & 0 \\
   &   & -7 & 0 & 14 & \\
\hline
   & 1 & 0 & -2 & 0 &
\end{array}
$$

The remaining polynomial is $x^2 - 2$. Its solutions are $\pm \sqrt{2}$.

Hence, the solutions are -7, $\pm \sqrt{2}$, and 4.

21 We first factor out x^2, leaving us with the equation $x^2(6x^3 + 19x^2 + x - 6) = 0$. The x^2 factor indicates that the number 0 is a zero of multiplicity two. We can now concentrate on solving the equation $6x^3 + 19x^2 + x - 6 = 0$.

$$\frac{1,\ 2,\ 3,\ 6}{\pm 1,\ \pm 2,\ \pm 3,\ \pm 6} \longrightarrow \pm 1,\ \pm 2,\ \pm 3,\ \pm 6,\ \pm \tfrac{1}{2},\ \pm \tfrac{3}{2},\ \pm \tfrac{1}{3},\ \pm \tfrac{2}{3},\ \pm \tfrac{1}{6}$$

$$
\begin{array}{r|rrrr}
-3 & 6 & 19 & 1 & -6 \\
   &   & -18 & -3 & 6 \\
\hline
   & 6 & 1 & -2 & 0
\end{array}
$$

$6x^2 + x - 2 = (3x + 2)(2x - 1)$ $\qquad\qquad\qquad$ ★ $-3,\ -\tfrac{2}{3},\ \tfrac{1}{2}$

23 $\dfrac{1,\ 3,\ 9,\ 27}{\pm 1,\ \pm 2,\ \pm 4,\ \pm 8} \longrightarrow$

$\pm 1,\ \pm 3,\ \pm 9,\ \pm 27,\ \pm\frac{1}{2},\ \pm\frac{1}{4},\ \pm\frac{1}{8},\ \pm\frac{3}{2},\ \pm\frac{3}{4},\ \pm\frac{3}{8},\ \pm\frac{9}{2},\ \pm\frac{9}{4},\ \pm\frac{9}{8},\ \pm\frac{27}{2},\ \pm\frac{27}{4},\ \pm\frac{27}{8}$

This is a tough one because only $-\frac{3}{4}$ is a solution.

$$
\begin{array}{r|rrrr}
-\frac{3}{4} & 8 & 18 & 45 & 27 \\
 & & -6 & -9 & -27 \\
\hline
 & 8 & 12 & 36 & 0
\end{array}
$$

$8x^2 + 12x + 36 = 0 \ \{\text{divide by } 4\} \Rightarrow 2x^2 + 3x + 9 = 0 \Rightarrow$

$x = \dfrac{-3 \pm \sqrt{9-72}}{4} = -\dfrac{3}{4} \pm \dfrac{1}{4}\sqrt{63}\,i = -\dfrac{3}{4} \pm \dfrac{3}{4}\sqrt{7}\,i$ $\bigstar\ -\dfrac{3}{4},\ -\dfrac{3}{4} \pm \dfrac{3}{4}\sqrt{7}\,i$

29 (a) $V(x) = x(20 - 2x)(30 - 2x) = 1000 \Rightarrow 4x^3 - 100x^2 + 600x - 1000 = 0 \Rightarrow$

$4(x - 5)\left[x - (10 - 5\sqrt{2})\right]\left[x - (10 + 5\sqrt{2})\right] = 0.$ The allowable range from

Exercise 33 of Section 4.1 was $0 < x < 10$, so discard $10 + 5\sqrt{2}$.

The two boxes having volume 1000 in^3 have dimensions

$[\![A]\!]\ 5 \times 10 \times 20$ and $[\![B]\!]\ (10 - 5\sqrt{2}) \times (10\sqrt{2}) \times (10 + 10\sqrt{2}).$

(b) The surface area function is

$S(x) = (20 - 2x)(30 - 2x) + 2(x)(20 - 2x) + 2(x)(30 - 2x) = -4x^2 + 600.$

$S(5) = 500$ and $S(10 - 5\sqrt{2}) = 400\sqrt{2} \approx 565.7$ so box $[\![A]\!]$ has less surface area.

31 (a) Let x denote one of the sides of the triangle and $x + 1$ its hypotenuse.

Using the Pythagorean theorem, the third side y is given by

$$x^2 + y^2 = (x + 1)^2 \Rightarrow y^2 = (x^2 + 2x + 1) - x^2 \Rightarrow y = \sqrt{2x + 1}.$$

Hence, the sides of the triangle are $x,\ \sqrt{2x + 1},$ and $x + 1$.

$A = \frac{1}{2}bh \Rightarrow 30 = \frac{1}{2}x\sqrt{2x+1} \Rightarrow 60 = x\sqrt{2x+1} \Rightarrow 60^2 = x^2(2x+1) \Rightarrow$

$$3600 = 2x^3 + x^2 \Rightarrow 2x^3 + x^2 - 3600 = 0.$$

(b) There is one sign change in $f(x) = 2x^3 + x^2 - 3600$.

By Descartes' rule of signs there is one positive real root.

Synthetically dividing 13 into f, we obtain

$$
\begin{array}{r|rrrr}
13 & 2 & 1 & 0 & -3600 \\
 & & 26 & 351 & 4563 \\
\hline
 & 2 & 27 & 351 & 963
\end{array}
$$

The numbers in the third row are nonnegative,

so 13 is an upper bound for the zeros of f.

(c) $2x^3 + x^2 - 3600 = 0 \Leftrightarrow (x-12)(2x^2 + 25x + 300) = 0$. The solutions of

$2x^2 + 25x + 300 = 0$ are $x = -\frac{25}{4} \pm \frac{5}{4}\sqrt{71}\,i$, and hence, $x = 12$ is the only real

solution. The legs of the triangle are 12 ft and 5 ft, and the hypotenuse is 13 ft.

$\boxed{33}$ (a) $\text{Volume}_{\text{total}} = \text{Volume}_{\text{cube}} + \text{Volume}_{\text{roof}}$

$$= x^3 + \tfrac{1}{2}bhx = x^3 + \tfrac{1}{2}(x)(6-x)(x) = x^3 + \tfrac{1}{2}x^2(6-x).$$

(b) $\text{Volume} = 80 \Rightarrow x^3 + \frac{1}{2}x^2(6-x) = 80 \Rightarrow x^3 + 3x^2 - \frac{1}{2}x^3 = 80 \Rightarrow \frac{1}{2}x^3 + 3x^2 = 80 \Rightarrow$

$x^3 + 6x^2 - 160 = 0 \Rightarrow (x-4)(x^2 + 10x + 40) = 0$. The length of the side is 4 ft.

$\boxed{35}$ $x^5 + 1.1x^4 - 3.21x^3 - 2.835x^2 + 2.7x + 0.62 = -1 \Leftrightarrow$

$x^5 + 1.1x^4 - 3.21x^3 - 2.835x^2 + 2.7x + 1.62 = 0$.

The graph of $y = x^5 + 1.1x^4 - 3.21x^3 - 2.835x^2 + 2.7x + 1.62$ intersects the x-axis

three times. The zeros at -1.5 and 1.2 have even multiplicity (since the graph is

tangent to the x-axis at these points) and the zero at -0.5 has odd multiplicity (since

the graph crosses the x-axis at this point). Since the equation has degree 5, the only

possibility is that the zeros at -1.5 and 1.2 have multiplicity 2 and the zero at -0.5

has multiplicity 1. Thus, the equation has no nonreal solutions.

$[-4.5, 4.5]$ by $[-3, 3]$

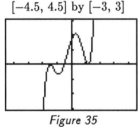

Xscl $= 1$

Yscl $= 1$

Figure 35

$[-4.5, 4.5]$ by $[-3, 3]$

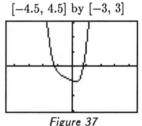

Xscl $= 1$

Yscl $= 1$

Figure 37

$\boxed{37}$ Graph $y = x^4 + 1.4x^3 + 0.44x^2 - 0.56x - 0.96$.

From the graph, zeros are located at -1.2 and 0.8. Using synthetic division,

$\dfrac{x^4 + 1.4x^3 + 0.44x^2 - 0.56x - 0.96}{x + 1.2} = x^3 + 0.2x^2 + 0.2x - 0.8$ and

$\dfrac{x^3 + 0.2x^2 + 0.2x - 0.8}{x - 0.8} = x^2 + x + 1$. The zeros of $x^2 + x + 1$ are $-\frac{1}{2} \pm \frac{\sqrt{3}}{2}i$.

Thus, the solutions to the equation are -1.2, 0.8, $-\frac{1}{2} \pm \frac{\sqrt{3}}{2}i$.

$\boxed{39}$ From the graph, we see that $D(h) = 0.4$ when $h \approx 10{,}200$.

Thus, the density of the atmosphere is 0.4 kg/m^3 at $10{,}200$ m.

$[0, 30{,}000]$ by $[0, 1.2]$

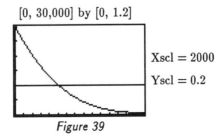

Xscl $= 2000$

Yscl $= 0.2$

Figure 39

4.5 Exercises

Note: We will use the guidelines listed in the text on page 266.

Below is a summary of these guidelines.

$$\text{Let } f(x) = \frac{a_n x^n + a_{n-1} x^{n-1} + \cdots + a_1 x + a_0}{b_k x^k + b_{k-1} x^{k-1} + \cdots + b_1 x + b_0}, \text{ where } a_n \neq 0 \text{ and } b_k \neq 0.$$

(1) Find the x-intercepts { the zeros of the numerator }.

(2) Find the vertical asymptotes { the zeros of the denominator }.

(3) Find the y-intercept { the ratio of constant terms, a_0/b_0 }.

(4) Find the horizontal or oblique asymptote.

    (a) If $n < k$, then $y = 0$ { the x-axis } is the horizontal asymptote.

    (b) If $n = k$,

        then $y = a_n/b_k$ { the ratio of leading coefficients } is the horizontal asymptote.

    (c) If $n > k$, then the asymptote is found by long division. The graph of f is

        asymptotic to $y = q(x)$, where $q(x)$ is the quotient of the division process.

(5) Find the intersection points of the function and the asymptote found in step 3. This will help us decide *how* the function approaches the asymptote.

(6) Sketch the graph by regions, where the regions are determined by the vertical asymptotes. We may use the sign of a particular function value to help us determine if the function is positive or negative. We will not plot any points since the purpose in this section is to determine the general shape of the graph and understand the general principles involved with asymptotes.

1 $f(x) = 4/x$ •

  (a) (1) There are no x-intercepts since the numerator is never equal to zero.

    (2) There is a vertical asymptote at $x = 0$ { the y-axis }.

    (3) There is no y-intercept since the function is undefined for $x = 0$.

    (4) The degree of the numerator is 0, which is less than the degree of the denominator, 1, so the horizontal asymptote is $y = 0$ { the x-axis }.

    (5) Setting the function equal to the value of the asymptote found in step 4 gives us $4/x = 0$, which has no solutions.

    (6) There is one vertical asymptote, and it separates the plane into 2 regions. For the region $x < 0$, $f(x) = 4/x < 0$—that is, the y-values are negative. This indicates that the graph is under the y-axis as in *Figure 1*. For the region $x > 0$, $f(x) > 0$, and the graph is above the x-axis.

  (b) The domain D is the set of all nonzero real numbers—that is, $\mathbb{R} - \{0\}$. The range R is equal to the same set of numbers.

  (c) The function is decreasing on $(-\infty, 0)$ and also on $(0, \infty)$.

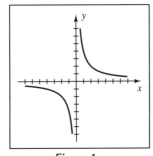

Figure 1

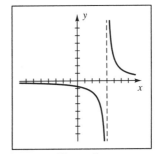

Figure 3

3 $f(x) = \dfrac{3}{x - 4} = 3\left(\dfrac{1}{x - 4}\right)$ •

    Rather than use the guidelines, we will think of this graph in terms of the shifting and stretching properties presented in Section 3.5. Consider the graph of $g(x) = 1/x$, which is similar to the graph in Exercise 1. Now $h(x) = 1/(x - 4)$ is just $g(x - 4)$, which shifts the graph of g to the right by 4 units, making $x = 4$ the vertical asymptote. The effect of the 3 in the numerator is to vertically stretch the graph of h by a factor of 3.

5 $f(x) = \dfrac{-3x}{x+2}$ •

(1) $-3x = 0 \Rightarrow x = 0$, the <u>only</u> x-intercept.

(2) $x + 2 = 0 \Rightarrow x = -2$, the only vertical asymptote.

(3) $f(0) = \frac{0}{2} = 0 \Rightarrow (0, 0)$ is the y-intercept. { We already knew this from step 1. }

(4) Degree of numerator $= 1 =$ degree of denominator $\Rightarrow$

$$y = \tfrac{-3}{1} \{ \text{ratio of leading coefficients} \} = -3 \text{ is the horizontal asymptote.}$$

(5) Function = asymptote $\Rightarrow f(x) = -3 \Rightarrow \dfrac{-3x}{x+2} = -3 \Rightarrow -3x = -3(x+2) \Rightarrow$

$-3x = -3x - 6 \Rightarrow 0 = -6$. This is a contradiction and indicates that there are <u>no</u> intersection points on the horizontal asymptote. Remember, the function can not intersect a vertical asymptote, but can intersect the horizontal asymptote.

(6) For the region $x < -2$, there are no x-intercepts and the graph must be below the horizontal asymptote $y = -3$. { If it was above $y = -3$, it would have to intersect the x-axis at some point. }

For the region $x > -2$, there is an x-intercept at 0 and the graph is above the horizontal asymptote. A common mistake is to confuse the x-axis with the horizontal asymptote. Remember, as $x \to \infty$ or $x \to -\infty$, $f(x)$ will get close to the horizontal or oblique asymptote, not the x-axis { unless the x-axis is the horizontal asymptote }.

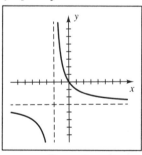

Figure 5

7 $f(x) = \dfrac{x-2}{x^2-x-6} = \dfrac{x-2}{(x+2)(x-3)}$ •

(1) $x - 2 = 0 \Rightarrow x = 2$, the only x-intercept

(2) $(x+2)(x-3) = 0 \Rightarrow x = -2$ and $x = 3$, the vertical asymptotes

(3) $f(0) = \frac{-2}{-6} = \frac{1}{3} \Rightarrow (0, \frac{1}{3})$ is the y-intercept

(4) Degree of numerator $= 1 < 2 = $ degree of denominator $\Rightarrow$

$y = 0$ is the horizontal asymptote.

(5) We have already solved $f(x) = 0$ in step 1.

Hence, we know that the function will cross the horizontal asymptote at $(2, 0)$.

(6) For $x < -2$, we have no information and will examine $f(-5)$ $\{-5$ is to the left of $-2\}$. Using the factored form of f and only the signs of the values, we

obtain $f(-5) = \dfrac{(-)}{(-)(-)} = \{$ some negative number $\} < 0$ since the combination

of 3 negative signs will be negative. Hence, the graph will be below the

horizontal asymptote.

For $-2 < x < 3$, we have an x-intercept at 2 and need to know what the function

does on each side of 2. Choosing 0 and 2.5 for test points, we check the sign

of $f(0)$ and $f(2.5)$ as above. $f(0) = \dfrac{(-)}{(+)(-)} > 0$ and $f(2.5) = \dfrac{(+)}{(+)(-)} < 0.$

Hence, the function will change signs from positive to negative as it passes

through 2.

For $x > 3$, $f(5) = \dfrac{(+)}{(+)(+)} > 0$ and the function is above the horizontal asymptote.

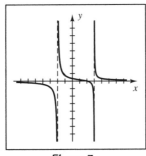

Figure 7

9 $f(x) = \dfrac{-4}{(x-2)^2}$ • Note that the function is always negative since it is the quotient

of a negative and a positive, provided $x \neq 2$.

$\boxed{11}$ $f(x) = \dfrac{x-3}{x^2-1} = \dfrac{x-3}{(x+1)(x-1)}$ •

(6) For $-1 < x < 1$, we have the y-intercept at 3. Since the function cannot intersect the x-axis {the only x-intercept is at 3}, the function remains positive and we have $f(x) \to \infty$ as $x \to 1^-$ and $f(x) \to \infty$ as $x \to -1^+$.

For $x > 1$, we have the x-intercept at 3 as a point to work with. The function will change sign at 3 since 3 is a zero of odd multiplicity {1}. If the function was going to merely touch the point $(3, 0)$ and "turn around", 3 would have to be a zero of even multiplicity. Hence, it suffices to find only one sign. We choose 2 as our test point. $f(2) = \dfrac{(-)}{(+)(+)} < 0$. The function values are negative for $1 < x < 3$ and positive for $x > 3$. Remember, eventually $f(x)$ will be extremely close to 0 as $x \to \infty$. Make sure your sketch reflects that fact.

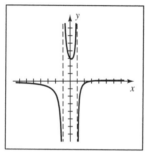

Figure 11

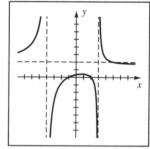

Figure 13

$\boxed{13}$ $f(x) = \dfrac{2x^2 - 2x - 4}{x^2 + x - 12} = \dfrac{2(x+1)(x-2)}{(x+4)(x-3)}$ • As a generalization, you should always factor the function first. For steps 1–6 of the guidelines, it is convenient to use the original function for steps 3, 4, and 5, and to use the factored form for steps 1, 2, and 6. After this problem, we will concentrate on steps 3 {when appropriate}, 5, and 6. You should be able to pick up the information for steps 1–4 by merely looking at both forms of the function {except for the oblique asymptote case}.

(1) $2(x+1)(x-2) = 0 \Rightarrow x = -1, 2$; the x-intercepts

(2) $(x+4)(x-3) = 0 \Rightarrow x = -4, 3$; the vertical asymptotes

(3) $f(0) = \dfrac{-4}{-12} = \dfrac{1}{3} \Rightarrow (0, \dfrac{1}{3})$ is the y-intercept

(4) Degree of numerator $= 2 =$ degree of denominator $\Rightarrow$

$y = \dfrac{2}{1} = 2$ is the horizontal asymptote.

(5) $\dfrac{2x^2 - 2x - 4}{x^2 + x - 12} = 2 \Rightarrow 2x^2 - 2x - 4 = 2(x^2 + x - 12) \Rightarrow$

$2x^2 - 2x - 4 = 2x^2 + 2x - 24 \Rightarrow 20 = 4x \Rightarrow x = 5$. Hence, the function will intersect the horizontal asymptote <u>only</u> at the point (5, 2).

(6) For $x < -4$, there are no x-intercepts so the function is above the horizontal asymptote.

For $-4 < x < 3$, the function must go through $(-1, 0)$, $(0, \frac{1}{3})$, and $(2, 0)$. It must go "down" by both vertical asymptotes because if it went "up", it would have to cross the horizontal asymptote. But we know this only occurs at the point (5, 2) and not in this region.

For $x > 3$, we know that the graph goes up as we get close to 3 because if it went down, there would have to be an x-intercept, but there isn't one. The function goes through (5, 2), but then turns around {before touching the x-axis} and gets very close to 2 as x increases.

Note: We will let $I(x, y)$ denote the intersection point found in step 5.

$\boxed{15}$ $f(x) = \dfrac{-x^2 - x + 6}{x^2 + 3x - 4} = \dfrac{-1(x + 3)(x - 2)}{(x + 4)(x - 1)}$ •

(5) $\dfrac{-x^2 - x + 6}{x^2 + 3x - 4} = -1 \Rightarrow -x^2 - x + 6 = -x^2 - 3x + 4 \Rightarrow$

$$2x = -2 \Rightarrow x = -1, \ I = (-1, -1)$$

(6) For $x < -4$, the function is below $y = -1$ since there are no x-intercepts.

For $-4 < x < 1$, the function passes through $(-3, 0)$, $(-1, -1)$, and $(0, -\frac{3}{2})$. These points are known from finding information in steps 1–5.

For $x > 1$, the function passes through $(2, 0)$ and is always above the horizontal asymptote since it only intersects the horizontal asymptote at $(-1, -1)$.

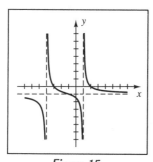

Figure 15

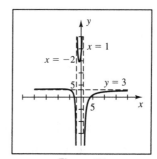

Figure 17

$\boxed{17}$ $f(x) = \dfrac{3x^2 - 3x - 36}{x^2 + x - 2} = \dfrac{3(x + 3)(x - 4)}{(x + 2)(x - 1)}$ •

(5) $\dfrac{3x^2 - 3x - 36}{x^2 + x - 2} = 3 \Rightarrow 3x^2 - 3x - 36 = 3x^2 + 3x - 6 \Rightarrow$

$$-30 = 6x \Rightarrow x = -5, \ I = (-5, 3)$$

(6) For $x < -2$, the function passes through $(-3, 0)$ and $(-5, 3)$ and stays above the horizontal asymptote, $y = 3$, as $x \to -\infty$.

(cont.)

For $-2 < x < 1$, we have the y-intercept at $(0, 18)$. Since the graph doesn't intersect the horizontal asymptote in this region, the function goes up as it gets close to each vertical asymptote.

For $x > 1$, we have the x-intercept $(4, 0)$ and the function stays below the horizontal asymptote.

$\boxed{19}$ $f(x) = \dfrac{-2x^2 + 10x - 12}{x^2 + x} = \dfrac{-2(x - 2)(x - 3)}{(x + 1)(x)}$ •

(2) Since $(x + 1)(x) = 0 \Rightarrow x = -1$ and 0,

we have the y-axis as a vertical asymptote, and hence, there is no y-intercept.

(5) $\dfrac{-2x^2 + 10x - 12}{x^2 + x} = -2 \Rightarrow -2x^2 + 10x - 12 = -2x^2 - 2x \Rightarrow$

$$12x = 12 \Rightarrow x = 1, \; I = (1, -2)$$

(6) For $-1 < x < 0$, $f(-\tfrac{1}{2}) = \dfrac{(-)(-)(-)}{(+)(-)} > 0$ and the function is above the x-axis.

For $x > 0$, the function goes through $(1, -2)$, $(2, 0)$, and $(3, 0)$ and gradually gets closer to $y = -3$ as $x \to \infty$.

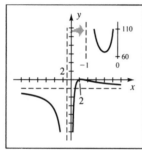

Figure 19

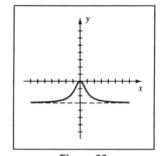

Figure 23

$\boxed{23}$ $f(x) = \dfrac{-3x^2}{x^2 + 1}$ • f is an even function

(2) There are no vertical asymptotes since $x^2 + 1 \neq 0$ for any real number.

(5) $\dfrac{-3x^2}{x^2 + 1} = -3 \Rightarrow -3x^2 = -3x^2 - 3 \Rightarrow 0 = -3 \Rightarrow$

there are no intersection points of the horizontal asymptote and the function.

(6) We have the x-intercept at $(0, 0)$ and the function must get close to the horizontal asymptote, $y = -3$, as $x \to \infty$ or $x \to -\infty$.

$\boxed{25}$ $f(x) = \dfrac{x^2 - x - 6}{x + 1} = \dfrac{(x+2)(x-3)}{x+1}$ •

(4) We could use long division to find the oblique asymptote, but since the denominator is of the from $x - c$, we will use synthetic division.

$$
\begin{array}{r|rrr}
-1 & 1 & -1 & -6 \\
   &   & -1 & 2 \\
\hline
   & 1 & -2 & -4
\end{array}
$$

The third row indicates that $f(x) = x - 2 - \dfrac{4}{x+1}$. The expression $\dfrac{4}{x+1} \to 0$ as $x \to \pm\infty$, so $y = x - 2$ is an oblique asymptote for f.

(5) $\dfrac{x^2 - x - 6}{x + 1} = x - 2 \Rightarrow x^2 - x - 6 = (x - 2)(x + 1) \Rightarrow$

$x^2 - x - 6 = x^2 - x - 2 \Rightarrow -6 = -2 \Rightarrow$ there are no intersection points of the oblique asymptote and the function.

(6) For $x < -1$, we have the x-intercept at $(-2,\ 0)$ and the function is above the oblique asymptote.

For $x > -1$, we have the points $(0,\ -6)$ and $(3,\ 0)$ and the function is below the oblique asymptote.

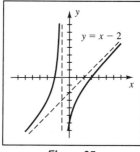

Figure 25

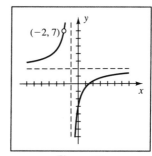

Figure 29

$\boxed{29}$ $f(x) = \dfrac{2x^2 + x - 6}{x^2 + 3x + 2} = \dfrac{(x+2)(2x-3)}{(x+2)(x+1)} = \dfrac{2x-3}{x+1}$ for $x \ne -2$. We must reduce the

function first and then *sketch the reduced function*. After sketching the reduced function, we need to remember to put a hole in the graph where the original function was undefined. In this case, the reduced function is $f(x) = \dfrac{2x-3}{x+1}$. We sketch this as

we did the other rational functions. To determine the value of y when $x = -2$,

substitute -2 into $\dfrac{2x-3}{x+1}$ to get 7. There is a hole in the graph at $(-2,\ 7)$.

31 $f(x) = \dfrac{x-1}{1-x^2}$

$= \dfrac{x-1}{(1+x)(1-x)}$

$= \dfrac{-1}{x+1}$ for $x \neq 1$; hole at $\left(1, -\tfrac{1}{2}\right)$

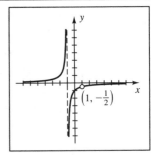

Figure 31

35 (a) The radius of the outside cylinder is $(r+0.5)$ ft and its height is $(h+1)$ ft.

Since the volume is 16π ft^3, we have $16\pi = \pi(r+0.5)^2(h+1) \Rightarrow$

$$h+1 = \frac{16\pi}{\pi(r+0.5)^2} \Rightarrow h = \frac{16}{(r+0.5)^2} - 1.$$

(b) $V(r) = \pi r^2 h = \pi r^2 \left[\dfrac{16}{(r+0.5)^2} - 1\right]$

(c) r and h must both be positive. $h > 0 \Rightarrow \dfrac{16}{(r+0.5)^2} - 1 > 0 \Rightarrow$

$16 > (r+0.5)^2 \Rightarrow |r+0.5| < 4 \Rightarrow -4.5 < r < 3.5$. The last inequality

combined with $r > 0$ means that the excluded values are $r \leq 0$ and $r \geq 3.5$.

37 (a) Since 5 gallons of water flow into the tank each minute, $V(t) = 50 + 5t$. Since

each additional gallon of water contains 0.1 lb of salt, $A(t) = 5(0.1)t = 0.5t$.

(b) $c(t) = \dfrac{A(t)}{V(t)} = \dfrac{0.5t}{50+5t} = \dfrac{t}{10t+100}$ lb/gal

(c) As $t \rightarrow \infty$, $c(t) \rightarrow 0.1$ lb. of salt per gal.

39 Assign $20x^2 + 80x + 72$ to Y_1, $10x^2 + 40x + 41$ to Y_2, and Y_1/Y_2 to Y_3.

Zoom-in around $(-2, -8)$ to confirm that this a low point and that there is not a

vertical asymptote at $x = -2$.

$[-9, 3]$ by $[-9, 3]$ $[0.7, 1.3]$ by $[0.8, 1.2]$

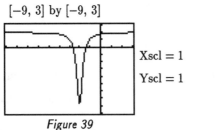
Xscl $= 1$

Yscl $= 1$

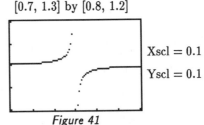
Xscl $= 0.1$

Yscl $= 0.1$

Figure 39   *Figure 41*

41 $f(x) = \dfrac{(x-1)^2}{(x-0.999)^2}$ • Note that the standard viewing rectangle gives the

horizontal line $y = 1$. *Figure 41* was obtained by using Dot mode. If Connected

mode is used, the calculator will draw a near-vertical line at $x = 0.999$.

Chapter 4 Review Exercises

[3] $f(x) = -\frac{1}{4}(x+2)(x-1)^2(x-3)$ has zeros at -2, 1 (multiplicity 2), and 3.

$$f(x) > 0 \text{ if } -2 < x < 1 \text{ or } 1 < x < 3, \ f(x) < 0 \text{ if } x < -2 \text{ or } x > 3.$$

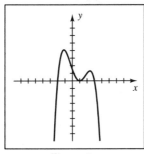

Figure 3

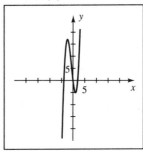

Figure 5

[5] $f(x) = x^3 + 2x^2 - 8x = x(x^2 + 2x - 8) = x(x+4)(x-2)$.

$$f(x) > 0 \text{ if } -4 < x < 0 \text{ or } x > 2, \ f(x) < 0 \text{ if } x < -4 \text{ or } 0 < x < 2.$$

[11] Synthetically dividing $f(x) = -4x^4 + 3x^3 - 5x^2 + 7x - 10$ by $x+2$, we have

$$
\begin{array}{r|rrrrr}
-2 & -4 & 3 & -5 & 7 & -10 \\
   &   & 8 & -22 & 54 & -122 \\
\hline
   & -4 & 11 & -27 & 61 & -132
\end{array}
$$

By the remainder theorem, $f(-2) = -132$.

[15] Since $-3 + 5i$ is a zero, so is $-3 - 5i$.

$$f(x) = a\big[x - (-3 + 5i)\big]\big[x - (-3 - 5i)\big](x+1) = a(x^2 + 6x + 34)(x+1).$$

$$f(1) = a(41)(2) \text{ and } f(1) = 4 \Rightarrow 82a = 4 \Rightarrow a = \tfrac{2}{41}.$$

Hence, $f(x) = \frac{2}{41}(x^2 + 6x + 34)(x+1)$.

[17] $f(x) = x^5(x+3)^2$

$\{$ leading coefficient is 1 $\}$

$\{$ 0 is a zero of multiplicity 5 $\}$

$\{$ −3 is a zero of multiplicity 2 $\}$

$= x^5(x^2 + 6x + 9)$

$= x^7 + 6x^6 + 9x^5$

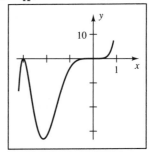

Figure 17

[20] $f(x) = x^6 + 2x^4 + x^2 = x^2(x^4 + 2x^2 + 1) = x^2(x^2 + 1)^2$

★ $0, \ \pm i$ (all have multiplicity 2)

$\boxed{22}$ (a) Let $f(x) = x^5 - 4x^3 + 6x^2 + x + 4$. Since there are 2 sign changes in $f(x)$ and 3

sign changes in $f(-x) = -x^5 + 4x^3 + 6x^2 - x + 4$, there are either

2 positive and 3 negative solutions;

2 positive, 1 negative, and 2 nonreal complex;

3 negative and 2 nonreal complex;

or   1 negative and 4 nonreal complex solutions.

(b) Upper bound is 2, lower bound is -3

$\boxed{25}$ $\dfrac{\text{positive divisors of 3}}{\text{divisors of 16}} = \dfrac{1,\ 3}{\pm 1,\ \pm 2,\ \pm 4,\ \pm 8,\ \pm 16} \longrightarrow$

$$\pm 1,\ \pm 3,\ \pm \tfrac{1}{2},\ \pm \tfrac{3}{2},\ \pm \tfrac{1}{4},\ \pm \tfrac{3}{4},\ \pm \tfrac{1}{8},\ \pm \tfrac{3}{8},\ \pm \tfrac{1}{16},\ \pm \tfrac{3}{16}$$

After a few attempts, we try $-\tfrac{1}{2}$.

$$
\begin{array}{r|rrrr}
-\tfrac{1}{2} & 16 & -20 & -8 & 3 \\
 & & -8 & 14 & -3 \\
\hline
 & 16 & -28 & 6 & 0
\end{array}
$$

$16x^2 - 28x + 6 = 0 \Rightarrow 8x^2 - 14x + 3 = 0 \Rightarrow (4x-1)(2x-3) = 0 \Rightarrow x = \tfrac{1}{4}, \tfrac{3}{2}$.

Thus, the solutions of $16x^3 - 20x^2 - 8x + 3 = 0$ are $x = -\tfrac{1}{2}, \tfrac{1}{4}, \tfrac{3}{2}$.

$\boxed{29}$ $f(x) = \dfrac{3x^2}{16 - x^2} = \dfrac{3x^2}{(4+x)(4-x)}$. $f(x) = -3 \Rightarrow 3x^2 = 3x^2 - 48 \Rightarrow 0 = -48$,

a contradiction $\Rightarrow$ the function does not intersect the horizontal asymptote.

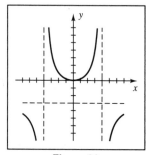

Figure 29

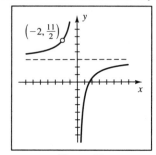

Figure 33

$\boxed{33}$ $f(x) = \dfrac{3x^2 + x - 10}{x^2 + 2x} = \dfrac{(3x-5)(x+2)}{x(x+2)} = \dfrac{3x-5}{x}$ if $x \neq -2$.

Substituting $x = -2$ into $\dfrac{3x-5}{x}$ yields $\tfrac{11}{2}$. Thus, we have a hole at $(-2, \tfrac{11}{2})$.

34 $f(x) = \dfrac{-2x^2 - 8x - 6}{x^2 - 6x + 8} = \dfrac{-2(x^2 + 4x + 3)}{(x-2)(x-4)} = \dfrac{-2(x+1)(x+3)}{(x-2)(x-4)}.$

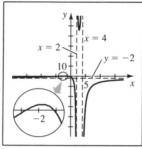

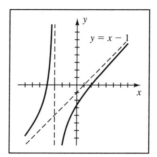

Figure 34   Figure 35

35 $f(x) = \dfrac{x^2 + 2x - 8}{x+3} = \dfrac{(x+4)(x-2)}{x+3} = x - 1 - \dfrac{5}{x+3}.$

$y = x - 1$ is an oblique asymptote.

38 (a) The volume of the cylinder is $\pi r^2 x$, where r is the radius of the cylinder and $x = \overline{AD} = \overline{BC}$ is the height. Edge AB has length $2\pi r$, since it is the circumference of the lower edge of the cylinder. Also, $\overline{AB}^2 + \overline{BC}^2 = l^2 \Rightarrow (2\pi r)^2 + x^2 = l^2 \Rightarrow 4\pi^2 r^2 = l^2 - x^2 \Rightarrow r^2 = \frac{1}{4\pi^2}(l^2 - x^2).$

Now, $V = \pi r^2 x = \pi\left[\frac{1}{4\pi^2}(l^2 - x^2)\right](x) = \frac{1}{4\pi}x(l^2 - x^2).$

(b) If $x > 0$, $V > 0$ if $l^2 - x^2 > 0$ or $l > x$. Thus, when $0 < x < l$, $V > 0$.

41 (a) S is the value that is changing. As S gets large, the value of $R = \dfrac{kS^n}{S^n + a^n}$ will approach the ratio of leading coefficients, $k/1$. Hence, an equation of the horizontal asymptote is $R = k$.

(b) k is the maximum rate at which the liver can remove alcohol from the bloodstream.

Chapter 5: Exponential and Logarithmic Functions

3 $3^{2x+3} = 3^{(x^2)} \Rightarrow 2x + 3 = x^2 \Rightarrow x^2 - 2x - 3 = 0 \Rightarrow (x-3)(x+1) = 0 \Rightarrow x = -1, 3$

5 We need to obtain the same base on each side of the equals sign, then we can apply

part (2) of the theorem about exponential functions being one-to-one.

$2^{-100x} = (0.5)^{x-4} \Rightarrow (2^{-1})^{100x} = \left(\frac{1}{2}\right)^{x-4} \Rightarrow \left(\frac{1}{2}\right)^{100x} = \left(\frac{1}{2}\right)^{x-4} \Rightarrow$

$$100x = x - 4 \Rightarrow 99x = -4 \Rightarrow x = -\frac{4}{99}$$

7 $4^{x-3} = 8^{4-x} \Rightarrow (2^2)^{x-3} = (2^3)^{4-x} \Rightarrow 2^{2x-6} = 2^{12-3x} \Rightarrow$

$$2x - 6 = 12 - 3x \Rightarrow 5x = 18 \Rightarrow x = \frac{18}{5}$$

9 (a) Let $g = f(x) = 2^x$ for reference purposes.

The graph of g goes through the points $(-1, \frac{1}{2})$, $(0, 1)$, and $(1, 2)$.

(b) $f(x) = -2^x$ •

Reflect g through the x-axis since f is just $-1 \cdot 2^x$. Do not confuse this function

with $(-2)^x$ — remember, the base is positive for exponential functions.

(c) $f(x) = 3 \cdot 2^x$ • vertically stretch g by a factor of 3

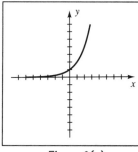

Figure 9(a)

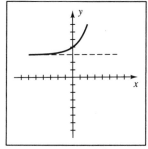

Figure 9(b)

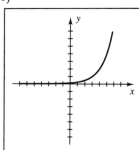

Figure 9(c)

(d) $f(x) = 2^{x+3}$ • shift g left 3 units since f is $g(x+3)$

(e) $f(x) = 2^x + 3$ • vertically shift g up 3 units

(f) $f(x) = 2^{x-3}$ • shift g right 3 units since f is $g(x-3)$

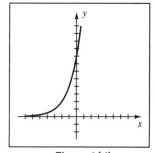

Figure 9(d)

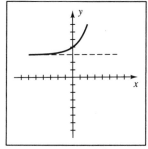

Figure 9(e)

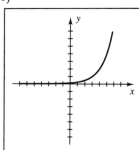

Figure 9(f)

(g) $f(x) = 2^x - 3$ •   vertically shift g down 3 units

(h) $f(x) = 2^{-x}$ •   reflect g through the y-axis since f is $g(-x)$

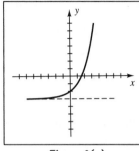

Figure 9(g)

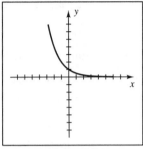

Figure 9(h)

(i) $f(x) = \left(\frac{1}{2}\right)^x$ • $\left(\frac{1}{2}\right)^x = (2^{-1})^x = 2^{-x}$, same graph as in part (h)

(j) $f(x) = 2^{3-x}$ • $2^{3-x} = 2^{-(x-3)}$, shift g right 3 units and reflect through the

line $x = 3$. Alternatively, $2^{3-x} = 2^3 2^{-x} = 8\left(\frac{1}{2}\right)^x$, vertically stretch $y = \left(\frac{1}{2}\right)^x$ (the

graph in part (i))by a factor of 8.

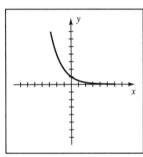

Figure 9(i)

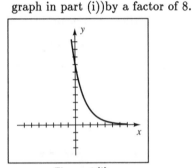

Figure 9(j)

$\boxed{13}$ $f(x) = -\left(\frac{1}{2}\right)^x + 4$ •   reflect $y = \left(\frac{1}{2}\right)^x$ through the x-axis and shift up 4 units

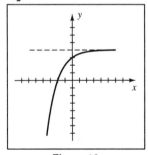

Figure 13

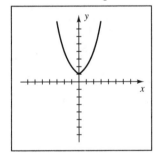

Figure 15

$\boxed{15}$ $f(x) = 2^{|x|} = \begin{cases} 2^x & \text{if } x \geq 0 \\ 2^{-x} & \text{if } x < 0 \end{cases} = \begin{cases} 2^x & \text{if } x \geq 0 \\ \left(\frac{1}{2}\right)^x & \text{if } x < 0 \end{cases}$

Use the portion of $y = 2^x$ with $x \geq 0$ and reflect it through the y-axis since f is even.

Note: For Exercises 17, 18, and 5 of the review exercises, refer to Example 5 in the text

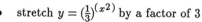

for the basic graph of $y = a^{-x^2} = (a^{-1})^{(x^2)} = \left(\frac{1}{a}\right)^{(x^2)}$, where $a > 1$.

$\boxed{17}$ $f(x) = 3^{1-x^2} = 3^1 3^{-x^2} = 3\left(\frac{1}{3}\right)^{(x^2)}$ •   stretch $y = \left(\frac{1}{3}\right)^{(x^2)}$ by a factor of 3

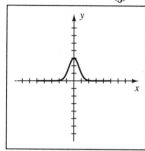

Figure 17

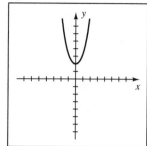

Figure 19

$\boxed{19}$ $f(x) = 3^x + 3^{-x}$ • Adding the functions $g(x) = 3^x$ and $h(x) = 3^{-x} = \left(\frac{1}{3}\right)^x$ together, we see that the y-intercept will be $(0, 2)$. If $x > 0$, f looks like $y = 3^x$ since 3^x dominates 3^{-x} (3^{-x} gets close to 0 and 3^x grows very large). If $x < 0$, f looks like $y = 3^{-x}$ since 3^{-x} dominates 3^x.

$\boxed{23}$ (a) 8:00 A.M. corresponds to $t = 1$ and $f(1) = 600\sqrt{3} \approx 1039$.

     10:00 A.M. corresponds to $t = 3$ and $f(3) = 600(3\sqrt{3}) = 1800\sqrt{3} \approx 3118$.

     11:00 A.M. corresponds to $t = 4$ and $f(4) = 600(9) = 5400$.

(b) The graph of f is an increasing exponential that passes through $(0, 600)$ and the

points in part (a).

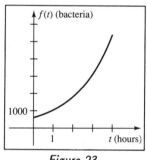

Figure 23

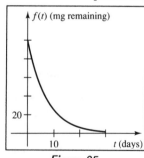

Figure 25

$\boxed{25}$ (a) $f(5) = 100(2)^{-1} = 50$ mg; $f(10) = 100(2)^{-2} = 25$ mg;

$$f(12.5) = 100(2)^{-2.5} = \frac{100}{4\sqrt{2}} = \frac{25}{2}\sqrt{2} \approx 17.7 \text{ mg}$$

(b) The end points are $(0, 100)$ and $(30, 1.5625)$.

$\boxed{27}$ A half-life of 1600 years means that when $t = 1600$, the amount remaining, $q(t)$, will be one-half the original amount—that is, $\frac{1}{2}q_0$. $q(t) = \frac{1}{2}q_0$ when $t = 1600 \Rightarrow$ $\frac{1}{2}q_0 = q_0 2^{k(1600)} \Rightarrow 2^{-1} = 2^{1600k} \Rightarrow 1600k = -1 \Rightarrow k = -\frac{1}{1600}$.

$\boxed{33}$ $P = 1000$, $r = 0.06$, and $n = 4 \Rightarrow A(t) = 1000\left(1 + \frac{0.06}{4}\right)^{4t} = 1000(1.015)^{4t}$

(a) $A(1) \approx \$1061.36$

(b) $A(2) \approx \$1126.49$

(c) $A(5) \approx \$1346.86$

(d) $A(10) \approx \$1814.02$

$\boxed{35}$ (a) Examine the pattern formed by the value y in the year n.

year (n)	value (y)
0	y_0
1	$(1-a)y_0 = y_1$
2	$(1-a)y_1 = (1-a)\left[(1-a)y_0\right] = (1-a)^2 y_0 = y_2$
3	$(1-a)y_2 = (1-a)\left[(1-a)^2 y_0\right] = (1-a)^3 y_0 = y_3$

(b) $s = (1-a)^T y_0 \Rightarrow (1-a)^T = s/y_0 \Rightarrow 1 - a = \sqrt[T]{s/y_0} \Rightarrow a = 1 - \sqrt[T]{s/y_0}$

$\boxed{37}$ (a) $r = 0.12$, $t = 30$, $L = 90{,}000 \Rightarrow k \approx 35.95$, $M \approx 925.75$

(b) (360 payments) $\times \$925.75 - \$90{,}000 = \$243{,}270$

$\boxed{43}$ Part (b) may be interpreted as doubling an investment at 8.5%.

(a) If $y = (1.085)^x$ and $x = 40$, then $y \approx 26.13$.   (b) If $y = 2$, then $x \approx 8.50$.

[0, 60] by [0, 40]   [−3, 3] by [−2, 2]

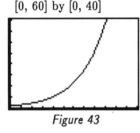

Xscl = 5

Yscl = 5

Figure 43

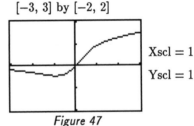

Xscl = 1

Yscl = 1

Figure 47

$\boxed{47}$ (a) f is not one-to-one since

the horizontal line $y = -0.1$ intersects the graph of f more than once.

(b) The only zero of f is $x = 0$.

$\boxed{51}$ Graph $y = 4(0.125)^{(0.25^x)}$. The line $y = k = 4$ is a horizontal asymptote for the

Gompertz function. The maximum number of sales of the product approaches k.

[0, 7.5] by [0, 5]   [0, 40] by [0, 200,000]

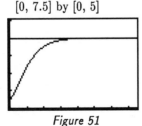

Xscl = 1

Yscl = 1

Figure 51

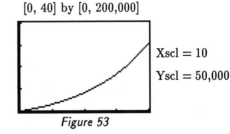

Xscl = 10

Yscl = 50,000

Figure 53

$\boxed{53}$ From the graph, we see that $A = 100{,}000$ when $n \approx 32.8$.

5.2 Exercises

Note: Examine Figure 13 in this section to reinforce the idea that $y = e^x$ is just a special case of $y = a^x$ with $a > 1$.

3 (a) $f(x) = e^{x+4}$ • shift $y = e^x$ left 4 units

(b) $f(x) = e^x + 4$ • shift $y = e^x$ up 4 units

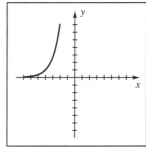

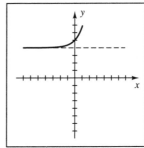

Figure 3(a)   Figure 3(b)

5 $A = Pe^{rt} = 1000e^{(0.0825)(5)} \approx \$1510.59$

7 $100,000 = Pe^{(0.11)(18)} \Rightarrow P = \dfrac{100,000}{e^{1.98}} \approx \$13,806.92$

9 $13,464 = 1000e^{(r)(20)} \Rightarrow e^{20r} = 13.464.$ From Table 2, $e^x \approx 13.464$ if $x = 2.60.$

Thus, $20r = 2.60$ and $r = 0.13$ or $13\%.$

11 $e^{(x^2)} = e^{7x-12} \Rightarrow x^2 = 7x - 12 \Rightarrow x^2 - 7x + 12 = 0 \Rightarrow (x-3)(x-4) = 0 \Rightarrow x = 3, 4$

15 $x^3(4e^{4x}) + 3x^2 e^{4x} = 0 \Rightarrow x^2 e^{4x}(4x + 3) = 0 \Rightarrow x = -\frac{3}{4}, 0.$ Note that $e^{4x} \neq 0.$

17 $\dfrac{(e^x + e^{-x})(e^x + e^{-x}) - (e^x - e^{-x})(e^x - e^{-x})}{(e^x + e^{-x})^2} =$

$$\frac{(e^{2x} + 2 + e^{-2x}) - (e^{2x} - 2 + e^{-2x})}{(e^x + e^{-x})^2} = \frac{4}{(e^x + e^{-x})^2}$$

21 2000 corresponds to $t = 2000 - 1980 = 20$; $N(20) = 227e^{(0.007)(20)} \approx 261.1$ million

23 $N(10) = N_0 e^{-2}.$ The percentage of the original number still alive after 10 years is

$$100 \times \left(\frac{N(10)}{N_0} \right) = 100e^{-2} \approx 13.5\%.$$

29 $x = 1 \Rightarrow y = 79.041 + 6.39 - e^{2.268} \approx 75.77$ cm.

$$x = 1 \Rightarrow R = 6.39 + 0.993e^{2.268} \approx 15.98 \text{ cm/yr.}$$

31 $2000 - 1971 = 29 \Rightarrow t = 29$ years. Using the continuously compounded interest formula with $P = 1.60$ and $r = 0.05$, we have $A = 1.60e^{(0.05)(29)} \approx \$6.82$ per hour.

33 (a) Note here that the amount of money invested is not of interest and that we are only concerned with the percent of growth.

$$\left(1 + \tfrac{0.07}{4}\right)^{4 \cdot 1} \approx 1.0719. \quad (1.0719 - 1) \times 100\% = 7.19\%$$

(b) $e^{(0.07)(1)} \approx 1.0725.$ $(1.0725 - 1) \times 100\% = 7.25\%$. The results indicate that we would receive an extra 0.06% in interest by investing our money in an account that is compounded continuously rather than quarterly. This is only an extra 6 cents on a $100 investment, but $600 extra on a $1,000,000 investment (actually $649.15 if the computations are carried beyond 0.01%).

35 It may be of interest to compare this graph with the graph of $y = (1.085)^x$ in Exercise 43 of §5.1. Both are compounding functions with $r = 8.5\%$.

Note that $e^{0.085x} = (e^{0.085})^x \approx (1.0887)^x > (1.085)^x$ for $x > 0$.

(a) If $y = e^{0.085x}$ and $x = 40$, then $y \approx 29.96$.   (b) If $y = 2$, then $x \approx 8.15$.

[0, 60] by [0, 40]

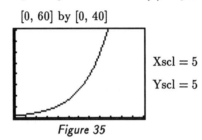

Xscl $= 5$

Yscl $= 5$

Figure 35

37 (a) As $x \to \infty$, $e^{-x} \to 0$ and f will resemble $\tfrac{1}{2}e^x$.

As $x \to -\infty$, $e^x \to 0$ and f will resemble $-\tfrac{1}{2}e^x$.

(b) At $x = 0$, $f(x) = 0$, and g will have a vertical asymptote since g is undefined (division by 0). As $x \to \infty$, $f(x) \to \infty$, and since the reciprocal of a large positive number is a small positive number, we have $g(x) \to 0$. As $x \to -\infty$, $f(x) \to -\infty$, and since the reciprocal of a large negative number is a small negative number, we have $g(x) \to 0$.

[−7.5, 7.5] by [−5, 5]

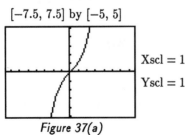

Xscl $= 1$

Yscl $= 1$

Figure 37(a)

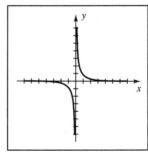

Figure 37(b)

39 (a) $f(x) = \dfrac{e^x - e^{-x}}{e^x + e^{-x}} = \dfrac{e^x - 1/e^x}{e^x + 1/e^x} \cdot \dfrac{e^x}{e^x} = \dfrac{e^{2x} - 1}{e^{2x} + 1}.$ At $x = 0$, $f(x) = 0$.   As $x \to \infty$,

$f(x) \to 1$ since the numerator and denominator are nearly the same number. As $x \to -\infty$, $f(x) \to -1$.

(b) At $x = 0$, we will have a vertical asymptote. As $x \to \infty$, $f(x) \to 1$, and since g is the reciprocal of f, $g(x) \to 1$. Similarly, as $x \to -\infty$, $g(x) \to -1$.

$[-4.5,\ 4.5]$ by $[-3,\ 3]$

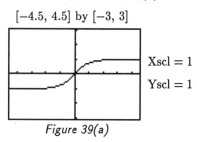

Xscl = 1

Yscl = 1

Figure 39(a)

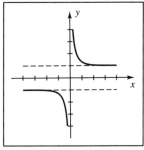

Figure 39(b)

41 The approximate coordinates of the points where the graphs of f and g intersect are $(-1.04,\ -0.92)$, $(2.11,\ 2.44)$, and $(8.51,\ 70.42)$. The region near the origin in *Figure 41(a)* is enhanced in *Figure 41(b)*. Thus, the solutions are $x \approx -1.04$, 2.11, and 8.51.

$[-3,\ 11]$ by $[-10,\ 80]$ $[-2.26,\ 3.34]$ by $[-7.14,\ 8.57]$

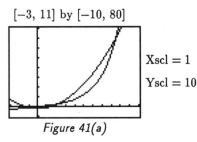

Xscl = 1

Yscl = 10

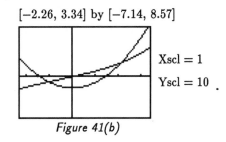

Xscl = 1

Yscl = 10 .

Figure 41(a)   *Figure 41(b)*

45 From the graph, we see that f has zeros at $x \approx 0.11$, 0.79, and 1.13.

$[-2,\ 2.5]$ by $[-1,\ 2]$ $[0,\ 200]$ by $[0,\ 8]$

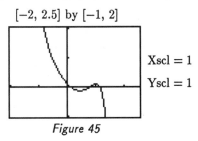

Xscl = 1

Yscl = 1

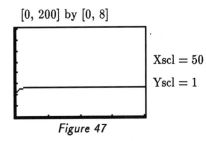

Xscl = 50

Yscl = 1

Figure 45   *Figure 47*

47 From the graph, there is a horizontal asymptote of $y \approx 2.71$.

f is approaching the value of e asymptotically.

53 (a) When $y = 0$ and $z = 0$, the equation becomes $C = \dfrac{2Q}{\pi v a b} e^{-h^2/(2b^2)}$. As h increases, the exponent becomes a larger *negative* value, and hence the concentration C decreases.

(b) When $z = 0$, the equation becomes $C = \dfrac{2Q}{\pi v a b} e^{-y^2/(2a^2)} e^{-h^2/(2b^2)}$.

As y increases, the concentration C decreases.

55 (a) Larger values of c cause F to decrease more rapidly. This indicates that the chip will fail sooner and be less reliable.

(b) $0.35 = 1 - e^{-0.125t} \Rightarrow e^{-0.125t} = 0.65$. From a graph of $Y_1 = e^{-0.125x}$, we see that $Y_1 = 0.65$ when $x \approx 3.45$ yr. Alternatively, you could graph $Y_1 = 1 - e^{-0.125x}$ and determine where Y_1 equals 0.35.

5.3 Exercises

Note: Exercises 1–4 are designed to familiarize the reader with the definition of $\log_a$ in this section. It is very important that you can generalize your understanding of this definition to the following case:

$$\log_{\text{base}}(\text{argument}) = \text{exponent} \quad \textit{is equivalent to} \quad (\text{base})^{\text{exponent}} = \text{argument}$$

1 (e) In this case, the *base* is 5, the *exponent* is $7t$, and the *argument* is $\dfrac{a+b}{a}$. Thus,

$$5^{7t} = \frac{a+b}{a} \quad \text{is equivalent to} \quad \log_5 \frac{a+b}{a} = 7t.$$

3 (e) In this case, the *base* is 2, the *argument* is m, and the *exponent* is $3x + 4$. Thus,

$$\log_2 m = 3x + 4 \quad \text{is equivalent to} \quad 2^{3x+4} = m.$$

7 In order to solve for t, we must isolate the expression containing t—in this case, that expression is the exponential a^{Ct}. $A = Ba^{Ct} + D \Rightarrow A - D = Ba^{Ct} \Rightarrow$

$$\frac{A-D}{B} = a^{Ct} \Rightarrow Ct = \log_a\left(\frac{A-D}{B}\right) \Rightarrow t = \frac{1}{C}\log_a\left(\frac{A-D}{B}\right).$$

The confusing step to most students in the above solution is $\frac{A-D}{B} = a^{Ct} \Rightarrow Ct = \log_a\left(\frac{A-D}{B}\right)$. This is similar to $y = a^x \Rightarrow \log_a y = x$, except x and y are more complicated expressions.

9 (a) Changing $10^5 = 100{,}000$ to logarithmic form gives us $\log_{10} 100{,}000 = 5$.

Since this is a common logarithm, we denote it as $\log 100{,}000 = 5$.

(e) Changing $e^{2t} = 3 - x$ to logarithmic form gives us $\log_e (3 - x) = 2t$.

Since this is a natural logarithm, we denote it as $\ln (3 - x) = 2t$.

$\boxed{11}$ (b) Remember that $\log x = 20t$ is the same as $\log_{10} x = 20t$.

Changing to exponential form, we have $10^{20t} = x$.

(d) Remember that $\ln w = 4 + 3x$ is the same as $\log_e w = 4 + 3x$.

Changing to exponential form, we have $e^{4+3x} = w$.

$\boxed{13}$ (c) Remember that you cannot take the logarithm, any base, of a negative number.

Hence, $\log_4(-2)$ is undefined.

(g) We will change the form of $\frac{1}{16}$ so that it can be written as an exponential expression with the same base as the logarithm—in this case, that base is 4.

$$\log_4 \tfrac{1}{16} = \log_4 4^{-2} = -2$$

$\boxed{15}$ Parts (a)–(f) are direct applications of the properties in the chart on page 299.

For part (g), we use a property of exponents that will enable us to use the

property $e^{\ln x} = x$. (g) $e^{2+\ln 3} = e^2 e^{\ln 3} = e^2(3) = 3e^2$

$\boxed{19}$ $\log_5(x-2) = \log_5(3x+7) \Rightarrow x - 2 = 3x + 7$ { since the logarithm function is one-to-one } $\Rightarrow 2x = -9 \Rightarrow x = -\frac{9}{2}$. We must check to make sure that all proposed solutions do not make any of the original expressions undefined. The value $x = -\frac{9}{2}$ is extraneous since it makes either of the original logarithm expressions undefined. Hence, there is no solution.

$\boxed{21}$ $\log x^2 = \log(-3x-2) \Rightarrow x^2 = -3x - 2 \Rightarrow x^2 + 3x + 2 = 0 \Rightarrow (x+1)(x+2) = 0 \Rightarrow$

$x = -1, -2$. Checking -1 and -2, we find that both are valid solutions.

$\boxed{25}$ $\log_9 x = \frac{3}{2} \Rightarrow x = 9^{3/2} = (9^{1/2})^3$ { remember, root first, power second } $= 3^3 = 27$

$\boxed{29}$ $e^{2\ln x} = 9 \Rightarrow (e^{\ln x})^2 = 9 \Rightarrow x^2 = 9 \Rightarrow x = \pm 3$; -3 is extraneous

$\boxed{31}$ (a) $f(x) = \log_4 x$ •   This graph has a vertical asymptote of $x = 0$ and

goes through $(\frac{1}{4}, -1)$, $(1, 0)$, and $(4, 1)$. For reference purposes, call this $F(x)$.

(b) $f(x) = -\log_4 x$ •   reflect F through the x-axis

(c) $f(x) = 2\log_4 x$ •   vertically stretch F by a factor of 2

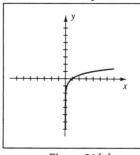

Figure 31(a)

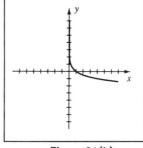

Figure 31(b)

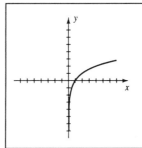

Figure 31(c)

(d) $f(x) = \log_4 (x + 2)$ •   shift F left 2 units

(e) $f(x) = (\log_4 x) + 2$ •   shift F up 2 units

(f) $f(x) = \log_4 (x - 2)$ •   shift F right 2 units

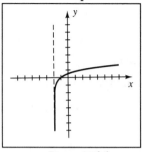 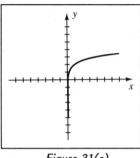 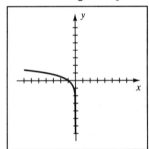

Figure 31(d)   *Figure 31(e)*   *Figure 31(f)*

(g) $f(x) = (\log_4 x) - 2$ •   shift F down 2 units

(h) $f(x) = \log_4 |x|$ •   include the reflection of F through the y-axis since x may

                                 be positive or negative, but $\log_4 |x|$ will give the same result

(i) $f(x) = \log_4 (-x)$ • x must be negative so that $-x$ is positive,

                                                     reflect F through the y-axis

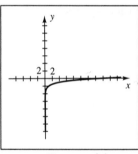 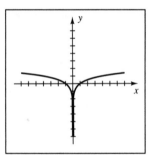

Figure 31(g)   *Figure 31(h)*   *Figure 31(i)*

(j) $f(x) = \log_4(3 - x) = \log_4[-(x - 3)]$ •   Shift F right 3 units and reflect through the line $x = 3$. It may be helpful to determine the domain of this function. We know that $3 - x$ must be positive for the function to be defined. Thus, $3 - x > 0 \Rightarrow 3 > x$, or, equivalently, $x < 3$.

(k) $f(x) = |\log_4 x|$ •

reflect points with negative y-coordinates through the x-axis

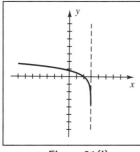 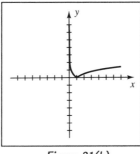 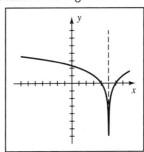

Figure 31(j)   *Figure 31(k)*   *Figure 35*

35 $f(x) = \log_2|x - 5|$ •   shift $y = \log_2|x|$ right 5 units

37   This is the basic "logarithm with base 2" graph, call it $F(x) = \log_2 x$.   ★ $f(x) = \log_2 x$

41   the reflection of F through the y-axis   ★ $f(x) = \log_2(-x)$

43 F appears to be stretched. Since the graph goes through $(2, 2)$ instead of $(2, 1)$ and $(4, 4)$ instead of $(4, 2)$, we might guess that the y-coordinates of F are doubled, that is, $f(x) = 2\log_2 x$. $f(x) \neq \log_2 x^2$ since the domain of f is $(0, \infty)$ and the domain of $g(x) = \log_2 x^2$ is $\mathbb{R} - \{0\}$.   ★ $f(x) = 2\log_2 x$

45   (a) $\log x = 3.6274 \Rightarrow x = 10^{3.6274} \approx 4240.333$, or 4240 to three significant figures

     (f) $\ln x = -1.6 \Rightarrow x = e^{-1.6} \approx 0.2019$, or 0.202

47 $q = q_0 (2)^{-t/1600} \Rightarrow \dfrac{q}{q_0} = 2^{-t/1600}$ {change to logarithm form} $\Rightarrow$

$$-\frac{t}{1600} = \log_2\left(\frac{q}{q_0}\right) \text{ \{multiply by } -1600\} \Rightarrow t = -1600\log_2\left(\frac{q}{q_0}\right)$$

53   We will find a general formula for α first.

$$I = 10^a I_0 \Rightarrow \alpha = 10\log\left(\frac{I}{I_0}\right) = 10\log\left(\frac{10^a I_0}{I_0}\right) = 10\left(\log 10^a\right) = 10(a) = 10a.$$

     Hence, for $10^1 I_0$, $10^3 I_0$, and $10^4 I_0$, the answers are: (a) 10   (b) 30   (c) 40

57   (a) $\ln W = \ln 2.4 + (1.84)h$ {change to exponential form} $\Rightarrow W = e^{[\ln 2.4 + (1.84)h]} \Rightarrow$

$$W = e^{\ln 2.4} e^{1.84h} \text{ \{since } e^x e^y = e^{x+y}\} \Rightarrow W = 2.4 e^{1.84h}$$

     (b) $h = 1.5 \Rightarrow W = 2.4 e^{(1.84)(1.5)} = 2.4 e^{2.76} \approx 37.92$ kg

$\boxed{59}$ (a) $10 = 14.7e^{-0.0000385h} \Rightarrow \frac{10}{14.7} = e^{-0.0000385h} \Rightarrow \ln\left(\frac{10}{14.7}\right) = -0.0000385h \Rightarrow$

$$h = -\frac{1}{0.0000385}\ln\left(\frac{10}{14.7}\right) \approx 10{,}007 \text{ ft.}$$

(b) At sea level, $h = 0$, and $p(0) = 14.7$. Setting $p(h)$ equal to $\frac{1}{2}(14.7)$,

and solving as in part (a), we have $h = -\frac{1}{0.0000385}\ln\left(\frac{1}{2}\right) \approx 18{,}004 \text{ ft.}$

$\boxed{61}$ (a) $t = 0 \Rightarrow W = 2600(1 - 0.51)^3 = 2600(0.49)^3 \approx 305.9 \text{ kg}$

(b) (1) From the graph, if $W = 1800$, t appears to be about 20.

(2) Solving the equation for t, we have $1800 = 2600\left(1 - 0.51e^{-0.075t}\right)^3 \Rightarrow$

$\frac{1800}{2600} = \left(1 - 0.51e^{-0.075t}\right)^3 \Rightarrow \sqrt[3]{\frac{9}{13}} = 1 - 0.51e^{-0.075t} \Rightarrow$

$e^{-0.075t} = \left(1 - \sqrt[3]{\frac{9}{13}}\right)\left(\frac{100}{51}\right)$ { call this A } $\Rightarrow (-0.075)t = \ln A \Rightarrow t \approx 19.8 \text{ yr.}$

$\boxed{65}$ Since the half-life is eight days, $A(t) = \frac{1}{2}A_0$ when $t = 8$.

Thus, $\frac{1}{2}A_0 = A_0 a^{-8} \Rightarrow a^{-8} = \frac{1}{2} \Rightarrow \frac{1}{a^8} = \frac{1}{2} \Rightarrow$

$$a^8 = 2 \Rightarrow a = 2^{1/8} \text{ { take the eighth root } } \approx 1.09.$$

$\boxed{67}$ (a) Since $\log P$ is an increasing function, increasing the population increases the

walking speed. Pedestrians have faster average walking speeds in large cities.

(b) $S = 5 \Rightarrow 5 = 0.05 + 0.86\log P \Rightarrow 4.95 = 0.86\log P \Rightarrow$

$$\frac{4.95}{0.86} = \log P \Rightarrow P = 10^{4.95/0.86} \approx 570{,}000$$

$\boxed{69}$ (a) $f(1) = \log 1 - 10^{-1} = 0 - \frac{1}{10} = -0.1 < 0$ and

$f(2) = \log 2 - 10^{-2} = \log 2 - \frac{1}{100} \approx 0.30 - 0.01 = 0.29 > 0.$

Thus, f assumes both positive and negative values on $[1, 2]$.

(b) $\log x - 10^{-x} = 0 \Rightarrow \log x = 10^{-x} \Rightarrow \log_{10} x = 10^{-x} \Rightarrow x = (10)^{10^{-x}}$

$x_1 = 1.5 \Rightarrow x_2 = (10)^{10^{-x_1}} = (10)^{10^{-1.5}} \approx 1.075531 \Rightarrow x_3 \approx 1.213492 \Rightarrow$

$x_4 \approx 1.151240 \Rightarrow x_5 \approx 1.176502 \Rightarrow x_6 \approx 1.165745 \Rightarrow x_7 \approx 1.170237.$

The zero is approximately 1.17. *Note:* If you are using a TI-81, type 1.5 and press the ENTER key to store 1.5 in the memory location ANS. Now type $10^{\char94}(10^{\char94}(-\text{Ans}))$ and then successively press the ENTER key to obtain the approximations for x_1, x_2, ...

3 $\log_a \dfrac{x^3 w}{y^2 z^4} = \log_a x^3 w - \log_a y^2 z^4 = \log_a x^3 + \log_a w - (\log_a y^2 + \log_a z^4) =$

$$3\log_a x + \log_a w - 2\log_a y - 4\log_a z$$

The most common mistake is to not have the minus sign in front of $4\log_a z$.

This error results from not having the parentheses in the correct place.

7 $\ln \sqrt[4]{\dfrac{x^7}{y^5 z}} = \ln x^{7/4} - \ln y^{5/4} z^{1/4} = \ln x^{7/4} - \ln y^{5/4} - \ln z^{1/4} = \frac{7}{4}\ln x - \frac{5}{4}\ln y - \frac{1}{4}\ln z$

As a generalization for exercises similar to those in 1–8, if the exponents on the variables are positive, then the sign in front of the individual logarithms will be positive if the variable was originally in the numerator and negative if the variable was originally in the denominator.

11 $2\log_a x + \frac{1}{3}\log_a (x-2) - 5\log_a (2x+3)$

$\quad = \log_a x^2 + \log_a (x-2)^{1/3} - \log_a (2x+3)^5$ {logarithm law (3)}

$\quad = \log_a x^2 \sqrt[3]{x-2} - \log_a (2x+3)^5$ {logarithm law (1)}

$\quad = \log_a \dfrac{x^2 \sqrt[3]{x-2}}{(2x+3)^5}$ {logarithm law (2)}

15 $\ln y^3 + \frac{1}{3}\ln (x^3 y^6) - 5\ln y = \ln y^3 + \ln (xy^2) - \ln y^5 = \ln\left[(xy^5)/y^5\right] = \ln x$

19 $2\log_3 x = 3\log_3 5 \Rightarrow \log_3 x^2 = \log_3 5^3 \Rightarrow x^2 = 125 \Rightarrow x = \pm 5\sqrt{5};$

$\qquad\qquad -5\sqrt{5}$ is extraneous since it would make $\log_3 x$ undefined

21 $\log x - \log (x+1) = 3\log 4 \Rightarrow \log \dfrac{x}{x+1} = \log 64 \Rightarrow$

$\qquad \dfrac{x}{x+1} = 64 \Rightarrow x = 64x + 64 \Rightarrow x = -\frac{64}{63};\ -\frac{64}{63}$ is extraneous, no solution

23 $\ln(-4-x) + \ln 3 = \ln(2-x) \Rightarrow \ln(-12-3x) = \ln(2-x) \Rightarrow -12-3x = 2-x \Rightarrow$

$2x = -14 \Rightarrow x = -7.$ Remember, the solution of a logarithmic equation may be negative—you must examine what happens to the original logarithm expressions. In this case, we have $\ln 3 + \ln 3 = \ln 9$, which is true.

27 $\log_3 (x+3) + \log_3 (x+5) = 1 \Rightarrow \log_3\left[(x+3)(x+5)\right] = 1 \Rightarrow \log_3 (x^2 + 8x + 15) = 1 \Rightarrow$

$x^2 + 8x + 15 = 3 \Rightarrow x^2 + 8x + 12 = 0 \Rightarrow (x+2)(x+6) = 0 \Rightarrow x = -6,\ -2;$

$\qquad\qquad\qquad\qquad\qquad -6$ is extraneous

31 $\ln x = 1 - \ln(x+2) \Rightarrow \ln x + \ln(x+2) = 1 \Rightarrow \ln\left[x(x+2)\right] = 1 \Rightarrow x^2 + 2x = e^1 \Rightarrow$

$x^2 + 2x - e = 0 \Rightarrow x = \dfrac{-2 \pm \sqrt{4+4e}}{2} = \dfrac{-2 \pm 2\sqrt{1+e}}{2} = -1 \pm \sqrt{1+e}.$

$x = -1 + \sqrt{1+e} \approx 0.93$ is a valid solution,

$\qquad\qquad\qquad\qquad\qquad$ but $x = -1 - \sqrt{1+e} \approx -2.93$ is extraneous.

33 $f(x) = \log_3(3x) = \log_3 3 + \log_3 x = \log_3 x + 1$ • shift $y = \log_3 x$ up 1 unit

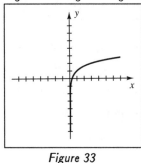

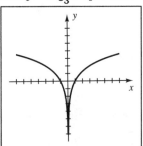

Figure 33   Figure 37

37 $f(x) = \log_3(x^2) = 2\log_3 x$ • Vertically stretch $y = \log_3 x$ by a factor of 2 and include its reflection through the y-axis since the domain of the original function, $f(x) = \log_3(x^2)$, is $\mathbb{R} - \{0\}$. Keep in mind that the laws of logarithms are established for positive real numbers, so that when we make the step $\log_3(x^2) = 2\log_3 x$, it is only for positive x.

41 $f(x) = \log_2 \sqrt{x} = \log_2 x^{1/2} = \frac{1}{2}\log_2 x$ •

vertically compress $y = \log_2 x$ by a factor of $1/(1/2) = 2$

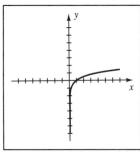

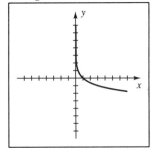

Figure 41   Figure 43

43 $f(x) = \log_3\left(\frac{1}{x}\right) = \log_3 x^{-1} = -\log_3 x$ • reflect $y = \log_3 x$ through the x-axis

45 The values of $F(x) = \log_2 x$ are doubled and the reflection of F through the y-axis is included. The domain of f is $\mathbb{R} - \{0\}$ and $f(x) = \log_2 x^2$.

47 $F(x) = \log_2 x$ is shifted up 3 units since $(1, 0)$ on F is $(1, 3)$ on the graph.

Hence, $f(x) = 3 + \log_2 x = \log_2 2^3 + \log_2 x = \log_2(8x)$.

49 $\log y = \log b - k \log x \Rightarrow \log y = \log b - \log x^k \Rightarrow \log y = \log \frac{b}{x^k} \Rightarrow y = \frac{b}{x^k}$

53 (a) $R(x_0) = a\log\left(\frac{x_0}{x_0}\right) = a\log 1 = a \cdot 0 = 0$

(b) $R(2x) = a\log\left(\frac{2x}{x_0}\right) = a\left[\log 2 + \log\left(\frac{x}{x_0}\right)\right] = a\log 2 + a\log\left(\frac{x}{x_0}\right) = R(x) + a\log 2$

55 $\ln I_0 - \ln I = kx \Rightarrow \ln\frac{I_0}{I} = kx \Rightarrow x = \frac{1}{k}\ln\frac{I_0}{I} = \frac{1}{0.39}\ln 1.12 \approx 0.29$ cm.

57 From the graph, the coordinates of the points of intersection are approximately

(1.01, 0.48) and (2.4, 0.86). $f(x) \geq g(x)$ on the intervals (0, 1.01] and [2.4, ∞).

[0, 6] by [−1, 3]

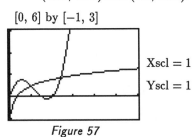

Xscl = 1

Yscl = 1

Figure 57

[0, 8] by [−1.67, 3.67]

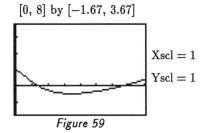

Xscl = 1

Yscl = 1

Figure 59

59 Graph $y = e^{-x} - 2 \log(1 + x^2) + 0.5x$ and estimate any x-intercepts. From the graph,

we see that the roots of the equation are approximately $x \approx 1.41, 6.59$.

5.5 Exercises

3 (a) $3^{4-x} = 5 \Rightarrow \log(3^{4-x}) = \log 5 \Rightarrow (4 - x)\log 3 = \log 5 \Rightarrow$

$4 - x = \dfrac{\log 5}{\log 3} \Rightarrow x = 4 - \dfrac{\log 5}{\log 3} \approx 2.54$. *Note:* The answer could also

be written as $4 - \dfrac{\log 5}{\log 3} = \dfrac{4 \log 3 - \log 5}{\log 3} = \dfrac{\log 81 - \log 5}{\log 3} = \dfrac{\log \frac{81}{5}}{\log 3}$.

(b) $3^{4-x} = 5 \Rightarrow 4 - x = \log_3 5 \Rightarrow x = 4 - \dfrac{\log 5}{\log 3} \approx 2.54$.

7 $\log_9 0.2 = \dfrac{\log 0.2}{\log 9} \left\{ \text{or, equivalently, } \dfrac{\ln 0.2}{\ln 9} \right\} \approx -0.7325$.

Note that either log or ln can be used here. You should get comfortable using both.

9 $\dfrac{\log_5 16}{\log_5 4} = \log_4 16 = \log_4 4^2 = 2$

13 The steps are similar to those in Example 4. $2^{2x-3} = 5^{x-2} \Rightarrow$

$\log(2^{2x-3}) = \log(5^{x-2}) \Rightarrow (2x - 3)\log 2 = (x - 2)\log 5 \Rightarrow$

$2x \log 2 - 3 \log 2 = x \log 5 - 2 \log 5 \Rightarrow 2x \log 2 - x \log 5 = 3 \log 2 - 2 \log 5 \Rightarrow$

$x(2 \log 2 - \log 5) = \log 2^3 - \log 5^2 \Rightarrow x = \dfrac{\log 8 - \log 25}{\log 4 - \log 5} \Rightarrow x = \dfrac{\log \frac{8}{25}}{\log \frac{4}{5}} \approx 5.11$

19 $\log(x^2 + 4) - \log(x + 2) = 2 + \log(x - 2) \Rightarrow \log\left(\dfrac{x^2 + 4}{x + 2}\right) - \log(x - 2) = 2 \Rightarrow$

$\log\left(\dfrac{x^2 + 4}{x^2 - 4}\right) = 2 \Rightarrow \dfrac{x^2 + 4}{x^2 - 4} = 10^2 \Rightarrow x^2 + 4 = 100x^2 - 400 \Rightarrow 404 = 99x^2 \Rightarrow$

$x = \pm\sqrt{\dfrac{404}{99}} = \pm\dfrac{2}{3}\sqrt{\dfrac{101}{11}} \approx \pm 2.02; \ -\dfrac{2}{3}\sqrt{\dfrac{101}{11}}$ is extraneous

21 See Example 5 for more detail concerning this type of exercise.

$5^x + 125(5^{-x}) = 30 \left\{ \text{multiply by } 5^x \right\} \Rightarrow$

$(5^x)^2 - 30(5^x) + 125 = 0 \left\{ \text{recognize as a quadratic in } 5^x \text{ and factor} \right\} \Rightarrow$

$(5^x - 5)(5^x - 25) = 0 \Rightarrow 5^x = 5, 25 \Rightarrow 5^x = 5^1, 5^2 \Rightarrow x = 1, 2$

☐23 $4^x - 3(4^{-x}) = 8$ { multiply by 4^x } $\Rightarrow$

$(4^x)^2 - 8(4^x) - 3 = 0$ { recognize as a quadratic in 4^x } $\Rightarrow$

$4^x = \dfrac{8 \pm \sqrt{76}}{2} = \dfrac{8 \pm 2\sqrt{19}}{2} = 4 \pm \sqrt{19}$.

Since since 4^x is positive and $4 - \sqrt{19}$ is negative, $4 - \sqrt{19}$ is discarded.

Continuing, $4^x = 4 + \sqrt{19} \Rightarrow x = \log_4 (4 + \sqrt{19}) = \dfrac{\log(4 + \sqrt{19})}{\log 4}$ { use the change of

base formula to approximate } ≈ 1.53

☐25 $\log(x^2) = (\log x)^2 \Rightarrow 2\log x = (\log x)^2 \Rightarrow (\log x)^2 - 2\log x = 0 \Rightarrow$

$(\log x)(\log x - 2) = 0 \Rightarrow \log x = 0, 2 \Rightarrow x = 10^0, 10^2 \Rightarrow x = 1$ or 100

☐27 Don't confuse $\log(\log x)$ with $(\log x)(\log x)$. The first expression is

the log of the log of x, whereas the second expression is the log of x times itself.

$\log(\log x) = 2 \Rightarrow \log x = 10^2 = 100 \Rightarrow x = 10^{100}$

☐29 $x^{\sqrt{\log x}} = 10^8$ { take the log of both sides } $\Rightarrow \log\left(x^{\sqrt{\log x}}\right) = \log 10^8 \Rightarrow$

$\sqrt{\log x}\,(\log x) = 8 \Rightarrow (\log x)^{1/2}(\log x)^1 = 8 \Rightarrow (\log x)^{3/2} = 8 \Rightarrow$

$\left[(\log x)^{3/2}\right]^{2/3} = (8)^{2/3} \Rightarrow \log x = (\sqrt[3]{8})^2 = 4 \Rightarrow x = 10{,}000$

Note: For Exercises 31–34 and 39–40 of the review exercises, let D denote the domain of

the function determined by the original equation, and R its range. These are then

the range and domain, respectively, of the equation listed in the answer.

☐31 $y = \dfrac{10^x + 10^{-x}}{2}$ { $D = \mathbb{R},\ R = [1, \infty)$ } $\Rightarrow$

$2y = 10^x + 10^{-x}$ { since $10^{-x} = \dfrac{1}{10^x}$, multiply by 10^x to eliminate denominator } $\Rightarrow$

$10^{2x} - 2y\,10^x + 1 = 0$ { treat as a quadratic in 10^x } $\Rightarrow$

$10^x = \dfrac{2y \pm \sqrt{4y^2 - 4}}{2} = y \pm \sqrt{y^2 - 1} \Rightarrow x = \log(y \pm \sqrt{y^2 - 1})$

☐33 $y = \dfrac{10^x - 10^{-x}}{10^x + 10^{-x}}$ { $D = \mathbb{R},\ R = (-1, 1)$ } $\Rightarrow y\,10^x + y\,10^{-x} = 10^x - 10^{-x} \Rightarrow$

$y\,10^{2x} + y = 10^{2x} - 1 \Rightarrow (y - 1)\,10^{2x} = -1 - y \Rightarrow$

$10^{2x} = \dfrac{-1 - y}{y - 1} \Rightarrow 2x = \log\left(\dfrac{1 + y}{1 - y}\right) \Rightarrow x = \tfrac{1}{2}\log\left(\dfrac{1 + y}{1 - y}\right)$

☐35 $y = \dfrac{e^x - e^{-x}}{2}$ { $D = R = \mathbb{R}$ } $\Rightarrow 2y = e^x - e^{-x} \Rightarrow$

$e^{2x} - 2y\,e^x - 1 = 0 \Rightarrow e^x = \dfrac{2y \pm \sqrt{4y^2 + 4}}{2} = y \pm \sqrt{y^2 + 1};$

$\sqrt{y^2 + 1} > y$, so $y - \sqrt{y^2 + 1} < 0$, but $e^x > 0$ and thus, $x = \ln(y + \sqrt{y^2 + 1})$

39 $f(x) = \log_2{(x+3)}$ • $x = 0 \Rightarrow$ y-intercept $= \log_2 3 = \dfrac{\log 3}{\log 2} \approx 1.5850$

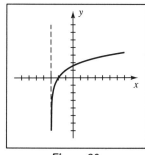

Figure 39

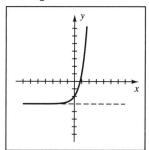

Figure 41

41 $f(x) = 4^x - 3$ • $y = 0 \Rightarrow 4^x = 3 \Rightarrow$ x-intercept $= \log_4 3 = \dfrac{\log 3}{\log 4} \approx 0.7925$

45 $[\text{H}^+] < 10^{-7} \Rightarrow \log{[\text{H}^+]} < \log 10^{-7}$ { since $\log$ is an increasing function } $\Rightarrow$

$\log{[\text{H}^+]} < -7 \Rightarrow -\log{[\text{H}^+]} > -(-7) \Rightarrow \text{pH} > 7$ for basic solutions;

similarly, $\text{pH} < 7$ for acidic solutions.

47 Solving $A = P(1 + \frac{r}{n})^{nt}$ for t with $A = 2P$, $r = 0.06$, and $n = 12$ yields

$2P = P(1 + \frac{0.06}{12})^{12t}$ { divide by P } $\Rightarrow 2 = (1.005)^{12t}$ { take the $\ln$ of both sides } $\Rightarrow$

$\ln 2 = \ln{(1.005)^{12t}} \Rightarrow \ln 2 = 12t \ln{(1.005)} \Rightarrow$

$$t = \dfrac{\ln 2}{12 \ln{(1.005)}} \approx 11.58 \text{ yr, or about 11 years and 7 months.}$$

49 50% of the light reaching a depth of 13 meters corresponds to the equation

$\frac{1}{2}I_0 = I_0 c^{13}$. Solving for c, we have $c^{13} = \frac{1}{2} \Rightarrow c = \sqrt[13]{\frac{1}{2}} = 2^{-1/13}$.

Now letting $I = 0.01\,I_0$, $c = 2^{-1/13}$, and using the formula from Example 6,

$$x = \dfrac{\log{(I/I_0)}}{\log c} = \dfrac{\log{[(0.01\,I_0)/I_0]}}{\log 2^{-1/13}} = \dfrac{\log 10^{-2}}{-\frac{1}{13}\log 2} = \dfrac{26}{\log 2} \approx 86.4 \text{ m.}$$

51 (a) $x^y = y^x \Rightarrow \ln{(x^y)} = \ln{(y^x)} \Rightarrow y \ln x = x \ln y \Rightarrow \dfrac{\ln x}{x} = \dfrac{\ln y}{y}$

(b) Note that $f(e) = \dfrac{\ln e}{e} = \dfrac{1}{e} \approx 0.37$ is the maximum value of f.

Any horizontal line $y = k$, with $0 < k < \frac{1}{e}$, will intersect the graph at the two

points $\left(x_1, \dfrac{\ln x_1}{x_1}\right)$ and $\left(x_2, \dfrac{\ln x_2}{x_2}\right)$, where $1 < x_1 < e$ { before the maximum } and

$x_2 > e$ { after the maximum }.

55 (a) $F = F_0(1-m)^t \Rightarrow (1-m)^t = \frac{F}{F_0} \Rightarrow \log(1-m)^t = \log\left(\frac{F}{F_0}\right) \Rightarrow$

$$t\log(1-m) = \log\left(\frac{F}{F_0}\right) \Rightarrow t = \frac{\log(F/F_0)}{\log(1-m)}$$

(b) Using part (a) with $F = \frac{1}{2}F_0$ and $m = 0.00005$,

$$t = \frac{\log(F/F_0)}{\log(1-m)} = \frac{\log(\frac{1}{2}F_0/F_0)}{\log(1-0.00005)} = \frac{\log\frac{1}{2}}{\log 0.99995} \approx 13{,}863 \text{ generations.}$$

57 (a) $t = 10 \Rightarrow h = \dfrac{120}{1+200e^{-2}} \approx 4.28$ ft

(b) $h = 50 \Rightarrow 50 = \dfrac{120}{1+200e^{-0.2t}} \Rightarrow 1 + 200e^{-0.2t} = \dfrac{120}{50} \Rightarrow 200e^{-0.2t} = \dfrac{12}{5} - 1 \Rightarrow$

$e^{-0.2t} = \dfrac{7}{5} \cdot \dfrac{1}{200} \Rightarrow e^{-0.2t} = 0.007 \Rightarrow -0.2t = \ln 0.007 \Rightarrow t = \dfrac{\ln 0.007}{-0.2} \approx 24.8$ yr

61 When $x = 0$, $y = c2^0 = c = 4$. Thus, $y = 4(2)^{kx}$. Similarly, $x = 1 \Rightarrow$ $y = 4(2)^k = 3.249 \Rightarrow k = \log_2\left(\frac{3.249}{4}\right) \approx -0.300$. Thus, $y = 4(2)^{-0.3x}$. Checking the remaining two points, we see that $x = 2 \Rightarrow y \approx 2.639$ and $x = 3 \Rightarrow y \approx 2.144$. The points all lie on the graph of $y = 4(2)^{-0.3x}$ to within three-decimal-place accuracy.

63 When $x = 0$, $y = c\log 10 = c = 1.5$. Thus, $y = 1.5\log(kx + 10)$. Similarly, $x = 1 \Rightarrow y = 1.5\log(k + 10) = 1.619 \Rightarrow k + 10 = 10^{1.619/1.5} \Rightarrow$ $k = 10^{1.619/1.5} - 10 \approx 2.004$. Thus, $y = 1.5\log(2.004x + 10)$. Checking the remaining two points, we see that $x = 2 \Rightarrow y \approx 1.720$, and $x = 3 \Rightarrow y \approx 1.807$. The points do not all lie on the graph of $y = c2^{kx}$ to within three-decimal-place accuracy.

67 From the graph, we see that the graphs of f and g intersect at three points. Their coordinates are approximately $(-0.32, 0.50)$, $(1.52, -1.33)$, and $(6.84, -6.65)$. The region near the origin in *Figure 67(a)* is enhanced in *Figure 67(b)*. Thus, $f(x) > g(x)$ on $(-\infty, -0.32)$ and $(1.52, 6.84)$.

[−5, 10] by [−8, 2]   [−1.53, 2.26] by [−2.92, 1.05]

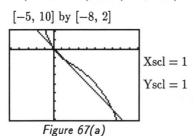

Xscl = 1

Yscl = 1

Figure 67(a)

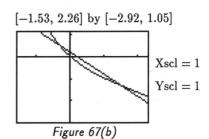

Xscl = 1

Yscl = 1

Figure 67(b)

Chapter 5 Review Exercises

$\boxed{3}$ $f(x) = \left(\frac{3}{2}\right)^{-x} = \left(\frac{2}{3}\right)^{x}$ • goes through $\left(-1, \frac{3}{2}\right)$, $(0, 1)$, and $\left(1, \frac{2}{3}\right)$

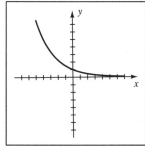

Figure 3

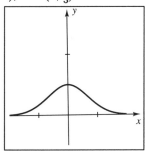

Figure 5

$\boxed{5}$ $f(x) = 3^{-x^2} = \left(3^{-1}\right)^{\left(x^2\right)} = \left(\frac{1}{3}\right)^{\left(x^2\right)}$ • see the note in §5.1 before Exercise 17

$\boxed{7}$ $f(x) = e^{x/2} = \left(e^{1/2}\right)^{x} \approx (1.65)^{x}$ • goes through $\left(-1, 1/\sqrt{e}\right)$, $(0, 1)$, and $\left(1, \sqrt{e}\right)$;

or approximately $(-1, 0.61)$, $(0, 1)$, and $(1, 1.65)$

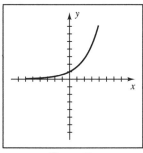

Figure 7

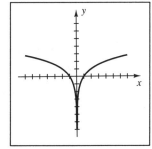

Figure 13

$\boxed{13}$ $f(x) = \log_4\left(x^2\right) = 2\log_4 x$ • stretch $y = \log_4 x$ by a factor of 2 and include its

reflection through the y-axis since the domain of the original function is $\mathbb{R} - \{0\}$

$\boxed{17}$ (a) $\log_2 \frac{1}{16} = \log_2 2^{-4} = -4$ (b) $\log_\pi 1 = 0$ (c) $\ln e = 1$

(d) $6^{\log_6 4} = 4$ (e) $\log 1{,}000{,}000 = \log 10^6 = 6$

(f) $10^{3\log 2} = 10^{\log 2^3} = 2^3 = 8$ (g) $\log_4 2 = \log_4 4^{1/2} = \frac{1}{2}$

$\boxed{19}$ $2^{3x-1} = \frac{1}{2} \Rightarrow 2^{3x-1} = 2^{-1} \Rightarrow 3x - 1 = -1 \Rightarrow 3x = 0 \Rightarrow x = 0$

$\boxed{23}$ $2\ln(x+3) - \ln(x+1) = 3\ln 2 \Rightarrow \ln\frac{(x+3)^2}{x+1} = \ln 2^3 \Rightarrow (x+3)^2 = 8(x+1) \Rightarrow$

$$x^2 + 6x + 9 = 8x + 8 \Rightarrow x^2 - 2x + 1 = 0 \Rightarrow (x-1)^2 = 0 \Rightarrow x = 1$$

$\boxed{27}$ $2^{5x+3} = 3^{2x+1} \Rightarrow \log\left(2^{5x+3}\right) = \log\left(3^{2x+1}\right) \Rightarrow (5x+3)\log 2 = (2x+1)\log 3 \Rightarrow$

$5x \log 2 + 3\log 2 = 2x \log 3 + \log 3 \Rightarrow 5x \log 2 - 2x \log 3 = \log 3 - 3\log 2 \Rightarrow$

$$x(5\log 2 - 2\log 3) = \log 3 - \log 2^3 \Rightarrow x = \frac{\log 3 - \log 8}{\log 32 - \log 9} = \frac{\log\frac{3}{8}}{\log\frac{32}{9}}$$

29 $\log_4 x = \sqrt[3]{\log_4 x} \Rightarrow \log_4 x = (\log_4 x)^{1/3} \Rightarrow (\log_4 x)^3 = \log_4 x \Rightarrow$

$(\log_4 x)^3 - \log_4 x = 0 \Rightarrow \log_4 x \left[(\log_4 x)^2 - 1\right] = 0 \Rightarrow$

$\log_4 x = 0$ or $\log_4 x = \pm 1 \Rightarrow x = 1$ or $x = 4, \frac{1}{4}$

31 $10^{2\log x} = 5 \Rightarrow 10^{\log x^2} = 5 \Rightarrow x^2 = 5 \Rightarrow x = \pm\sqrt{5}; -\sqrt{5}$ is extraneous

35 (a) $\log x^2 = \log(6 - x) \Rightarrow x^2 = 6 - x \Rightarrow x^2 + x - 6 = 0 \Rightarrow (x + 3)(x - 2) = 0 \Rightarrow$

$x = -3, 2$

(b) $2\log x = \log(6 - x) \Rightarrow \log x^2 = \log(6 - x)$, which is the equation in part (a).

This equation has the same solutions provided they are in the domain.

But -3 is extraneous, so 2 is the only solution.

39 $y = \dfrac{1}{10^x + 10^{-x}} \left\{ D = \mathbb{R}, R = (0, \frac{1}{2}] \right\} \Rightarrow y\, 10^x + y\, 10^{-x} = 1 \Rightarrow$

$y\, 10^{2x} + y = 10^x \Rightarrow y\, 10^{2x} - 10^x + y = 0 \Rightarrow 10^x = \dfrac{1 \pm \sqrt{1 - 4y^2}}{2y} \Rightarrow$

$x = \log\left(\dfrac{1 \pm \sqrt{1 - 4y^2}}{2y}\right)$

45 (a) $N = 64(0.5)^{t/8} = 64\left[(0.5)^{1/8}\right]^t \approx 64(0.917)^t$

(b) $N = \frac{1}{2}N_0 \Rightarrow \frac{1}{2}N_0 = N_0(0.5)^{t/8} \Rightarrow$

$(\frac{1}{2})^1 = (\frac{1}{2})^{t/8} \Rightarrow 1 = t/8 \Rightarrow t = 8$ days

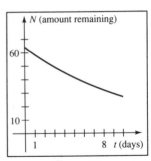

Figure 45

47 (a) Using $A = Pe^{rt}$ with $A = \$35,000$, $P = \$10,000$, and $r = 11\%$, we have

$35,000 = 10,000e^{0.11t} \Rightarrow e^{0.11t} = 3.5 \Rightarrow 0.11t = \ln 3.5 \Rightarrow t = \frac{1}{0.11}\ln 3.5 \approx 11.39$ yr.

(b) $2 \cdot 10,000 = 10,000e^{0.11t} \Rightarrow e^{0.11t} = 2 \Rightarrow 0.11t = \ln 2 \Rightarrow t \approx 6.30$ yr

49 (a) $\alpha = 10\log\left(\dfrac{I}{I_0}\right) \Rightarrow \dfrac{\alpha}{10} = \log\left(\dfrac{I}{I_0}\right) \Rightarrow 10^{\alpha/10} = \dfrac{I}{I_0} \Rightarrow I = I_0 10^{\alpha/10}$

(b) Let $I(\alpha)$ be the intensity corresponding to α decibels.

$I(\alpha + 1) = I_0 10^{(\alpha + 1)/10} = I_0 10^{\alpha/10} 10^{1/10} = I(\alpha)\, 10^{1/10} \approx 1.26\, I(\alpha),$

which represents a 26% increase in $I(\alpha)$.

$\boxed{51}$ $R = 2.3 \log{(A + 3000)} - 5.1 \Rightarrow R + 5.1 = 2.3 \log{(A + 3000)} \Rightarrow$

$\dfrac{R + 5.1}{2.3} = \log{(A + 3000)} \Rightarrow$

$A + 3000 = 10^{(R + 5.1)/2.3}$ { change to exponential form } $\Rightarrow A = 10^{(R + 5.1)/2.3} - 3000$

$\boxed{55}$ Substituting $v = 0$ and $m = m_1 + m_2$ in $v = -a \ln{m} + b$ yields

$0 = -a \ln{(m_1 + m_2)} + b$. Thus, $b = a \ln{(m_1 + m_2)}$. At burnout, $m = m_1$,

and hence, $v = -a \ln{m_1} + b\ = -a \ln{m_1} + a \ln{(m_1 + m_2)}$

$$= a\left[\ln{(m_1 + m_2)} - \ln{m_1}\right] \Rightarrow v = a \ln\left(\frac{m_1 + m_2}{m_1}\right)$$

$\boxed{57}$ (a) $\log{E} = 11.4 + (1.5)R \Rightarrow E = 10^{11.4 + 1.5R}$ { merely change the form }

(b) $R = 8.4 \Rightarrow E = 10^{11.4 + 1.5(8.4)} = 10^{24}$ ergs

$\boxed{61}$ (a) $x = 4\% \Rightarrow T = -8310 \ln{(0.04)} \approx 26{,}749$ yr.

(b) $T = 10{,}000 \Rightarrow 10{,}000 = -8310 \ln{x} \Rightarrow -\dfrac{10{,}000}{8310} = \ln{x} \Rightarrow$

$$x = e^{-1000/831} \approx 0.30, \text{ or } 30\%.$$

Chapter 6: The Trigonometric Functions

Note: Exercises 1 and 3: The answers listed are the smallest (in magnitude) two positive coterminal angles and two negative coterminal angles.

$\boxed{1}$ (a) $120° + 1(360°) = 480°$, $120° + 2(360°) = 840°$;

       $120° - 1(360°) = -240°$, $120° - 2(360°) = -600°$

    (b) $135° + 1(360°) = 495°$, $135° + 2(360°) = 855°$;

       $135° - 1(360°) = -225°$, $135° - 2(360°) = -585°$

    (c) $-30° + 1(360°) = 330°$, $-30° + 2(360°) = 690°$;

       $-30° - 1(360°) = -390°$, $-30° - 2(360°) = -750°$

$\boxed{3}$ (a) $620° - 1(360°) = 260°$, $620° + 1(360°) = 980°$;

       $620° - 2(360°) = -100°$, $620° - 3(360°) = -460°$

    (b) $\frac{5\pi}{6} + 1(2\pi) = \frac{5\pi}{6} + \frac{12\pi}{6} = \frac{17\pi}{6}$, $\frac{5\pi}{6} + 2(2\pi) = \frac{5\pi}{6} + \frac{24\pi}{6} = \frac{29\pi}{6}$;

       $\frac{5\pi}{6} - 1(2\pi) = \frac{5\pi}{6} - \frac{12\pi}{6} = -\frac{7\pi}{6}$, $\frac{5\pi}{6} - 2(2\pi) = \frac{5\pi}{6} - \frac{24\pi}{6} = -\frac{19\pi}{6}$

    (c) $-\frac{\pi}{4} + 1(2\pi) = -\frac{\pi}{4} + \frac{8\pi}{4} = \frac{7\pi}{4}$, $-\frac{\pi}{4} + 2(2\pi) = -\frac{\pi}{4} + \frac{16\pi}{4} = \frac{15\pi}{4}$;

       $-\frac{\pi}{4} - 1(2\pi) = -\frac{\pi}{4} - \frac{8\pi}{4} = -\frac{9\pi}{4}$, $-\frac{\pi}{4} - 2(2\pi) = -\frac{\pi}{4} - \frac{16\pi}{4} = -\frac{17\pi}{4}$

$\boxed{5}$ (a) $90° - 5°17'34'' = 84°42'26''$ (b) $90° - 32.5° = 57.5°$

$\boxed{7}$ (a) $180° - 48°51'37'' = 131°8'23''$ (b) $180° - 136.42° = 43.58°$

Note: Multiply each degree measure by $\frac{\pi}{180}$ to obtain the listed radian measure.

$\boxed{9}$ (a) $150° \cdot \frac{\pi}{180} = \frac{5 \cdot 30\pi}{6 \cdot 30} = \frac{5\pi}{6}$ (b) $-60° \cdot \frac{\pi}{180} = -\frac{60\pi}{3 \cdot 60} = -\frac{\pi}{3}$

    (c) $225° \cdot \frac{\pi}{180} = \frac{5 \cdot 45\pi}{4 \cdot 45} = \frac{5\pi}{4}$

Note: Multiply each radian measure by $\frac{180}{\pi}$ to obtain the listed degree measure.

$\boxed{15}$ (a) $-\frac{7\pi}{2} \cdot \left(\frac{180}{\pi}\right)° = -\left(\frac{7 \cdot 90 \cdot 2\pi}{2\pi}\right)° = -630°$ (b) $7\pi \cdot \left(\frac{180}{\pi}\right)° = (7 \cdot 180)° = 1260°$

    (c) $\frac{\pi}{9} \cdot \left(\frac{180}{\pi}\right)° = \left(\frac{20 \cdot 9\pi}{9\pi}\right)° = 20°$

$\boxed{17}$ *Note:* Some calculators can easily change radians to degrees, minutes, and seconds by pressing a couple keys. Check your calculator manual for this feature.

We first convert 2 radians to degrees. $2 \cdot \left(\frac{180}{\pi}\right)° \approx 114.59156° = 114° + 0.59156°$.

We now use the decimal portion, $0.59156°$, and convert it to minutes.

Since $60' = 1°$, we have $0.59156° = 0.59156(60') = 35.4936'$.

Using the decimal portion, $0.4936'$, we convert it to seconds.

Since $60'' = 1'$, we have $0.4936' = 0.4936(60'') \approx 30''$. $\therefore$ 2 radians $\approx 114°35'30''$

$\boxed{21}$ Since $1' = \left(\frac{1}{60}\right)°$, $41' = \left(\frac{41}{60}\right)°$. Thus, $37°41' = \left(37 + \frac{41}{60}\right)° \approx 37.6833°$.

$\boxed{23}$ Since $1'' = \left(\frac{1}{3600}\right)°$, $27'' = \left(\frac{27}{3600}\right)°$. Thus, $115°26'27'' = \left(115 + \frac{26}{60} + \frac{27}{3600}\right)° \approx 115.4408°$.

25 We have $63°$ and a portion of one more degree. Since $1° = 60'$,

$0.169° = 0.169\,(60') = 10.14'$. We now have $10'$ and a portion of one more minute.

Since $1' = 60''$, $0.14' = 0.14\,(60'') = 8.4''$. $\therefore 63.169° \approx 63°10'8''$

29 We will use the formula for the length of a circular arc.

$$s = r\theta \Rightarrow r = \tfrac{s}{\theta} = \tfrac{10}{4} = 2.5 \text{ cm.}$$

31 (a) $s = r\theta = 8 \cdot (45 \cdot \tfrac{\pi}{180}) = 8 \cdot \tfrac{\pi}{4} = 2\pi \approx 6.28$ cm

(b) $A = \tfrac{1}{2}r^2\theta = \tfrac{1}{2}(8)^2(\tfrac{\pi}{4}) = 8\pi \approx 25.13$ cm^2

33 (a) Remember that θ is measured in radians. $s = r\theta \Rightarrow \theta = \tfrac{s}{r} = \tfrac{7}{4} = 1.75$ radians.

Converting to degrees, we have $\tfrac{7}{4} \cdot (\tfrac{180}{\pi})° = (\tfrac{315}{\pi})° \approx 100.27°$.

(b) $A = \tfrac{1}{2}r^2\theta = \tfrac{1}{2}(4)^2(\tfrac{7}{4}) = 14$ cm^2

35 (a) A measure of $50°$ is equivalent to $(50 \cdot \tfrac{\pi}{180})$ radians. The radius is one-half of the

diameter. Thus, $s = r\theta = (\tfrac{1}{2} \cdot 16)(50 \cdot \tfrac{\pi}{180}) = 8 \cdot \tfrac{5\pi}{18} = \tfrac{20\pi}{9} \approx 6.98$ m.

(b) $A = \tfrac{1}{2}r^2\theta = \tfrac{1}{2}(8)^2(\tfrac{5\pi}{18}) = \tfrac{80\pi}{9} \approx 27.93$ m^2

37 radius $= \tfrac{1}{2} \cdot 8000$ miles $= 4000$ miles

(a) $s = r\theta = 4000\,(60 \cdot \tfrac{\pi}{180}) = \tfrac{4000\pi}{3} \approx 4189$ miles

(b) $s = r\theta = 4000\,(45 \cdot \tfrac{\pi}{180}) = 1000\pi \approx 3142$ miles

(c) $s = r\theta = 4000\,(30 \cdot \tfrac{\pi}{180}) = \tfrac{2000\pi}{3} \approx 2094$ miles

(d) $s = r\theta = 4000\,(10 \cdot \tfrac{\pi}{180}) = \tfrac{2000\pi}{9} \approx 698$ miles

(e) $s = r\theta = 4000\,(1 \cdot \tfrac{\pi}{180}) = \tfrac{200\pi}{9} \approx 70$ miles

39 $\theta = \tfrac{s}{r} = \tfrac{500}{4000} = \tfrac{1}{8}$ radian; $(\tfrac{1}{8})(\tfrac{180}{\pi})° = (\tfrac{45}{2\pi})° \approx 7°10'$

41 23 hours, 56 minutes, and 4 seconds $= 23(60)^2 + 56(60) + 4 = 86,164$ sec.

Since the earth turns through 2π radians in 86,164 seconds,

it rotates through $\tfrac{2\pi}{86,164} \approx 7.29 \times 10^{-5}$ radians in one second.

43 (a) $\left(40 \; \dfrac{\text{revolutions}}{\text{minute}}\right)\left(2\pi \; \dfrac{\text{radians}}{\text{revolution}}\right) = 80\pi \; \dfrac{\text{radians}}{\text{minute}}$.

Note: Remember to write out and "cancel" the units if you are unsure about

what units your answer is measured in.

(b) The distance that a point on the circumference travels is

$$s = r\theta = (5 \text{ in.}) \cdot 80\pi = 400\pi \text{ in.}$$

Hence, its linear speed is

$$400\pi \text{ in./min} = \tfrac{100\pi}{3} \text{ ft/min} \; \{\, 400\pi \cdot \tfrac{1}{12} \,\} \approx 104.72 \text{ ft/min.}$$

45 (a) As in Exercise 43, we take the number of

revolutions per minute times the number of radians per revolution and

obtain $(33\tfrac{1}{3})(2\pi) = \tfrac{200\pi}{3}$ and $45(2\pi) = 90\pi$.

(b) $s = r\theta = (\frac{1}{2} \cdot 12)(\frac{200\pi}{3}) = 400\pi$ in. Linear speed $= 400\pi$ in./min $= \frac{100\pi}{3}$ ft/min.

$s = r\theta = (\frac{1}{2} \cdot 7)(90\pi) = 315\pi$ in. Linear speed $= 315\pi$ in./min $= \frac{105\pi}{4}$ ft/min.

47 (a) The distance that the cargo is lifted is equal to the arc length that the cable is

moved through. $s = r\theta = (\frac{1}{2} \cdot 3)(\frac{7\pi}{4}) = \frac{21\pi}{8} \approx 8.25$ ft.

(b) $s = r\theta \Rightarrow d = (\frac{1}{2} \cdot 3)\theta \Rightarrow \theta = (\frac{2}{3}d)$ radians. For example, to lift the cargo 6 feet,

the winch must rotate $\frac{2}{3} \cdot 6 = 4$ radians, or about 229°.

49 $\text{Area}_{\text{small}} = \frac{1}{2}r^2\theta = \frac{1}{2}(\frac{1}{2} \cdot 18)^2 \cdot (\frac{2\pi}{6}) = \frac{27\pi}{2}$. $\text{Area}_{\text{large}} = \frac{1}{2}(\frac{1}{2} \cdot 26)^2 \cdot (\frac{2\pi}{8}) = \frac{169\pi}{8}$.

$\text{Cost}_{\text{small}} = \text{Area}_{\text{small}} \div \text{Cost} = \frac{27\pi}{2} \div 2 \approx 21.21$ in.^{2}/dollar.

$\text{Cost}_{\text{large}} = \text{Area}_{\text{large}} \div \text{Cost} = \frac{169\pi}{8} \div 3 \approx 22.12$ in.^{2}/dollar.

The large slice provides slightly more pizza per dollar.

51 $\frac{40 \text{ miles}}{\text{hour}} = \frac{40 \text{ miles}}{\text{hour}} \cdot \frac{1 \text{ hour}}{3600 \text{ seconds}} \cdot \frac{5280 \text{ feet}}{\text{mile}} \cdot \frac{12 \text{ inches}}{\text{foot}} = \frac{704 \text{ inches}}{\text{second}}$.

The circumference of the wheel is $2\pi(14)$ inches. The back sprocket then rotates

$\frac{704 \text{ inches}}{\text{second}} \cdot \frac{1 \text{ revolution}}{28\pi \text{ inches}} = \frac{704 \text{ revolutions}}{28\pi \text{ second}}$ or $\frac{704}{28\pi} \cdot 2\pi = \frac{352}{7}$ radians per second.

The front sprocket's angular speed is given by { from Exercise 50 }

$\theta_1 = \frac{r_2\theta_2}{r_1} = \frac{2 \cdot \frac{352}{7}}{5} = \frac{704}{35} \approx 20.114$ radians per second

or 3.2 revolutions per second or 192.08 revolutions per minute.

6.2 Exercises

Note: Answers are in the order *sin, cos, tan, cot, sec, csc* for any exercises that require

the values of the six trigonometric functions.

1 Using the definition of the trigonometric functions,

$$\sin\theta = \frac{\text{opp}}{\text{hyp}} = \frac{4}{5}, \quad \cos\theta = \frac{\text{adj}}{\text{hyp}} = \frac{3}{5}, \quad \text{and} \quad \tan\theta = \frac{\text{opp}}{\text{adj}} = \frac{4}{3}.$$

We now use the reciprocal identities to find the values of the other trigonometric

functions:

$$\cot\theta = \frac{1}{\tan\theta} = \frac{3}{4}, \quad \sec\theta = \frac{1}{\cos\theta} = \frac{5}{3}, \quad \text{and} \quad \csc\theta = \frac{1}{\sin\theta} = \frac{5}{4}.$$

3 Using the Pythagorean theorem, $(\text{adj})^2 + (\text{opp})^2 = (\text{hyp})^2 \Rightarrow$

$(\text{adj})^2 = (\text{hyp})^2 - (\text{opp})^2 \Rightarrow \text{adj} = \sqrt{(\text{hyp})^2 - (\text{opp})^2} = \sqrt{5^2 - 2^2} = \sqrt{21}.$

$$\bigstar \frac{2}{5}, \frac{\sqrt{21}}{5}, \frac{2}{\sqrt{21}}, \frac{\sqrt{21}}{2}, \frac{5}{\sqrt{21}}, \frac{5}{2}$$

$\boxed{5}$ Using the Pythagorean theorem, hyp $= \sqrt{(\text{adj})^2 + (\text{opp})^2} = \sqrt{a^2 + b^2}$.

$$\star \; \frac{a}{\sqrt{a^2+b^2}}, \; \frac{b}{\sqrt{a^2+b^2}}, \; \frac{a}{b}, \frac{b}{a}, \; \frac{\sqrt{a^2+b^2}}{b}, \; \frac{\sqrt{a^2+b^2}}{a}$$

$\boxed{9}$ Since we want to find the value of the hypotenuse x, we need to use a trigonometric function which relates x to two given parts of the triangle—in this case, the angle $30°$ and the opposite side of length 4. The sine function relates the opposite side and the hypotenuse. Hence, $\sin 30° = \frac{4}{x} \Rightarrow \frac{1}{2} = \frac{4}{x} \Rightarrow x = 8$. The tangent function relates the opposite and the adjacent side. Hence, $\tan 30° = \frac{4}{y} \Rightarrow \frac{\sqrt{3}}{3} = \frac{4}{y} \Rightarrow y = 4\sqrt{3}$.

$\boxed{13}$ $\sin 60° = \frac{x}{8} \Rightarrow \frac{\sqrt{3}}{2} = \frac{x}{8} \Rightarrow x = 4\sqrt{3}$ and $\cos 60° = \frac{y}{8} \Rightarrow \frac{1}{2} = \frac{y}{8} \Rightarrow y = 4$.

Note: It may help to sketch a triangle as shown for Exercises 15 and 19.

   Use the Pythagorean theorem to find the remaining side.

$\boxed{15}$ $(\text{adj})^2 + (\text{opp})^2 = (\text{hyp})^2 \Rightarrow (\text{adj})^2 + 3^2 = 5^2 \Rightarrow \text{adj} = \sqrt{25-9} = 4.$ $\star \; \frac{3}{5}, \frac{4}{5}, \frac{3}{4}, \frac{4}{3}, \frac{5}{4}, \frac{5}{3}$

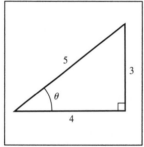

Figure 15

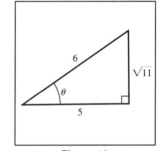

Figure 19

$\boxed{17}$ $12^2 + 5^2 = (\text{hyp})^2 \Rightarrow \text{hyp} = \sqrt{144+25} = 13.$ $\star \; \frac{5}{13}, \frac{12}{13}, \frac{5}{12}, \frac{12}{5}, \frac{13}{12}, \frac{13}{5}$

$\boxed{19}$ $5^2 + (\text{opp})^2 = 6^2 \Rightarrow \text{opp} = \sqrt{36-25} = \sqrt{11}.$ $\star \; \frac{\sqrt{11}}{6}, \frac{5}{6}, \frac{\sqrt{11}}{5}, \frac{5}{\sqrt{11}}, \frac{6}{5}, \frac{6}{\sqrt{11}}$

$\boxed{21}$ (a) Using the *degree* mode on a calculator, $\sin 73°20' \approx 0.958$.

   (b) Using the *radian* mode on a calculator, $\cos 0.68 \approx 0.778$.

$\boxed{25}$ (a) First compute $\cos 67°50'$, obtaining 0.3773. Now use the reciprocal key, usually labeled as either $\boxed{1/x}$ or $\boxed{x^{-1}}$, to obtain $\sec 67°50' \approx 2.650$.

   (b) As in part (a), since $\sin 0.32 \approx 0.3146$, we have $\csc 0.32 \approx 3.179$.

$\boxed{27}$ (a) Using the degree mode, calculate $\cos^{-1}(0.8620)$ to obtain $30.46°$ to the nearest one-hundredth of a degree.

   (b) Using the answer from part (a), subtract 30 and multiply that result by 60 to obtain $30°27'$ to the nearest minute.

$\boxed{31}$ (a) $\sin \theta = 0.4217 \Rightarrow \theta = \sin^{-1}(0.4217) \approx 24.94°$ (b) $24.94° \approx 24°57'$

33 (a) After entering 4.246, use $\boxed{1/x}$ and then $\boxed{\text{INV}}\ \boxed{\text{COS}}$, or, equivalently, $\boxed{\text{COS}^{-1}}$. Similarly, for the cosecant function use $\boxed{1/x}$ and then $\boxed{\text{INV}}\ \boxed{\text{SIN}}$ and for the cotangent function use $\boxed{1/x}$ and then $\boxed{\text{INV}}\ \boxed{\text{TAN}}$.

$\sec\theta = 4.246 \Rightarrow \cos\theta = \frac{1}{4.246} \Rightarrow \theta = \cos^{-1}\left(\frac{1}{4.246}\right) \approx 76.38°.$

(b) $76.38° \approx 76°23'$

35 $\cot\theta = \frac{\cos\theta}{\sin\theta}$ { cotangent identity } $= \frac{\sqrt{1-\sin^2\theta}}{\sin\theta}$ { $\sin^2\theta + \cos^2\theta = 1$ }

37 $\sec\theta = \frac{1}{\cos\theta}$ { reciprocal identity } $= \frac{1}{\sqrt{1-\sin^2\theta}}$

39 One solution is $\sin\theta = \sqrt{1-\cos^2\theta} = \sqrt{1-\frac{1}{\sec^2\theta}} = \frac{\sqrt{\sec^2\theta - 1}}{\sec\theta}$.

Alternatively, $\sin\theta = \frac{\sin\theta/\cos\theta}{1/\cos\theta} = \frac{\tan\theta}{\sec\theta} = \frac{\sqrt{\sec^2\theta - 1}}{\sec\theta}$ { $1 + \tan^2\theta = \sec^2\theta$ }.

41 $\cos\theta\,\sec\theta = \cos\theta\,(1/\cos\theta)$ { reciprocal identity } $= 1$

43 $\sin\theta\,\sec\theta = \sin\theta\,(1/\cos\theta) = \sin\theta/\cos\theta = \tan\theta$ { tangent identity }

45 $\frac{\csc\theta}{\sec\theta} = \frac{1/\sin\theta}{1/\cos\theta}$ { reciprocal identities } $= \frac{\cos\theta}{\sin\theta} = \cot\theta$ { cotangent identity }

47 $(1+\cos\theta)(1-\cos\theta) = 1 - \cos^2\theta = \sin^2\theta$ { Pythagorean identity }

49 $\cos^2\theta\,(\sec^2\theta - 1) = \cos^2\theta\,(\tan^2\theta)$ { Pythagorean identity }

$= \cos^2\theta \cdot \frac{\sin^2\theta}{\cos^2\theta}$ { tangent identity } $= \sin^2\theta$

51 $\frac{\sin\theta}{\csc\theta} + \frac{\cos\theta}{\sec\theta} = \frac{\sin\theta}{1/\sin\theta} + \frac{\cos\theta}{1/\cos\theta} = \sin^2\theta + \cos^2\theta = 1$

53 $(1+\sin\theta)(1-\sin\theta) = 1 - \sin^2\theta$ { multiply as a difference of squares } $= \cos^2\theta = \frac{1}{\sec^2\theta}$

55 $\sec\theta - \cos\theta = \frac{1}{\cos\theta} - \cos\theta = \frac{1-\cos^2\theta}{\cos\theta} = \frac{\sin^2\theta}{\cos\theta} = \frac{\sin\theta}{\cos\theta}\cdot\sin\theta = \tan\theta\,\sin\theta$

57 $(\cot\theta + \csc\theta)(\tan\theta - \sin\theta)$

$= \cot\theta\,\tan\theta - \cot\theta\,\sin\theta + \csc\theta\,\tan\theta - \csc\theta\,\sin\theta$ { multiply binomials }

$= \frac{1}{\tan\theta}\,\tan\theta - \frac{\cos\theta}{\sin\theta}\,\sin\theta + \frac{1}{\sin\theta}\,\frac{\sin\theta}{\cos\theta} - \frac{1}{\sin\theta}\,\sin\theta$

{ reciprocal, cotangent, and tangent identities }

$= 1 - \cos\theta + \frac{1}{\cos\theta} - 1$ { cancel terms }

$= -\cos\theta + \sec\theta = \sec\theta - \cos\theta$

59 $\sec^2\theta\,\csc^2\theta = (1+\tan^2\theta)(1+\cot^2\theta)$ { Pythagorean identities }

$= 1 + \tan^2\theta + \cot^2\theta + 1$ { multiply binomials }

$= (1+\tan^2\theta) + (\cot^2\theta + 1)$ { group terms }

$= \sec^2\theta + \csc^2\theta$ { Pythagorean identities }

61 $\log \csc \theta = \log\left(\dfrac{1}{\sin \theta}\right)$ { reciprocal identity }

$\qquad\qquad = \log 1 - \log \sin \theta$ { property of logarithms }

$\qquad\qquad = 0 - \log \sin \theta$ { $\log 1 = 0$ }

$\qquad\qquad = -\log \sin \theta$

63 $\ln I_0 - \ln I = kx \sec \theta \Rightarrow \ln\dfrac{I_0}{I} = kx \sec \theta$ { property of logarithms } $\Rightarrow$

$\qquad x = \dfrac{1}{k \sec \theta} \ln\dfrac{I_0}{I}$ { solve for x }

$\qquad\quad = \dfrac{1}{1.88 \sec 12°} \ln 1.72$ { substitute and approximate } ≈ 0.28 cm.

65 (a) Since $\cos \theta$ and $\sin \phi$ are both less than or equal to 1, $R = R_0 \cos \theta \sin \phi$, the solar
radiation R will equal its maximum R_0 when $\cos \theta = \sin \phi = 1$. This occurs when
$\theta = 0°$ and $\phi = 90°$, and corresponds to when the sun is just rising in the east.

(b) The sun located in the southeast corresponds to $\phi = 45°$.

$$\text{percentage of } R_0 = \frac{\text{amount of } R_0}{R_0} = \frac{R_0 \cos 60° \sin 45°}{R_0} = \frac{1}{2} \cdot \frac{\sqrt{2}}{2} = \frac{\sqrt{2}}{4} \approx 35\%.$$

6.3 Exercises

Note: The missing values are found in terms of the given values.

         We could also use proportions to find the remaining parts.

1   Since α is given and $\gamma = 90°$, we can easily find β. $\beta = 90° - \alpha = 90° - 30° = 60°$.

To find a, we will relate it to the given parts, α and b, using the tangent function.

$$\tan \alpha = \tfrac{a}{b} \Rightarrow a = b \tan \alpha = 20 \tan 30° = 20(\tfrac{1}{3}\sqrt{3}) = \tfrac{20}{3}\sqrt{3}.$$

$$\sec \alpha = \tfrac{c}{b} \Rightarrow c = b \sec \alpha = 20 \sec 30° = 20(\tfrac{2}{3}\sqrt{3}) = \tfrac{40}{3}\sqrt{3}.$$

3 $\alpha = 90° - \beta = 90° - 45° = 45°$.

$\qquad \cos \beta = \tfrac{a}{c} \Rightarrow a = c \cos \beta = 30 \cos 45° = 30(\tfrac{1}{2}\sqrt{2}) = 15\sqrt{2}.$ $b = a$ in a $45° - 45° - 90°$ $\triangle$.

5 $\tan \alpha = \tfrac{a}{b} = \tfrac{5}{5} = 1 \Rightarrow \alpha = 45°$. $\beta = 90° - \alpha = 90° - 45° = 45°$.   Using the

$\qquad$ Pythagorean theorem, $a^2 + b^2 = c^2 \Rightarrow c = \sqrt{a^2 + b^2} = \sqrt{25 + 25} = \sqrt{50} = 5\sqrt{2}$.

7 $\cos \alpha = \tfrac{b}{c} = \dfrac{5\sqrt{3}}{10\sqrt{3}} = \tfrac{1}{2} \Rightarrow \alpha = 60°$. $\beta = 90° - \alpha = 90° - 60° = 30°$.

$$a = \sqrt{c^2 - b^2} = \sqrt{300 - 75} = \sqrt{225} = 15.$$

9 $\beta = 90° - \alpha = 90° - 37° = 53°$. $\tan \alpha = \tfrac{a}{b} \Rightarrow a = b \tan \alpha = 24 \tan 37° \approx 18$.

$$\sec \alpha = \tfrac{c}{b} \Rightarrow c = b \sec \alpha = 24 \sec 37° \approx 30.$$

$\boxed{11}$ $\alpha = 90° - \beta = 90° - 71°51' = 18°9'$. $\cot \beta = \frac{a}{b} \Rightarrow a = b \cot \beta = 240.0 \cot 71°51' \approx 78.7$.

$$\csc \beta = \frac{c}{b} \Rightarrow c = b \csc \beta = 240.0 \csc 71°51' \approx 252.6.$$

$\boxed{13}$ $\tan \alpha = \frac{a}{b} = \frac{25}{45} \Rightarrow \alpha = \tan^{-1} \frac{25}{45} \approx 29°$. $\beta = 90° - \alpha \approx 90° - 29° = 61°$.

$$c = \sqrt{a^2 + b^2} = \sqrt{25^2 + 45^2} = \sqrt{625 + 2025} = \sqrt{2650} \approx 51.$$

$\boxed{15}$ $\cos \alpha = \frac{b}{c} = \frac{2.1}{5.8} \Rightarrow \alpha = \cos^{-1} \frac{21}{58} \approx 69°$. $\beta = 90° - \alpha \approx 90° - 69° = 21°$.

$$a = \sqrt{c^2 - b^2} = \sqrt{(5.8)^2 - (2.1)^2} = \sqrt{33.64 - 4.41} = \sqrt{29.23} \approx 5.4.$$

Note: Refer to Figures 17 and 18 in the text for the labeling of the sides and angles.

$\boxed{17}$ We need to find a relationship involving b, c, and α. We want angle α with its adjacent side b and hypotenuse c. The cosine or secant are the functions of α that involve b and c. We choose the cosine since it is easier to solve for b { b is in the numerator }. $\cos \alpha = \frac{b}{c} \Rightarrow b = c \cos \alpha$.

$\boxed{19}$ We want angle β with its adjacent side a and opposite side b. The tangent or cotangent are the functions of β that involve a and b. $\cot \beta = \frac{a}{b} \Rightarrow a = b \cot \beta$.

$\boxed{21}$ We want angle α with its opposite side a and hypotenuse c. The sine or cosecant are the functions of α that involve a and c. $\csc \alpha = \frac{c}{a} \Rightarrow c = a \csc \alpha$.

$\boxed{23}$ $a^2 + b^2 = c^2 \Rightarrow b^2 = c^2 - a^2 \Rightarrow b = \sqrt{c^2 - a^2}$

$\boxed{25}$ Let h denote the height of the tree.

$$\tan \theta = \frac{\text{opp}}{\text{adj}} \Rightarrow \tan 60° = \frac{h}{200} \Rightarrow h = 200 \tan 60° = 200\sqrt{3} \approx 346.4 \text{ ft.}$$

$\boxed{27}$ Let h denote the height of the kite and $x = h - 4$. $\sin 60° = \frac{x}{500} \Rightarrow$

$$x = 500 \sin 60° = 500(\tfrac{1}{2}\sqrt{3}) = 250\sqrt{3}. \quad h = x + 4 = 250\sqrt{3} + 4 \approx 437 \text{ ft.}$$

$\boxed{29}$ $\sin 10° = \frac{5000}{x} \Rightarrow x = \frac{5000}{\sin 10°} \Rightarrow$

$x = 5000 \csc 10° \approx 28{,}793.85$, or $28{,}800$ ft.

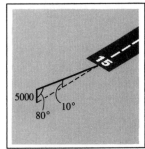

Figure 29

$\boxed{33}$ The 10,000 feet would represent the hypotenuse in a triangle depicting this

information. Let h denote the altitude. $\sin 75° = \frac{h}{10{,}000} \Rightarrow h \approx 9659$ ft.

35 (a) The bridge section is 75 feet long. Using the right triangle with the 75 foot section as its hypotenuse and $(d-15)$ as the side opposite the 35° angle, we have

$$\sin 35° = \frac{d-15}{75} \Rightarrow d = 75\sin 35° + 15 \approx 58 \text{ ft.}$$

(b) Let x be the horizontal distance from the end of a bridge section to a point directly underneath the end of the section. $\cos 35° = \frac{x}{75} \Rightarrow x = 75\cos 35°$.

The distance between the ends of the two sections is (total distance) −

(the 2 horizontal distances under the bridge sections) $= 150 - 2x \approx 27 \text{ ft.}$

39 Let D denote the position of the duck and t the number of seconds required for a direct hit. The duck will move $(7t)$ cm. and the bullet will travel $(25t)$ cm.

$$\sin \varphi = \frac{\overline{AD}}{\overline{OD}} = \frac{7t}{25t} \Rightarrow \sin \varphi = \frac{7}{25} \Rightarrow \varphi \approx 16.3°.$$

41 The central angle of a section of the Pentagon has measure $\frac{360°}{5} = 72°$.

Bisecting that angle, we have an angle of 36° whose opposite side is $\frac{921}{2}$.

The height h is given by $\tan 36° = \frac{\frac{921}{2}}{h} \Rightarrow h = \frac{921}{2\tan 36°}$.

$$\text{Area} = 5(\tfrac{1}{2}bh) = 5(\tfrac{1}{2})(921)\left(\frac{921}{2\tan 36°}\right) \approx 1{,}459{,}379 \text{ ft}^2.$$

43 The diagonal of the base is $\sqrt{8^2 + 6^2} = 10$. $\tan \theta = \frac{4}{10} \Rightarrow \theta \approx 21.8°$

45 $\cot 53°30' = \frac{x}{h} \Rightarrow x = h\cot 53°30'$. $\cot 26°50' = \frac{x+25}{h} \Rightarrow$

$x + 25 = h\cot 26°50' \Rightarrow x = h\cot 26°50' - 25$.

We now have two expressions for x.

We will set these equal to each other and solve for h.

Thus, $h\cot 53°30' = h\cot 26°50' - 25 \Rightarrow$

$25 = h\cot 26°50' - h\cot 53°30' \Rightarrow$

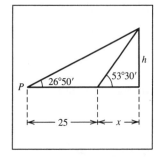

Figure 45

$h = \dfrac{25}{\cot 26°50' - \cot 53°30'} \approx 20.2 \text{ m.}$

47 When the angle of elevation is 19°20', $\tan 19°20' = \frac{h_1}{110} \Rightarrow h_1 = 110\tan 19°20'$.

When the angle of elevation is 31°50', $\tan 31°50' = \frac{h_2}{110} \Rightarrow h_2 = 110\tan 31°50'$.

The change in elevation is $h_2 - h_1 \approx 68.29 - 38.59 = 29.7 \text{ km.}$

49 The distance from the spacelab to the center of the earth is $(380 + r)$ miles.

$$\sin 65.8° = \frac{r}{r+380} \Rightarrow r \sin 65.8° + 380 \sin 65.8° = r \Rightarrow r - r \sin 65.8° = 380 \sin 65.8° \Rightarrow$$

$$r(1 - \sin 65.8°) = 380 \sin 65.8° \Rightarrow r = \frac{380 \sin 65.8°}{1 - \sin 65.8°} \approx 3944 \text{ mi}.$$

51 Let d be the distance traveled. $\tan 42° = \frac{10{,}000}{d} \Rightarrow d = 10{,}000 \cot 42°$.

Converting to mi/hr, we have $\dfrac{10{,}000 \cot 42° \text{ ft}}{1 \text{ minute}} \cdot \dfrac{60 \text{ minutes}}{1 \text{ hour}} \cdot \dfrac{1 \text{ mile}}{5280 \text{ ft}} \approx 126$ mi/hr.

53 (a) As in Exercise 49, there is a right angle formed on the earth's surface.

Bisecting angle θ and forming a right triangle, we have

$$\cos \frac{\theta}{2} = \frac{R}{R+a} = \frac{4000}{26{,}300}. \text{ Thus, } \frac{\theta}{2} \approx 81.25° \Rightarrow \theta \approx 162.5°.$$

The percentage of the equator that is within signal range is $\frac{162.5°}{360°} \times 100 \approx 45\%$.

(b) Each satellite has a signal range of more than $120°$,

and thus all 3 will cover all points on the equator.

55 Let $x = h - c$. $\sin \alpha = \frac{x}{d} \Rightarrow x = d \sin \alpha$. $h = x + c = d \sin \alpha + c$.

57 Let x denote the distance from the base of the tower to the closer point.

$$\cot \beta = \frac{x}{h} \Rightarrow x = h \cot \beta. \quad \cot \alpha = \frac{x+d}{h} \Rightarrow x + d = h \cot \alpha \Rightarrow x = h \cot \alpha - d.$$

Thus, $h \cot \beta = h \cot \alpha - d \Rightarrow d = h \cot \alpha - h \cot \beta \Rightarrow d = h(\cot \alpha - \cot \beta) \Rightarrow$

$$h = \frac{d}{\cot \alpha - \cot \beta}.$$

59 When the angle of elevation is α, $\tan \alpha = \frac{h_1}{d} \Rightarrow h_1 = d \tan \alpha$.

When the angle of elevation is β, $\tan \beta = \frac{h_2}{d} \Rightarrow h_2 = d \tan \beta$.

$$h = h_2 - h_1 = d \tan \beta - d \tan \alpha = d(\tan \beta - \tan \alpha).$$

61 The bearing from P to A is $90° - 20° = 70°$ east of north and is denoted by N70°E.

The bearing from P to B is $40°$ west of north and is denoted by N40°W.

The bearing from P to C is $90° - 75° = 15°$ west of south and is denoted by S15°W.

The bearing from P to D is $25°$ east of south and is denoted by S25°E.

63 (a) The first ship travels 2 hours @ 24 mi/hr for a

distance of 48 miles. The second ship travels

$1\frac{1}{2}$ hours @ 18 mi/hr for a distance of 27 miles.

The paths form a right triangle with legs of 48 miles

and 27 miles. The distance between the two ships is

$$\sqrt{27^2 + 48^2} \approx 55 \text{ miles}.$$

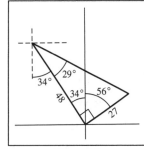

Figure 63

(b) The angle between the side of length 48 and the

hypotenuse is found by solving $\tan\alpha = \frac{27}{48}$ for α. Now

$\alpha \approx 29°$, so the second ship is approximately $29° + 34° = $ S63°E of the first ship.

$\boxed{65}$ 30 minutes @ 360 mi/hr = 180 miles.   45 minutes @ 360 mi/hr = 270 miles.

$227° - 137° = 90°$, and hence, the plane's flight forms a right triangle with legs 180

miles and 270 miles. The distance from A to the airplane is equal to the hypotenuse

of the triangle—that is, $\sqrt{180^2 + 270^2} \approx 324.5$ mi.

6.4 Exercises

$\boxed{1}$ For the point $P(x, y)$, we let r denote the distance from the origin to P. Applying

the definition of trigonometric functions of any angle with $x = 4$, $y = -3$, and

$r = \sqrt{x^2 + y^2} = \sqrt{4^2 + (-3)^2} = 5$, we obtain the following:

$$\sin\theta = \frac{y}{r} = \frac{-3}{5} = -\frac{3}{5} \qquad\qquad \cos\theta = \frac{x}{r} = \frac{4}{5}$$

$$\tan\theta = \frac{y}{x} = \frac{-3}{4} = -\frac{3}{4} \qquad\qquad \cot\theta = \frac{x}{y} = \frac{4}{-3} = -\frac{4}{3}$$

$$\sec\theta = \frac{r}{x} = \frac{5}{4} \qquad\qquad \csc\theta = \frac{r}{y} = \frac{5}{-3} = -\frac{5}{3}$$

$\boxed{3}$ $x = -2$ and $y = -5 \Rightarrow r = \sqrt{x^2 + y^2} = \sqrt{(-2)^2 + (-5)^2} = \sqrt{29}$.

$$\sin\theta = \frac{y}{r} = \frac{-5}{\sqrt{29}}, \text{ or } -\frac{5}{29}\sqrt{29} \qquad \cos\theta = \frac{x}{r} = \frac{-2}{\sqrt{29}}, \text{ or } -\frac{2}{29}\sqrt{29}$$

$$\tan\theta = \frac{y}{x} = \frac{-5}{-2} = \frac{5}{2} \qquad\qquad \cot\theta = \frac{x}{y} = \frac{-2}{-5} = \frac{2}{5}$$

$$\sec\theta = \frac{r}{x} = \frac{\sqrt{29}}{-2}, \text{ or } -\frac{1}{2}\sqrt{29} \qquad \csc\theta = \frac{r}{y} = \frac{\sqrt{29}}{-5} = -\frac{1}{5}\sqrt{29}$$

Note: In the following exercises, we will only find the values of x, y, and r. The above

definitions can then be used to find the values of the trigonometric functions of θ.

These values are listed in the usual order in the answer.

$\boxed{5}$ Since the terminal side of θ is in QII, choose x to be negative.

If $x = -1$, then $y = 4$ and $(-1, 4)$ is a point on the terminal side of θ.

$x = -1$ and $y = 4 \Rightarrow r = \sqrt{(-1)^2 + 4^2} = \sqrt{17}$.   ★ $\frac{4}{\sqrt{17}}, -\frac{1}{\sqrt{17}}, -4, -\frac{1}{4}, -\sqrt{17}, \frac{\sqrt{17}}{4}$

$\boxed{7}$ Remember that $y = mx$ is an equation of a line that passes through the origin and

has slope m. Thus, an equation of the line is $y = \frac{4}{3}x$.

If $x = 3$, then $y = 4$ and $(3, 4)$ is a point on the terminal side of θ.

$x = 3$ and $y = 4 \Rightarrow r = \sqrt{3^2 + 4^2} = 5$.   ★ $\frac{4}{5}, \frac{3}{5}, \frac{4}{3}, \frac{3}{4}, \frac{5}{3}, \frac{5}{4}$

⑨ $2y - 7x + 2 = 0 \Leftrightarrow y = \frac{7}{2}x - 1$. Thus, the slope of the given line is $\frac{7}{2}$.

An equation of the line through the origin with that slope is $y = \frac{7}{2}x$.

If $x = -2$, then $y = -7$ and $(-2, -7)$ is a point on the terminal side of θ.

$x = -2$ and $y = -7 \Rightarrow r = \sqrt{(-2)^2 + (-7)^2} = \sqrt{53}$.

$$\bigstar \; -\frac{7}{\sqrt{53}}, \; -\frac{2}{\sqrt{53}}, \; \frac{7}{2}, \; \frac{2}{7}, \; -\frac{\sqrt{53}}{2}, \; -\frac{\sqrt{53}}{7}$$

⑪ *Note:* U denotes *undefined.*

(a) For $\theta = 90°$, choose $x = 0$ and $y = 1$. $r = 1$. $\bigstar$ 1, 0, U, 0, U, 1

(b) For $\theta = 0°$, choose $x = 1$ and $y = 0$. $r = 1$. $\bigstar$ 0, 1, 0, U, 1, U

(c) For $\theta = \frac{7\pi}{2}$, choose $x = 0$ and $y = -1$. $r = 1$. $\bigstar$ −1, 0, U, 0, U, −1

(d) For $\theta = 3\pi$, choose $x = -1$ and $y = 0$. $r = 1$. $\bigstar$ 0, −1, 0, U, −1, U

⑬ (a) $\cos\theta > 0 \; \{x > 0\}$ implies that the terminal side of θ is in QI or QIV.

   $\sin\theta < 0 \; \{y < 0\}$ implies that the terminal side of θ is in QIII or QIV.

   Thus, θ must be in quadrant IV to satisfy both conditions.

(b) $\sin\theta < 0 \Rightarrow \theta$ is in QIII or QIV. $\cot\theta > 0 \Rightarrow \theta$ is in QI or QIII. $\therefore \theta$ is in QIII.

(c) $\csc\theta > 0 \Rightarrow \theta$ is in QI or QII. $\sec\theta < 0 \Rightarrow \theta$ is in QII or QIII. $\therefore \theta$ is in QII.

(d) $\sec\theta < 0 \Rightarrow \theta$ is in QII or QIII. $\tan\theta > 0 \Rightarrow \theta$ is in QI or QIII. $\therefore \theta$ is in QIII.

Note: Let θ_C denote the coterminal angle of θ such that $0° \le \theta_C < 360°$ { or $0 \le \theta_C < 2\pi$ }.

The following formulas are then used in the solutions.

(1) If θ_C is in QI, then $\theta_R = \theta_C$.

(2) If θ_C is in QII, then $\theta_R = 180° - \theta_C$ { or $\pi - \theta_C$ }.

(3) If θ_C is in QIII, then $\theta_R = \theta_C - 180°$ { or $\theta_C - \pi$ }.

(4) If θ_C is in QIV, then $\theta_R = 360° - \theta_C$ { or $2\pi - \theta_C$ }.

⑮ (a) Since $240°$ is in QIII, $\theta_R = 240° - 180° = 60°$.

(b) Since $340°$ is in QIV, $\theta_R = 360° - 340° = 20°$.

(c) $\theta_C = -202° + 1(360°) = 158° \in$ QII. $\theta_R = 180° - 158° = 22°$.

(d) $\theta_C = -660° + 2(360°) = 60° \in$ QI. $\theta_R = 60°$.

⑰ (a) Since $\frac{3\pi}{4}$ is in QII, $\theta_R = \pi - \frac{3\pi}{4} = \frac{\pi}{4}$.

(b) Since $\frac{4\pi}{3}$ is in QIII, $\theta_R = \frac{4\pi}{3} - \pi = \frac{\pi}{3}$.

(c) $\theta_C = -\frac{\pi}{6} + 1(2\pi) = \frac{11\pi}{6} \in$ QIV. $\theta_R = 2\pi - \frac{11\pi}{6} = \frac{\pi}{6}$.

(d) $\theta_C = \frac{9\pi}{4} - 1(2\pi) = \frac{\pi}{4} \in$ QI. $\theta_R = \frac{\pi}{4}$.

⑲ (a) Since $\frac{\pi}{2} < 3 < \pi$, θ is in QII and $\theta_R = \pi - 3 \approx 0.14$, or $8.1°$.

(b) $\theta_C = -2 + 1(2\pi) = 2\pi - 2 \approx 4.28$.

   Since $\pi < 4.28 < \frac{3\pi}{2}$, θ_C is in QIII and $\theta_R = (2\pi - 2) - \pi = \pi - 2 \approx 1.14$, or $65.4°$.

(c) Since $\frac{3\pi}{2} < 5.5 < 2\pi$, θ is in QIV and $\theta_R = 2\pi - 5.5 \approx 0.78$, or $44.9°$.

(d) The number of revolutions formed by θ is $\frac{100}{2\pi} \approx 15.92$, so

$\theta_C = 100 - 15(2\pi) = 100 - 30\pi \approx 5.75$. Since $\frac{3\pi}{2} < 5.75 < 2\pi$,

θ_C is in QIV and $\theta_R = 2\pi - (100 - 30\pi) = 32\pi - 100 \approx 0.53$, or $30.4°$.

Alternatively, if your calculator is capable of computing trigonometric functions

of large values, then computing $\sin^{-1}(\sin 100) \approx -0.53 \Rightarrow \theta_R = 0.53$, or $30.4°$.

Note: For the following problems, we use the theorem on reference angles before

evaluating.

21 (a) $\sin\frac{2\pi}{3} = \sin\frac{\pi}{3}$ { since the sine is positive in QII

and $\frac{\pi}{3}$ is the reference angle for $\frac{2\pi}{3}$ } $= \frac{\sqrt{3}}{2}$

(b) $\sin\left(-\frac{5\pi}{4}\right) = \sin\frac{3\pi}{4}$ { since $\frac{3\pi}{4}$ is coterminal with $-\frac{5\pi}{4}$ } $=$

$\sin\frac{\pi}{4}$ { since the sine is positive in QII and $\frac{\pi}{4}$ is the reference angle for $\frac{3\pi}{4}$ } $= \frac{\sqrt{2}}{2}$

23 (a) $\cos 150° = -\cos 30°$ { since the cosine is negative in QII } $= -\frac{\sqrt{3}}{2}$

(b) $\cos(-60°) = \cos 300° = \cos 60°$ { since the cosine is positive in QIV } $= \frac{1}{2}$

25 (a) $\tan\frac{5\pi}{6} = -\tan\frac{\pi}{6}$ { since the tangent is negative in QII } $= -\frac{\sqrt{3}}{3}$

(b) $\tan\left(-\frac{\pi}{3}\right) = \tan\frac{5\pi}{3} = -\tan\frac{\pi}{3}$ { since the tangent is negative in QIV } $= -\sqrt{3}$

27 (a) $\cot 120° = -\cot 60°$ { since the cotangent is negative in QII } $= -\frac{\sqrt{3}}{3}$

(b) $\cot(-150°) = \cot 210° = \cot 30°$ { since the cotangent is positive in QIII } $= \sqrt{3}$

29 (a) $\sec\frac{2\pi}{3} = -\sec\frac{\pi}{3}$ { since the secant is negative in QII } $= -2$

(b) $\sec\left(-\frac{\pi}{6}\right) = \sec\frac{11\pi}{6} = \sec\frac{\pi}{6}$ { since the secant is positive in QIV } $= \frac{2}{\sqrt{3}}$

31 (a) $\csc 240° = -\csc 60°$ { since the cosecant is negative in QIII } $= -\frac{2}{\sqrt{3}}$

(b) $\csc(-330°) = \csc 30° = 2$

33 (a) $\sin 98°10' \approx 0.9899$ (b) $\cos 623.7° \approx -0.1097$ (c) $\tan 3 \approx -0.1425$

(d) $\cot 231°40' \approx 0.7907$ (e) $\sec 1175.1° \approx -11.2493$ (f) $\csc 0.82 \approx 1.3677$

35 (a) Use the degree mode. $\sin\theta = -0.5640 \Rightarrow \theta = \sin^{-1}(-0.5640) \approx -34.3° \Rightarrow$

$\theta_R \approx 34.3°$. Since the sine is negative in QIII and QIV, we use θ_R in those

quadrants. $180° + 34.3° = \underline{214.3°}$ and $360° - 34.3° = \underline{325.7°}$

(b) $\cos\theta = 0.7490 \Rightarrow \theta = \cos^{-1}(0.7490) \approx 41.5°$. $\theta_R \approx 41.5°$, QI: $41.5°$, QIV: $318.5°$

(c) $\tan\theta = 2.798 \Rightarrow \theta = \tan^{-1}(2.798) \approx 70.3°$. $\theta_R \approx 70.3°$, QI: $70.3°$, QIII: $250.3°$

(d) $\cot\theta = -0.9601 \Rightarrow \tan\theta = -\frac{1}{0.9601} \Rightarrow \theta = \tan^{-1}(-\frac{1}{0.9601}) \approx -46.2°$.

$\theta_R \approx 46.2°$, QII: $133.8°$, QIV: $313.8°$

(e) $\sec\theta = -1.116 \Rightarrow \cos\theta = -\frac{1}{1.116} \Rightarrow \theta = \cos^{-1}(-\frac{1}{1.116}) \approx 153.6°$.

$\theta_R \approx 180° - 153.6° = 26.4°$, QII: $153.6°$, QIII: $206.4°$

(f) $\csc\theta = 1.485 \Rightarrow \sin\theta = \frac{1}{1.485} \Rightarrow \theta = \sin^{-1}(\frac{1}{1.485}) \approx 42.3°$.

$\theta_R \approx 42.3°$, QI: $42.3°$, QII: $137.7°$

$\boxed{37}$ (a) Use the radian mode. $\sin\theta = 0.4195 \Rightarrow \theta = \sin^{-1}(0.4195) \approx 0.43$.

$\theta_R \approx 0.43$ is one answer. Since the sine is positive in QI and QII,

we also use the reference angle for θ in quadrant II. QII: $\pi - 0.43 \approx 2.71$

(b) $\cos\theta = -0.1207 \Rightarrow \theta = \cos^{-1}(-0.1207) \approx 1.69$ is one answer. Since 1.69 is in QII,

$\theta_R \approx \pi - 1.69 \approx 1.45$. The cosine is also negative in QIII. QIII: $\pi + 1.45 \approx 4.59$

(c) $\tan\theta = -3.2504 \Rightarrow \theta = \tan^{-1}(-3.2504) \approx -1.27 \Rightarrow \theta_R \approx 1.27$.

QII: $\pi - 1.27 \approx 1.87$, QIV: $2\pi - 1.27 \approx 5.01$

(d) $\cot\theta = 2.6815 \Rightarrow \tan\theta = \frac{1}{2.6815} \Rightarrow \theta = \tan^{-1}(\frac{1}{2.6815}) \approx 0.36 \Rightarrow$

$\theta_R \approx 0.36$ is one answer. QIII: $\pi + 0.36 \approx 3.50$

(e) $\sec\theta = 1.7452 \Rightarrow \cos\theta = \frac{1}{1.7452} \Rightarrow \theta = \cos^{-1}(\frac{1}{1.7452}) \approx 0.96 \Rightarrow$

$\theta_R \approx 0.96$ is one answer. QIV: $2\pi - 0.96 \approx 5.32$

(f) $\csc\theta = -4.8521 \Rightarrow \sin\theta = -\frac{1}{4.8521} \Rightarrow \theta = \sin^{-1}(-\frac{1}{4.8521}) \approx -0.21 \Rightarrow \theta_R \approx 0.21$.

QIII: $\pi + 0.21 \approx 3.35$, QIV: $2\pi - 0.21 \approx 6.07$

$\boxed{39}$ $\sqrt{\sec^2\theta - 1} = \sqrt{\tan^2\theta}$ { Pythagorean identity } $= |\tan\theta|$ $\left\{ \sqrt{x^2} = |x| \right\} =$

$-\tan\theta$ since $\tan\theta < 0$ if $\pi/2 < \theta < \pi$.

$\boxed{41}$ $\sqrt{1 + \tan^2\theta} = \sqrt{\sec^2\theta}$ { Pythagorean identity } $= |\sec\theta|$ $\left\{ \sqrt{x^2} = |x| \right\} =$

$\sec\theta$ since $\sec\theta > 0$ if $3\pi/2 < \theta < 2\pi$.

$\boxed{43}$ $\sqrt{\sin^2(\theta/2)} = |\sin(\theta/2)|$ $\left\{ \sqrt{x^2} = |x| \right\} =$

$-\sin(\theta/2)$ since $\sin(\theta/2) < 0$ if $2\pi < \theta < 4\pi$ { $\pi < \theta/2 < 2\pi$ }.

$\boxed{45}$ $\sin\theta = \frac{b}{c} \Rightarrow \sin 60° = \frac{b}{18} \Rightarrow b = 18\sin 60° = 18 \cdot \frac{\sqrt{3}}{2} = 9\sqrt{3} \approx 15.6$.

$\cos\theta = \frac{a}{c} \Rightarrow \cos 60° = \frac{a}{18} \Rightarrow a = 18\cos 60° = 18 \cdot \frac{1}{2} = 9$.

The hand is located at $(9, 9\sqrt{3})$.

6.5 Exercises

$\boxed{1}$ (a) $\sin(-30°) = -\sin 30°$ { since $\sin(-t) = -\sin t$ } $= -\frac{1}{2}$

(b) $\cos(-\frac{3\pi}{4}) = \cos\frac{3\pi}{4}$ { since $\cos(-t) = \cos t$ } $= -\frac{\sqrt{2}}{2}$

(c) $\tan(-45°) = -\tan 45°$ { since $\tan(-t) = -\tan t$ } $= -1$

$\boxed{3}$ (a) $\cot(-\frac{3\pi}{4}) = -\cot\frac{3\pi}{4}$ { since $\cot(-t) = -\cot t$ } $= -(-1) = 1$

(b) $\sec(-150°) = \sec 150°$ { since $\sec(-t) = \sec t$ } $= -\sec 30° = -\frac{2}{\sqrt{3}}$

(c) $\csc(-\frac{7\pi}{6}) = -\csc\frac{7\pi}{6}$ { since $\csc(-t) = -\csc t$ } $= -(-\csc\frac{\pi}{6}) = 2$

$\boxed{5}$ $\sin(-t)\sec(-t) = (-\sin t)\sec t$ { formulas for negatives }

$= (-\sin t)(1/\cos t)$ { reciprocal identity }

$= -\tan t$ { tangent identity }

$\boxed{7}$ $\dfrac{\cot(-t)}{\csc(-t)} = \dfrac{-\cot t}{-\csc t}$ { formulas for negatives }

$\qquad\qquad = \dfrac{\cos t/\sin t}{1/\sin t}$ { cotangent identity and reciprocal identity }

$\qquad\qquad = \cos t$ { simplify }

$\boxed{9}$ $\dfrac{1}{\cos(-t)} - \tan(-t)\sin(-t) = \dfrac{1}{\cos t} - (-\tan t)(-\sin t)$ { formulas for negatives }

$\qquad\qquad\qquad\qquad\qquad = \dfrac{1}{\cos t} - \dfrac{\sin t}{\cos t}\sin t$ { tangent identity }

$\qquad\qquad\qquad\qquad\qquad = \dfrac{1 - \sin^2 t}{\cos t}$ { combine terms }

$\qquad\qquad\qquad\qquad\qquad = \dfrac{\cos^2 t}{\cos t}$ { Pythagorean identity }

$\qquad\qquad\qquad\qquad\qquad = \cos t$ { cancel $\cos t$ }

$\boxed{11}$ (a) Using Figure 40 in the text, we see that as t gets close to 0 through numbers greater than 0 (from the *right* of 0), $\sin t$ approaches 0.

(b) As t approaches $-\frac{\pi}{2}$ through numbers less than $-\frac{\pi}{2}$ (from the *left* of $-\frac{\pi}{2}$), $\sin t$ approaches -1.

$\boxed{13}$ (a) Using Figure 42 in the text, we see that as t gets close to $\frac{\pi}{4}$ through numbers greater than $\frac{\pi}{4}$ (from the *right* of $\frac{\pi}{4}$), $\cos t$ approaches $\frac{\sqrt{2}}{2}$. Note that the value $\frac{\sqrt{2}}{2}$ is in the table containing specific values of the cosine function.

(b) As t approaches π through numbers less than π (from the *left* of π), $\cos t$ approaches -1.

$\boxed{15}$ (a) Using Figure 45 in the text, we see that as t gets close to $\frac{\pi}{4}$ through numbers greater than $\frac{\pi}{4}$ (from the *right* of $\frac{\pi}{4}$), $\tan t$ approaches 1.

(b) As t approaches $\frac{\pi}{2}$ through numbers greater than $\frac{\pi}{2}$ (from the *right* of $\frac{\pi}{2}$), $\tan t$ is approaching the vertical asymptote at $t = \frac{\pi}{2}$. $\tan t$ is *decreasing* without bound, and we use the notation $\underline{\tan t \to -\infty}$ to denote this.

$\boxed{17}$ (a) Using Figure 48 in the text, we see that as t gets close to $-\frac{\pi}{4}$ through numbers less than $-\frac{\pi}{4}$ (from the *left* of $-\frac{\pi}{4}$), $\cot t$ approaches -1.

(b) As t approaches 0 through numbers greater than 0 (from the *right* of 0), $\cot t$ is approaching the vertical asymptote at $t = 0$ { the y-axis }. $\cot t$ is *increasing* without bound, and we use the notation $\underline{\cot t \to \infty}$ to denote this.

$\boxed{19}$ (a) Using Figure 47 in the text, we see that as t gets close to $\frac{\pi}{2}$ through numbers less than $\frac{\pi}{2}$ (from the *left* of $\frac{\pi}{2}$), $\sec t \to \infty$.

(b) As t approaches $\frac{\pi}{4}$ through numbers greater than $\frac{\pi}{4}$ (from the *right* of $\frac{\pi}{4}$), $\sec t$ approaches $\sqrt{2}$. Recall that $\cos\frac{\pi}{4} = \frac{1}{\sqrt{2}}$ and that $\sec t = \frac{1}{\cos t}$.

$\boxed{21}$ (a) Using Figure 46 in the text, we see that as t gets close to 0 through numbers less than 0 (from the *left* of 0), $\csc t \to -\infty$.

(b) As t approaches $\frac{\pi}{2}$ through numbers greater than $\frac{\pi}{2}$ (from the *right* of $\frac{\pi}{2}$), $\csc t$ approaches 1.

$\boxed{23}$ Refer to Figure 40 and the accompanying table. We see that $\sin\frac{3\pi}{2} = -1$.

Since the period of the sine is 2π, the second value in $[0, 4\pi]$ is $\frac{3\pi}{2} + 2\pi = \frac{7\pi}{2}$.

$\boxed{25}$ Recall that $\sin\frac{\pi}{6} = \frac{1}{2}$ and $\sin\frac{5\pi}{6} = \frac{1}{2}$. Since the period of the sine is 2π,

other values in $[0, 4\pi]$ are $\frac{\pi}{6} + 2\pi = \frac{13\pi}{6}$ and $\frac{5\pi}{6} + 2\pi = \frac{17\pi}{6}$.

$\boxed{27}$ Refer to Figure 42 and the accompanying table. We see that $\cos 0 = \cos 2\pi = 1$.

Since the period of the cosine is 2π, the other value in $[0, 4\pi]$ is $2\pi + 2\pi = 4\pi$.

$\boxed{29}$ Refer to Figure 42 and the accompanying table. We see that $\cos\frac{\pi}{4} = \cos\frac{7\pi}{4} = \frac{\sqrt{2}}{2}$.

Since the period of the cosine is 2π,

other values in $[0, 4\pi]$ are $\frac{\pi}{4} + 2\pi = \frac{9\pi}{4}$ and $\frac{7\pi}{4} + 2\pi = \frac{15\pi}{4}$.

$\boxed{31}$ Refer to Figure 45. In the interval $(-\frac{\pi}{2}, \frac{\pi}{2})$, $\tan t = 1$ only if $t = \frac{\pi}{4}$. Since the period of the tangent is π, the desired value in the interval $(\frac{\pi}{2}, \frac{3\pi}{2})$ is $\frac{\pi}{4} + \pi = \frac{5\pi}{4}$.

$\boxed{33}$ Refer to Figure 45. In the interval $(-\frac{\pi}{2}, \frac{\pi}{2})$, $\tan t = 0$ only if $t = 0$. Since the period of the tangent is π, the desired value in the interval $(\frac{\pi}{2}, \frac{3\pi}{2})$ is $0 + \pi = \pi$.

$\boxed{35}$ $y = \sin t$; $[-2\pi, 2\pi]$; $a = \frac{1}{2}$ • Refer to Figure 40. $\sin t = \frac{1}{2} \Rightarrow t = \frac{\pi}{6}$ and $\frac{5\pi}{6}$.

Also, $\frac{\pi}{6} - 2\pi = -\frac{11\pi}{6}$ and $\frac{5\pi}{6} - 2\pi = -\frac{7\pi}{6}$. $\sin t > \frac{1}{2}$ when the graph is *above* the horizontal line $y = \frac{1}{2}$. $\sin t < \frac{1}{2}$ when the graph is *below* the horizontal line $y = \frac{1}{2}$.

★ (a) $-\frac{11\pi}{6}, -\frac{7\pi}{6}, \frac{\pi}{6}, \frac{5\pi}{6}$ (b) $-\frac{11\pi}{6} < t < -\frac{7\pi}{6}$ and $\frac{\pi}{6} < t < \frac{5\pi}{6}$

(c) $-2\pi \le t < -\frac{11\pi}{6}$, $-\frac{7\pi}{6} < t < \frac{\pi}{6}$, and $\frac{5\pi}{6} < t \le 2\pi$

$\boxed{37}$ $y = \cos t$; $[-2\pi, 2\pi]$; $a = -\frac{1}{2}$ • Refer to Figure 42. $\cos t = -\frac{1}{2} \Rightarrow t = \frac{2\pi}{3}$ and $\frac{4\pi}{3}$.

Also, $\frac{2\pi}{3} - 2\pi = -\frac{4\pi}{3}$ and $\frac{4\pi}{3} - 2\pi = -\frac{2\pi}{3}$. $\cos t > -\frac{1}{2}$ when the graph is *above* the horizontal line $y = -\frac{1}{2}$. $\cos t < -\frac{1}{2}$ when the graph is *below* the horizontal line $y = -\frac{1}{2}$.

★ (a) $-\frac{4\pi}{3}, -\frac{2\pi}{3}, \frac{2\pi}{3}, \frac{4\pi}{3}$ (b) $-2\pi \le t < -\frac{4\pi}{3}$, $-\frac{2\pi}{3} < t < \frac{2\pi}{3}$, and $\frac{4\pi}{3} < t \le 2\pi$

(c) $-\frac{4\pi}{3} < t < -\frac{2\pi}{3}$ and $\frac{2\pi}{3} < t < \frac{4\pi}{3}$

39 $y = 2 + \sin t$ •   Shift $y = \sin t$ up 2 units.

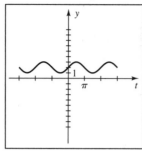

Figure 39

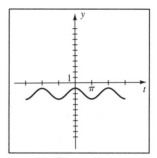

Figure 41

41 $y = \cos t - 2$ •   Shift $y = \cos t$ down 2 units.

43 $y = 1 + \tan t$ •   Shift $y = \tan t$ up 1 unit.

Since $1 + \tan\left(-\frac{\pi}{4}\right) = 1 + (-1) = 0$, and the period of the tangent is π,

there are t-intercepts at $t = -\frac{\pi}{4} + \pi n$.

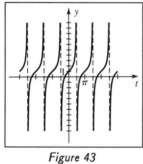

Figure 43

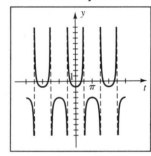

Figure 45

45 $y = \sec t - 2$ •

Shift $y = \sec t$ down 2 units, t-intercepts are at $t = \frac{\pi}{3} + 2\pi n, \frac{5\pi}{3} + 2\pi n$.

47 (a) As we move from left to right, the function increases { goes up } on the

intervals $[-2\pi, -\frac{3\pi}{2})$, $(-\frac{3\pi}{2}, -\pi]$, $[0, \frac{\pi}{2})$, $(\frac{\pi}{2}, \pi]$.

(b) As we move from left to right, the function decreases { goes down } on the

intervals $[-\pi, -\frac{\pi}{2})$, $(-\frac{\pi}{2}, 0]$, $[\pi, \frac{3\pi}{2})$, $(\frac{3\pi}{2}, 2\pi]$.

49 (a) The tangent function increases on *all* intervals on which it is defined.

Between -2π and 2π,

these intervals are $[-2\pi, -\frac{3\pi}{2})$, $(-\frac{3\pi}{2}, -\frac{\pi}{2})$, $(-\frac{\pi}{2}, \frac{\pi}{2})$, $(\frac{\pi}{2}, \frac{3\pi}{2})$, and $(\frac{3\pi}{2}, 2\pi]$.

(b) The tangent function is *never* decreasing on any interval for which it is defined.

51 This is good advice.

53 Graph $y = \sin(t^2)$ and $y = 0.5$ on the same coordinate plane. From the graph,

we see that $\sin(t^2)$ assumes the value of 0.5 at $x \approx \pm 0.72, \pm 1.62, \pm 2.61, \pm 2.98$.

$$\text{Xscl} = 0.785 \approx \tfrac{\pi}{4}.$$

$[-3.14, 3.14]$ by $[-2.09, 2.09]$

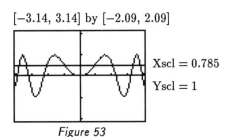

Xscl = 0.785

Yscl = 1

Figure 53

$[-6.28, 6.28]$ by $[-5.19, 3.19]$

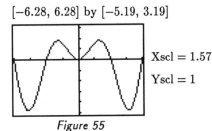

Xscl = 1.57

Yscl = 1

Figure 55

55 We see that the graph of $y = t \sin t$ assumes a maximum value of approximately

$$1.82 \text{ at } t \approx \pm 2.03, \text{ and a minimum value of } -4.81 \text{ at } t \approx \pm 4.91.$$

57 As $t \to 0^+$, $f(t) = \dfrac{1 - \cos t}{t} \to 0$.

$[-1, 1]$ by $[-0.67, 0.67]$

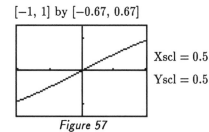

Xscl = 0.5

Yscl = 0.5

Figure 57

$[-2, 2]$ by $[-1.33, 1.33]$

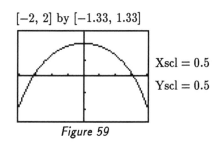

Xscl = 0.5

Yscl = 0.5

Figure 59

59 As $t \to 0^+$, $f(t) = t \cot t \to 1$.

61 As $t \to 0^+$, $f(t) = \dfrac{\tan t}{t} \to 1$.

$[-1.5, 1.5]$ by $[0, 3]$

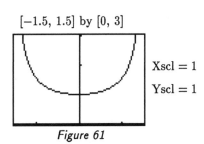

Xscl = 1

Yscl = 1

Figure 61

6.6 Exercises

Note: Exercises 1 & 3: We will refer to $y = \sin x$ as just $\sin x$ ($y = \cos x$ as $\cos x$, etc.).

For the form $y = a \sin bx$, the amplitude is $|a|$ and the period is $\dfrac{2\pi}{|b|}$.

These are merely listed in the answer along with the values of the x-intercepts.

Let n denote any integer.

⚀ (a) $y = 4\sin x$ • Vertically stretch $\sin x$ by a factor of 4. The x-intercepts are not affected by a vertical stretch or compression. ★ 4, 2π, x-int. @ πn

(b) $y = \sin 4x$ • Horizontally compress $\sin x$ by a factor of 4. The x-intercepts are affected by a horizontal stretch or compression by the same factor—that is, a horizontal compression by a factor of k will move the x-intercepts of $\sin x$ from πn to $\frac{\pi}{k}n$, and a horizontal stretch by a factor of k will move the x-intercepts of $\sin x$ from πn to $k\pi n$. ★ 1, $\frac{\pi}{2}$, x-int. @ $\frac{\pi}{4}n$

(c) $y = \frac{1}{4}\sin x$ • Vertically compress $\sin x$ by a factor of 4. ★ $\frac{1}{4}$, 2π, x-int. @ πn

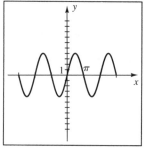

Figure 1(a)

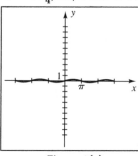

Figure 1(b)

Figure 1(c)

(d) $y = \sin\frac{1}{4}x$ • Horizontally stretch $\sin x$ by a factor of 4. ★ 1, 8π, x-int. @ $4\pi n$

(e) $y = 2\sin\frac{1}{4}x$ • Vertically stretch the graph in part (d) by a factor of 2. ★ 2, 8π, x-int. @ $4\pi n$

(f) $y = \frac{1}{2}\sin 4x$ • Vertically compress the graph in part (b) by a factor of 2. ★ $\frac{1}{2}$, $\frac{\pi}{2}$, x-int. @ $\frac{\pi}{4}n$

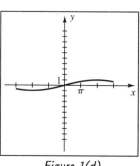

Figure 1(d)

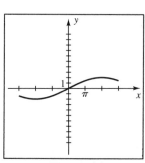

Figure 1(e)

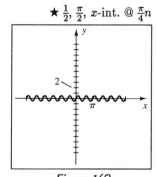

Figure 1(f)

(g) $y = -4\sin x$ • Reflect the graph in part (a) through the x-axis.

★ 4, 2π, x-int. @ πn

(h) $y = \sin(-4x) = -\sin 4x$ using a formula for negatives. Reflect the graph in part

(b) through the x-axis. ★ 1, $\frac{\pi}{2}$, x-int. @ $\frac{\pi}{4}n$

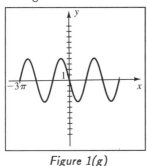

Figure 1(g)

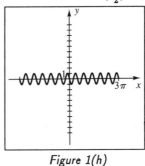

Figure 1(h)

[3] (a) $y = 3\cos x$ • Vertically stretch $\cos x$ by a factor of 3.

★ 3, 2π, x-int. @ $\frac{\pi}{2} + \pi n$

(b) $y = \cos 3x$ • Horizontally compress $\cos x$ by a factor of 3. The x-intercepts are affected by a horizontal stretch or compression by the same factor—that is, horizontal compression by a factor of k will move the x-intercepts of $\cos x$ from $\frac{\pi}{2} + \pi n$ to $\frac{\pi}{2k} + \frac{\pi}{k}n$, and a horizontal stretch by a factor of k will move the x-intercepts of $\cos x$ from $\frac{\pi}{2} + \pi n$ to $\frac{k\pi}{2} + k\pi n$. ★ 1, $\frac{2\pi}{3}$, x-int. @ $\frac{\pi}{6} + \frac{\pi}{3}n$

(c) $y = \frac{1}{3}\cos x$ • Vertically compress $\cos x$ by a factor of 3.

★ $\frac{1}{3}$, 2π, x-int. @ $\frac{\pi}{2} + \pi n$

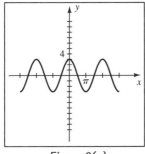

Figure 3(a)

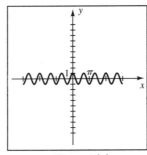

Figure 3(b)

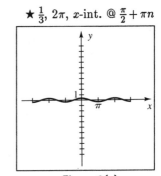

Figure 3(c)

(d) $y = \cos\frac{1}{3}x$ • Horizontally stretch $\cos x$ by a factor of 3.

★ $1, 6\pi$, x-int. @ $\frac{3\pi}{2} + 3\pi n$

(e) $y = 2\cos\frac{1}{3}x$ • Vertically stretch the graph in part (d) by a factor of 2.

★ $2, 6\pi$, x-int. @ $\frac{3\pi}{2} + 3\pi n$

(f) $y = \frac{1}{2}\cos 3x$ • Vertically compress the graph in part (b) by a factor of 2.

★ $\frac{1}{2}, \frac{2\pi}{3}$, x-int. @ $\frac{\pi}{6} + \frac{\pi}{3}n$

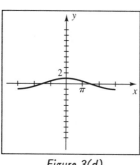

Figure 3(d)

Figure 3(e)

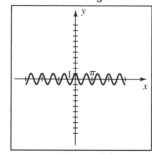

Figure 3(f)

(g) $y = -3\cos x$ • Reflect the graph in part (a) through the x-axis.

★ $3, 2\pi$, x-int. @ $\frac{\pi}{2} + \pi n$

(h) $y = \cos(-3x) = \cos 3x$ using a formula for negatives. This is the same as the graph in part (b).

★ $1, \frac{2\pi}{3}$, x-int. @ $\frac{\pi}{6} + \frac{\pi}{3}n$

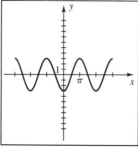

Figure 3(g)

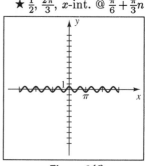

Figure 3(h)

Note: We will write $y = a\sin(bx + c)$ in the form $y = a\sin\left[b\left(x + \frac{c}{b}\right)\right]$. From this form we have the amplitude, $|a|$, the period, $\frac{2\pi}{|b|}$, and the phase shift, $-\frac{c}{b}$. We will also list the interval that corresponds to $[0, 2\pi]$ for the sine functions and to $[-\frac{\pi}{2}, \frac{3\pi}{2}]$ for the cosine functions. The work to determine those intervals is shown in each exercise.

$\boxed{5}$ $y = \sin\left(x - \frac{\pi}{2}\right)$ • $0 \le x - \frac{\pi}{2} \le 2\pi \Rightarrow \frac{\pi}{2} \le x \le \frac{5\pi}{2}$

Phase shift $= -\left(-\frac{\pi}{2}\right) = \frac{\pi}{2}.$

★ $1, 2\pi, \frac{\pi}{2}, \left[\frac{\pi}{2}, \frac{5\pi}{2}\right]$

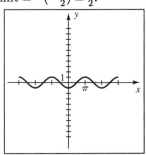

Figure 5

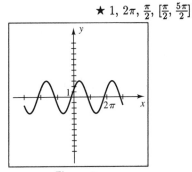

Figure 7

$\boxed{7}$ $y = 3\sin\left(x + \frac{\pi}{6}\right)$ • $0 \le x + \frac{\pi}{6} \le 2\pi \Rightarrow -\frac{\pi}{6} \le x \le \frac{11\pi}{6}$

Phase shift $= -\left(\frac{\pi}{6}\right) = -\frac{\pi}{6}.$

★ $3, 2\pi, -\frac{\pi}{6}, \left[-\frac{\pi}{6}, \frac{11\pi}{6}\right]$

$\boxed{9}$ $y = \cos\left(x + \frac{\pi}{2}\right)$ • $-\frac{\pi}{2} \le x + \frac{\pi}{2} \le \frac{3\pi}{2} \Rightarrow -\pi \le x \le \pi$

Phase shift $= -\left(\frac{\pi}{2}\right) = -\frac{\pi}{2}.$

★ $1, 2\pi, -\frac{\pi}{2}, [-\pi, \pi]$

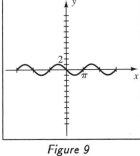

Figure 9

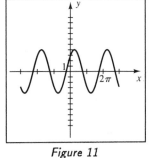

Figure 11

$\boxed{11}$ $y = 4\cos\left(x - \frac{\pi}{4}\right)$ • $-\frac{\pi}{2} \le x - \frac{\pi}{4} \le \frac{3\pi}{2} \Rightarrow -\frac{\pi}{4} \le x \le \frac{7\pi}{4}$

Phase shift $= -\left(-\frac{\pi}{4}\right) = \frac{\pi}{4}.$

★ $4, 2\pi, \frac{\pi}{4}, \left[-\frac{\pi}{4}, \frac{7\pi}{4}\right]$

13 $y = \sin(2x - \pi) + 1 = \sin\left[2\left(x - \frac{\pi}{2}\right)\right] + 1.$ •   Period $= \frac{2\pi}{|2|} = \pi.$ The normal range

of the sine, -1 to 1, is affected by the "$+1$" at the end of the equation. It shifts

the graph up 1 unit and the resulting range is 0 to 2. It may be easiest to graph

$y = \sin\left[2\left(x - \frac{\pi}{2}\right)\right]$ {1 period of a sine wave with end points at $\frac{\pi}{2}$ and $\frac{3\pi}{2}$} and then

make a vertical shift of 1 unit up to complete the graph of $y = \sin\left[2\left(x - \frac{\pi}{2}\right)\right] + 1.$

$0 \leq 2x - \pi \leq 2\pi \Rightarrow \pi \leq 2x \leq 3\pi \Rightarrow \frac{\pi}{2} \leq x \leq \frac{3\pi}{2}$ ★ $1, \pi, \frac{\pi}{2}, \left[\frac{\pi}{2}, \frac{3\pi}{2}\right]$

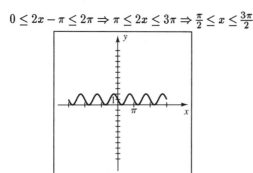

Figure 13   Figure 15

15 $y = -\cos(3x + \pi) - 2 = -\cos\left[3\left(x + \frac{\pi}{3}\right)\right] - 2$ •   The "$-$" in front of cos has the

effect of reflecting the graph of $y = \cos(3x + \pi)$ through the x-axis. The "-2" at the

end of the equation lowers the range 2 units to -3 to -1. Period $= \frac{2\pi}{|3|} = \frac{2\pi}{3}$, phase

shift $= -\left(\frac{\pi}{3}\right) = -\frac{\pi}{3}.$ $-\frac{\pi}{2} \leq 3x + \pi \leq \frac{3\pi}{2} \Rightarrow -\frac{3\pi}{2} \leq 3x \leq \frac{\pi}{2} \Rightarrow -\frac{\pi}{2} \leq x \leq \frac{\pi}{6}$

★ $1, \frac{2\pi}{3}, -\frac{\pi}{3}, \left[-\frac{\pi}{2}, \frac{\pi}{6}\right]$

17 $y = -2\sin(3x - \pi) = -2\sin\left[3\left(x - \frac{\pi}{3}\right)\right].$ •   Amplitude $= |-2| = 2.$ The negative

before the "2" has the effect of reflecting the graph of $y = 2\sin(3x - \pi)$ through the

x-axis.   Period $= \frac{2\pi}{|3|} = \frac{2\pi}{3},$ phase   shift $= -\left(-\frac{\pi}{3}\right) = \frac{\pi}{3},$ $0 \leq 3x - \pi \leq 2\pi \Rightarrow$

$\pi \leq 3x \leq 3\pi \Rightarrow \frac{\pi}{3} \leq x \leq \pi.$ ★ $2, \frac{2\pi}{3}, \frac{\pi}{3}, \left[\frac{\pi}{3}, \pi\right]$

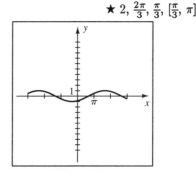

Figure 17   Figure 19

19 $y = \sin\left(\frac{1}{2}x - \frac{\pi}{3}\right) = \sin\left[\frac{1}{2}\left(x - \frac{2\pi}{3}\right)\right].$ •   Period $= \frac{2\pi}{|1/2|} = 4\pi.$

$0 \leq \frac{1}{2}x - \frac{\pi}{3} \leq 2\pi \Rightarrow \frac{\pi}{3} \leq \frac{1}{2}x \leq \frac{7\pi}{3} \Rightarrow \frac{2\pi}{3} \leq x \leq \frac{14\pi}{3}$ ★ $1, 4\pi, \frac{2\pi}{3}, \left[\frac{2\pi}{3}, \frac{14\pi}{3}\right]$

$\boxed{21}$ $y = 6\sin\pi x$ • Period $= \dfrac{2\pi}{|\pi|} = 2.$ $0 \le \pi x \le 2\pi \Rightarrow 0 \le x \le 2$ ★ $6, 2, 0, [0, 2]$

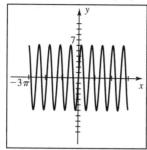

Figure 21

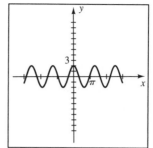

Figure 23

$\boxed{23}$ $y = 2\cos\dfrac{\pi}{2}x$ • Period $= \dfrac{2\pi}{|\pi/2|} = 4.$ $-\dfrac{\pi}{2} \le \dfrac{\pi}{2}x \le \dfrac{3\pi}{2} \Rightarrow -1 \le x \le 3$

★ $2, 4, 0, [-1, 3]$

$\boxed{25}$ $y = \dfrac{1}{2}\sin 2\pi x$ • $0 \le 2\pi x \le 2\pi \Rightarrow 0 \le x \le 1$ ★ $\dfrac{1}{2}, 1, 0, [0, 1]$

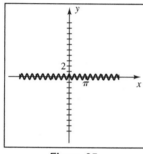

Figure 25

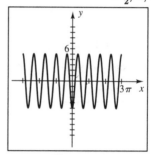

Figure 27

$\boxed{27}$ $y = 5\sin\left(3x - \dfrac{\pi}{2}\right) = 5\sin\left[3\left(x - \dfrac{\pi}{6}\right)\right].$ •   ★ $5, \dfrac{2\pi}{3}, \dfrac{\pi}{6}, \left[\dfrac{\pi}{6}, \dfrac{5\pi}{6}\right]$

$0 \le 3x - \dfrac{\pi}{2} \le 2\pi \Rightarrow \dfrac{\pi}{2} \le 3x \le \dfrac{5\pi}{2} \Rightarrow \dfrac{\pi}{6} \le x \le \dfrac{5\pi}{6}.$ To graph this function, draw

one period of a sine wave with end points at $\dfrac{\pi}{6}$ and $\dfrac{5\pi}{6}$ and an amplitude of 5.

$\boxed{29}$ $y = 3\cos\left(\dfrac{1}{2}x - \dfrac{\pi}{4}\right) = 3\cos\left[\dfrac{1}{2}\left(x - \dfrac{\pi}{2}\right)\right].$ •

$-\dfrac{\pi}{2} \le \dfrac{1}{2}x - \dfrac{\pi}{4} \le \dfrac{3\pi}{2} \Rightarrow -\dfrac{\pi}{4} \le \dfrac{1}{2}x \le \dfrac{7\pi}{4} \Rightarrow -\dfrac{\pi}{2} \le x \le \dfrac{7\pi}{2}$ ★ $3, 4\pi, \dfrac{\pi}{2}, \left[-\dfrac{\pi}{2}, \dfrac{7\pi}{2}\right]$

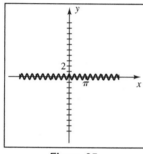

Figure 29

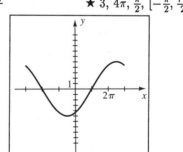

Figure 31

$\boxed{31}$ $y = -5\cos\left(\dfrac{1}{3}x + \dfrac{\pi}{6}\right) = -5\cos\left[\dfrac{1}{3}\left(x + \dfrac{\pi}{2}\right)\right].$ •

$-\dfrac{\pi}{2} \le \dfrac{1}{3}x + \dfrac{\pi}{6} \le \dfrac{3\pi}{2} \Rightarrow -\dfrac{2\pi}{3} \le \dfrac{1}{3}x \le \dfrac{4\pi}{3} \Rightarrow -2\pi \le x \le 4\pi$ ★ $5, 6\pi, -\dfrac{\pi}{2}, [-2\pi, 4\pi]$

33 $y = 3\cos(\pi x + 4\pi) = 3\cos[\pi(x+4)].$ •

$-\frac{\pi}{2} \le \pi x + 4\pi \le \frac{3\pi}{2} \Rightarrow -\frac{9\pi}{2} \le \pi x \le -\frac{5\pi}{2} \Rightarrow -\frac{9}{2} \le x \le -\frac{5}{2}$ ★ $3, 2, -4, [-\frac{9}{2}, -\frac{5}{2}]$

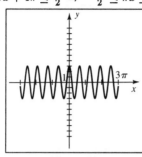

Figure 33

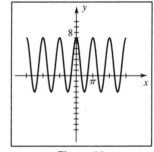

Figure 35

35 $y = -\sqrt{2}\sin\left(\frac{\pi}{2}x - \frac{\pi}{4}\right) = -\sqrt{2}\sin\left[\frac{\pi}{2}\left(x - \frac{1}{2}\right)\right].$ •

$0 \le \frac{\pi}{2}x - \frac{\pi}{4} \le 2\pi \Rightarrow \frac{\pi}{4} \le \frac{\pi}{2}x \le \frac{9\pi}{4} \Rightarrow \frac{1}{2} \le x \le \frac{9}{2}$ ★ $\sqrt{2}, 4, \frac{1}{2}, [\frac{1}{2}, \frac{9}{2}]$

37 $y = -2\sin(2x - \pi) + 3 = -2\sin[2(x - \frac{\pi}{2})] + 3.$ •

$0 \le 2x - \pi \le 2\pi \Rightarrow \pi \le 2x \le 3\pi \Rightarrow \frac{\pi}{2} \le x \le \frac{3\pi}{2}.$ The amplitude of 2 makes the normal sine range of -1 to 1 change to -2 to 2. The "$+3$" at the end of the equation raises the graph up 3 units and the range is 1 to 5.   ★ $2, \pi, \frac{\pi}{2}, [\frac{\pi}{2}, \frac{3\pi}{2}]$

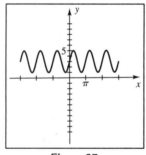

Figure 37

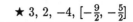

Figure 39

39 $y = 5\cos(2x + 2\pi) + 2 = 5\cos[2(x + \pi)] + 2.$ •

$-\frac{\pi}{2} \le 2x + 2\pi \le \frac{3\pi}{2} \Rightarrow -\frac{5\pi}{2} \le 2x \le -\frac{\pi}{2} \Rightarrow -\frac{5\pi}{4} \le x \le -\frac{\pi}{4}$ ★ $5, \pi, -\pi, [-\frac{5\pi}{4}, -\frac{\pi}{4}]$

41 (a) The amplitude a is 4 and the period {from $-\pi$ to π} is 2π.

       The phase shift is the first negative zero that occurs before a maximum, $-\pi$.

  (b) Period $= \frac{2\pi}{b} \Rightarrow 2\pi = \frac{2\pi}{b} \Rightarrow b = 1$. The least number c is the phase shift, $-\pi$.

                Hence, $y = 4\sin[1(x - (-\pi))]$, or, equivalently, $y = 4\sin(x + \pi)$.

43 (a) The amplitude a is 2 and the period {from -3 to 1} is 4.

       The phase shift is the first negative zero that occurs before a maximum, -3.

  (b) Period $= \frac{2\pi}{b} \Rightarrow 4 = \frac{2\pi}{b} \Rightarrow b = \frac{\pi}{2}.$

                Hence, $y = 2\sin[\frac{\pi}{2}(x - (-3))]$, or, equivalently, $y = 2\sin(\frac{\pi}{2}x + \frac{3\pi}{2}).$

[45] In the first second, there are 2 complete cycles.

Hence 1 cycle is completed in $\frac{1}{2}$ second and thus, the period is $\frac{1}{2}$.

Also, the period is $\frac{2\pi}{b}$. Equating these expressions yields $\frac{2\pi}{b} = \frac{1}{2} \Rightarrow b = 4\pi$.

[47] We first note that $\frac{1}{2}$ period takes place in $\frac{1}{4}$ second and thus 1 period in $\frac{1}{2}$ second.

As in Exercise 45, $\frac{2\pi}{b} = \frac{1}{2} \Rightarrow b = 4\pi$. Since the maximum flow rate is 8 liters/minute,

the amplitude is 8. $a = 8$ and $b = 4\pi \Rightarrow y = 8\sin 4\pi t$.

[49] $f(t) = \frac{1}{2}\cos\left[\frac{\pi}{6}\left(t - \frac{11}{2}\right)\right]$, amplitude $= \frac{1}{2}$, period $= \frac{2\pi}{\pi/6} = 12$, phase shift $= \frac{11}{2}$

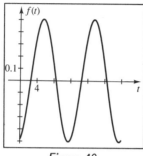

Figure 49

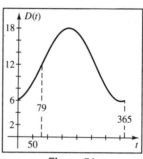

Figure 51

[51] $D(t) = 6\sin\left[\frac{2\pi}{365}(t - 79)\right] + 12$, amplitude $= 6$, period $= \frac{2\pi}{2\pi/365} = 365$,

phase shift $= 79$, range $= \underline{12 - 6}$ to $\underline{12 + 6}$ or 6 to 18

Note: Exer. 53–56: The period is 24 hours. Thus, $24 = \frac{2\pi}{b} \Rightarrow b = \frac{\pi}{12}$.

[53] A high of $10\,°\text{C}$ and a low of $-10\,°\text{C}$ imply that $d = \dfrac{\text{high} + \text{low}}{2} = \dfrac{10 + (-10)}{2} = 0$ and

$a = \text{high} - \text{average} = 10 - 0 = 10$. The average temperature of $0\,°\text{C}$ will occur 6 hours

$\{$one-half of 12$\}$ after the low at 4 A.M., which corresponds to $t = 10$. Letting this

correspond to the first zero of the sine function, we have

$$f(t) = 10\sin\left[\frac{\pi}{12}(t - 10)\right] + 0 = 10\sin\left(\frac{\pi}{12}t - \frac{5\pi}{6}\right) \text{ with } a = 10,\ b = \frac{\pi}{12},\ c = -\frac{5\pi}{6},\ d = 0.$$

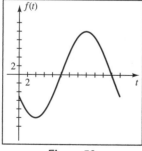

Figure 53

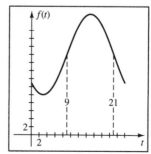

Figure 55

55 A high of 30°C and a low of 10°C imply that $d = \dfrac{30 + 10}{2} = 20$ and $a = 30 - 20 = 10$.

The average temperature of 20°C at 9 A.M. corresponds to $t = 9$.

Letting this correspond to the first zero of the sine function, we have

$f(t) = 10 \sin\left[\frac{\pi}{12}(t - 9)\right] + 20 = 10 \sin\left(\frac{\pi}{12}t - \frac{3\pi}{4}\right) + 20$ with

$a = 10, \ b = \frac{\pi}{12}, \ c = -\frac{3\pi}{4}, \ d = 20$. See *Figure 55*.

57 As $x \to 0^-$ or as $x \to 0^+$,

y oscillates between -1 and 1 and does not approach a unique value.

[−2, 2] by [−1.33, 1.33]

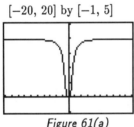

Xscl = 0.5

Yscl = 0.5

Figure 57

[−2, 2] by [−0.33, 2.33]

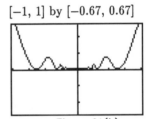

Xscl = 0.5

Yscl = 0.5

Figure 59

59 As $x \to 0^-$ or as $x \to 0^+$, y appears to approach 2.

61 From the graph, we see that there is a horizontal asymptote of $y = 4$.

[−20, 20] by [−1, 5]

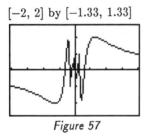

Xscl = 2

Yscl = 1

Figure 61(a)

[−1, 1] by [−0.67, 0.67]

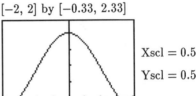

Xscl = 0.25

Yscl = 0.25

Figure 61(b)

63 $\cos 3x \geq \frac{1}{2}x - \sin x \Leftrightarrow \cos 3x - \frac{1}{2}x + \sin x \geq 0$. Graph $y = \cos 3x - \frac{1}{2}x + \sin x$.

From the graph, the x-intercepts occur at $x \approx -1.63, -0.45, 0.61, 1.49, 2.42$.

Thus, $\cos 3x \geq \frac{1}{2}x - \sin x$ on $[-\pi, -1.63] \cup [-0.45, 0.61] \cup [1.49, 2.42]$.

[−3.14, 3.14] by [−2.09, 2.09]

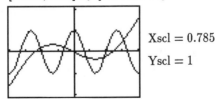

Xscl = 0.785

Yscl = 1

Figure 63

6.7 Exercises

Note: If $y = a\tan(bx + c)$ or $y = a\cot(bx + c)$,

then the periods for the tangent and cotangent graphs are $\pi / |b|$.

If $y = a\sec(bx + c)$ or $y = a\csc(bx + c)$,

then the periods for the secant and cosecant graphs are $2\pi / |b|$.

$\boxed{1}$ $y = 4\tan x$ • Vertically stretch $\tan x$ by a factor of 4. The *x-intercepts* of $\tan x$ and $\cot x$ are not affected by vertically stretching or compressing their graphs. The *vertical asymptotes* of $\tan x$, $\cot x$, $\sec x$, and $\csc x$ are not affected by vertically stretching or compressing their graphs. ★ π

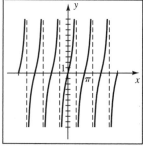

Figure 1

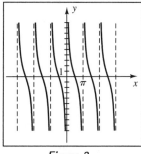

Figure 3

$\boxed{3}$ $y = 3\cot x$ • Vertically stretch $\cot x$ by a factor of 3. ★ π

$\boxed{5}$ $y = 2\csc x$ • Vertically stretch $\csc x$ by a factor of 2.

Note that there is now a minimum value of 2 at $x = \frac{\pi}{2}$.

The range of this function is $(-\infty, -2] \cup [2, \infty)$, or $|y| \geq 2$. ★ 2π

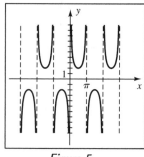

Figure 5

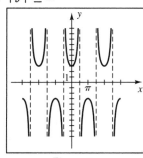

Figure 7

$\boxed{7}$ $y = 3\sec x$ • Vertically stretch $\sec x$ by a factor of 3.

Note that there is now a minimum value of 3 at $x = 0$.

The range of this function is $(-\infty, -3] \cup [3, \infty)$ or $|y| \geq 3$. ★ 2π

Note: The vertical asymptotes of each function are denoted by *VA* @ $x =$. The work to determine two consecutive vertical asymptotes is shown for each exercise. For the tangent and secant functions, the region from $-\frac{\pi}{2}$ to $\frac{\pi}{2}$ is used. For the cotangent and cosecant functions, the region from 0 to π is used.

$\boxed{9}$ $y = \tan\left(x - \frac{\pi}{4}\right)$ • Shift $\tan x$ right $\frac{\pi}{4}$ units, *VA* @ $x = -\frac{\pi}{4} + \pi n$. Note that the asymptotes remain π units apart. $-\frac{\pi}{2} \le x - \frac{\pi}{4} \le \frac{\pi}{2} \Rightarrow -\frac{\pi}{4} \le x \le \frac{3\pi}{4}$ ★ π

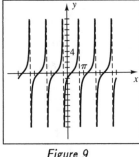

Figure 9

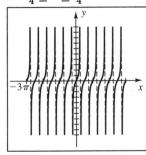

Figure 11

$\boxed{11}$ $y = \tan 2x$ • Horizontally compress $\tan x$ by a factor of 2, *VA* @ $x = -\frac{\pi}{4} + \frac{\pi}{2}n$. Note that the asymptotes are only $\frac{\pi}{2}$ units apart. $-\frac{\pi}{2} \le 2x \le \frac{\pi}{2} \Rightarrow -\frac{\pi}{4} \le x \le \frac{\pi}{4}$ ★ $\frac{\pi}{2}$

$\boxed{13}$ $y = \tan\frac{1}{4}x$ • Horizontally stretch $\tan x$ by a factor of 4, *VA* @ $x = -2\pi + 4\pi n$. Note that the asymptotes are 4π units apart. $-\frac{\pi}{2} \le \frac{1}{4}x \le \frac{\pi}{2} \Rightarrow -2\pi \le x \le 2\pi$ ★ 4π

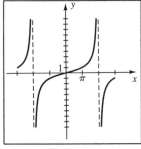

Figure 13

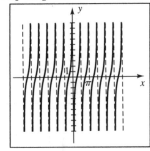

Figure 15

$\boxed{15}$ $y = 2\tan\left(2x + \frac{\pi}{2}\right) = 2\tan\left[2\left(x + \frac{\pi}{4}\right)\right]$. •

The phase shift is $-\frac{\pi}{4}$, the period is $\frac{\pi}{2}$, and we have a vertical stretching factor of 2.

$-\frac{\pi}{2} \le 2x + \frac{\pi}{2} \le \frac{\pi}{2} \Rightarrow -\pi \le 2x \le 0 \Rightarrow -\frac{\pi}{2} \le x \le 0$, *VA* @ $x = \frac{\pi}{2}n$ ★ $\frac{\pi}{2}$

$\boxed{17}$ $y = -\frac{1}{4}\tan\left(\frac{1}{2}x + \frac{\pi}{3}\right) = -\frac{1}{4}\tan\left[\frac{1}{2}\left(x + \frac{2\pi}{3}\right)\right].$ • Note that the "−" in front of the $\frac{1}{4}$

reflects the graph through the x-axis. This changes the appearance of a tangent

graph to that of a cotangent graph {increasing to decreasing}.

$-\frac{\pi}{2} \le \frac{1}{2}x + \frac{\pi}{3} \le \frac{\pi}{2} \Rightarrow -\frac{5\pi}{6} \le \frac{1}{2}x \le \frac{\pi}{6} \Rightarrow -\frac{5\pi}{3} \le x \le \frac{\pi}{3},$ VA @ $x = -\frac{5\pi}{3} + 2\pi n$ ★ 2π

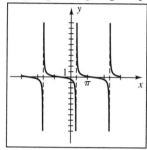

Figure 17

Figure 19

$\boxed{19}$ $y = \cot\left(x - \frac{\pi}{2}\right)$ • $0 \le x - \frac{\pi}{2} \le \pi \Rightarrow \frac{\pi}{2} \le x \le \frac{3\pi}{2},$ VA @ $x = \frac{\pi}{2} + \pi n$ ★ π

$\boxed{21}$ $y = \cot 2x$ • $0 \le 2x \le \pi \Rightarrow 0 \le x \le \frac{\pi}{2},$ VA @ $x = \frac{\pi}{2} n$ ★ $\frac{\pi}{2}$

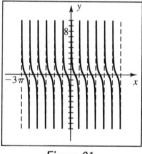

Figure 21

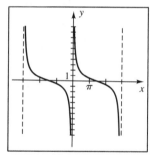

Figure 23

$\boxed{23}$ $y = \cot\frac{1}{3}x$ • $0 \le \frac{1}{3}x \le \pi \Rightarrow 0 \le x \le 3\pi,$ VA @ $x = 3\pi n$ ★ 3π

$\boxed{25}$ $y = 2\cot\left(2x + \frac{\pi}{2}\right) = 2\cot\left[2\left(x + \frac{\pi}{4}\right)\right].$ •

$0 \le 2x + \frac{\pi}{2} \le \pi \Rightarrow -\frac{\pi}{2} \le 2x \le \frac{\pi}{2} \Rightarrow -\frac{\pi}{4} \le x \le \frac{\pi}{4},$ VA @ $x = -\frac{\pi}{4} + \frac{\pi}{2} n$ ★ $\frac{\pi}{2}$

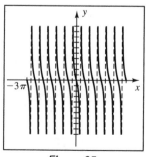

Figure 25

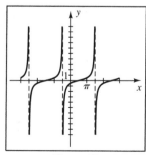

Figure 27

$\boxed{27}$ $y = -\frac{1}{2}\cot\left(\frac{1}{2}x + \frac{\pi}{4}\right) = -\frac{1}{2}\cot\left[\frac{1}{2}\left(x + \frac{\pi}{2}\right)\right].$ •

$0 \le \frac{1}{2}x + \frac{\pi}{4} \le \pi \Rightarrow -\frac{\pi}{4} \le \frac{1}{2}x \le \frac{3\pi}{4} \Rightarrow -\frac{\pi}{2} \le x \le \frac{3\pi}{2},$ VA @ $x = -\frac{\pi}{2} + 2\pi n$ ★ 2π

29 $y = \sec\left(x - \frac{\pi}{2}\right)$ •   Note that this is the same graph as the graph of $y = \csc x$.

$-\frac{\pi}{2} \le x - \frac{\pi}{2} \le \frac{\pi}{2} \Rightarrow 0 \le x \le \pi$, $VA @ x = \pi n$ ★ 2π

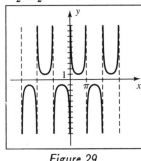

Figure 29

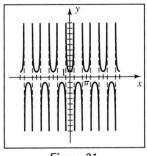

Figure 31

31 $y = \sec 2x$ •   Note that the asymptotes move closer together by a factor of 2.

$-\frac{\pi}{2} \le 2x \le \frac{\pi}{2} \Rightarrow -\frac{\pi}{4} \le x \le \frac{\pi}{4}$, $VA @ x = -\frac{\pi}{4} + \frac{\pi}{2}n$ ★ π

33 $y = \sec\frac{1}{3}x$ •   Note that the asymptotes move farther apart by a factor of 3,

just as they did with the graphs of tan and cot.

$-\frac{\pi}{2} \le \frac{1}{3}x \le \frac{\pi}{2} \Rightarrow -\frac{3\pi}{2} \le x \le \frac{3\pi}{2}$, $VA @ x = -\frac{3\pi}{2} + 3\pi n$ ★ 6π

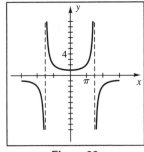

Figure 33

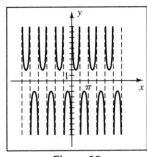

Figure 35

35 $y = 2\sec\left(2x - \frac{\pi}{2}\right) = 2\sec\left[2\left(x - \frac{\pi}{4}\right)\right]$.   •

$-\frac{\pi}{2} \le 2x - \frac{\pi}{2} \le \frac{\pi}{2} \Rightarrow 0 \le 2x \le \pi \Rightarrow 0 \le x \le \frac{\pi}{2}$, $VA @ x = \frac{\pi}{2}n$ ★ π

37 $y = -\frac{1}{3}\sec\left(\frac{1}{2}x + \frac{\pi}{4}\right) = -\frac{1}{3}\sec\left[\frac{1}{2}\left(x + \frac{\pi}{2}\right)\right]$.   •

$-\frac{\pi}{2} \le \frac{1}{2}x + \frac{\pi}{4} \le \frac{\pi}{2} \Rightarrow -\frac{3\pi}{4} \le \frac{1}{2}x \le \frac{\pi}{4} \Rightarrow -\frac{3\pi}{2} \le x \le \frac{\pi}{2}$, $VA @ x = -\frac{3\pi}{2} + 2\pi n$ ★ 4π

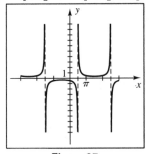

Figure 37

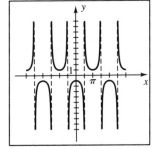

Figure 39

39 $y = \csc\left(x - \frac{\pi}{2}\right)$ • $0 \le x - \frac{\pi}{2} \le \pi \Rightarrow \frac{\pi}{2} \le x \le \frac{3\pi}{2}$, $VA @ x = \frac{\pi}{2} + \pi n$ ★ 2π

$\boxed{41}$ $y = \csc 2x$ • $0 \le 2x \le \pi \Rightarrow 0 \le x \le \frac{\pi}{2}$, VA @ $x = \frac{\pi}{2}n$ ★ π

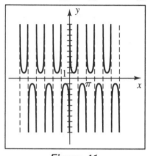

Figure 41

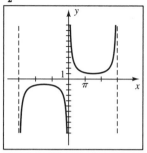

Figure 43

$\boxed{43}$ $y = \csc\frac{1}{3}x$ • $0 \le \frac{1}{3}x \le \pi \Rightarrow 0 \le x \le 3\pi$, VA @ $x = 3\pi n$ ★ 6π

$\boxed{45}$ $y = 2\csc\left(2x + \frac{\pi}{2}\right) = 2\csc\left[2\left(x + \frac{\pi}{4}\right)\right]$. •

$0 \le 2x + \frac{\pi}{2} \le \pi \Rightarrow -\frac{\pi}{2} \le 2x \le \frac{\pi}{2} \Rightarrow -\frac{\pi}{4} \le x \le \frac{\pi}{4}$, VA @ $x = -\frac{\pi}{4} + \frac{\pi}{2}n$ ★ π

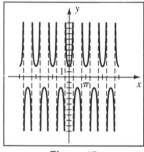

Figure 45

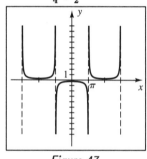

Figure 47

$\boxed{47}$ $y = -\frac{1}{4}\csc\left(\frac{1}{2}x + \frac{\pi}{2}\right) = -\frac{1}{4}\csc\left[\frac{1}{2}(x + \pi)\right]$. •

$0 \le \frac{1}{2}x + \frac{\pi}{2} \le \pi \Rightarrow -\frac{\pi}{2} \le \frac{1}{2}x \le \frac{\pi}{2} \Rightarrow -\pi \le x \le \pi$, VA @ $x = -\pi + 2\pi n$ ★ 4π

$\boxed{49}$ $y = \tan\frac{\pi}{2}x$ • Horizontally stretch $\tan x$ by a factor of $2/\pi$, VA @ $x = -1 + 2n$.

$-\frac{\pi}{2} \le \frac{\pi}{2}x \le \frac{\pi}{2} \Rightarrow -1 \le x \le 1$ ★ 2

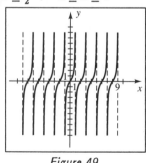

Figure 49

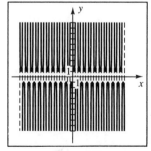

Figure 51

$\boxed{51}$ $y = \csc 2\pi x$ • $0 \le 2\pi x \le \pi \Rightarrow 0 \le x \le \frac{1}{2}$, VA @ $x = \frac{1}{2}n$ ★ 1

$\boxed{53}$ Reflecting the graph of $y = \cot x$ about the x-axis, which is $y = -\cot x$, gives us the graph of $y = \tan\left(x + \frac{\pi}{2}\right)$. If we shift this graph to the left (or right), we will obtain the graph of $y = \tan x$. Thus, one equation is $y = -\cot\left(x + \frac{\pi}{2}\right)$.

55 $y = |\sin x|$ • Reflect the negative values of $y = \sin x$ through the x-axis. In general, when sketching the graph of $y = |f(x)|$, reflect the negative values of $f(x)$ through the x-axis. The absolute value does not affect the nonnegative values.

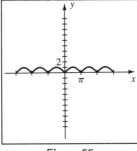

Figure 55

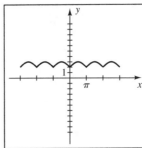

Figure 57

57 $y = |\sin x| + 2$ • Shift $y = |\sin x|$ up 2 units.

59 $y = -|\cos x| + 1$ • Similar to Exercise 55, we first reflect the negative values of $y = \cos x$ through the x-axis. The "$-$" in front of $|\cos x|$ has the effect of reflecting $y = |\cos x|$ through the x-axis. Finally, we shift that graph 1 unit up to obtain the graph of $y = -|\cos x| + 1$.

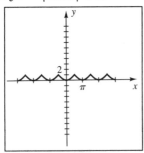

Figure 59

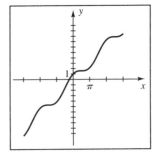

Figure 61

61 $y = x + \cos x$ • The value of $\cos x$ is between -1 and 1—adding this relatively small amount to the value of x has the effect of oscillating the graph about the line $y = x$.

63 $y = 2^{-x} \cos x$ • This graph is similar to the graph in Example 8.

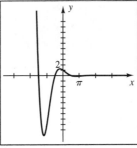

Figure 63

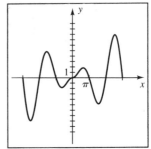

Figure 65

65 $y = |x| \sin x$ • See *Figure 65*.   The graph will coincide with the graph of

$y = |x|$ if $\sin x = 1$—that is, if $x = \frac{\pi}{2} + 2\pi n$. The graph will coincide with the graph

of $y = -|x|$ if $\sin x = -1$—that is, if $x = \frac{3\pi}{2} + 2\pi n$.

67 The damping factor of $y = e^{-x/4} \sin 4x$ is $e^{-x/4}$.

$[-6.28, 6.28]$ by $[-4.19, 4.19]$ $[-3.14, 3.14]$ by $[-4, 4]$

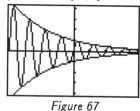
$Xscl = 1.57$
$Yscl = 1$

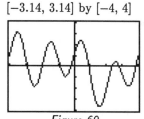
$Xscl = 0.785$
$Yscl = 1$

Figure 67   *Figure 69*

69 From the graph, we see that the maximum occurs at the approximate coordinates

$(-2.76, 3.09)$, and the minimum occurs at the approximate coordinates $(1.23, -3.68)$.

71 From the graph, we see that f is increasing and one-to-one between

$a \approx -0.70$ and $b \approx 0.12$.   Thus, the interval is approximately $[-0.70, 0.12]$.

$[-2, 2]$ by $[-1.33, 1.33]$ $[-3.14, 3.14]$ by $[-2.09, 2.09]$

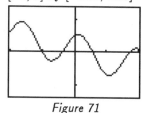
$Xscl = 1$
$Yscl = 1$

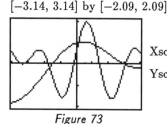
$Xscl = 0.785$
$Yscl = 1$

Figure 71   *Figure 73*

73 $\cos(2x - 1) + \sin 3x \geq \sin\frac{1}{3}x + \cos x \Leftrightarrow \cos(2x - 1) + \sin 3x - \sin\frac{1}{3}x - \cos x \geq 0$.

Graph $y = \cos(2x - 1) + \sin 3x - \sin\frac{1}{3}x - \cos x$.

From the graph, the x-intercepts occur at $x \approx -1.31, 0.11, 0.95, 2.39$.

Thus, $\cos(2x - 1) + \sin 3x \geq \sin\frac{1}{3}x + \cos x$ on $[-\pi, -1.31] \cup [0.11, 0.95] \cup [2.39, \pi]$.

75 (a) The damping factor of $S = A_0 e^{-\alpha z} \sin(kt - \alpha z)$ is $A_0 e^{-\alpha z}$.

(b) The phase shift at depth z_0 can be found by solving the equation $kt - \alpha z_0 = 0$ for

t.   Doing so gives us $kt = \alpha z_0$, and hence, $t = \frac{\alpha}{k} z_0$.

(c) At the surface, $z = 0$.   Hence, $S = A_0 \sin kt$ and the amplitude at the surface is

A_0. $\text{Amplitude}_{\text{wave}} = \frac{1}{2}\text{Amplitude}_{\text{surface}} \Rightarrow$

$$A_0 e^{-\alpha z} = \frac{1}{2}A_0 \Rightarrow e^{-\alpha z} = \frac{1}{2} \Rightarrow -\alpha z = \ln\frac{1}{2} \Rightarrow z = \frac{-\ln 2}{-\alpha} = \frac{\ln 2}{\alpha}.$$

6.8 Exercises

1 (a) $\omega = \left(\frac{2\pi \,\text{radians}}{\text{revolution}}\right) \cdot \left(\frac{100\,\text{revolutions}}{\text{minute}}\right) = 200\pi$ radians/minute

  (b) The diameter of the wheel is $40\,\text{cm}$ so its radius is $20\,\text{cm}$. Using the formulas next to text Figure 78, we have $x = 20\cos 200\pi t$ cm and $y = 20\sin 200\pi t$ cm.

3 Amplitude, $10\,\text{cm}$; period $= \frac{2\pi}{6\pi} = \frac{1}{3}$ sec; frequency $= \frac{6\pi}{2\pi} = 3$ oscillations/sec. The point is at the origin at $t = 0$. It moves upward with decreasing speed, reaching the point with coordinate 10 when $6\pi t = \frac{\pi}{2}$ or $t = \frac{1}{12}$. It then reverses direction and moves downward, gaining speed until it reaches the origin when $6\pi t = \pi$ or $t = \frac{1}{6}$. It continues downward with decreasing speed, reaching the point with coordinate -10 when $6\pi t = \frac{3\pi}{2}$ or $t = \frac{1}{4}$. It then reverses direction and moves upward with increasing speed, returning to the origin when $6\pi t = 2\pi$ or $t = \frac{1}{3}$ to complete one oscillation. Another approach is to simply model this movement in terms of proportions of the sine curve. For one period, the sine increases for $\frac{1}{4}$ period, decreases for $\frac{1}{2}$ period, and increases for its last $\frac{1}{4}$ period.

5 Amplitude, $4\,\text{cm}$; period $= \frac{2\pi}{3\pi/2} = \frac{4}{3}$ sec; frequency $= \frac{3\pi/2}{2\pi} = \frac{3}{4}$ oscillation/sec. The point is at $d = 4$ when $t = 0$. It then decreases in height until $\frac{3\pi}{2}t = \pi$ or $t = \frac{2}{3}$ where it obtains a minimum of $d = -4$. It then reverses direction and increases to a height of $d = 4$ when $\frac{3\pi}{2}t = 2\pi$ or $t = \frac{4}{3}$ to complete one oscillation.

7 Period $= 3 \Rightarrow \frac{2\pi}{\omega} = 3 \Rightarrow \omega = \frac{2\pi}{3}$. Amplitude $= 5 \Rightarrow a = 5$. $d = 5\cos\frac{2\pi}{3}t$

9 See the discussion near text Figure 82.

  For $I = 20\sin\left(360\pi t - \frac{\pi}{4}\right)$, we have $\phi = \frac{\pi}{4}$ and $\omega = 360\pi$.

  Since $\phi > 0$, the current lags the electromotive force (emf)

  by $\frac{\phi}{\omega} = \frac{\pi/4}{360\pi} = \frac{1}{1440}$ second.

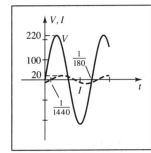

Figure 9

11 (a) 2 revolutions per minute $= 4\pi$ radians per minute or $\frac{4\pi}{60} = \frac{\pi}{15}$ radian/sec

  (b) period $= 30 \Rightarrow \frac{2\pi}{\omega} = 30 \Rightarrow \omega = \frac{\pi}{15}$. When $t = 0$, $\cos\frac{\pi}{15}t$ is at a maximum and we will subtract $50\cos\frac{\pi}{15}t$ from the height of the center of the wheel $\{\,60\ \text{ft}\,\}$.

  $h(t) = 60 - 50\cos\frac{\pi}{15}t$ or equivalently, $h(t) = 60 + 50\sin\left(\frac{\pi}{15}t - \frac{\pi}{2}\right)$ since

$$\cos\tfrac{\pi}{15}t = \sin\left(\tfrac{\pi}{2} - \tfrac{\pi}{15}t\right) = \sin\left[-\left(\tfrac{\pi}{15}t - \tfrac{\pi}{2}\right)\right] = -\sin\left(\tfrac{\pi}{15}t - \tfrac{\pi}{2}\right).$$

13 (a) $y = a \sin bt = 8 \sin \frac{\pi}{6} t \Rightarrow a = 8$ and $b = \frac{\pi}{6}$.

Thus, the wave has an amplitude of 8 ft and its period is $\frac{2\pi}{b} = \frac{2\pi}{\pi/6} = 12$ min.

(b) The wave moves 21 km every 12 min.

The wave's velocity is $\frac{21}{12} = 1.75$ km/min, or, 105 km/hr.

15 A high of 12 feet and a low of 3 feet imply that we have an average height of

$\frac{12+3}{2} = 7.5$ ft with an amplitude of $7.5 - 3 = 4.5$. We have minimums at 8 A.M.

and 8 P.M. so the period is 12 hours. $12 = \frac{2\pi}{b} \Rightarrow b = \frac{\pi}{6}$. There is a maximum

halfway between the minimums, at $t = 14$ { 2:00 P.M. }. The average before a high

occurs halfway between the minimum at $t = 8$ and the maximum at $t = 14$. This is

at $t = 11$ and we want this to correspond to the first zero of the sine curve.

Combining all of the above, we have $y = (4.5)\sin\left[\frac{\pi}{6}(t-11)\right] + 7.5$.

Chapter 6 Review Exercises

1 $330° \cdot \frac{\pi}{180} = \frac{11 \cdot 30\pi}{6 \cdot 30} = \frac{11\pi}{6}$; $\qquad\qquad 405° \cdot \frac{\pi}{180} = \frac{9 \cdot 45\pi}{4 \cdot 45} = \frac{9\pi}{4}$

2 $\frac{9\pi}{2} \cdot \left(\frac{180}{\pi}\right)° = \left(\frac{9 \cdot 90 \cdot 2\pi}{2\pi}\right)° = 810°$; $\qquad -\frac{2\pi}{3} \cdot \left(\frac{180}{\pi}\right)° = -\left(\frac{2 \cdot 60 \cdot 3\pi}{3\pi}\right)° = -120°$;

3 (a) $\theta = \frac{s}{r} = \frac{20\,\text{cm}}{2\,\text{m}} = \frac{20\,\text{cm}}{2\,(100)\,\text{cm}} = 0.1\,\text{radian}$

(b) $A = \frac{1}{2}r^2\theta = \frac{1}{2}(2)^2(0.1) = 0.2\,\text{m}^2$

4 (a) $s = r\theta = (15 \cdot \frac{1}{2})(70 \cdot \frac{\pi}{180}) = \frac{35\pi}{12} \approx 9.16\,\text{cm}$

(b) $A = \frac{1}{2}r^2\theta = \frac{1}{2}(15 \cdot \frac{1}{2})^2(70 \cdot \frac{\pi}{180}) = \frac{175\pi}{16} \approx 34.4\,\text{cm}^2$

5 $\sin 60° = \frac{9}{x} \Rightarrow \frac{\sqrt{3}}{2} = \frac{9}{x} \Rightarrow x = 6\sqrt{3}$; $\tan 60° = \frac{9}{y} \Rightarrow \sqrt{3} = \frac{9}{y} \Rightarrow y = 3\sqrt{3}$

7 $1 + \tan^2\theta = \sec^2\theta \Rightarrow \tan^2\theta = \sec^2\theta - 1 \Rightarrow \tan\theta = \sqrt{\sec^2\theta - 1}$

9 $\text{opp} = \sqrt{\text{hyp}^2 - \text{adj}^2} = \sqrt{7^2 - 4^2} = \sqrt{33}$. $\qquad \bigstar\ \frac{\sqrt{33}}{7}, \frac{4}{7}, \frac{\sqrt{33}}{4}, \frac{4}{\sqrt{33}}, \frac{7}{4}, \frac{7}{\sqrt{33}}$

10 (b) $\cot t = \frac{\cos t}{\sin t} = \frac{\csc t}{\sec t} \Rightarrow -\frac{3}{2} = \frac{\sqrt{13}/2}{\sec t} \Rightarrow \sec t = -\frac{\sqrt{13}}{3}$;

the other values are just the reciprocals.

11 (a) $\sec\theta < 0 \Rightarrow$ terminal side of θ is in QII or QIII.

$\sin\theta > 0 \Rightarrow$ terminal side of θ is in QI or QII. Hence, θ is in QII.

(b) $\cot\theta > 0 \Rightarrow$ terminal side of θ is in QI or QIII.

$\csc\theta < 0 \Rightarrow$ terminal side of θ is in QIII or QIV. Hence, θ is in QIII.

12 (a) $t = -\frac{9\pi}{8} \Rightarrow \theta_C = \frac{7\pi}{8}$ and $t_R = \pi - \frac{7\pi}{8} = \frac{\pi}{8}$.

(b) $\theta = 892° \Rightarrow \theta_C = 172°$ and $\theta_R = 180° - 172° = 8°$.

$\boxed{13}$ (b) For $\theta = -\frac{5\pi}{4}$, choose $x = -1$ and $y = 1$. $r = \sqrt{2}$.

$$\star \text{ (b) } \frac{\sqrt{2}}{2}, \; -\frac{\sqrt{2}}{2}, \; -1, \; -1, \; -\sqrt{2}, \; \sqrt{2}$$

(d) For $\theta = \frac{11\pi}{6}$, choose $x = \sqrt{3}$ and $y = -1$. $r = 2$.

$$\star \text{ (d) } -\frac{1}{2}, \; \frac{\sqrt{3}}{2}, \; -\frac{\sqrt{3}}{3}, \; -\sqrt{3}, \; \frac{2}{\sqrt{3}}, \; -2$$

$\boxed{14}$ (a) $\cos 225° = -\cos 45° = -\dfrac{\sqrt{2}}{2}$ (b) $\tan 150° = -\tan 30° = -\dfrac{\sqrt{3}}{3}$

     (c) $\sin\left(-\frac{\pi}{6}\right) = -\sin\frac{\pi}{6} = -\frac{1}{2}$ (d) $\sec\frac{4\pi}{3} = -\sec\frac{\pi}{3} = -2$

     (e) $\cot\frac{7\pi}{4} = -\cot\frac{\pi}{4} = -1$ (f) $\csc 300° = -\csc 60° = -\dfrac{2}{\sqrt{3}}$

$\boxed{15}$ (a) $x = 30$ and $y = -40 \Rightarrow r = \sqrt{30^2 + (-40)^2} = 50$. $\star$ (a) $-\frac{4}{5}, \frac{3}{5}, -\frac{4}{3}, -\frac{3}{4}, \frac{5}{3}, -\frac{5}{4}$

     (b) $2x + 3y + 6 = 0 \Leftrightarrow y = -\frac{2}{3}x - 2$, so the slope of the given line is $-\frac{2}{3}$.

         The line through the origin with that slope is $y = -\frac{2}{3}x$.

         If $x = -3$, then $y = 2$ and $(-3,\,2)$ is a point on the terminal side of θ.

$$x = -3 \text{ and } y = 2 \Rightarrow r = \sqrt{(-3)^2 + 2^2} = \sqrt{13}.$$

$$\star \text{ (b) } \frac{2}{\sqrt{13}}, \; -\frac{3}{\sqrt{13}}, \; -\frac{2}{3}, \; -\frac{3}{2}, \; -\frac{\sqrt{13}}{3}, \; \frac{\sqrt{13}}{2}$$

     (c) For $\theta = -90°$, choose $x = 0$ and $y = -1$. r is 1. $\star$ (c) $-1, 0, U, 0, U, -1$

$\boxed{16}$ $\sin\theta = -0.7604 \Rightarrow \theta = \sin^{-1}(-0.7604) \approx -49.5° \Rightarrow \theta_R \approx 49.5°$.

     Since the sine is negative in QIII and QIV, and the secant is positive in QIV,

         we want the fourth-quadrant angle having $\theta_R = 49.5°$. $360° - 49.5° = 310.5°$

$\boxed{17}$ α and β are complementary, so $\alpha = 90° - \beta = 90° - 60° = 30°$.

     $\cot\beta = \frac{a}{b} \Rightarrow a = b\cot\beta = 40\cot 60° = 40(\frac{1}{3}\sqrt{3}) \approx 23$.

     $\csc\beta = \frac{c}{b} \Rightarrow c = b\csc\beta = 40\csc 60° = 40(\frac{2}{3}\sqrt{3}) \approx 46$.

$\boxed{19}$ $\tan\alpha = \frac{a}{b} = \frac{62}{25} \Rightarrow \alpha \approx 68°$. $\beta = 90° - \alpha \approx 90° - 68° = 22°$.

$$c = \sqrt{a^2 + b^2} = \sqrt{62^2 + 25^2} = \sqrt{3844 + 625} = \sqrt{4469} \approx 67.$$

$\boxed{21}$ $\sin\theta\,(\csc\theta - \sin\theta)$ $= \sin\theta\csc\theta - \sin^2\theta$ { multiply terms }

                        $= \sin\theta \cdot \dfrac{1}{\sin\theta} - \sin^2\theta$ { reciprocal identity }

                        $= 1 - \sin^2\theta$ { simplify }

                        $= \cos^2\theta$ { Pythagorean identity }

$\boxed{23}$ $(\cos^2\theta - 1)(\tan^2\theta + 1) = (\cos^2\theta - 1)(\sec^2\theta)$ { Pythagorean identity }

                        $= \cos^2\theta\sec^2\theta - \sec^2\theta$ { multiply terms }

                        $= 1 - \sec^2\theta$ { reciprocal identity }

$\boxed{25}$ $\dfrac{1+\tan^2\theta}{\tan^2\theta} = \dfrac{1}{\tan^2\theta} + \dfrac{\tan^2\theta}{\tan^2\theta}$ { split up the fraction }

$= \cot^2\theta + 1$ { reciprocal identity, simplify }

$= \csc^2\theta$ { Pythagorean identity }

$\boxed{27}$ $\dfrac{\cot\theta - 1}{1 - \tan\theta} = \dfrac{\dfrac{\cos\theta}{\sin\theta} - 1}{1 - \dfrac{\sin\theta}{\cos\theta}}$ { put in terms of sines and cosines }

$= \dfrac{\dfrac{\cos\theta - \sin\theta}{\sin\theta}}{\dfrac{\cos\theta - \sin\theta}{\cos\theta}}$ $\left\{ \begin{array}{l} \text{make the numerator and the} \\ \text{denominator each a single fraction} \end{array} \right\}$

$= \dfrac{(\cos\theta - \sin\theta)\cos\theta}{(\cos\theta - \sin\theta)\sin\theta}$ { simplify a complex fraction }

$= \dfrac{\cos\theta}{\sin\theta}$ { cancel like term }

$= \cot\theta$ { cotangent identity }

$\boxed{29}$ $\dfrac{\tan(-\theta) + \cot(-\theta)}{\tan\theta} = \dfrac{-\tan\theta - \cot\theta}{\tan\theta}$ { formulas for negatives }

$= -\dfrac{\tan\theta}{\tan\theta} - \dfrac{\cot\theta}{\tan\theta}$ { split up fraction }

$= -1 - \cot^2\theta$ { simplify, reciprocal identity }

$= -(1 + \cot^2\theta)$ { factor out -1 }

$= -\csc^2\theta$ { Pythagorean identity }

$\boxed{31}$ $y = 5\cos x$ •   Vertically stretch $\cos x$ by a factor of 5.   ★ 5, 2π, x-int. @ $\frac{\pi}{2} + \pi n$

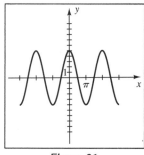

Figure 31

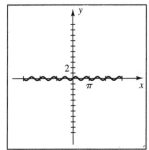

Figure 33

$\boxed{33}$ $y = \frac{1}{3}\sin 3x$ •   Horizontally compress $\sin x$ by a factor of 3 and vertically compress

that graph by a factor of 3.   ★ $\frac{1}{3}$, $\frac{2\pi}{3}$, x-int. @ $\frac{\pi}{3}n$

[35] $y = -3\cos\frac{1}{2}x$ • Horizontally stretch $\cos x$ by a factor of 2, vertically stretch that graph by a factor of 3, and reflect that graph through the x-axis.

★ 3, 4π, x-int. @ $\pi + 2\pi n$

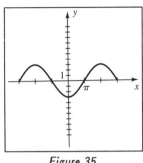

Figure 35

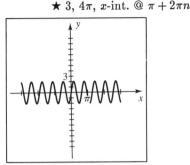

Figure 37

[37] $y = 2\sin \pi x$ • Horizontally compress $\sin x$ by a factor of π, and then vertically stretch that graph by a factor of 2.   ★ 2, 2, x-int. @ n

Note: Let a denote the amplitude and p the period.

[39] (a) $a = |-1.43| = 1.43$. We have $\frac{3}{4}$ of a period from $(0, 0)$ to $(1.5, -1.43)$.

$$\tfrac{3}{4}p = 1.5 \Rightarrow p = 2.$$

(b) Since the period p is given by $\frac{2\pi}{b}$, we can solve for b, obtaining $b = \frac{2\pi}{P}$.

Hence, $b = \frac{2\pi}{P} = \frac{2\pi}{2} = \pi$ and consequently, $y = 1.43\sin \pi x$.

[41] (a) Since the y-intercept is -3, $a = |-3| = 3$. The second positive x-intercept is π, so $\frac{3}{4}$ of a period occurs from $(0, -3)$ to $(\pi, 0)$. Thus, $\frac{3}{4}p = \pi \Rightarrow p = \frac{4\pi}{3}$.

(b) $b = \frac{2\pi}{P} = \frac{2\pi}{4\pi/3} = \frac{3}{2}$, $y = -3\cos\frac{3}{2}x$.

[43] $y = 2\sin\left(x - \frac{2\pi}{3}\right)$ • $0 \le x - \frac{2\pi}{3} \le 2\pi \Rightarrow \frac{2\pi}{3} \le x \le \frac{8\pi}{3}$.

There are x-intercepts at $x = \frac{2\pi}{3} + \pi n$.

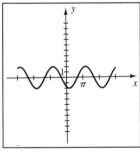

Figure 43

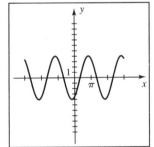

Figure 45

[45] $y = -4\cos\left(x + \frac{\pi}{6}\right)$ • $-\frac{\pi}{2} \le x + \frac{\pi}{6} \le \frac{3\pi}{2} \Rightarrow -\frac{2\pi}{3} \le x \le \frac{4\pi}{3}$.

There are x-intercepts at $x = -\frac{2\pi}{3} + \pi n$.

$\boxed{47}$ $y = 2\tan\left(\frac{1}{2}x - \pi\right) = 2\tan\left[\frac{1}{2}(x - 2\pi)\right]$. •

$-\frac{\pi}{2} \le \frac{1}{2}x - \pi \le \frac{\pi}{2} \Rightarrow \frac{\pi}{2} \le \frac{1}{2}x \le \frac{3\pi}{2} \Rightarrow \pi \le x \le 3\pi$, VA @ $x = \pi + 2\pi n$

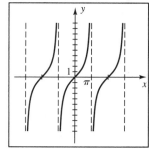

Figure 47

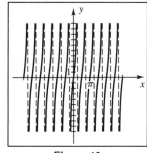

Figure 49

$\boxed{49}$ $y = -4\cot\left(2x - \frac{\pi}{2}\right) = -4\cot\left[2\left(x - \frac{\pi}{4}\right)\right]$. •

$0 \le 2x - \frac{\pi}{2} \le \pi \Rightarrow \frac{\pi}{2} \le 2x \le \frac{3\pi}{2} \Rightarrow \frac{\pi}{4} \le x \le \frac{3\pi}{4}$, VA @ $x = \frac{\pi}{4} + \frac{\pi}{2}n$

$\boxed{51}$ $y = \sec\left(\frac{1}{2}x + \pi\right) = \sec\left[\frac{1}{2}(x + 2\pi)\right]$. •

$-\frac{\pi}{2} \le \frac{1}{2}x + \pi \le \frac{\pi}{2} \Rightarrow -\frac{3\pi}{2} \le \frac{1}{2}x \le -\frac{\pi}{2} \Rightarrow -3\pi \le x \le -\pi$, VA @ $x = -3\pi + 2\pi n$

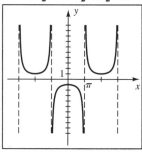

Figure 51

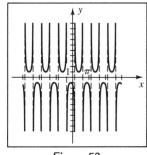

Figure 53

$\boxed{53}$ $y = \csc\left(2x - \frac{\pi}{4}\right) = \csc\left[2\left(x - \frac{\pi}{8}\right)\right]$. •

$0 \le 2x - \frac{\pi}{4} \le \pi \Rightarrow \frac{\pi}{4} \le 2x \le \frac{5\pi}{4} \Rightarrow \frac{\pi}{8} \le x \le \frac{5\pi}{8}$, VA @ $x = \frac{\pi}{8} + \frac{\pi}{2}n$

$\boxed{55}$ (a) $\left(\frac{545 \text{ rev}}{1 \text{ min}}\right)\left(\frac{2\pi \text{ rad}}{1 \text{ rev}}\right)\left(\frac{1 \text{ min}}{60 \text{ sec}}\right) = \frac{109\pi}{6}$ rad/sec ≈ 57 rad/sec

(b) $d = 22.625$ ft $\Rightarrow C = \pi d = 22.625\pi$ ft.

$$\left(\frac{22.625\pi \text{ ft}}{1 \text{ rev}}\right)\left(\frac{545 \text{ rev}}{1 \text{ min}}\right)\left(\frac{1 \text{ mile}}{5280 \text{ ft}}\right)\left(\frac{60 \text{ min}}{1 \text{ hour}}\right) \approx 440.2 \text{ mi/hr}$$

$\boxed{57}$ The angle φ has an adjacent side of $\frac{1}{2}(230$ m), or 115 m. $\tan\varphi = \frac{147}{115} \Rightarrow \varphi \approx 52°$.

$\boxed{59}$ The depth of the cone is 4 inches and its slant height is 5 inches. A right triangle is formed by a cross section of the cone with sides 4, 5, and r. Thus, $4^2 + r^2 = 5^2 \Rightarrow$ $r = 3$ inches. The circumference of the rim of the cone is $2\pi r = 2\pi(3) = 6\pi$.

This circumference is the arc length from A to B on the circle.

Using the formula for arc length, $\theta = \frac{s}{r} = \frac{6\pi}{5}$ radians $= 216°$.

61 (a) Let h denote the height of the building and x the distance between the two buildings. $\tan 59° = \dfrac{h-50}{x}$ and $\tan 62° = \dfrac{h}{x}$. We now solve both of these equations for h, giving us $h = x \tan 59° + 50$ and $h = x \tan 62°$. Setting these expressions equal to each other and solving for x, we have

$x \tan 62° = x \tan 59° + 50 \Rightarrow x \tan 62° - x \tan 59° = 50 \Rightarrow$

$$x(\tan 62° - \tan 59°) = 50 \Rightarrow x = \frac{50}{\tan 62° - \tan 59°} \approx 231.0 \text{ ft.}$$

(b) From part (a), $h = x \tan 62° \approx 434.5$ ft.

63 (a) Extend the two boundary lines for h { call these l_{top} and l_{bottom} } to the right until they intersect a line l extended down from the front edge of the building. Let x denote the distance from the intersection of the incline and l_{bottom} to l and y the distance on l from l_{top} to the lower left corner of the building.

$\cos \alpha = \dfrac{x}{d} \Rightarrow x = d \cos \alpha.$ $\sin \alpha = \dfrac{h+y}{d} \Rightarrow h + y = d \sin \alpha \Rightarrow y = d \sin \alpha - h.$

$\tan \theta = \dfrac{y+T}{x} \Rightarrow y + T = x \tan \theta \Rightarrow$

$$\begin{aligned}
T = x \tan \theta - y &= (d \cos \alpha) \tan \theta - (d \sin \alpha - h) \\
&= d \cos \alpha \tan \theta - d \sin \alpha + h \\
&= h + (d \cos \alpha \tan \theta - d \sin \alpha) = h + d(\cos \alpha \tan \theta - \sin \alpha).
\end{aligned}$$

(b) $T = 6 + 50 (\cos 15° \tan 31.4° - \sin 15°) \approx 6 + 50 (0.3308) \approx 22.54$ ft.

65 (a) Let $x = \overline{PT}$ and $y = \overline{QT}$. Now $x^2 + d^2 = y^2$, $h = x \sin \alpha$, and $h = y \sin \beta$.

$$d^2 = y^2 - x^2 = \left(\frac{h}{\sin \beta}\right)^2 - \left(\frac{h}{\sin \alpha}\right)^2 = \frac{h^2}{\sin^2 \beta} - \frac{h^2}{\sin^2 \alpha} = \frac{h^2 \sin^2 \alpha - h^2 \sin^2 \beta}{\sin^2 \beta \sin^2 \alpha} =$$

$$\frac{h^2(\sin^2 \alpha - \sin^2 \beta)}{\sin^2 \alpha \sin^2 \beta} \Rightarrow h^2 = \frac{d^2 \sin^2 \alpha \sin^2 \beta}{\sin^2 \alpha - \sin^2 \beta} \Rightarrow h = \frac{d \sin \alpha \sin \beta}{\sqrt{\sin^2 \alpha - \sin^2 \beta}}.$$

(b) $\alpha = 30°$, $\beta = 20°$, and $d = 10 \Rightarrow h = \dfrac{10 \sin 30° \sin 20°}{\sqrt{\sin^2 30° - \sin^2 20°}} \approx 4.69$ miles.

67 (a) The side opposite angle θ is one-half the length of the base, $\frac{1}{2}x$. $\sin \theta = \dfrac{\frac{1}{2}x}{a} \Rightarrow$ $x = 2a \sin \theta$. The area of each face is the area of a triangle, and S is the total area of the four faces. The area of one face is $\frac{1}{2}(\text{base})(\text{height}) = \frac{1}{2}xa$.

$$S = 4(\tfrac{1}{2}ax) = 2ax = 2a(2a \sin \theta) = 4a^2 \sin \theta.$$

(b) $\cos \theta = \dfrac{y}{a} \Rightarrow y = a \cos \theta.$ The volume of a pyramid is one-third times the area of its base times its height. Hence,

$$V = \tfrac{1}{3}(\text{base area})(\text{height}) = \tfrac{1}{3}x^2 y = \tfrac{1}{3}(2a \sin \theta)^2 (a \cos \theta) = \tfrac{4}{3}a^3 \sin^2 \theta \cos \theta.$$

$\boxed{69}$ $y = 1 - 1\cos\left(\frac{1}{2}\pi x/10\right) = -\cos\left(\frac{\pi}{20}x\right) + 1$. To obtain the range for y, we will start with the range for x, perform operations on these values, and try to obtain the expression for y. For $0 \le x \le 10$, $0 \le \frac{\pi}{20}x \le \frac{\pi}{2}$ { multiply by $\frac{\pi}{20}$ }, $1 \ge \cos\left(\frac{\pi}{20}x\right) \ge 0$ { take the cosine of all 3 parts }, $-1 \le -\cos\left(\frac{\pi}{20}x\right) \le 0$ { multiply by -1 }, $0 \le -\cos\left(\frac{\pi}{20}x\right) + 1 \le 1$ { add 1 }, which is equivalent to $0 \le y \le 1$.

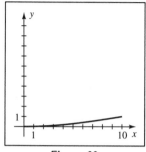

Figure 69

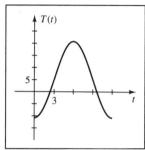

Figure 71

$\boxed{71}$ (a) $p = \frac{2\pi}{\pi/6} = 12$ months; range $= \underline{5 - 15.8}$ to $\underline{5 + 15.8}$, or, equivalently,

-10.8 to 20.8; phase shift $= 3$

(b) The highest temperature will occur when the argument of the sine is $\frac{\pi}{2}$ since this is when the sine function has a maximum. $\frac{\pi}{6}(t - 3) = \frac{\pi}{2} \Rightarrow t - 3 = 3 \Rightarrow$

$t = 6$ months. This is July 1st and the temperature is $20.8\,°\text{C}$, or $69.44\,°\text{F}$.

$\boxed{73}$ (a) The cork is in simple harmonic motion. At $t = 0$, its height is $s(0) = 12 + \cos 0 = 12 + 1 = 13$ ft. It decreases until $t = 1$, reaching a minimum of $s(1) = 12 + \cos \pi = 12 + (-1) = 11$ ft. It then increases, reaching a maximum of 13 ft at $t = 2$.

(b) From part (a), the cork is rising for $1 \le t \le 2$.

Chapter 7: Analytic Trigonometry

1 $\csc\theta - \sin\theta = \dfrac{1}{\sin\theta} - \sin\theta = \dfrac{1-\sin^2\theta}{\sin\theta} = \dfrac{\cos^2\theta}{\sin\theta} = \dfrac{\cos\theta}{\sin\theta}\cdot\cos\theta = \cot\theta\;\cos\theta$

3 $\dfrac{\sec^2 u - 1}{\sec^2 u} = 1 - \dfrac{1}{\sec^2 u}\;\{\text{split up fraction}\} = 1 - \cos^2 u = \sin^2 u$

5 $\dfrac{\csc^2\theta}{1+\tan^2\theta} = \dfrac{\csc^2\theta}{\sec^2\theta} = \dfrac{1/\sin^2\theta}{1/\cos^2\theta} = \dfrac{\cos^2\theta}{\sin^2\theta} = \left(\dfrac{\cos\theta}{\sin\theta}\right)^2 = \cot^2\theta$

7 $\dfrac{1+\cos t}{\sin t} + \dfrac{\sin t}{1+\cos t} = \dfrac{(1+\cos t)^2 + \sin^2 t}{\sin t\,(1+\cos t)}$ $\{\text{combine fractions}\}$

$$= \dfrac{1+2\cos t + \cos^2 t + \sin^2 t}{\sin t\,(1+\cos t)}\quad\{\text{expand}\}$$

$$= \dfrac{2+2\cos t}{\sin t\,(1+\cos t)}\qquad\{\cos^2 t + \sin^2 t = 1\}$$

$$= \dfrac{2(1+\cos t)}{\sin t\,(1+\cos t)}\qquad\{\text{factor out 2}\}$$

$$= 2\csc t\qquad\qquad\{\text{cancel like term}\}$$

9 $\dfrac{1}{1-\cos\gamma} + \dfrac{1}{1+\cos\gamma} = \dfrac{1+\cos\gamma + 1 - \cos\gamma}{1-\cos^2\gamma} = \dfrac{2}{\sin^2\gamma} = 2\csc^2\gamma$

11 $(\sec u - \tan u)(\csc u + 1) = \left(\dfrac{1}{\cos u} - \dfrac{\sin u}{\cos u}\right)\left(\dfrac{1}{\sin u} + 1\right)$ $\{\text{change to sines and cosines}\}$

$$= \left(\dfrac{1-\sin u}{\cos u}\right)\left(\dfrac{1+\sin u}{\sin u}\right)\qquad\{\text{combine fractions}\}$$

$$= \dfrac{1-\sin^2 u}{\cos u\,\sin u}\qquad\qquad\{\text{multiply terms}\}$$

$$= \dfrac{\cos^2 u}{\cos u\,\sin u} = \dfrac{\cos u}{\sin u} = \cot u$$

13 $\csc^4 t - \cot^4 t = (\csc^2 t)^2 - (\cot^2 t)^2$ $\{\text{recognize as the diff. of 2 squares}\}$

$$= (\csc^2 t + \cot^2 t)(\csc^2 t - \cot^2 t)\quad\{\text{factor}\}$$

$$= (\csc^2 t + \cot^2 t)(1)\qquad\qquad\{\text{Pythagorean id., } 1+\cot^2 t = \csc^2 t\}$$

$$= \csc^2 t + \cot^2 t$$

[15] The first step in the following verification is to multiply the denominator, $1 - \sin\beta$, by its conjugate, $1 + \sin\beta$. This procedure will give us the difference of two squares, $1 - \sin^2\beta$, which is equal to $\cos^2\beta$. This step is often helpful when simplifying trigonometric expressions because manipulation of the Pythagorean identities often allows us to reduce the resulting expression to a single term.

$$\frac{\cos\beta}{1-\sin\beta} = \frac{\cos\beta}{1-\sin\beta}\cdot\frac{1+\sin\beta}{1+\sin\beta} = \frac{\cos\beta\,(1+\sin\beta)}{1-\sin^2\beta} = \frac{\cos\beta\,(1+\sin\beta)}{\cos^2\beta} = \frac{1+\sin\beta}{\cos\beta} =$$

$$\frac{1}{\cos\beta} + \frac{\sin\beta}{\cos\beta} = \sec\beta + \tan\beta$$

[17] $\dfrac{\tan^2 x}{\sec x+1} = \dfrac{\sec^2 x-1}{\sec x+1} = \dfrac{(\sec x+1)(\sec x-1)}{\sec x+1} = \sec x - 1 = \dfrac{1}{\cos x} - 1 = \dfrac{1-\cos x}{\cos x}$

[19] $\dfrac{\cot u-1}{\cot u+1} = \dfrac{\dfrac{1}{\tan u}-1}{\dfrac{1}{\tan u}+1} = \dfrac{\dfrac{1-\tan u}{\tan u}}{\dfrac{1+\tan u}{\tan u}} = \dfrac{1-\tan u}{1+\tan u}$

[21] $\sin^4 r - \cos^4 r = (\sin^2 r - \cos^2 r)(\sin^2 r + \cos^2 r) = (\sin^2 r - \cos^2 r)(1) = \sin^2 r - \cos^2 r$

[23] $\tan^4 k - \sec^4 k = (\tan^2 k - \sec^2 k)(\tan^2 k + \sec^2 k)$ { factor as the diff. of 2 squares }

$\qquad = (-1)(\sec^2 k - 1 + \sec^2 k)$ { Pythagorean identity }

$\qquad = (-1)(2\sec^2 k - 1) = 1 - 2\sec^2 k$

[25] $(\sec t + \tan t)^2 = \left(\dfrac{1}{\cos t} + \dfrac{\sin t}{\cos t}\right)^2 = \left(\dfrac{1+\sin t}{\cos t}\right)^2 = \dfrac{(1+\sin t)^2}{\cos^2 t} =$

$$\frac{(1+\sin t)^2}{1-\sin^2 t} = \frac{(1+\sin t)^2}{(1+\sin t)(1-\sin t)} = \frac{1+\sin t}{1-\sin t}$$

[27] $(\sin^2\theta + \cos^2\theta)^3 = (1)^3 = 1$

[29] $\dfrac{1+\csc\beta}{\cot\beta+\cos\beta} = \dfrac{1+\dfrac{1}{\sin\beta}}{\dfrac{\cos\beta}{\sin\beta}+\cos\beta} = \dfrac{\dfrac{\sin\beta+1}{\sin\beta}}{\dfrac{\cos\beta+\cos\beta\sin\beta}{\sin\beta}} = \dfrac{\sin\beta+1}{\cos\beta\,(1+\sin\beta)} = \dfrac{1}{\cos\beta} = \sec\beta$

[31] As demonstrated in the following verification, it may be beneficial to try to combine expressions rather than expand them. $(\csc t - \cot t)^4(\csc t + \cot t)^4 =$

$$\left[(\csc t - \cot t)(\csc t + \cot t)\right]^4 = (\csc^2 t - \cot^2 t)^4 = (1)^4 = 1$$

[33] $\text{RS} = \dfrac{\tan\alpha+\tan\beta}{1-\tan\alpha\,\tan\beta} = \dfrac{\dfrac{\sin\alpha}{\cos\alpha}+\dfrac{\sin\beta}{\cos\beta}}{1-\dfrac{\sin\alpha}{\cos\alpha}\cdot\dfrac{\sin\beta}{\cos\beta}} = \dfrac{\dfrac{\sin\alpha\cos\beta+\cos\alpha\sin\beta}{\cos\alpha\cos\beta}}{\dfrac{\cos\alpha\cos\beta-\sin\alpha\sin\beta}{\cos\alpha\cos\beta}} =$

$$\frac{\sin\alpha\cos\beta+\cos\alpha\sin\beta}{\cos\alpha\cos\beta-\sin\alpha\sin\beta} = \text{LS}$$

Note: We could obtain the RS by dividing

the numerator and denominator of the LS by $(\cos\alpha\,\cos\beta)$.

35 $\dfrac{\tan\alpha}{1+\sec\alpha}+\dfrac{1+\sec\alpha}{\tan\alpha}=\dfrac{\tan^2\alpha+(1+\sec\alpha)^2}{(1+\sec\alpha)\tan\alpha}=\dfrac{(\sec^2\alpha-1)+(1+2\sec\alpha+\sec^2\alpha)}{(1+\sec\alpha)\tan\alpha}=$

$$\dfrac{2\sec^2\alpha+2\sec\alpha}{(1+\sec\alpha)\tan\alpha}=\dfrac{2\sec\alpha\,(\sec\alpha+1)\cot\alpha}{1+\sec\alpha}=\dfrac{2}{\cos\alpha}\cdot\dfrac{\cos\alpha}{\sin\alpha}=\dfrac{2}{\sin\alpha}=2\csc\alpha$$

37 $\dfrac{1}{\tan\beta+\cot\beta}=\dfrac{1}{\dfrac{\sin\beta}{\cos\beta}+\dfrac{\cos\beta}{\sin\beta}}=\dfrac{1}{\dfrac{\sin^2\beta+\cos^2\beta}{\cos\beta\,\sin\beta}}=\dfrac{1}{\dfrac{1}{\cos\beta\,\sin\beta}}=\sin\beta\,\cos\beta$

39 $\sec\theta+\csc\theta-\cos\theta-\sin\theta=\dfrac{1}{\cos\theta}+\dfrac{1}{\sin\theta}-\cos\theta-\sin\theta$

$$=\left(\dfrac{1}{\cos\theta}-\cos\theta\right)+\left(\dfrac{1}{\sin\theta}-\sin\theta\right)$$

$$=\dfrac{1-\cos^2\theta}{\cos\theta}+\dfrac{1-\sin^2\theta}{\sin\theta}$$

$$=\dfrac{\sin^2\theta}{\cos\theta}+\dfrac{\cos^2\theta}{\sin\theta}$$

$$=\left(\sin\theta\cdot\dfrac{\sin\theta}{\cos\theta}\right)+\left(\cos\theta\cdot\dfrac{\cos\theta}{\sin\theta}\right)=\sin\theta\,\tan\theta+\cos\theta\,\cot\theta$$

41 $\text{RS}=\sec^4\phi-4\tan^2\phi=(\sec^2\phi)^2-4\tan^2\phi=(1+\tan^2\phi)^2-4\tan^2\phi=$

$$(1+2\tan^2\phi+\tan^4\phi)-4\tan^2\phi=1-2\tan^2\phi+\tan^4\phi=(1-\tan^2\phi)^2=\text{LS}$$

43 $\dfrac{\cot(-t)+\tan(-t)}{\cot t}=\dfrac{-\cot t-\tan t}{\cot t}=-\dfrac{\cot t}{\cot t}-\dfrac{\tan t}{\cot t}=-(1+\tan^2 t)=-\sec^2 t$

45 $\log 10^{\tan t}=\log_{10}10^{\tan t}=\tan t$, since $\log_a a^x=x$

47 $\ln\cot x=\ln(\cot x)=\ln(\tan x)^{-1}=-\ln(\tan x)\ \{\text{since }\ln a^x=x\ln a\}=-\ln\tan x$

49 $\ln|\sec\theta+\tan\theta|\ =\ln\left|\dfrac{(\sec\theta+\tan\theta)(\sec\theta-\tan\theta)}{\sec\theta-\tan\theta}\right|$ {multiply by the conjugate}

$$=\ln\left|\dfrac{\sec^2\theta-\tan^2\theta}{\sec\theta-\tan\theta}\right| \qquad \{\text{simplify}\}$$

$$=\ln\left|\dfrac{1}{\sec\theta-\tan\theta}\right| \qquad \{\text{Pythagorean identity}\}$$

$$=\ln\dfrac{|1|}{|\sec\theta-\tan\theta|} \qquad \left\{\left|\dfrac{a}{b}\right|=\dfrac{|a|}{|b|}\right\}$$

$$=\ln|1|-\ln|\sec\theta-\tan\theta| \qquad \left\{\ln\dfrac{a}{b}=\ln a-\ln b\right\}$$

$$=-\ln|\sec\theta-\tan\theta| \qquad \{\ln 1=0\}$$

51 $\cos^2 t=1-\sin^2 t\Rightarrow\cos t=\pm\sqrt{1-\sin^2 t}$. Since the given equation is

$\cos t=+\sqrt{1-\sin^2 t}$, we may choose any t such that $\cos t<0$.

Using $t=\pi$, LS $=\cos\pi=-1$. RS $=\sqrt{1-\sin^2\pi}=1$. Since $-1\neq 1$, LS $\neq$ RS.

[53] $\sqrt{\sin^2 t} = |\sin t| = \pm \sin t$. Hence, choose any t such that $\sin t < 0$.

Using $t = \frac{3\pi}{2}$, LS $= \sqrt{(-1)^2} = 1$. RS $= \sin \frac{3\pi}{2} = -1$. Since $1 \neq -1$, LS $\neq$ RS.

[55] $(\sin \theta + \cos \theta)^2 = \sin^2\theta + 2\sin \theta \cos \theta + \cos^2\theta$. Since the right side of the given

equation is only $\sin^2\theta + \cos^2\theta$, we may choose any θ such that $2\sin \theta \cos \theta \neq 0$.

Using $\theta = \frac{\pi}{4}$, LS $= (\frac{1}{2}\sqrt{2} + \frac{1}{2}\sqrt{2})^2 = (\sqrt{2})^2 = 2$.

$$RS = (\tfrac{1}{2}\sqrt{2})^2 + (\tfrac{1}{2}\sqrt{2})^2 = \tfrac{1}{2} + \tfrac{1}{2} = 1. \text{ Since } 2 \neq 1, \text{ LS} \neq \text{RS.}$$

[57] $\cos(-t) = -\cos t$ • Since $\cos(-t) = \cos t$, we may choose any t such that

$\cos t \neq -\cos t$—that is, any t such that $\cos t \neq 0$. Using $t = \pi$, LS $= \cos(-\pi) = -1$.

$$RS = -\cos \pi = -(-1) = 1. \text{ Since } -1 \neq 1, \text{ LS} \neq \text{RS.}$$

[59] Don't confuse $\cos(\sec t) = 1$ with $\cos t \cdot \sec t = 1$. The former is true if $\sec t = 2\pi$ or

an integer multiple of 2π. The latter is true for any value of t as long as $\sec t$ is

defined. Choose any t such that $\sec t \neq 2\pi n$.

$$\text{Using } t = \tfrac{\pi}{4}, \text{ LS} = \cos(\sec \tfrac{\pi}{4}) = \cos \sqrt{2} \neq 1 = \text{RS.}$$

[61] $\sin^2 t - 4\sin t - 5 = 0 \Rightarrow (\sin t - 5)(\sin t + 1) = 0 \Rightarrow \sin t = 5$ or $\sin t = -1$.

Since $\sin t$ cannot equal 5, we may choose any t such that $\sin t \neq -1$.

$$\text{Using } t = \pi, \text{ LS} = -5 \neq 0 = \text{RS.}$$

Note: Exer. 63–66: Use $\sqrt{a^2 - x^2} = a \cos \theta$ because

$$\sqrt{a^2 - x^2} = \sqrt{a^2 - a^2 \sin^2\theta} = \sqrt{a^2(1 - \sin^2\theta)} = \sqrt{a^2 \cos^2\theta} = |a| \, |\cos \theta| = a \cos \theta$$

$$\text{since } \cos \theta > 0 \text{ if } -\tfrac{\pi}{2} < \theta < \tfrac{\pi}{2} \text{ and } a > 0.$$

[63] $(a^2 - x^2)^{3/2} = (\sqrt{a^2 - x^2})^3 = (a \cos \theta)^3 \{ \text{see above note} \} = a^3 \cos^3\theta$

[65] $\dfrac{x^2}{\sqrt{a^2 - x^2}} = \dfrac{a^2 \sin^2\theta}{a \cos \theta} = a \cdot \dfrac{\sin \theta}{\cos \theta} \cdot \sin \theta = a \tan \theta \sin \theta$

Note: Exer. 67–70: Use $\sqrt{a^2 + x^2} = a \sec \theta$ because

$$\sqrt{a^2 + x^2} = \sqrt{a^2 + a^2 \tan^2\theta} = \sqrt{a^2(1 + \tan^2\theta)} = \sqrt{a^2 \sec^2\theta} = |a| \, |\sec \theta| =$$

$$a \sec \theta \text{ since } \sec \theta > 0 \text{ if } -\tfrac{\pi}{2} < \theta < \tfrac{\pi}{2} \text{ and } a > 0.$$

[67] $\sqrt{a^2 + x^2} = a \sec \theta \{ \text{see above note} \}$

[69] $\dfrac{1}{x^2 + a^2} = \dfrac{1}{(\sqrt{a^2 + x^2})^2} = \dfrac{1}{(a \sec \theta)^2} = \dfrac{1}{a^2 \sec^2\theta} = \dfrac{1}{a^2} \cos^2\theta$

Note: Exer. 71–74: Use $\sqrt{x^2 - a^2} = a \tan \theta$ because

$$\sqrt{x^2 - a^2} = \sqrt{a^2 \sec^2\theta - a^2} = \sqrt{a^2(\sec^2\theta - 1)} = \sqrt{a^2 \tan^2\theta} = |a| \, |\tan \theta| =$$

$$a \tan \theta \text{ since } \tan \theta > 0 \text{ if } 0 < \theta < \tfrac{\pi}{2} \text{ and } a > 0.$$

[71] $\sqrt{x^2 - a^2} = a \tan \theta \{ \text{see above note} \}$

[73] $x^3 \sqrt{x^2 - a^2} = (a^3 \sec^3\theta)(a \tan \theta) = a^4 \sec^3\theta \tan \theta$

75 The graph of $f(x) = \dfrac{\sin^2 x - \sin^4 x}{(1 - \sec^2 x)\cos^4 x}$ appears to be that of the horizontal line

$y = g(x) = -1$. Verifying this identity, we have

$$\frac{\sin^2 x - \sin^4 x}{(1 - \sec^2 x)\cos^4 x} = \frac{\sin^2 x(1 - \sin^2 x)}{-\tan^2 x \, \cos^4 x} = \frac{\sin^2 x \, \cos^2 x}{-(\sin^2 x/\cos^2 x)\cos^4 x} = \frac{\sin^2 x \, \cos^2 x}{-\sin^2 x \, \cos^2 x} = -1.$$

77 The graph of $f(x) = \sec x \,(\sin x \, \cos x + \cos^2 x) - \sin x$ appears to be that of

$y = g(x) = \cos x$. Verifying this identity, we have

$$\sec x \,(\sin x \, \cos x + \cos^2 x) - \sin x = \sec x \, \cos x \,(\sin x + \cos x) - \sin x$$

$$= (\sin x + \cos x) - \sin x = \cos x.$$

7.2 Exercises

1 In $[0,\, 2\pi)$, $\sin x = -\dfrac{\sqrt{2}}{2}$ only if $x = \dfrac{5\pi}{4},\, \dfrac{7\pi}{4}$. All solutions would include these angles

plus all angles coterminal with them. Hence, $x = \dfrac{5\pi}{4} + 2\pi n,\, \dfrac{7\pi}{4} + 2\pi n$.

3 $\tan\theta = \sqrt{3} \Rightarrow \theta = \dfrac{\pi}{3} + \pi n$. *Note:* This solution could be written as $\dfrac{\pi}{3} + 2\pi n$ and

$\dfrac{4\pi}{3} + 2\pi n$. Since the period of the tangent (and the cotangent) is π, we use the

abbreviated form $\dfrac{\pi}{3} + \pi n$ to describe all solutions. We will use "πn" for solutions of

exercises involving the tangent and cotangent functions.

7 $\sin x = \dfrac{\pi}{2}$ has no solution since $\dfrac{\pi}{2} > 1$, which is not in the range $[-1,\, 1]$. Your

calculator will give some kind of error message if you attempt to find a solution to

this problem or any similar type problem.

9 $\cos\theta = \dfrac{1}{\sec\theta}$ is true for all values *for which the equation is defined.*

★ All θ except $\theta = \dfrac{\pi}{2} + \pi n$

11 $2\cos 2\theta - \sqrt{3} = 0 \Rightarrow \cos 2\theta = \dfrac{\sqrt{3}}{2}$ $\{\, 2\theta$ is just an angle — so we solve this equation

for 2θ and then divide those solutions by $2\,\} \Rightarrow$

$$2\theta = \frac{\pi}{6} + 2\pi n,\, \frac{11\pi}{6} + 2\pi n \Rightarrow \theta = \frac{\pi}{12} + \pi n,\, \frac{11\pi}{12} + \pi n$$

13 $\sqrt{3} \tan\frac{1}{3}t = 1 \Rightarrow \tan\frac{1}{3}t = \dfrac{1}{\sqrt{3}} \Rightarrow \frac{1}{3}t = \dfrac{\pi}{6} + \pi n \Rightarrow t = \dfrac{\pi}{2} + 3\pi n$

15 $\sin\left(\theta + \frac{\pi}{4}\right) = \frac{1}{2} \Rightarrow \theta + \frac{\pi}{4} = \frac{\pi}{6} + 2\pi n,\, \frac{5\pi}{6} + 2\pi n \Rightarrow \theta = -\frac{\pi}{12} + 2\pi n,\, \frac{7\pi}{12} + 2\pi n$

17 $\sin\left(2x - \frac{\pi}{3}\right) = \frac{1}{2} \Rightarrow 2x - \frac{\pi}{3} = \frac{\pi}{6} + 2\pi n,\, \frac{5\pi}{6} + 2\pi n \Rightarrow 2x = \frac{\pi}{2} + 2\pi n,\, \frac{7\pi}{6} + 2\pi n \Rightarrow$

$$x = \frac{\pi}{4} + \pi n,\, \frac{7\pi}{12} + \pi n$$

21 $\tan^2 x = 1 \Rightarrow \tan x = \pm 1 \Rightarrow x = \frac{\pi}{4} + \pi n,\, \frac{3\pi}{4} + \pi n$, or simply $\frac{\pi}{4} + \frac{\pi}{2}n$

23 $(\cos\theta - 1)(\sin\theta + 1) = 0 \Rightarrow (\cos\theta - 1) = 0$ or $(\sin\theta + 1) = 0 \Rightarrow$

$$\cos\theta = 1 \text{ or } \sin\theta = -1 \Rightarrow \theta = 2\pi n \text{ or } \theta = \frac{3\pi}{2} + 2\pi n$$

$\boxed{25}$ $\sec^2\alpha - 4 = 0 \Rightarrow \sec^2\alpha = 4 \Rightarrow \sec\alpha = \pm 2 \Rightarrow$

$$\alpha = \tfrac{\pi}{3} + 2\pi n,\ \tfrac{5\pi}{3} + 2\pi n,\ \tfrac{2\pi}{3} + 2\pi n,\ \tfrac{4\pi}{3} + 2\pi n,\ \text{or simply } \tfrac{\pi}{3} + \pi n,\ \tfrac{2\pi}{3} + \pi n$$

$\boxed{29}$ $\cot^2 x - 3 = 0 \Rightarrow \cot^2 x = 3 \Rightarrow \cot x = \pm\sqrt{3} \Rightarrow x = \tfrac{\pi}{6} + \pi n,\ \tfrac{5\pi}{6} + \pi n$

$\boxed{31}$ $(2\sin\theta + 1)(2\cos\theta + 3) = 0 \Rightarrow \sin\theta = -\tfrac{1}{2}$ or $\sin\theta = -\tfrac{3}{2} \Rightarrow$

$$\theta = \tfrac{7\pi}{6} + 2\pi n,\ \tfrac{11\pi}{6} + 2\pi n\ \{\sin\theta = -\tfrac{3}{2} \text{ has no solutions}\}$$

$\boxed{33}$ $\sin 2x\,(\csc 2x - 2) = 0 \Rightarrow 1 - 2\sin 2x = 0 \Rightarrow \sin 2x = \tfrac{1}{2} \Rightarrow$

$$2x = \tfrac{\pi}{6} + 2\pi n,\ \tfrac{5\pi}{6} + 2\pi n \Rightarrow x = \tfrac{\pi}{12} + \pi n,\ \tfrac{5\pi}{12} + \pi n$$

$\boxed{35}$ $\cos\left(2x - \tfrac{\pi}{4}\right) = 0 \Rightarrow 2x - \tfrac{\pi}{4} = \tfrac{\pi}{2} + \pi n \Rightarrow 2x = \tfrac{3\pi}{4} + \pi n \Rightarrow x = \tfrac{3\pi}{8} + \tfrac{\pi}{2}n.$

x will be in the interval $[0, 2\pi)$ if $n = 0$, 1, 2, or 3. Thus, $x = \tfrac{3\pi}{8},\ \tfrac{7\pi}{8},\ \tfrac{11\pi}{8},\ \tfrac{15\pi}{8}$.

$\boxed{37}$ $2 - 8\cos^2 t = 0 \Rightarrow \cos^2 t = \tfrac{1}{4} \Rightarrow \cos t = \pm\tfrac{1}{2} \Rightarrow t = \tfrac{\pi}{3},\ \tfrac{2\pi}{3},\ \tfrac{4\pi}{3},\ \tfrac{5\pi}{3}$

$\boxed{39}$ $2\sin^2 u = 1 - \sin u \Rightarrow 2\sin^2 u + \sin u - 1 = 0 \Rightarrow (2\sin u - 1)(\sin u + 1) = 0 \Rightarrow$

$$\sin u = \tfrac{1}{2},\ -1 \Rightarrow u = \tfrac{\pi}{6},\ \tfrac{5\pi}{6},\ \tfrac{3\pi}{2}$$

$\boxed{41}$ $\tan^2 x\,\sin x = \sin x \Rightarrow \tan^2 x\,\sin x - \sin x = 0 \Rightarrow \sin x\,(\tan^2 x - 1) = 0 \Rightarrow \sin x = 0$ or

$\tan x = \pm 1 \Rightarrow x = 0,\ \pi,\ \tfrac{\pi}{4},\ \tfrac{3\pi}{4},\ \tfrac{5\pi}{4},\ \tfrac{7\pi}{4}$. *Note:* A common mistake is to divide both

sides of the given equation by $\sin x$—doing so results in losing the solutions for

$\sin x = 0$.

$\boxed{43}$ $2\cos^2\gamma + \cos\gamma = 0 \Rightarrow \cos\gamma\,(2\cos\gamma + 1) = 0 \Rightarrow \cos\gamma = 0,\ -\tfrac{1}{2} \Rightarrow \gamma = \tfrac{\pi}{2},\ \tfrac{3\pi}{2},\ \tfrac{2\pi}{3},\ \tfrac{4\pi}{3}$

$\boxed{45}$ $\sin^2\theta + \sin\theta - 6 = 0 \Rightarrow (\sin\theta + 3)(\sin\theta - 2) = 0 \Rightarrow \sin\theta = -3,\ 2.$

There are *no solutions* for either equation.

$\boxed{47}$ $1 - \sin t = \sqrt{3}\cos t$ $\bullet$ Square both sides to obtain an equation in either sin or cos.

$(1 - \sin t)^2 = (\sqrt{3}\cos t)^2 \Rightarrow 1 - 2\sin t + \sin^2 t = 3\cos^2 t \Rightarrow$

$\sin^2 t - 2\sin t + 1 = 3(1 - \sin^2 t) \Rightarrow 4\sin^2 t - 2\sin t - 2 = 0 \Rightarrow 2\sin^2 t - \sin t - 1 = 0 \Rightarrow$

$(2\sin t + 1)(\sin t - 1) = 0 \Rightarrow \sin t = -\tfrac{1}{2},\ 1 \Rightarrow t = \tfrac{7\pi}{6},\ \tfrac{11\pi}{6},\ \tfrac{\pi}{2}$. Since each side of the

equation was squared, the solutions must be checked in the original equation.

Checking $\tfrac{7\pi}{6}$, we have $\text{LS} = 1 - \sin\tfrac{7\pi}{6} = 1 - (-\tfrac{1}{2}) = \tfrac{3}{2}$ and $\text{RS} = \sqrt{3}\cos\tfrac{7\pi}{6} =$

$\sqrt{3}\left(-\tfrac{\sqrt{3}}{2}\right) = -\tfrac{3}{2}$. Since $\text{LS} \neq \text{RS}$, $\tfrac{7\pi}{6}$ is an extraneous solution. Checking $\tfrac{11\pi}{6}$, we

have $\text{LS} = \tfrac{3}{2}$ and $\text{RS} = \tfrac{3}{2}$. Since $\text{LS} = \text{RS}$, $\tfrac{11\pi}{6}$ is a valid solution. Similarly, $\tfrac{\pi}{2}$ is a

valid solution and our solution is $\tfrac{\pi}{2}$ and $\tfrac{11\pi}{6}$.

$\boxed{49}$ $\cos\alpha + \sin\alpha = 1 \Rightarrow \cos\alpha = 1 - \sin\alpha\ \{\text{square both sides}\} \Rightarrow$

$\cos^2\alpha = 1 - 2\sin\alpha + \sin^2\alpha$

$\{\text{change } \cos^2\alpha \text{ to } 1 - \sin^2\alpha \text{ to obtain an equation involving only } \sin\alpha\} \Rightarrow$

$1 - \sin^2\alpha = 1 - 2\sin\alpha + \sin^2\alpha \Rightarrow 2\sin^2\alpha - 2\sin\alpha = 0 \Rightarrow$

$2\sin\alpha\,(\sin\alpha - 1) = 0 \Rightarrow \sin\alpha = 0,\ 1 \Rightarrow \alpha = 0,\ \pi,\ \tfrac{\pi}{2}.$ π is an extraneous solution.

$\boxed{51}$ $2\tan t - \sec^2 t = 0 \Rightarrow 2\tan t - (1 + \tan^2 t) = 0 \Rightarrow \tan^2 t - 2\tan t + 1 = 0 \Rightarrow$

$$(\tan t - 1)^2 = 0 \Rightarrow \tan t = 1 \Rightarrow t = \tfrac{\pi}{4}, \tfrac{5\pi}{4}$$

$\boxed{53}$ $\cot\alpha + \tan\alpha = \csc\alpha\sec\alpha \Rightarrow \dfrac{\cos\alpha}{\sin\alpha} + \dfrac{\sin\alpha}{\cos\alpha} = \dfrac{1}{\sin\alpha\cos\alpha} \Rightarrow$

$\dfrac{\cos^2\alpha + \sin^2\alpha}{\sin\alpha\cos\alpha} = \dfrac{1}{\sin\alpha\cos\alpha}$. This is an identity and is true for *all numbers in* $[0, 2\pi)$

$\quad$ *except* 0, $\tfrac{\pi}{2}$, π, and $\tfrac{3\pi}{2}$ since these values make the original equation undefined.

$\boxed{55}$ $2\sin^3 x + \sin^2 x - 2\sin x - 1 = 0$ { factor by grouping since there are four terms } $\Rightarrow$

$\sin^2 x (2\sin x + 1) - 1(2\sin x + 1) = 0 \Rightarrow$

$$(\sin^2 x - 1)(2\sin x + 1) = 0 \Rightarrow \sin x = \pm 1, -\tfrac{1}{2} \Rightarrow x = \tfrac{\pi}{2}, \tfrac{3\pi}{2}, \tfrac{7\pi}{6}, \tfrac{11\pi}{6}$$

$\boxed{57}$ $2\tan t\csc t + 2\csc t + \tan t + 1 = 0 \Rightarrow 2\csc t(\tan t + 1) + 1(\tan t + 1) \Rightarrow$

$(2\csc t + 1)(\tan t + 1) = 0 \Rightarrow \csc t = -\tfrac{1}{2}$ or $\tan t = -1 \Rightarrow t = \tfrac{3\pi}{4}, \tfrac{7\pi}{4}$ { since $\csc t \neq -\tfrac{1}{2}$ }

$\boxed{59}$ $\sin^2 t - 4\sin t + 1 = 0 \Rightarrow \sin t = \dfrac{4 \pm \sqrt{12}}{2} = 2 \pm \sqrt{3}$.

$(2 + \sqrt{3}) > 1$ is not in the range of the sine, so $\sin t = 2 - \sqrt{3} \Rightarrow$

$$t = 15°30' \text{ or } 164°30' \text{ { to the nearest ten minutes }}$$

$\boxed{61}$ $\tan^2\theta + 3\tan\theta + 2 = 0 \Rightarrow (\tan\theta + 1)(\tan\theta + 2) = 0 \Rightarrow$

$$\tan\theta = -1, -2 \Rightarrow \theta = 135°, 315°, 116°30', 296°30'$$

$\boxed{63}$ $12\sin^2 u - 5\sin u - 2 = 0 \Rightarrow (3\sin u - 2)(4\sin u + 1) = 0 \Rightarrow \sin u = \tfrac{2}{3}, -\tfrac{1}{4} \Rightarrow$

$$u = 41°50', 138°10', 194°30', 345°30'$$

$\boxed{65}$ The top of the wave will be above the sea wall when its height is greater than 12.5.

$y > 12.5 \Rightarrow 25\cos\tfrac{\pi}{15}t > 12.5 \Rightarrow \cos\tfrac{\pi}{15}t > \tfrac{1}{2} \Rightarrow$

{ To visualize this step, it may help to look at a unit circle and draw a vertical line

$\quad$ through 0.5 on the x-axis — $x > 0.5$ is the same as $\cos\tfrac{\pi}{15}t > 0.5$. }

$-\tfrac{\pi}{3} < \tfrac{\pi}{15}t < \tfrac{\pi}{3} \Rightarrow -5 < t < 5$ { multiply by $\tfrac{15}{\pi}$ } $\Rightarrow$

$\quad y > 12.5$ for about $5 - (-5) = 10$ minutes of each 30-minute period.

$\boxed{67}$ $I = \tfrac{1}{2}I_M$ and $D = 12 \Rightarrow \tfrac{1}{2}I_M = I_M \sin^3\tfrac{\pi}{12}t \Rightarrow \sin^3\tfrac{\pi}{12}t = \tfrac{1}{2} \Rightarrow \sin\tfrac{\pi}{12}t = \sqrt[3]{\tfrac{1}{2}} \Rightarrow$

$\tfrac{\pi}{12}t \approx 0.9169$ and 2.2247 { $\pi - 0.9169 \approx 2.2247$ is the reference angle for 0.9169 in

$$\text{QII. } \} \Rightarrow t \approx 3.50 \text{ and } t \approx 8.50$$

$\boxed{69}$ 75% of the maximum intensity $= 0.75\,I_M$

(a) $I > 0.75\,I_M \Rightarrow I_M \sin^3\tfrac{\pi}{12}t > 0.75\,I_M \Rightarrow \sin^3\tfrac{\pi}{12}t > \tfrac{3}{4} \Rightarrow \sin\tfrac{\pi}{12}t > \sqrt[3]{\tfrac{3}{4}} \Rightarrow$

$$1.1398 < \tfrac{\pi}{12}t < 2.0018 \Rightarrow 4.3538 < t < 7.6462, \text{ or approximately 3.29 hours.}$$

(b) $I > 0.75\,I_M \Rightarrow I_M \sin^2\tfrac{\pi}{12}t > 0.75\,I_M \Rightarrow \sin^2\tfrac{\pi}{12}t > \tfrac{3}{4} \Rightarrow \sin\tfrac{\pi}{12}t > \tfrac{1}{2}\sqrt{3} \Rightarrow$

$$\tfrac{\pi}{3} < \tfrac{\pi}{12}t < \tfrac{2\pi}{3} \Rightarrow 4 < t < 8, \text{ or 4 hours.}$$

71 (a) $N(t) = 1000\cos\frac{\pi}{5}t + 4000$, amplitude $= 1000$,

period $= \frac{2\pi}{\pi/5} = 10$ years

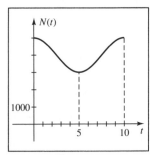

Figure 71

(b) $N > 4500 \Rightarrow 1000\cos\frac{\pi}{5}t + 4000 > 4500 \Rightarrow$

$\cos\frac{\pi}{5}t > \frac{1}{2} \Rightarrow \{$The cosine function is greater

than $\frac{1}{2}$ on $[0, \frac{\pi}{3})$ and $(\frac{5\pi}{3}, 2\pi].\}$

$0 \le \frac{\pi}{5}t < \frac{\pi}{3}$ and $\frac{5\pi}{3} < \frac{\pi}{5}t \le 2\pi \Rightarrow$

$0 \le t < \frac{5}{3}$ and $\frac{25}{3} < t \le 10$

73 $\frac{1}{2} + \cos x = 0 \Rightarrow \cos x = -\frac{1}{2} \Rightarrow x = -\frac{4\pi}{3},\ -\frac{2\pi}{3},\ \frac{2\pi}{3},$ and $\frac{4\pi}{3}$ $\{$for x in $[-2\pi, 2\pi]\}$

for A, B, C, and D, respectively. The corresponding y values are found by using

$y = \frac{1}{2}x + \sin x$ with each of the above values. The points are:

$A(-\frac{4\pi}{3}, -\frac{2\pi}{3} + \frac{1}{2}\sqrt{3})$, $B(-\frac{2\pi}{3}, -\frac{\pi}{3} - \frac{1}{2}\sqrt{3})$, $C(\frac{2\pi}{3}, \frac{\pi}{3} + \frac{1}{2}\sqrt{3})$, and $D(\frac{4\pi}{3}, \frac{2\pi}{3} - \frac{1}{2}\sqrt{3})$

75 $I(t) = k \Rightarrow 20\sin(60\pi t - 6\pi) = -10 \Rightarrow \sin(60\pi t - 6\pi) = -\frac{1}{2} \Rightarrow 60\pi t_1 - 6\pi = \frac{7\pi}{6} + 2\pi n$

or $60\pi t_2 - 6\pi = \frac{11\pi}{6} + 2\pi n$ $\{$We use t_1 and t_2 to distinguish between angles

coterminal with $\frac{7\pi}{6}$ and those coterminal with $\frac{11\pi}{6}.\} \Rightarrow 60t_1 = \frac{43}{6} + 2n$ or

$60t_2 = \frac{47}{6} + 2n \Rightarrow t_1 = \frac{43}{360} + \frac{1}{30}n$ or $t_2 = \frac{47}{360} + \frac{1}{30}n$. We must now find the smallest

positive value of t_1 or t_2. $t_1 > 0 \Rightarrow \frac{1}{30}n > -\frac{43}{360} \Rightarrow n > -\frac{43}{12} \approx -3.58$. The last result

indicates that n must be an integer in the set $\{-3, -2, -1, \ldots\}$. If $n = -3$, then

$t_1 = \frac{7}{360}$. Greater values of n yield greater values of t_1. Similarly, for t_2,

$t_2 > 0 \Rightarrow \frac{1}{30}n > -\frac{47}{360} \Rightarrow n > -\frac{47}{12}$. If $n = -3$, then $t_1 = \frac{11}{360}$. Thus, the smallest exact

value of t for which $I(t) = -10$ is $t = \frac{7}{360}$ sec.

77 Graph $y = \cos x$ and $y = 0.3$ on the same coordinate plane. See *Figure 77*.

The points of intersection are located at $x \approx 1.27$, 5.02, and $\cos x$ is less than 0.3

between these values. Therefore, $\cos x \ge 0.3$ on $[0, 1.27] \cup [5.02, 2\pi]$.

$[0, 6.28]$ by $[-2.09, 2.09]$ $[0, 6.28]$ by $[-2.09, 2.09]$

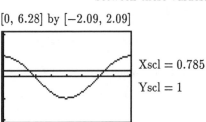

Xscl $= 0.785$

Yscl $= 1$

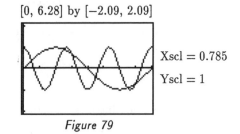

Xscl $= 0.785$

Yscl $= 1$

Figure 77   Figure 79

79 Graph $y = \cos 3x$ and $y = \sin x$ on the same coordinate plane. See *Figure 79*.

The points of intersection are located at $x \approx 0.39$, 1.96, 2.36, 3.53, 5.11, 5.50.

From the graph, we see that $\cos 3x$ is less than $\sin x$ on

$(0.39, 1.96) \cup (2.36, 3.53) \cup (5.11, 5.50)$.

81 Graph $y = \sin 2x$ and $y = 2 - x^2$. From the graph, we see that there are two points of intersection. The x-coordinates of these points are $x \approx -1.48,\ 1.08$.

[−3.14, 3.14] by [−2.09, 2.09]

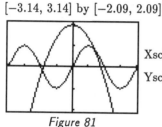

Xscl = 0.785

Yscl = 1

Figure 81

[−3.14, 3.14] by [−2.09, 2.09]

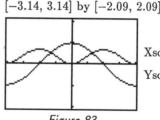

Xscl = 0.785

Yscl = 1

Figure 83

83 Graph $y = \ln(1 + \sin^2 x)$ and $y = \cos x$. From the graph, we see that there are two points of intersection. The x-coordinates of these points are $x \approx \pm 1.00$.

87 (a) $g = 9.8 \Rightarrow 9.8 = 9.8066(1 - 0.00264 \cos 2\theta) \Rightarrow 0.00264 \cos 2\theta = 1 - \frac{9.8}{9.8066} \Rightarrow$

$$\cos 2\theta = \frac{0.0066}{(9.8066)(0.00264)} \Rightarrow 2\theta \approx 75.2° \Rightarrow \theta \approx 37.6°$$

(b) At the equator, $g_0 = 9.8066(1 - 0.00264 \cos 0°) = 9.8066(0.99736)$. Since the weight W of a person on the earth's surface is directly proportional to the force of gravity, we have $W = kg$.

At $\theta = 0°$, $W = kg \Rightarrow 150 = kg_0 \Rightarrow k = \frac{150}{g_0} \Rightarrow W = \frac{150}{g_0} g$.

$$W = 150.5 \Rightarrow 150.5 = \frac{150}{g_0} g \Rightarrow 150.5 = \frac{150 \cdot 9.8066(1 - 0.00264 \cos 2\theta)}{9.8066(0.99736)} \Rightarrow$$

$$150.5 = \frac{150}{0.99736}(1 - 0.00264 \cos 2\theta) \Rightarrow 0.00264 \cos 2\theta = 1 - \frac{150.5(0.99736)}{150} \Rightarrow$$

$$\cos 2\theta \approx -0.2593 \Rightarrow 2\theta \approx 105.0° \Rightarrow \theta \approx 52.5°.$$

7.3 Exercises

1 *Note:* Use the cofunction formulas with $\left(\frac{\pi}{2} - u\right)$ for the argument if you're working with radian measure, $(90° - u)$ if you're working with degree measure.

(a) $\sin 46°37' = \cos(90° - 46°37') = \cos 43°23'$

(b) $\cos 73°12' = \sin(90° - 73°12') = \sin 16°48'$

(c) $\tan \frac{\pi}{6} = \cot\left(\frac{\pi}{2} - \frac{\pi}{6}\right) = \cot\left(\frac{3\pi}{6} - \frac{\pi}{6}\right) = \cot \frac{2\pi}{6} = \cot \frac{\pi}{3}$

(d) $\sec 17.28° = \csc(90° - 17.28°) = \csc 72.72°$

3 (a) $\cos \frac{7\pi}{20} = \sin\left(\frac{\pi}{2} - \frac{7\pi}{20}\right) = \sin \frac{3\pi}{20}$ (b) $\sin \frac{1}{4} = \cos\left(\frac{\pi}{2} - \frac{1}{4}\right) = \cos\left(\frac{2\pi - 1}{4}\right)$

(c) $\tan 1 = \cot\left(\frac{\pi}{2} - 1\right) = \cot\left(\frac{\pi - 2}{2}\right)$ (d) $\csc 0.53 = \sec\left(\frac{\pi}{2} - 0.53\right)$

5 (a) $\cos \frac{\pi}{4} + \cos \frac{\pi}{6} = \frac{\sqrt{2}}{2} + \frac{\sqrt{3}}{2} = \frac{\sqrt{2} + \sqrt{3}}{2}$

(b) $\cos \frac{5\pi}{12} = \cos\left(\frac{\pi}{4} + \frac{\pi}{6}\right) = \cos \frac{\pi}{4} \cos \frac{\pi}{6} - \sin \frac{\pi}{4} \sin \frac{\pi}{6} = \frac{\sqrt{2}}{2} \cdot \frac{\sqrt{3}}{2} - \frac{\sqrt{2}}{2} \cdot \frac{1}{2} = \frac{\sqrt{6} - \sqrt{2}}{4}$

$\boxed{7}$ (a) $\tan 60° + \tan 225° = \sqrt{3} + 1$

     (b) $\tan 285° = \tan(60° + 225°) =$

$$\frac{\tan 60° + \tan 225°}{1 - \tan 60° \tan 225°} = \frac{\sqrt{3}+1}{1-(\sqrt{3})(1)} \cdot \frac{1+\sqrt{3}}{1+\sqrt{3}} = \frac{4+2\sqrt{3}}{-2} = -2 - \sqrt{3}$$

$\boxed{9}$ (a) $\sin\frac{3\pi}{4} - \sin\frac{\pi}{6} = \frac{\sqrt{2}}{2} - \frac{1}{2} = \frac{\sqrt{2}-1}{2}$

     (b) $\sin\frac{7\pi}{12} = \sin\left(\frac{3\pi}{4} - \frac{\pi}{6}\right) = \sin\frac{3\pi}{4}\cos\frac{\pi}{6} - \cos\frac{3\pi}{4}\sin\frac{\pi}{6} = \frac{\sqrt{2}}{2} \cdot \frac{\sqrt{3}}{2} - \left(-\frac{\sqrt{2}}{2}\right) \cdot \frac{1}{2} = \frac{\sqrt{6}+\sqrt{2}}{4}$

$\boxed{11}$ Since the expression is of the form "cos cos plus sin sin", we recognize it as the subtraction formula for the cosine.

$$\cos 48° \cos 23° + \sin 48° \sin 23° = \cos(48° - 23°) = \cos 25°$$

$\boxed{13}$ $\cos 10° \sin 5° - \sin 10° \cos 5° = \sin(5° - 10°) = \sin(-5°)$

$\boxed{15}$ Since we have angle arguments of 2, 3, and -2, we want to change one of them so that we can apply one of the formulas. We recognize that $\cos 2 = \cos(-2)$ and this is probably the simplest change. $\cos 3 \sin(-2) - \cos 2 \sin 3 =$

$$\sin(-2)\cos 3 - \cos(-2)\sin 3 = \sin(-2 - 3) = \sin(-5)$$

$\boxed{17}$ See *Figure 17* for a drawing of angles α and β.

     (a) $\sin(\alpha + \beta) = \sin\alpha\cos\beta + \cos\alpha\sin\beta = \frac{3}{5} \cdot \frac{15}{17} + \frac{4}{5} \cdot \frac{8}{17} = \frac{77}{85}$

     (b) $\cos(\alpha + \beta) = \cos\alpha\cos\beta - \sin\alpha\sin\beta = \frac{4}{5} \cdot \frac{15}{17} - \frac{3}{5} \cdot \frac{8}{17} = \frac{36}{85}$

     (c)   Since the sine and cosine of $(\alpha + \beta)$ are positive, $(\alpha + \beta)$ is in QI.

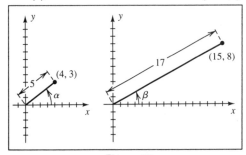 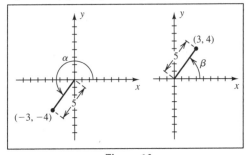

Figure 17   *Figure 19*

$\boxed{19}$ See *Figure 19* for a drawing of angles α and β.

     (a) $\sin(\alpha + \beta) = \sin\alpha\cos\beta + \cos\alpha\sin\beta = \left(-\frac{4}{5}\right) \cdot \frac{3}{5} + \left(-\frac{3}{5}\right) \cdot \frac{4}{5} = -\frac{24}{25}$

     (b) $\tan(\alpha + \beta) = \frac{\tan\alpha + \tan\beta}{1 - \tan\alpha\tan\beta} = \frac{\frac{4}{3} + \frac{4}{3}}{1 - \frac{4}{3} \cdot \frac{4}{3}} \cdot \frac{9}{9} = \frac{12 + 12}{9 - 16} = -\frac{24}{7}$

     (c)   Since the sine and tangent of $(\alpha + \beta)$ are negative, $(\alpha + \beta)$ is in QIV.

$\boxed{21}$ (a) $\sin(\alpha - \beta) = \sin\alpha\cos\beta - \cos\alpha\sin\beta =$

$$\left(-\frac{\sqrt{21}}{5}\right) \cdot \left(-\frac{3}{5}\right) - \left(-\frac{2}{5}\right) \cdot \left(-\frac{4}{5}\right) = \frac{3\sqrt{21} - 8}{25} \approx 0.23$$

(b) $\cos(\alpha - \beta) = \cos\alpha\,\cos\beta + \sin\alpha\,\sin\beta =$

$$\left(-\tfrac{2}{5}\right)\cdot\left(-\tfrac{3}{5}\right) + \left(-\frac{\sqrt{21}}{5}\right)\cdot\left(-\tfrac{4}{5}\right) = \frac{4\sqrt{21}+6}{25} \approx 0.97$$

(c) Since the sine and cosine of $(\alpha - \beta)$ are positive, $(\alpha - \beta)$ is in QI.

$\boxed{23}$ $\sin(\theta + \pi) = \sin\theta\,\cos\pi + \cos\theta\,\sin\pi = \sin\theta\,(-1) + \cos\theta\,(0) = -\sin\theta$

$\boxed{25}$ $\sin\left(x - \frac{5\pi}{2}\right) = \sin x\,\cos\frac{5\pi}{2} - \cos x\,\sin\frac{5\pi}{2} = \sin x\,(0) - \cos x\,(1) = -\cos x$

$\boxed{27}$ $\cos(\theta - \pi) = \cos\theta\,\cos\pi + \sin\theta\,\sin\pi = \cos\theta\,(-1) + \sin\theta\,(0) = -\cos\theta$

$\boxed{29}$ $\cos\left(x + \frac{3\pi}{2}\right) = \cos x\,\cos\frac{3\pi}{2} - \sin x\,\sin\frac{3\pi}{2} = \cos x\,(0) - \sin x\,(-1) = \sin x$

$\boxed{31}$ $\tan\left(x - \frac{\pi}{2}\right) = \dfrac{\sin\left(x - \frac{\pi}{2}\right)}{\cos\left(x - \frac{\pi}{2}\right)} = \dfrac{\sin x\,\cos\frac{\pi}{2} - \cos x\,\sin\frac{\pi}{2}}{\cos x\,\cos\frac{\pi}{2} + \sin x\,\sin\frac{\pi}{2}} = \dfrac{-\cos x}{\sin x} = -\cot x$

$\boxed{33}$ The tangent of a sum formula won't work here since $\tan\frac{\pi}{2}$ is undefined.

We will use a cofunction identity to verify the identity.

$$\tan\left(\theta + \tfrac{\pi}{2}\right) = \cot\left[\tfrac{\pi}{2} - \left(\theta + \tfrac{\pi}{2}\right)\right] = \cot\left(-\theta\right) = -\cot\theta$$

Alternatively, we could also write $\tan\left(\theta + \frac{\pi}{2}\right)$ as $\dfrac{\sin\left(\theta + \frac{\pi}{2}\right)}{\cos\left(\theta + \frac{\pi}{2}\right)}$ and then simplify.

$\boxed{37}$ $\tan\left(u + \frac{\pi}{4}\right) = \dfrac{\tan u + \tan\frac{\pi}{4}}{1 - \tan u\,\tan\frac{\pi}{4}} = \dfrac{\tan u + 1}{1 - \tan u\,(1)} = \dfrac{1 + \tan u}{1 - \tan u}$

$\boxed{39}$ $\cos(u + v) + \cos(u - v) = (\cos u\,\cos v - \sin u\,\sin v) + (\cos u\,\cos v + \sin u\,\sin v) =$

$$2\cos u\,\cos v$$

$\boxed{41}$ $\sin(u + v)\cdot\sin(u - v) = (\sin u\,\cos v + \cos u\,\sin v)\cdot(\sin u\,\cos v - \cos u\,\sin v)$

$$\{\text{addition and subtraction formulas for the sine}\}$$

$$= \sin^2 u\,\cos^2 v - \cos^2 u\,\sin^2 v$$

$$\{\text{recognize as the difference of two squares}\}$$

$$= \sin^2 u\,(1 - \sin^2 v) - (1 - \sin^2 u)\sin^2 v$$

$$\{\text{change to terms only involving sine}\}$$

$$= \sin^2 u - \sin^2 u\,\sin^2 v - \sin^2 v + \sin^2 u\,\sin^2 v = \sin^2 u - \sin^2 v$$

$\boxed{43}$ $\dfrac{1}{\cot\alpha - \cot\beta} = \dfrac{1}{\dfrac{\cos\alpha}{\sin\alpha} - \dfrac{\cos\beta}{\sin\beta}} = \dfrac{1}{\dfrac{\cos\alpha\,\sin\beta - \cos\beta\,\sin\alpha}{\sin\alpha\,\sin\beta}} = \dfrac{\sin\alpha\,\sin\beta}{\sin(\beta - \alpha)}$

$\boxed{45}$ $\sin(u + v + w) = \sin[(u + v) + w]$

$$= \sin(u + v)\cos w + \cos(u + v)\sin w$$

$$= (\sin u\,\cos v + \cos u\,\sin v)\cos w + (\cos u\,\cos v - \sin u\,\sin v)\sin w$$

$$= \sin u\,\cos v\,\cos w + \cos u\,\sin v\,\cos w + \cos u\,\cos v\,\sin w - \sin u\,\sin v\,\sin w$$

$\boxed{47}$ The question usually asked here is "why divide by $\sin u \sin v$?" Since we know the

form we want to end up with, we need to "force" the term "$-\sin u \sin v$" to equal

"-1", hence divide all terms by "$\sin u \sin v$."

$$\cot{(u+v)} = \frac{\cos{(u+v)}}{\sin{(u+v)}} = \frac{(\cos u \cos v - \sin u \sin v)(1/\sin u \sin v)}{(\sin u \cos v + \cos u \sin v)(1/\sin u \sin v)} = \frac{\cot u \cot v - 1}{\cot v + \cot u}$$

$\boxed{49}$ $\sin{(u-v)} = \sin{[u+(-v)]} = \sin u \cos{(-v)} + \cos u \sin{(-v)} = \sin u \cos v - \cos u \sin v$

$\boxed{51}$ $\dfrac{f(x+h) - f(x)}{h} = \dfrac{\cos{(x+h)} - \cos x}{h} = \dfrac{\cos x \cos h - \sin x \sin h - \cos x}{h} =$

$$\frac{\cos x \cos h - \cos x}{h} - \frac{\sin x \sin h}{h} = \cos x \left(\frac{\cos h - 1}{h}\right) - \sin x \left(\frac{\sin h}{h}\right)$$

$\boxed{53}$ $\sin 4t \cos t = \sin t \cos 4t \Rightarrow \sin 4t \cos t - \sin t \cos 4t = 0 \Rightarrow \sin{(4t - t)} = 0 \Rightarrow$

$$\sin 3t = 0 \Rightarrow 3t = \pi n \Rightarrow t = \tfrac{\pi}{3}n. \text{ In } [0, \pi), t = 0, \tfrac{\pi}{3}, \tfrac{2\pi}{3}.$$

$\boxed{55}$ $\cos 5t \cos 2t = -\sin 5t \sin 2t \Rightarrow \cos 5t \cos 2t + \sin 5t \sin 2t = 0 \Rightarrow$

$$\cos{(5t - 2t)} = 0 \Rightarrow \cos 3t = 0 \Rightarrow 3t = \tfrac{\pi}{2} + \pi n \Rightarrow t = \tfrac{\pi}{6} + \tfrac{\pi}{3}n. \text{ In } [0, \pi), t = \tfrac{\pi}{6}, \tfrac{\pi}{2}, \tfrac{5\pi}{6}.$$

$\boxed{57}$ $\tan 2t + \tan t = 1 - \tan 2t \tan t \Rightarrow \dfrac{\tan 2t + \tan t}{1 - \tan 2t \tan t} = 1 \Rightarrow \tan{(2t + t)} = 1 \Rightarrow$

$$\tan 3t = 1 \Rightarrow 3t = \tfrac{\pi}{4} + \pi n \Rightarrow t = \tfrac{\pi}{12} + \tfrac{\pi}{3}n. \text{ In } [0, \pi), t = \tfrac{\pi}{12}, \tfrac{5\pi}{12}, \tfrac{3\pi}{4}.$$

However, $\tan 2t$ is undefined if $t = \tfrac{3\pi}{4}$, so exclude this value of t.

$\boxed{59}$ (a) $f(x) = \sqrt{3} \cos 2x + \sin 2x$ • $A = \sqrt{(\sqrt{3})^2 + 1^2} = 2$. $\tan C = \tfrac{1}{\sqrt{3}} \Rightarrow C = \tfrac{\pi}{6}$.

$$f(x) = 2 \cos{(2x - \tfrac{\pi}{6})} = 2 \cos{[2(x - \tfrac{\pi}{12})]}$$

(b) amplitude $= 2$, period $= \tfrac{2\pi}{2} = \pi$, phase shift $= \tfrac{\pi}{12}$

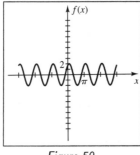

Figure 59

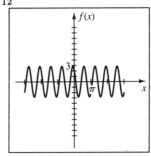

Figure 61

$\boxed{61}$ (a) $f(x) = 2 \cos 3x - 2 \sin 3x$ • $A = \sqrt{2^2 + 2^2} = 2\sqrt{2}$. $\tan C = \tfrac{-2}{2} = -1 \Rightarrow$

$$C = -\tfrac{\pi}{4}. \quad f(x) = 2\sqrt{2} \cos{(3x + \tfrac{\pi}{4})} = 2\sqrt{2} \cos{[3(x + \tfrac{\pi}{12})]}$$

(b) amplitude $= 2\sqrt{2}$, period $= \tfrac{2\pi}{3}$, phase shift $= -\tfrac{\pi}{12}$

$\boxed{63}$ $y = 50 \sin 60\pi t + 40 \cos 60\pi t$ • $A = \sqrt{50^2 + 40^2} = 10\sqrt{41}$. $\tan C = \tfrac{50}{40} \Rightarrow$

$$C = \tan^{-1} \tfrac{5}{4} \approx 0.8961. \quad y = 10\sqrt{41} \cos{(60\pi t - \tan^{-1} \tfrac{5}{4})} \approx 10\sqrt{41} \cos{(60\pi t - 0.8961)}.$$

65 (a) $y = 2\cos t + 3\sin t$; $A = \sqrt{2^2 + 3^2} = \sqrt{13}$; $\tan C = \frac{3}{2} \Rightarrow C \approx 0.98$;

$$y = \sqrt{13}\cos(t - C); \text{ amplitude} = \sqrt{13}, \text{ period} = \frac{2\pi}{1} = 2\pi$$

(b) $y = 0 \Rightarrow \sqrt{13}\cos(t - C) = 0 \Rightarrow t - C = \frac{\pi}{2} + \pi n \Rightarrow$

$$t = C + \frac{\pi}{2} + \pi n \approx 2.5536 + \pi n \text{ for every nonnegative integer } n.$$

67 (a) $p(t) = A\sin\omega t + B\sin(\omega t + \tau)$

$$= A\sin\omega t + B(\sin\omega t \cos\tau + \cos\omega t \sin\tau)$$

$$= A\underline{\sin\omega t} + B\cos\tau \underline{\sin\omega t} + B\sin\tau \underline{\cos\omega t}$$

$$= (B\sin\tau)\cos\omega t + (A + B\cos\tau)\sin\omega t$$

$$= a\cos\omega t + b\sin\omega t \text{ with } a = B\sin\tau \text{ and } b = A + B\cos\tau$$

(b) $C^2 = (B\sin\tau)^2 + (A + B\cos\tau)^2$

$$= B^2\sin^2\tau + A^2 + 2AB\cos\tau + B^2\cos^2\tau$$

$$= A^2 + B^2(\sin^2\tau + \cos^2\tau) + 2AB\cos\tau$$

$$= A^2 + B^2 + 2AB\cos\tau$$

69 (a) $C^2 = A^2 + B^2 + 2AB\cos\tau \le A^2 + B^2 + 2AB$, since $\cos\tau \le 1$ and

$$A > 0, \ B > 0. \text{ Thus, } C^2 \le (A + B)^2, \text{ and hence } C \le A + B.$$

(b) $C = A + B$ if $\cos\tau = 1$, or $\tau = 0, 2\pi$.

(c) Constructive interference will occur if $C > A$. $C > A \Rightarrow C^2 > A^2 \Rightarrow$

$$A^2 + B^2 + 2AB\cos\tau > A^2 \Rightarrow B^2 + 2AB\cos\tau > 0 \Rightarrow B(B + 2A\cos\tau) > 0.$$

Since $B > 0$, the product will be positive if $B + 2A\cos\tau > 0$, i.e., $\cos\tau > -\frac{B}{2A}$.

71 Graph $y = 3\sin 2t + 2\sin(4t + 1)$. Constructive interference will occur when $y > 3$ or

$y < -3$. From the graph, we see that this occurs on the intervals

$$(-2.97, -2.69), \ (-1.00, -0.37), \ (0.17, 0.46), \text{ and } (2.14, 2.77).$$

$$[-3.14, 3.14] \text{ by } [-5, 5]$$

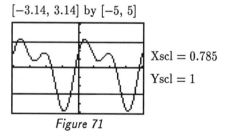

Xscl $= 0.785$

Yscl $= 1$

Figure 71

$\boxed{1}$　From *Figure 1*, $\sin\theta = \frac{4}{5}$ and $\cos\theta = \frac{3}{5}$. Thus, $\sin 2\theta = 2\sin\theta\cos\theta = 2(\frac{4}{5})(\frac{3}{5}) = \frac{24}{25}$.

$$\cos 2\theta = \cos^2\theta - \sin^2\theta = (\tfrac{3}{5})^2 - (\tfrac{4}{5})^2 = -\tfrac{7}{25}. \quad \tan 2\theta = \frac{\sin 2\theta}{\cos 2\theta} = \frac{24/25}{-7/25} = -\frac{24}{7}.$$

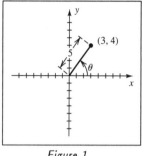

Figure 1

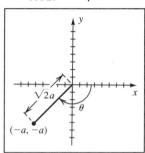

Figure 3　　　　　　　　Figure 7

$\boxed{3}$　From *Figure 3*, $\sin\theta = \sqrt{8}/3 = \frac{2}{3}\sqrt{2}$ and $\cos\theta = -\frac{1}{3}$.

Thus, $\sin 2\theta = 2\sin\theta\cos\theta = 2(\frac{2}{3}\sqrt{2})(-\frac{1}{3}) = -\frac{4}{9}\sqrt{2}$.

$\cos 2\theta = \cos^2\theta - \sin^2\theta = (-\frac{1}{3})^2 - (\frac{2}{3}\sqrt{2})^2 = \frac{1}{9} - \frac{8}{9} = -\frac{7}{9}$.

$$\tan 2\theta = \frac{\sin 2\theta}{\cos 2\theta} = \frac{-4\sqrt{2}/9}{-7/9} = \frac{4\sqrt{2}}{7}.$$

$\boxed{5}$　$\sec\theta = \frac{5}{4} \Rightarrow \cos\theta = \frac{4}{5}$. θ acute implies that $\frac{\theta}{2}$ is acute, so all 6 trigonometric functions of $\frac{\theta}{2}$ are positive and we use the "$+$" sign for the sine and cosine.

$$\sin\frac{\theta}{2} = \sqrt{\frac{1-\cos\theta}{2}} = \sqrt{\frac{1-\frac{4}{5}}{2}} = \sqrt{\frac{\frac{1}{5}}{2}} = \sqrt{\frac{1}{10}\cdot\frac{10}{10}} = \frac{\sqrt{10}}{10}.$$

$$\cos\frac{\theta}{2} = \sqrt{\frac{1+\cos\theta}{2}} = \sqrt{\frac{1+\frac{4}{5}}{2}} = \sqrt{\frac{\frac{9}{5}}{2}} = \sqrt{\frac{9}{10}\cdot\frac{10}{10}} = \frac{3\sqrt{10}}{10}.$$

$$\tan\frac{\theta}{2} = \frac{\sin\frac{\theta}{2}}{\cos\frac{\theta}{2}} = \frac{\sqrt{10}/10}{3\sqrt{10}/10} = \frac{1}{3}.$$

$\boxed{7}$　From *Figure 7* $(a > 0)$, $\cos\theta = -\dfrac{a}{\sqrt{2}\,a} = -\dfrac{\sqrt{2}}{2}$ and $\frac{\theta}{2}$ is in QIV.

$$\sin\frac{\theta}{2} = -\sqrt{\frac{1-\cos\theta}{2}} = -\sqrt{\frac{1+\sqrt{2}/2}{2}} = -\sqrt{\frac{2+\sqrt{2}}{4}} = -\frac{1}{2}\sqrt{2+\sqrt{2}}.$$

$$\cos\frac{\theta}{2} = \sqrt{\frac{1+\cos\theta}{2}} = \sqrt{\frac{1-\sqrt{2}/2}{2}} = \sqrt{\frac{2-\sqrt{2}}{4}} = \frac{1}{2}\sqrt{2-\sqrt{2}}.$$

$$\tan\frac{\theta}{2} = \frac{1-\cos\theta}{\sin\theta} = \frac{1+\sqrt{2}/2}{-\sqrt{2}/2}\cdot\frac{2}{2} = \frac{2+\sqrt{2}}{-\sqrt{2}} = -\sqrt{2}-1.$$

9 (a) $\cos 67°30' = \sqrt{\dfrac{1 + \cos 135°}{2}} = \sqrt{\dfrac{1 - \sqrt{2}/2}{2}} = \sqrt{\dfrac{2 - \sqrt{2}}{4}} = \frac{1}{2}\sqrt{2 - \sqrt{2}}.$

(b) $\sin 15° = \sqrt{\dfrac{1 - \cos 30°}{2}} = \sqrt{\dfrac{1 - \sqrt{3}/2}{2}} = \sqrt{\dfrac{2 - \sqrt{3}}{4}} = \frac{1}{2}\sqrt{2 - \sqrt{3}}.$

(c) $\tan\frac{3\pi}{8} = \dfrac{1 - \cos\frac{3\pi}{4}}{\sin\frac{3\pi}{4}} = \dfrac{1 + \sqrt{2}/2}{\sqrt{2}/2} \cdot \dfrac{2}{2} = \dfrac{2 + \sqrt{2}}{\sqrt{2}} = \sqrt{2} + 1.$

13 We recognize the product of the sine and the cosine as being of the form of the right side of the double-angle formula for the sine, and apply that formula "in reverse."

$$4\sin\frac{x}{2} \cos\frac{x}{2} = 2 \cdot 2\sin\frac{x}{2}\cos\frac{x}{2} = 2\sin\left(2 \cdot \frac{x}{2}\right) = 2\sin x$$

15 $(\sin t + \cos t)^2 = \sin^2 t + 2\sin t\,\cos t + \cos^2 t = (\sin^2 t + \cos^2 t) + (2\sin t\,\cos t) =$

$$1 + \sin 2t$$

17 $\sin 3u = \sin(2u + u)$

$\qquad = \sin 2u\,\cos u + \cos 2u\,\sin u$

$\qquad = (2\sin u\,\cos u)\cos u + (1 - 2\sin^2 u)\sin u$

$\qquad = 2\sin u\,\cos^2 u + \sin u - 2\sin^3 u$

$\qquad = 2\sin u(1 - \sin^2 u) + \sin u - 2\sin^3 u$

$\qquad = 2\sin u - 2\sin^3 u + \sin u - 2\sin^3 u$

$\qquad = 3\sin u - 4\sin^3 u$

$\qquad = \sin u(3 - 4\sin^2 u)$

19 $\cos 4\theta = \cos(2 \cdot 2\theta) = 2\cos^2 2\theta - 1 = 2(2\cos^2\theta - 1)^2 - 1 =$

$$2(4\cos^4\theta - 4\cos^2\theta + 1) - 1 = 8\cos^4\theta - 8\cos^2\theta + 1$$

21 $\sin^4 t = (\sin^2 t)^2 \quad = \left(\dfrac{1 - \cos 2t}{2}\right)^2$

$\qquad\qquad\qquad = \frac{1}{4}(1 - 2\cos 2t + \cos^2 2t)$

$\qquad\qquad\qquad = \frac{1}{4} - \frac{1}{2}\cos 2t + \frac{1}{4}\left(\dfrac{1 + \cos 4t}{2}\right)$

$\qquad\qquad\qquad = \frac{1}{4} - \frac{1}{2}\cos 2t + \frac{1}{8} + \frac{1}{8}\cos 4t$

$\qquad\qquad\qquad = \frac{3}{8} - \frac{1}{2}\cos 2t + \frac{1}{8}\cos 4t$

23 We do not have a formula for $\sec 2\theta$, so we will write $\sec 2\theta$ in terms of $\cos 2\theta$ in order to apply the double-angle formula for the cosine.

$$\sec 2\theta = \frac{1}{\cos 2\theta} = \frac{1}{2\cos^2\theta - 1} = \frac{1}{2\left(\dfrac{1}{\sec^2\theta}\right) - 1} = \frac{1}{\dfrac{2 - \sec^2\theta}{\sec^2\theta}} = \frac{\sec^2\theta}{2 - \sec^2\theta}$$

25 We need to match the arguments of the trigonometric functions involved—that is, either write both of them in terms of $2t$ or in terms of $4t$. Converting $\cos 4t$ to an expression with $2t$ as the angle argument gives us

$$2\sin^2 2t + \cos 4t = 2\sin^2 2t + \cos(2 \cdot 2t) = 2\sin^2 2t + (1 - 2\sin^2 2t) = 1.$$

Alternatively, if we write both arguments in terms of $4t$, we have

$$2\sin^2 2t + \cos 4t = 2 \cdot \frac{1 - \cos 4t}{2} + \cos 4t = 1 - \cos 4t + \cos 4t = 1.$$

27 $\tan 3u = \tan(2u + u) = \dfrac{\tan 2u + \tan u}{1 - \tan 2u \tan u}$

$$= \frac{\dfrac{2\tan u}{1 - \tan^2 u} + \tan u}{1 - \dfrac{2\tan u}{1 - \tan^2 u} \cdot \tan u}$$

$$= \frac{\dfrac{2\tan u + \tan u - \tan^3 u}{1 - \tan^2 u}}{\dfrac{1 - \tan^2 u - 2\tan^2 u}{1 - \tan^2 u}}$$

$$= \frac{3\tan u - \tan^3 u}{1 - 3\tan^2 u}$$

$$= \frac{\tan u\left(3 - \tan^2 u\right)}{1 - 3\tan^2 u}$$

29 $\cos^4 \dfrac{\theta}{2} = \left(\cos^2 \dfrac{\theta}{2}\right)^2 = \left(\dfrac{1 + \cos\theta}{2}\right)^2 = \dfrac{1 + 2\cos\theta + \cos^2\theta}{4} = \dfrac{1}{4} + \dfrac{1}{2}\cos\theta + \dfrac{1}{4}\left(\dfrac{1 + \cos 2\theta}{2}\right) =$

$$\tfrac{1}{4} + \tfrac{1}{2}\cos\theta + \tfrac{1}{8} + \tfrac{1}{8}\cos 2\theta = \tfrac{3}{8} + \tfrac{1}{2}\cos\theta + \tfrac{1}{8}\cos 2\theta$$

33 $\sin 2t + \sin t = 0 \Rightarrow 2\sin t \cos t + \sin t = 0 \Rightarrow \sin t(2\cos t + 1) = 0 \Rightarrow$

$$\sin t = 0 \text{ or } \cos t = -\tfrac{1}{2} \Rightarrow t = 0,\ \pi \text{ or } \tfrac{2\pi}{3},\ \tfrac{4\pi}{3}$$

35 $\cos u + \cos 2u = 0 \Rightarrow \cos u + 2\cos^2 u - 1 = 0 \Rightarrow 2\cos^2 u + \cos u - 1 = 0 \Rightarrow$

$$(2\cos u - 1)(\cos u + 1) = 0 \Rightarrow \cos u = \tfrac{1}{2},\ -1 \Rightarrow u = \tfrac{\pi}{3},\ \tfrac{5\pi}{3},\ \pi$$

37 A first approach uses the concept that if $\tan\alpha = \tan\beta$, then $\alpha = \beta + \pi n$.

$\tan 2x = \tan x \Rightarrow 2x = x + \pi n \Rightarrow x = \pi n \Rightarrow x = 0,\ \pi.$

Another approach is: $\tan 2x = \tan x \Rightarrow \dfrac{\sin 2x}{\cos 2x} = \dfrac{\sin x}{\cos x} \Rightarrow \sin 2x \cos x = \sin x \cos 2x \Rightarrow$

$$\sin 2x \cos x - \sin x \cos 2x = 0 \Rightarrow \sin(2x - x) = 0 \Rightarrow \sin x = 0 \Rightarrow x = 0,\ \pi.$$

39 $\sin\tfrac{1}{2}u + \cos u = 1 \Rightarrow \sin\tfrac{1}{2}u + \cos\left[2 \cdot \left(\tfrac{1}{2}u\right)\right] = 1 \Rightarrow \sin\tfrac{1}{2}u + \left(1 - 2\sin^2\tfrac{1}{2}u\right) = 1 \Rightarrow$

$$\sin\tfrac{1}{2}u - 2\sin^2\tfrac{1}{2}u = 0 \Rightarrow \sin\tfrac{1}{2}u\left(1 - 2\sin\tfrac{1}{2}u\right) = 0 \Rightarrow \sin\tfrac{1}{2}u = 0,\ \tfrac{1}{2} \Rightarrow$$

$$\tfrac{1}{2}u = 0,\ \tfrac{\pi}{6},\ \tfrac{5\pi}{6} \Rightarrow u = 0,\ \tfrac{\pi}{3},\ \tfrac{5\pi}{3}$$

$\boxed{41}$ $\sqrt{a^2+b^2}\sin(u+v) = \sqrt{a^2+b^2}\sin u\cos v + \sqrt{a^2+b^2}\cos u\sin v = a\sin u + b\cos u$

{ equate coefficients of $\sin u$ and $\cos u$ } $\Rightarrow a = \sqrt{a^2+b^2}\cos v$ and $b = \sqrt{a^2+b^2}\sin v$

$\Rightarrow \cos v = \dfrac{a}{\sqrt{a^2+b^2}}$ and $\sin v = \dfrac{b}{\sqrt{a^2+b^2}}$. Since $0 < u < \frac{\pi}{2}$, $\sin u > 0$ and $\cos v > 0$.

Now $a > 0$ and $b > 0$ combine with the above to imply that $\cos v > 0$ and $\sin v > 0$.

Thus, $0 < v < \frac{\pi}{2}$.

$\boxed{43}$ (a) $\cos 2x + 2\cos x = 0 \Rightarrow 2\cos^2 x + 2\cos x - 1 = 0 \Rightarrow$

$\cos x = \dfrac{-2 \pm \sqrt{12}}{4} = \dfrac{-1 \pm \sqrt{3}}{2} \approx 0.366 \left\{ \cos x \ne \dfrac{-1-\sqrt{3}}{2} < -1 \right\}$.

Thus, $x \approx 1.20$ and 5.09.

(b) $\sin 2x + \sin x = 0 \Rightarrow 2\sin x\cos x + \sin x = 0 \Rightarrow \sin x\,(2\cos x + 1) = 0 \Rightarrow \sin x = 0$ or

$\cos x = -\frac{1}{2} \Rightarrow x = 0,\ \pi,\ 2\pi$ or $\frac{2\pi}{3},\ \frac{4\pi}{3}$. $P(\frac{2\pi}{3}, -1.5)$, $Q(\pi, -1)$, $R(\frac{4\pi}{3}, -1.5)$

$\boxed{45}$ (a) $\cos 3x - 3\cos x = 0 \Rightarrow 4\cos^3 x - 3\cos x - 3\cos x = 0 \Rightarrow$

$4\cos^3 x - 6\cos x = 0 \Rightarrow 2\cos x\,(2\cos^2 x - 3) = 0 \Rightarrow \cos x = 0,\ \pm\sqrt{3/2} \Rightarrow$

$x = -\frac{3\pi}{2},\ -\frac{\pi}{2},\ \frac{\pi}{2},\ \frac{3\pi}{2}$ $\{ \cos x \ne \pm\sqrt{3/2}$ since $\sqrt{3/2} > 1 \}$

(b) $\sin 3x - \sin x = 0 \Rightarrow 3\sin x - 4\sin^3 x - \sin x = 0 \Rightarrow 4\sin^3 x - 2\sin x = 0 \Rightarrow$

$2\sin x\,(2\sin^2 x - 1) = 0 \Rightarrow \sin x = 0,\ \pm 1/\sqrt{2} \Rightarrow$

$x = 0,\ \pm\pi,\ \pm 2\pi,\ \pm\frac{\pi}{4},\ \pm\frac{3\pi}{4},\ \pm\frac{5\pi}{4},\ \pm\frac{7\pi}{4}$

$\boxed{47}$ (a) Let $y = \overline{BC}$. Form a right triangle with hypotenuse y, side opposite θ, 20, and

side adjacent θ, x. $\sin\theta = \dfrac{20}{y} \Rightarrow y = \dfrac{20}{\sin\theta}$. $\cos\theta = \dfrac{x}{y} \Rightarrow x = y\cos\theta = \dfrac{20\cos\theta}{\sin\theta}$.

Now $d = (40 - x) + y = 40 - \dfrac{20\cos\theta}{\sin\theta} + \dfrac{20}{\sin\theta} = 20\left(\dfrac{1-\cos\theta}{\sin\theta}\right) + 40 = 20\tan\frac{\theta}{2} + 40$.

(b) $50 = 20\tan\frac{\theta}{2} + 40 \Rightarrow \tan\frac{\theta}{2} = \frac{1}{2} \Rightarrow \dfrac{1-\cos\theta}{\sin\theta} = \frac{1}{2} \Rightarrow 2 - 2\cos\theta = \sin\theta \Rightarrow$

$4 - 8\cos\theta + 4\cos^2\theta = \sin^2\theta = 1 - \cos^2\theta \Rightarrow 5\cos^2\theta - 8\cos\theta + 3 = 0 \Rightarrow$

$(5\cos\theta - 3)(\cos\theta - 1) = 0 \Rightarrow \cos\theta = \frac{3}{5},\ 1$. { $\cos\theta = 1 \Rightarrow \theta = 0$ and 0 is

extraneous }. $\cos\theta = \frac{3}{5} \Rightarrow \sin\theta = \frac{4}{5}$ and $y = \dfrac{20}{4/5} = 25$. $\cos\theta = \dfrac{x}{y}$ and

$\cos\theta = \frac{3}{5} \Rightarrow \dfrac{x}{25} = \frac{3}{5} \Rightarrow x = 15$, which means that B would be 25 miles from A.

$\boxed{49}$ (a) From Example 8, the area A of a cross section is

$A = \frac{1}{2}(\text{side})^2(\text{sine of included angle}) = \frac{1}{2}(\frac{1}{2})^2\sin\theta = \frac{1}{8}\sin\theta$.

The volume $V = (\text{length of gutter})(\text{area of cross section}) = 20(\frac{1}{8}\sin\theta) = \frac{5}{2}\sin\theta$.

(b) $V = 2 \Rightarrow \frac{5}{2}\sin\theta = 2 \Rightarrow \sin\theta = \frac{4}{5} \Rightarrow \theta \approx 53.13°$.

51 (a) Let $y = \overline{DB}$ and x denote the distance from D to the midpoint of $\overline{BC}$.

$$\sin\frac{\theta}{2} = \frac{b/2}{y} \Rightarrow y = \frac{b}{2} \cdot \frac{1}{\sin(\theta/2)} \text{ and } \tan\frac{\theta}{2} = \frac{b/2}{x} \Rightarrow x = \frac{b}{2} \cdot \frac{\cos(\theta/2)}{\sin(\theta/2)}.$$

$$l = (a - x) + y = a - \frac{b}{2} \cdot \frac{\cos(\theta/2)}{\sin(\theta/2)} + \frac{b}{2} \cdot \frac{1}{\sin(\theta/2)} = a + \frac{b}{2} \cdot \frac{1 - \cos(\theta/2)}{\sin(\theta/2)} =$$

$$a + \frac{b}{2} \tan\left(\frac{\theta/2}{2}\right) = a + \frac{b}{2} \tan\frac{\theta}{4}.$$

(b) $a = 10$ mm, $b = 6$ mm, and $\theta = 156° \Rightarrow l = 10 + 3\tan 39° \approx 12.43$ mm.

7.5 Exercises

Note: We will reference the product-to-sum formulas as [P1]–[P4] and the sum-to-product formulas as [S1]–[S4] in the order they appear in the text. The formulas $\cos(-kx) = \cos kx$ and $\sin(-kx) = -\sin kx$ will be used without mention.

1. $\sin 7t \sin 3t = $ [P4] $\frac{1}{2}[\cos(7t - 3t) - \cos(7t + 3t)] = \frac{1}{2}\cos 4t - \frac{1}{2}\cos 10t$

3. $\cos 6u \cos(-4u) = $ [P3] $\frac{1}{2}\left\{\cos[6u + (-4u)] + \cos[6u - (-4u)]\right\} = \frac{1}{2}\cos 2u + \frac{1}{2}\cos 10u$

5. $2\sin 9\theta \cos 3\theta = $ [P1] $2 \cdot \frac{1}{2}[\sin(9\theta + 3\theta) + \sin(9\theta - 3\theta)] = \sin 12\theta + \sin 6\theta$

7. $3\cos x \sin 2x = $ [P2] $3 \cdot \frac{1}{2}[\sin(x + 2x) - \sin(x - 2x)] = \frac{3}{2}\sin 3x - \frac{3}{2}\sin(-x) =$

$$\frac{3}{2}\sin 3x + \frac{3}{2}\sin x$$

9. $\sin 6\theta + \sin 2\theta = $ [S1] $2\sin\frac{6\theta + 2\theta}{2} \cos\frac{6\theta - 2\theta}{2} = 2\sin 4\theta \cos 2\theta$

11. $\cos 5x - \cos 3x = $ [S4] $-2\sin\frac{5x + 3x}{2} \sin\frac{5x - 3x}{2} = -2\sin 4x \sin x$

13. $\sin 3t - \sin 7t = $ [S2] $2\cos\frac{3t + 7t}{2} \sin\frac{3t - 7t}{2} = 2\cos 5t \sin(-2t) = -2\cos 5t \sin 2t$

15. $\cos x + \cos 2x = $ [S3] $2\cos\frac{x + 2x}{2} \cos\frac{x - 2x}{2} = 2\cos\frac{3}{2}x \cos\left(-\frac{1}{2}x\right) = 2\cos\frac{3}{2}x \cos\frac{1}{2}x$

17. $\dfrac{\sin 4t + \sin 6t}{\cos 4t - \cos 6t} = \dfrac{[S1]\ 2\sin 5t \cos(-t)}{[S4]\ -2\sin 5t \sin(-t)} = \dfrac{\cos t}{\sin t} = \cot t$

19. $\dfrac{\sin u + \sin v}{\cos u + \cos v} = \dfrac{[S1]\ 2\sin\frac{1}{2}(u + v) \cos\frac{1}{2}(u - v)}{[S3]\ 2\cos\frac{1}{2}(u + v) \cos\frac{1}{2}(u - v)} = \dfrac{\sin\frac{1}{2}(u + v)}{\cos\frac{1}{2}(u + v)} = \tan\frac{1}{2}(u + v)$

21. $\dfrac{\sin u - \sin v}{\sin u + \sin v} = \dfrac{[S2]\ 2\cos\frac{1}{2}(u + v) \sin\frac{1}{2}(u - v)}{[S1]\ 2\sin\frac{1}{2}(u + v) \cos\frac{1}{2}(u - v)} = \cot\frac{1}{2}(u + v) \tan\frac{1}{2}(u - v) = \dfrac{\tan\frac{1}{2}(u - v)}{\tan\frac{1}{2}(u + v)}$

23 Since the arguments on the right side are all even multiples of x $(2x, 4x,$ and $6x)$, we begin by grouping the terms with the odd multiples of x together, and operate on them with a product-to-sum formula, which will convert these expressions to expressions with even multiples of x.

$$4\cos x \cos 2x \sin 3x = 2\cos 2x \left(2\sin 3x \cos x\right)$$
$$= 2\cos 2x \left([\text{P1}] \sin 4x + \sin 2x\right)$$
$$= (2\cos 2x \sin 4x) + (2\cos 2x \sin 2x)$$
$$= \big[[\text{P2}] \sin 6x - \sin (-2x)\big] + \big([\text{P2}] \sin 4x - \sin 0\big)$$
$$= \sin 2x + \sin 4x + \sin 6x$$

25 $(\sin ax)(\cos bx) = [\text{P1}] \frac{1}{2}\big[\sin (ax + bx) + \sin (ax - bx)\big] = \frac{1}{2}\sin\big[(a+b)x\big] + \frac{1}{2}\sin\big[(a-b)x\big]$

27 $\sin 5t + \sin 3t = 0 \Rightarrow [\text{S1}]\ 2\sin 4t \cos t = 0 \Rightarrow \sin 4t = 0$ or $\cos t = 0 \Rightarrow$
$$4t = \pi n \text{ or } t = \frac{\pi}{2} + \pi n \Rightarrow t = \frac{\pi}{4}n \ \{\text{which includes } t = \frac{\pi}{2} + \pi n\}$$

29 $\cos x = \cos 3x \Rightarrow \cos x - \cos 3x = 0 \Rightarrow [\text{S4}]\ -2\sin 2x \sin (-x) = 0 \Rightarrow$
$$\sin 2x = 0 \text{ or } \sin x = 0 \Rightarrow 2x = \pi n \text{ or } x = \pi n \Rightarrow x = \frac{\pi}{2}n \ \{\text{which includes } x = \pi n\}$$

31 $\cos 3x + \cos 5x = \cos x \Rightarrow [\text{S3}]\ 2\cos 4x \cos (-x) - \cos x = 0 \Rightarrow$
$$\cos x \left(2\cos 4x - 1\right) = 0 \Rightarrow \cos x = 0 \text{ or } \cos 4x = \frac{1}{2} \Rightarrow$$
$$x = \frac{\pi}{2} + \pi n \text{ or } 4x = \frac{\pi}{3} + 2\pi n, \frac{5\pi}{3} + 2\pi n \Rightarrow x = \frac{\pi}{2} + \pi n, \frac{\pi}{12} + \frac{\pi}{2}n, \frac{5\pi}{12} + \frac{\pi}{2}n$$

33 $\sin 2x - \sin 5x = 0 \Rightarrow [\text{S2}]\ 2\cos\frac{7}{2}x \sin (-\frac{3}{2}x) = 0 \Rightarrow \frac{7}{2}x = \frac{\pi}{2} + \pi n \text{ or } \frac{3}{2}x = \pi n \Rightarrow$
$$x = \frac{\pi}{7} + \frac{2\pi}{7}n \text{ or } x = \frac{2\pi}{3}n$$

35 $\cos x + \cos 3x = 0 \Rightarrow [\text{S3}]\ 2\cos 2x \cos (-x) = 0 \Rightarrow \cos 2x = 0 \text{ or } \cos x = 0 \Rightarrow$
$$2x = \frac{\pi}{2} + \pi n \text{ or } x = \frac{\pi}{2} + \pi n \Rightarrow x = \frac{\pi}{4} + \frac{\pi}{2}n \text{ or } x = \frac{\pi}{2} + \pi n \Rightarrow$$
$$x = \frac{\pi}{4}, \frac{3\pi}{4}, \frac{5\pi}{4}, \frac{7\pi}{4}, \frac{\pi}{2}, \frac{3\pi}{2} \text{ for } 0 \le x \le 2\pi$$

37 $\sin 3x - \sin x = 0 \Rightarrow [\text{S2}]\ 2\cos 2x \sin x = 0 \Rightarrow \cos 2x = 0 \text{ or } \sin x = 0 \Rightarrow$
$$2x = \frac{\pi}{2} + \pi n \text{ or } x = \pi n \Rightarrow x = \frac{\pi}{4} + \frac{\pi}{2}n \text{ or } x = \pi n \Rightarrow$$
$$x = 0, \ \pm \pi, \ \pm 2\pi, \ \pm\frac{\pi}{4}, \ \pm\frac{3\pi}{4}, \ \pm\frac{5\pi}{4}, \ \pm\frac{7\pi}{4} \text{ for } -2\pi \le x \le 2\pi.$$

39 $f(x) = \sin\left(\frac{\pi n}{l}x\right)\cos\left(\frac{k\pi n}{l}t\right)$

$$= [\text{P1}] \frac{1}{2}\left[\sin\left(\frac{\pi n}{l}x + \frac{k\pi n}{l}t\right) + \sin\left(\frac{\pi n}{l}x - \frac{k\pi n}{l}t\right)\right]$$

$$= \frac{1}{2}\left[\sin\frac{\pi n}{l}(x + kt) + \sin\frac{\pi n}{l}(x - kt)\right]$$

$$= \frac{1}{2}\sin\frac{\pi n}{l}(x + kt) + \frac{1}{2}\sin\frac{\pi n}{l}(x - kt)$$

41 (a) Estimating the x-intercepts, we have $x \approx 0$, ± 1.05, ± 1.57, ± 2.09, ± 3.14.

(b) $\sin 4x + \sin 2x = 2 \sin 3x \cos x = 0 \Rightarrow \sin 3x = 0$ or $\cos x = 0$.

$\sin 3x = 0 \Rightarrow 3x = \pi n \Rightarrow x = \frac{\pi}{3}n \Rightarrow x = 0$, $\pm \frac{\pi}{3}$, $\pm \frac{2\pi}{3}$, $\pm \pi$. $\cos x = 0 \Rightarrow x = \pm \frac{\pi}{2}$.

The x-intercepts are 0, $\pm \frac{\pi}{3}$, $\pm \frac{\pi}{2}$, $\pm \frac{2\pi}{3}$, $\pm \pi$.

$[-3.14, 3.14]$ by $[-2.09, 2.09]$

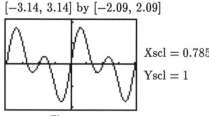

$\text{Xscl} = 0.785$

$\text{Yscl} = 1$

Figure 41

43 Graphing on an interval of $[-\pi, \pi]$ gives us a figure that resembles a tangent function. It appears that there is a vertical asymptote at about 0.78. Recognizing that this value is about $\frac{\pi}{4}$, we might make the conjecture that we have "halved" the period of the tangent and that the graph of $f(x) = \frac{\sin x + \sin 2x + \sin 3x}{\cos x + \cos 2x + \cos 3x}$ appears to be that of $y = g(x) = \tan 2x$. Verifying this identity, we have

$$\frac{\sin x + \sin 2x + \sin 3x}{\cos x + \cos 2x + \cos 3x} = \frac{\sin 2x + (\sin 3x + \sin x)}{\cos 2x + (\cos 3x + \cos x)} = \frac{\sin 2x + 2 \sin 2x \cos x}{\cos 2x + 2 \cos 2x \cos x} =$$

$$\frac{\sin 2x (1 + 2 \cos x)}{\cos 2x (1 + 2 \cos x)} = \frac{\sin 2x}{\cos 2x} = \tan 2x.$$

7.6 Exercises

1 (a) $\sin^{-1}\left(-\frac{\sqrt{2}}{2}\right) = -\frac{\pi}{4}$ since $\sin\left(-\frac{\pi}{4}\right) = -\frac{\sqrt{2}}{2}$ and $-\frac{\pi}{2} \leq -\frac{\pi}{4} \leq \frac{\pi}{2}$

(b) $\cos^{-1}\left(-\frac{1}{2}\right) = \frac{2\pi}{3}$ since $\cos \frac{2\pi}{3} = -\frac{1}{2}$ and $0 \leq \frac{2\pi}{3} \leq \pi$

(c) $\tan^{-1}(-\sqrt{3}) = -\frac{\pi}{3}$ since $\tan\left(-\frac{\pi}{3}\right) = -\sqrt{3}$ and $-\frac{\pi}{2} < -\frac{\pi}{3} < \frac{\pi}{2}$

3 (a) $\arcsin \frac{\sqrt{3}}{2} = \frac{\pi}{3}$ since $\sin \frac{\pi}{3} = \frac{\sqrt{3}}{2}$ and $-\frac{\pi}{2} \leq \frac{\pi}{3} \leq \frac{\pi}{2}$

(b) $\arccos \frac{\sqrt{2}}{2} = \frac{\pi}{4}$ since $\cos \frac{\pi}{4} = \frac{\sqrt{2}}{2}$ and $0 \leq \frac{\pi}{4} \leq \pi$

(c) $\arctan \frac{1}{\sqrt{3}} = \frac{\pi}{6}$ since $\tan \frac{\pi}{6} = \frac{1}{\sqrt{3}}$ and $-\frac{\pi}{2} < \frac{\pi}{6} < \frac{\pi}{2}$

5 (a) $\sin^{-1} \frac{\pi}{3}$ is <u>not defined</u> since $\frac{\pi}{3} > 1$, i.e., $\frac{\pi}{3} \notin [-1, 1]$

(b) $\cos^{-1} \frac{\pi}{2}$ is <u>not defined</u> since $\frac{\pi}{2} > 1$, i.e., $\frac{\pi}{2} \notin [-1, 1]$

(c) $\tan^{-1} 1 = \frac{\pi}{4}$ since $\tan \frac{\pi}{4} = 1$ and $-\frac{\pi}{2} < \frac{\pi}{4} < \frac{\pi}{2}$

Note: Exercises 7–10 refer to the boxed properties of $\sin^{-1}$, $\cos^{-1}$, and $\tan^{-1}$.

7. (a) $\sin\left[\arcsin\left(-\frac{3}{10}\right)\right] = -\frac{3}{10}$ since $-1 \le -\frac{3}{10} \le 1$

   (b) $\cos\left(\arccos\frac{1}{2}\right) = \frac{1}{2}$ since $-1 \le \frac{1}{2} \le 1$

   (c) $\tan\left(\arctan 14\right) = 14$ since $\tan\left(\arctan x\right) = x$ for every x

9. (a) $\sin^{-1}\left(\sin\frac{\pi}{3}\right) = \frac{\pi}{3}$ since $-\frac{\pi}{2} \le \frac{\pi}{3} \le \frac{\pi}{2}$ (b) $\cos^{-1}\left[\cos\left(\frac{5\pi}{6}\right)\right] = \frac{5\pi}{6}$ since $0 \le \frac{5\pi}{6} \le \pi$

   (c) $\tan^{-1}\left[\tan\left(-\frac{\pi}{6}\right)\right] = -\frac{\pi}{6}$ since $-\frac{\pi}{2} < -\frac{\pi}{6} < \frac{\pi}{2}$

11. (a) $\arcsin\left(\sin\frac{5\pi}{4}\right) = \arcsin\left(-\frac{\sqrt{2}}{2}\right) = -\frac{\pi}{4}$

    (b) $\arccos\left(\cos\frac{5\pi}{4}\right) = \arccos\left(-\frac{\sqrt{2}}{2}\right) = \frac{3\pi}{4}$ (c) $\arctan\left(\tan\frac{7\pi}{4}\right) = \arctan\left(-1\right) = -\frac{\pi}{4}$

13. (a) $\sin\left[\cos^{-1}\left(-\frac{1}{2}\right)\right] = \sin\frac{2\pi}{3} = \frac{\sqrt{3}}{2}$ (b) $\cos\left(\tan^{-1} 1\right) = \cos\frac{\pi}{4} = \frac{\sqrt{2}}{2}$

    (c) $\tan\left[\sin^{-1}\left(-1\right)\right] = \tan\left(-\frac{\pi}{2}\right)$, which is <u>not defined</u>.

15. (a) Let $\theta = \sin^{-1}\frac{2}{3}$. From *Figure 15(a)*, $\cot\left(\sin^{-1}\frac{2}{3}\right) = \cot\theta = \frac{x}{y} = \frac{\sqrt{5}}{2}$.

    (b) Let $\theta = \tan^{-1}\left(-\frac{3}{5}\right)$. From *Figure 15(b)*, $\sec\left[\tan^{-1}\left(-\frac{3}{5}\right)\right] = \sec\theta = \frac{r}{x} = \frac{\sqrt{34}}{5}$.

    (c) Let $\theta = \cos^{-1}\left(-\frac{1}{4}\right)$. From *Figure 15(c)*, $\csc\left[\cos^{-1}\left(-\frac{1}{4}\right)\right] = \csc\theta = \frac{r}{y} = \frac{4}{\sqrt{15}}$.

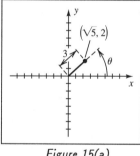

Figure 15(a)

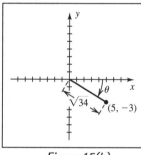

Figure 15(b)

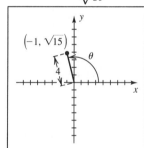

Figure 15(c)

17. (a) $\sin\left(\arcsin\frac{1}{2} + \arccos 0\right) = \sin\left(\frac{\pi}{6} + \frac{\pi}{2}\right) = \sin\frac{2\pi}{3} = \frac{\sqrt{3}}{2}$.

    (b) Remember, $\arctan\left(-\frac{3}{4}\right)$ and $\arcsin\frac{4}{5}$ are <u>angles</u>. To abbreviate this solution,

    we let $\alpha = \arctan\left(-\frac{3}{4}\right)$ and $\beta = \arcsin\frac{4}{5}$. Using the difference identity for the

    cosine and figures as in Exercises 15 and 16, we have $\cos\left[\arctan\left(-\frac{3}{4}\right) - \arcsin\frac{4}{5}\right]$

    $$= \cos\left(\alpha - \beta\right) = \cos\alpha\,\cos\beta + \sin\alpha\,\sin\beta = \frac{4}{5}\cdot\frac{3}{5} + \left(-\frac{3}{5}\right)\cdot\frac{4}{5} = 0.$$

    (c) Let $\alpha = \arctan\frac{4}{3}$ and $\beta = \arccos\frac{8}{17}$. $\tan\left(\arctan\frac{4}{3} + \arccos\frac{8}{17}\right) =$

    $$\tan\left(\alpha + \beta\right) = \frac{\tan\alpha + \tan\beta}{1 - \tan\alpha\,\tan\beta} = \frac{\frac{4}{3} + \frac{15}{8}}{1 - \frac{4}{3}\cdot\frac{15}{8}}\cdot\frac{24}{24} = \frac{32 + 45}{24 - 60} = -\frac{77}{36}.$$

[19] (a) We first recognize this expression as being of the form $\sin(twice\ an\ angle)$, where $\arccos\left(-\frac{3}{5}\right)$ is the angle. Let $\alpha = \arccos\left(-\frac{3}{5}\right)$. It may help to draw a figure as in Exercise 15 to determine that α is a second quadrant angle and that $\sin\alpha = \frac{4}{5}$.

Applying the double-angle formula for the sine gives us

$$\sin\left[2\arccos\left(-\tfrac{3}{5}\right)\right] = \sin 2\alpha = 2\sin\alpha\,\cos\alpha = 2\left(\tfrac{4}{5}\right)\left(-\tfrac{3}{5}\right) = -\tfrac{24}{25}.$$

(b) Let $\alpha = \sin^{-1}\frac{15}{17}$. Thus, $\cos\alpha = \frac{8}{17}$ and we apply the double-angle formula for the cosine. $\cos\left(2\sin^{-1}\frac{15}{17}\right) = \cos 2\alpha = \cos^2\alpha - \sin^2\alpha = \left(\tfrac{8}{17}\right)^2 - \left(\tfrac{15}{17}\right)^2 = -\tfrac{161}{289}.$

(c) Let $\alpha = \tan^{-1}\frac{3}{4}$. Thus, $\tan\alpha = \frac{3}{4}$ and we apply the double-angle formula for the tangent. $\tan\left(2\tan^{-1}\frac{3}{4}\right) = \tan 2\alpha = \dfrac{2\tan\alpha}{1-\tan^2\alpha} = \dfrac{2\cdot\frac{3}{4}}{1-\left(\frac{3}{4}\right)^2}\cdot\dfrac{16}{16} = \dfrac{24}{16-9} = \dfrac{24}{7}.$

[21] (a) We first recognize this expression as being of the form $\sin(one\text{-}half\ an\ angle)$, where $\sin^{-1}\left(-\frac{7}{25}\right)$ is the angle. Let $\alpha = \sin^{-1}\left(-\frac{7}{25}\right)$. Hence, α is a fourth quadrant angle and $\cos\alpha = \frac{24}{25}$. $-\frac{\pi}{2} < \alpha < 0 \Rightarrow -\frac{\pi}{4} < \frac{1}{2}\alpha < 0$. We need to know the sign of $\sin\frac{1}{2}\alpha$ in order to correctly apply the half-angle formula for the sine. Since $-\frac{\pi}{4} < \frac{1}{2}\alpha < 0$ { QIV }, $\sin\frac{1}{2}\alpha < 0$.

$$\sin\left[\tfrac{1}{2}\sin^{-1}\left(-\tfrac{7}{25}\right)\right] = \sin\tfrac{1}{2}\alpha = -\sqrt{\frac{1-\cos\alpha}{2}} = -\sqrt{\frac{1-\frac{24}{25}}{2}} = -\sqrt{\frac{1}{50}\cdot\frac{2}{2}} = -\tfrac{1}{10}\sqrt{2}.$$

(b) Let $\alpha = \tan^{-1}\frac{8}{15}$. $0 < \alpha < \frac{\pi}{2} \Rightarrow 0 < \frac{1}{2}\alpha < \frac{\pi}{4}$ and $\cos\frac{1}{2}\alpha > 0$.

Applying the half-angle formula for the cosine,

$$\cos\left(\tfrac{1}{2}\tan^{-1}\tfrac{8}{15}\right) = \cos\tfrac{1}{2}\alpha = \sqrt{\frac{1+\cos\alpha}{2}} = \sqrt{\frac{1+\frac{15}{17}}{2}} = \sqrt{\frac{16}{17}\cdot\frac{17}{17}} = \tfrac{4}{17}\sqrt{17}.$$

(c) Let $\alpha = \cos^{-1}\frac{3}{5}$. Applying the half-angle formula for the tangent,

$$\tan\left(\tfrac{1}{2}\cos^{-1}\tfrac{3}{5}\right) = \tan\tfrac{1}{2}\alpha = \frac{1-\cos\alpha}{\sin\alpha} = \frac{1-\frac{3}{5}}{\frac{4}{5}} = \frac{1}{2}.$$

[23] We recognize this expression as being of the form $\sin(some\ angle)$. The angle is *the angle whose tangent is* α. The easiest way to picture this angle is to consider it as being the angle whose ratio of opposite side to adjacent side is x to 1. Let $\alpha = \tan^{-1} x$. From *Figure 23*, $\sin\left(\tan^{-1} x\right) = \sin\alpha = \dfrac{x}{\sqrt{x^2+1}}$.

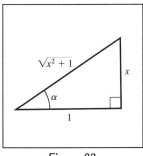

Figure 23

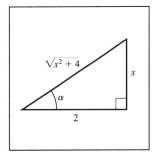

Figure 25

$\boxed{25}$ Let $\alpha = \sin^{-1}\dfrac{x}{\sqrt{x^2+4}}$. From *Figure 25*, $\sec\left(\sin^{-1}\dfrac{x}{\sqrt{x^2+4}}\right) = \sec\alpha = \dfrac{\sqrt{x^2+4}}{2}$.

$\boxed{27}$ Let $\alpha = \sin^{-1}x$. From *Figure 27*,

$$\sin\left(2\sin^{-1}x\right) = \sin 2\alpha = 2\sin\alpha\cos\alpha = 2\cdot\frac{x}{1}\cdot\frac{\sqrt{1-x^2}}{1} = 2x\sqrt{1-x^2}.$$

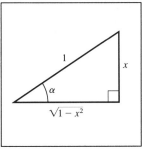

Figure 27

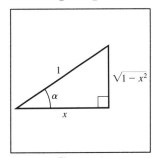

Figure 29

$\boxed{29}$ Let $\alpha = \arccos x$. $0 \le \alpha \le \pi \Rightarrow 0 \le \frac{1}{2}\alpha \le \frac{\pi}{2}$.

Thus $\cos\frac{1}{2}\alpha > 0$ and we use the "+" in the half-angle formula for the cosine.

$$\cos\left(\tfrac{1}{2}\arccos x\right) = \cos\tfrac{1}{2}\alpha = \sqrt{\frac{1+\cos\alpha}{2}} = \sqrt{\frac{1+x}{2}}.$$

$\boxed{31}$ (a) See text Figure 19. As $x \to -1^+$, $\sin^{-1}x \to \underline{\ -\frac{\pi}{2}\ }$.

(b) See text Figure 21. As $x \to 1^-$, $\cos^{-1}x \to \underline{\ 0\ }$.

(c) See text Figure 23. As $x \to \infty$, $\tan^{-1}x \to \underline{\ \frac{\pi}{2}\ }$.

$\boxed{33}$ $y = \sin^{-1} 2x$ •   Horizontally compress $y = \sin^{-1} x$ by a factor of 2.

Note that the domain changes from $[-1, 1]$ to $[-\frac{1}{2}, \frac{1}{2}]$.

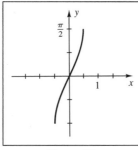

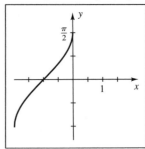

Figure 33   *Figure 35*

$\boxed{35}$ $y = \sin^{-1}(x + 1)$ •   Shift $y = \sin^{-1} x$ left 1 unit.

The domain changes from $[-1, 1]$ to $[-2, 0]$.

$\boxed{37}$ $y = \cos^{-1} \frac{1}{2}x$ •   Horizontally stretch $y = \cos^{-1} x$ by a factor of 2.

The domain changes from $[-1, 1]$ to $[-2, 2]$.

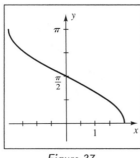

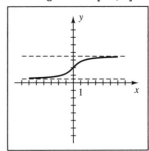

Figure 37   *Figure 39*

$\boxed{39}$ $y = 2 + \tan^{-1} x$ •   Shift $y = \tan^{-1} x$ up 2 units. The range changes from

$(-\frac{\pi}{2}, \frac{\pi}{2})$ to $(2 - \frac{\pi}{2}, 2 + \frac{\pi}{2})$, which is approximately $(0.43, 3.57)$.

$\boxed{41}$ If $\alpha = \arccos x$, then $\cos \alpha = x$, where $0 \le \alpha \le \pi$.

Hence, $y = \sin(\arccos x) = \sin \alpha = \sqrt{1 - \cos^2 \alpha} = \sqrt{1 - x^2}$.

Thus, we have the graph of the semicircle $y = \sqrt{1 - x^2}$

on the interval $[-1, 1]$.

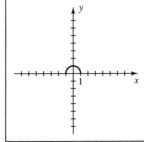

Figure 41

$\boxed{43}$ (a) Since the domain of the arcsine function is $[-1, 1]$, we know that $x - 3$ must be

in that interval. Thus, $-1 \le x - 3 \le 1 \Rightarrow 2 \le x \le 4$.

(b) The range of the arcsine function is $[-\frac{\pi}{2}, \frac{\pi}{2}]$.

Thus, $-\frac{\pi}{2} \le \sin^{-1}(x - 3) \le \frac{\pi}{2} \Rightarrow -\frac{\pi}{4} \le \frac{1}{2}\sin^{-1}(x - 3) \le \frac{\pi}{4} \Rightarrow -\frac{\pi}{4} \le y \le \frac{\pi}{4}$.

(c) $y = \frac{1}{2}\sin^{-1}(x - 3) \Rightarrow 2y = \sin^{-1}(x - 3) \Rightarrow \sin 2y = x - 3 \Rightarrow x = \sin 2y + 3$

45 (a) $-1 \le \frac{2}{3}x \le 1 \Rightarrow -\frac{3}{2} \le x \le \frac{3}{2}$

   (b) $0 \le \cos^{-1}\frac{2}{3}x \le \pi \Rightarrow 0 \le 4\cos^{-1}\frac{2}{3}x \le 4\pi \Rightarrow 0 \le y \le 4\pi$

   (c) $y = 4\cos^{-1}\frac{2}{3}x \Rightarrow \frac{1}{4}y = \cos^{-1}\frac{2}{3}x \Rightarrow \cos\frac{1}{4}y = \frac{2}{3}x \Rightarrow x = \frac{3}{2}\cos\frac{1}{4}y$

47 $y = -3 - \sin x \Rightarrow y + 3 = -\sin x \Rightarrow -(y+3) = \sin x \Rightarrow x = \sin^{-1}(-y-3)$

49 $y = 15 - 2\cos x \Rightarrow 2\cos x = 15 - y \Rightarrow \cos x = \frac{1}{2}(15-y) \Rightarrow x = \cos^{-1}\left[\frac{1}{2}(15-y)\right]$

51 $\frac{\sin x}{3} = \frac{\sin y}{4} \Rightarrow \sin x = \frac{3}{4}\sin y$. Since $0 < y < \pi$, we know that $\sin y > 0$. Thus, solving

$\sin x = \left(\frac{3}{4}\sin y\right)$ is similar to solving $\sin x = a$, where $0 < a < 1$. Remember that when

solving equations of this form, there are two solutions, a first quadrant angle and a

second quadrant angle. The reference angle for x is $x_R = \sin^{-1}(\frac{3}{4}\sin y)$, where

$0 < \frac{3}{4}\sin y \le \frac{3}{4} < 1$. If $0 < x < \frac{\pi}{2}$, then $x = x_R$. If $\frac{\pi}{2} < x < \pi$, then $x = \pi - x_R$.

53 $\cos^2 x + 2\cos x - 1 = 0 \Rightarrow \cos x = -1 \pm \sqrt{2} \approx 0.4142,\ -2.4142$.

   Since $-2.4142 < -1$, $x = \cos^{-1}(-1+\sqrt{2}) \approx 1.1437$ is one answer.

$$x = 2\pi - \cos^{-1}(-1+\sqrt{2}) \approx 2\pi - 1.1437 \approx 5.1395 \text{ is the other.}$$

57 $15\cos^4 x - 14\cos^2 x + 3 = 0 \Rightarrow (5\cos^2 x - 3)(3\cos^2 x - 1) = 0 \Rightarrow \cos^2 x = \frac{3}{5}, \frac{1}{3} \Rightarrow$

$\cos x = \pm\frac{1}{5}\sqrt{15},\ \pm\frac{1}{3}\sqrt{3} \Rightarrow x = \cos^{-1}(\pm\frac{1}{5}\sqrt{15}),\ \cos^{-1}(\pm\frac{1}{3}\sqrt{3})$.

$\cos^{-1}\frac{1}{5}\sqrt{15} \approx 0.6847,\ \cos^{-1}(-\frac{1}{5}\sqrt{15}) \approx 2.4569,$

$$\cos^{-1}\frac{1}{3}\sqrt{3} \approx 0.9553,\ \cos^{-1}(-\frac{1}{3}\sqrt{3}) \approx 2.1863$$

59 $6\sin^3\theta + 18\sin^2\theta - 5\sin\theta - 15 = 0 \Rightarrow 6\sin^2\theta(\sin\theta + 3) - 5(\sin\theta + 3) = 0 \Rightarrow$

$$(6\sin^2\theta - 5)(\sin\theta + 3) = 0 \Rightarrow \sin\theta = \pm\frac{1}{6}\sqrt{30} \Rightarrow \theta = \sin^{-1}(\pm\frac{1}{6}\sqrt{30}) \approx \pm1.1503$$

61 (a) $S = 4,\ D = 3.5,\ d = 1 \Rightarrow M = \frac{S}{2}\left(1 - \frac{2}{\pi}\tan^{-1}\frac{d}{D}\right) = \frac{4}{2}\left(1 - \frac{2}{\pi}\tan^{-1}\frac{1}{3.5}\right) \approx 1.65$ m

   (b) $d = 4 \Rightarrow M = \frac{S}{2}\left(1 - \frac{2}{\pi}\tan^{-1}\frac{d}{D}\right) = \frac{4}{2}\left(1 - \frac{2}{\pi}\tan^{-1}\frac{4}{3.5}\right) \approx 0.92$ m

   (c) $d = 10 \Rightarrow M = \frac{S}{2}\left(1 - \frac{2}{\pi}\tan^{-1}\frac{d}{D}\right) = \frac{4}{2}\left(1 - \frac{2}{\pi}\tan^{-1}\frac{10}{3.5}\right) \approx 0.43$ m

63 (a) Let β denote the angle by the sailboat with opposite side d and hypotenuse k.

   Now $\sin\beta = \frac{d}{k} \Rightarrow \beta = \sin^{-1}\frac{d}{k}$. Using alternate interior angles,

   we see that $\alpha + \beta = \theta$. Thus, $\alpha = \theta - \beta = \theta - \sin^{-1}\frac{d}{k}$.

   (b) $d = 50,\ k = 210,$ and $\theta = 53.4° \Rightarrow \alpha = 53.4° - \sin^{-1}\frac{50}{210} \approx 39.63°$, or $40°$.

Note: The following is a general outline that can be used for verifying trigonometric identities involving inverse trigonometric functions.

    (1) Define angles and their ranges—make sure the range of values for one side of the equation is equal to the range of values for the other side.

    (2) Choose a trigonometric function T that is one-to-one on the range of values listed in part (1).

    (3) Show that $T(\text{LS}) = T(\text{RS})$. Note that $T(\text{LS}) = T(\text{RS}) \not\Rightarrow \text{LS} = \text{RS}$.

    (4) Conclude that since T is one-to-one on the range of values, LS = RS.

65 Let $\alpha = \sin^{-1} x$ and $\beta = \tan^{-1}\dfrac{x}{\sqrt{1-x^2}}$ with $-\frac{\pi}{2} < \alpha < \frac{\pi}{2}$ and $-\frac{\pi}{2} < \beta < \frac{\pi}{2}$.

Thus, $\sin\alpha = x$ and $\sin\beta = x$. Since the sine function is one-to-one on $\left(-\frac{\pi}{2}, \frac{\pi}{2}\right)$,

$$\text{we have } \alpha = \beta\text{—that is, } \sin^{-1}x = \tan^{-1}\frac{x}{\sqrt{1-x^2}}.$$

67 Let $\alpha = \arcsin(-x)$ and $\beta = \arcsin x$ with $-\frac{\pi}{2} \le \alpha \le \frac{\pi}{2}$ and $-\frac{\pi}{2} \le \beta \le \frac{\pi}{2}$.

Thus, $\sin\alpha = -x$ and $\sin\beta = x$. Consequently, $\sin\alpha = -\sin\beta = \sin(-\beta)$.

Since the sine function is one-to-one on $\left[-\frac{\pi}{2}, \frac{\pi}{2}\right]$,

$$\text{we have } \alpha = -\beta\text{—that is, } \arcsin(-x) = -\arcsin x.$$

69 Let $\alpha = \arctan x$ and $\beta = \arctan(1/x)$.

Since $x > 0$, we have $0 < \alpha < \frac{\pi}{2}$ and $0 < \beta < \frac{\pi}{2}$, and hence $0 < \alpha + \beta < \pi$.

Thus, $\tan(\alpha + \beta) = \dfrac{\tan\alpha + \tan\beta}{1 - \tan\alpha \tan\beta} = \dfrac{x + (1/x)}{1 - x \cdot (1/x)} = \dfrac{x + (1/x)}{0}$.

Since the denominator is 0, $\tan(\alpha + \beta)$ is undefined and hence $\alpha + \beta = \frac{\pi}{2}$ since $\frac{\pi}{2}$ is the only value between 0 and π for which the tangent is undefined.

71 The domain of $\sin^{-1}(x - 1)$ is $[0, 2]$ and the domain of $\cos^{-1}\frac{1}{2}x$ is $[-2, 2]$.

The domain of f is the intersection of $[0, 2]$ and $[-2, 2]$, i.e., $[0, 2]$.

    From the graph, we see that the function is increasing and its range is $\left[-\frac{\pi}{2}, \pi\right]$.

$[-3, 6]$ by $[-2, 4]$ $[-3, 3]$ by $[-2, 2]$

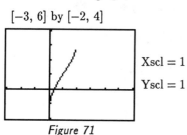
    Xscl = 1
    Yscl = 1

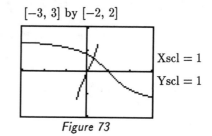
    Xscl = 1
    Yscl = 1

Figure 71   *Figure 73*

73 Graph $y = \sin^{-1} 2x$ and $y = \tan^{-1}(1 - x)$.

    From the graph, we see that there is one solution at $x \approx 0.29$.

[75] Make the assignments $Y_1 = \sin^{-1}(\sin x/1.52)$, $Y_2 = x - Y_1$, $Y_3 = x + Y_1$, and $Y_4 = .5((\sin Y_2)^2/(\sin Y_3)^2 + (\tan Y_2)^2/(\tan Y_3)^2)$. Now turn *off* Y_1, Y_2, and Y_3—leaving only Y_4 *on* to graph. From the graph, we see that when $f(\theta) = 0.2$, $\theta \approx 1.25$, or approximately $72°$.

$$[0, 1.57] \text{ by } [0, 1.05]$$

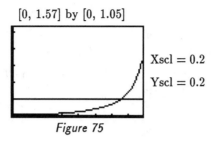

Xscl $= 0.2$

Yscl $= 0.2$

Figure 75

Chapter 7 Review Exercises

[3] $\dfrac{(\sec^2\theta - 1)\cot\theta}{\tan\theta\,\sin\theta + \cos\theta} = \dfrac{(\tan^2\theta)\cot\theta}{\dfrac{\sin\theta}{\cos\theta}\cdot\sin\theta + \cos\theta}$ { Pythagorean and tangent identities }

$$= \dfrac{\tan\theta\,(\tan\theta\,\cot\theta)}{\dfrac{\sin^2\theta}{\cos\theta} + \cos\theta} \quad \{\text{combine terms}\}$$

$$= \dfrac{\tan\theta}{\dfrac{\sin^2\theta + \cos^2\theta}{\cos\theta}} \quad \{\text{reciprocal identity, common denominator}\}$$

$$= \dfrac{\sin\theta\,/\cos\theta}{1/\cos\theta} \quad \{\text{Pythagorean and tangent identities}\}$$

$$= \sin\theta \quad \{\text{simplify}\}$$

[5] $\dfrac{1}{1+\sin t} = \dfrac{1}{1+\sin t}\cdot\dfrac{1-\sin t}{1-\sin t}$ $\left\{ \begin{array}{c} \text{multiply the numerator and the denominator} \\ \text{by conjugate of the denominator} \end{array} \right\}$

$$= \dfrac{1-\sin t}{1-\sin^2 t} \quad \{\text{the difference of two squares}\}$$

$$= \dfrac{1-\sin t}{\cos^2 t} \quad \{\text{Pythagorean identity}\}$$

$$= \dfrac{1-\sin t}{\cos t}\cdot\dfrac{1}{\cos t} \quad \{\text{break up since we want } \sec t \text{ on the right side}\}$$

$$= \left(\dfrac{1}{\cos t} - \dfrac{\sin t}{\cos t}\right)\cdot\sec t \quad \{\text{split up fraction}\}$$

$$= (\sec t - \tan t)\sec t \quad \{\text{reciprocal and tangent identities}\}$$

$\boxed{6}$ $\dfrac{\sin(\alpha - \beta)}{\cos(\alpha + \beta)} = \dfrac{\sin\alpha\cos\beta - \cos\alpha\sin\beta}{\cos\alpha\cos\beta - \sin\alpha\sin\beta}$ $\left\{ \begin{array}{l} \text{subtraction formula for the sine} \\ \text{addition formula for the cosine} \end{array} \right\}$

The first term in the denominator is $\cos\alpha\cos\beta$, but looking ahead, we see that the first term in the denominator of the expression we want to obtain is 1. Hence, we will divide both the numerator and the denominator by $\cos\alpha\cos\beta$.

$$= \frac{(\sin\alpha\cos\beta - \cos\alpha\sin\beta) \,/\, \cos\alpha\cos\beta}{(\cos\alpha\cos\beta - \sin\alpha\sin\beta) \,/\, \cos\alpha\cos\beta}$$

$$= \frac{\dfrac{\sin\alpha\cos\beta}{\cos\alpha\cos\beta} - \dfrac{\cos\alpha\sin\beta}{\cos\alpha\cos\beta}}{\dfrac{\cos\alpha\cos\beta}{\cos\alpha\cos\beta} - \dfrac{\sin\alpha\sin\beta}{\cos\alpha\cos\beta}} = \frac{\dfrac{\sin\alpha}{\cos\alpha} - \dfrac{\sin\beta}{\cos\beta}}{1 - \dfrac{\sin\alpha}{\cos\alpha}\cdot\dfrac{\sin\beta}{\cos\beta}} = \frac{\tan\alpha - \tan\beta}{1 - \tan\alpha\tan\beta}$$

$\boxed{7}$ $\quad \tan 2u = \dfrac{2\tan u}{1 - \tan^2 u}$ $\qquad$ { apply the double-angle formula for the tangent }

$$= \frac{2\cdot\dfrac{1}{\cot u}}{1 - \dfrac{1}{\cot^2 u}} \qquad \text{\{ put in terms of cot since it appears on the right side \}}$$

$$= \frac{\dfrac{2}{\cot u}}{\dfrac{\cot^2 u - 1}{\cot^2 u}} \qquad \text{\{ combine into one fraction \}}$$

$$= \frac{2\cot u}{\cot^2 u - 1} \qquad \text{\{ simplify complex fraction \}}$$

$$= \frac{2\cot u}{(\csc^2 u - 1) - 1} \qquad \text{\{ Pythagorean identity \}}$$

$$= \frac{2\cot u}{\csc^2 u - 2} \qquad \text{\{ simplify \}}$$

$\boxed{10}$ $\text{LS} = \dfrac{\sin u + \sin v}{\csc u + \csc v} = \dfrac{\sin u + \sin v}{\dfrac{1}{\sin u} + \dfrac{1}{\sin v}} = \dfrac{\sin u + \sin v}{\dfrac{\sin v + \sin u}{\sin u \sin v}} = \sin u \sin v$

At this point, there is no apparent "next step." Thus, we will stop working with the left side and try to simplify the right side to the same expression, $\sin u \sin v$.

$\text{RS} = \dfrac{1 - \sin u \sin v}{-1 + \csc u \csc v} = \dfrac{1 - \sin u \sin v}{-1 + \dfrac{1}{\sin u \sin v}} = \dfrac{1 - \sin u \sin v}{\dfrac{1 - \sin u \sin v}{\sin u \sin v}} = \sin u \sin v$

Since the LS and RS equal the same expression and the steps are reversible,

$\hfill$ the identity is verified.

[12] $\dfrac{\cos \gamma}{1 - \tan \gamma} + \dfrac{\sin \gamma}{1 - \cot \gamma} = \dfrac{\cos \gamma}{1 - \dfrac{\sin \gamma}{\cos \gamma}} + \dfrac{\sin \gamma}{1 - \dfrac{\cos \gamma}{\sin \gamma}} = \dfrac{\cos \gamma}{\dfrac{\cos \gamma - \sin \gamma}{\cos \gamma}} + \dfrac{\sin \gamma}{\dfrac{\sin \gamma - \cos \gamma}{\sin \gamma}} =$

$\dfrac{\cos^2\gamma}{\cos \gamma - \sin \gamma} + \dfrac{\sin^2\gamma}{\sin \gamma - \cos \gamma} = \dfrac{\cos^2\gamma}{\cos \gamma - \sin \gamma} - \dfrac{\sin^2\gamma}{-(\sin \gamma - \cos \gamma)} =$

$\dfrac{\cos^2\gamma}{\cos \gamma - \sin \gamma} - \dfrac{\sin^2\gamma}{\cos \gamma - \sin \gamma} = \dfrac{\cos^2\gamma - \sin^2\gamma}{\cos \gamma - \sin \gamma} = \dfrac{(\cos \gamma + \sin \gamma)(\cos \gamma - \sin \gamma)}{\cos \gamma - \sin \gamma} =$

$$\cos \gamma + \sin \gamma$$

[13] $\dfrac{\cos (-t)}{\sec (-t) + \tan (-t)} = \dfrac{\cos t}{\sec t + (-\tan t)} = \dfrac{\cos t}{\sec t - \tan t} = \dfrac{\cos t}{\dfrac{1}{\cos t} - \dfrac{\sin t}{\cos t}} = \dfrac{\cos t}{\dfrac{1 - \sin t}{\cos t}} =$

$$\dfrac{\cos^2 t}{1 - \sin t} = \dfrac{1 - \sin^2 t}{1 - \sin t} = \dfrac{(1 - \sin t)(1 + \sin t)}{1 - \sin t} = 1 + \sin t$$

[15] In the following solution, we could multiply both the numerator and the denominator by *either* the conjugate of the numerator *or* the conjugate of the denominator. Since the numerator on the right side looks like the numerator on the left side, we'll change the denominator.

$$\sqrt{\dfrac{1 - \cos t}{1 + \cos t}} = \sqrt{\dfrac{(1 - \cos t)}{(1 + \cos t)} \cdot \dfrac{(1 - \cos t)}{(1 - \cos t)}} = \sqrt{\dfrac{(1 - \cos t)^2}{1 - \cos^2 t}} = \sqrt{\dfrac{(1 - \cos t)^2}{\sin^2 t}} =$$

$$\dfrac{\sqrt{(1 - \cos t)^2}}{\sqrt{\sin^2 t}} = \dfrac{|1 - \cos t|}{|\sin t|} = \dfrac{1 - \cos t}{|\sin t|}, \text{ since } (1 - \cos t) \geq 0.$$

[19] We need to break down the angle argument of 4β into terms with only β as their argument. We can do this by using either the double-angle formula for the sine or the addition formula for the sine.

$$\tfrac{1}{4}\sin 4\beta = \tfrac{1}{4}\sin (2 \cdot 2\beta) = \tfrac{1}{4}(2 \sin 2\beta \, \cos 2\beta) = \tfrac{1}{2}(2 \sin \beta \, \cos \beta)(\cos^2\beta - \sin^2\beta) =$$

$$\sin \beta \, \cos^3\beta - \cos \beta \, \sin^3\beta$$

[20] $\tan \tfrac{1}{2}\theta = \dfrac{1 - \cos \theta}{\sin \theta}$ { apply the half-angle formula for the tangent }

$\quad = \dfrac{1}{\sin \theta} - \dfrac{\cos \theta}{\sin \theta}$ { split up the fraction }

$\quad = \csc \theta - \cot \theta$ { reciprocal and cotangent identities }

[22] Let $\alpha = \arctan x$ and $\beta = \arctan \dfrac{2x}{1 - x^2}$.

Because $-1 < x < 1$, $\arctan (-1) < \arctan (x) < \arctan (1)$, and hence $-\tfrac{\pi}{4} < \alpha < \tfrac{\pi}{4}$.

Thus, $\tan \alpha = x$ and $\tan \beta = \dfrac{2x}{1 - x^2} = \dfrac{2 \tan \alpha}{1 - \tan^2\alpha} = \tan 2\alpha$.

Since the tangent function is one-to-one on $(-\tfrac{\pi}{2}, \tfrac{\pi}{2})$, we have $\beta = 2\alpha$ or, equivalently,

$$\alpha = \tfrac{1}{2}\beta. \text{ In terms of } x, \text{ we have } \arctan x = \tfrac{1}{2}\arctan \dfrac{2x}{1 - x^2}.$$

$\boxed{24}$ $2\cos\alpha + \tan\alpha = \sec\alpha \Rightarrow 2\cos\alpha + \frac{\sin\alpha}{\cos\alpha} = \frac{1}{\cos\alpha} \Rightarrow \{\text{multiply by the lcd, } \cos\alpha\}$

$2\cos^2\alpha + \sin\alpha = 1 \Rightarrow 2(1-\sin^2\alpha) + \sin\alpha = 1 \Rightarrow 2 - 2\sin^2\alpha + \sin\alpha = 1 \Rightarrow$

$2\sin^2\alpha - \sin\alpha - 1 = 0 \Rightarrow (2\sin\alpha + 1)(\sin\alpha - 1) = 0 \Rightarrow \sin\alpha = -\frac{1}{2}, 1 \Rightarrow \alpha = \frac{7\pi}{6}, \frac{11\pi}{6}, \frac{\pi}{2}.$

Checking these values in the original equation, we see that $\tan\frac{\pi}{2}$ is undefined so

exclude $\frac{\pi}{2}$ and our solution is $\alpha = \frac{7\pi}{6}, \frac{11\pi}{6}.$

$\boxed{25}$ $\sin\theta = \tan\theta \Rightarrow \sin\theta - \frac{\sin\theta}{\cos\theta} = 0 \Rightarrow \sin\theta\left(1 - \frac{1}{\cos\theta}\right) = 0 \Rightarrow \sin\theta = 0$ or $1 = \frac{1}{\cos\theta} \Rightarrow$

$\sin\theta = 0$ or $\cos\theta = 1 \Rightarrow \theta = 0, \pi$ or $\theta = 0 \Rightarrow \theta = 0, \pi$

$\boxed{27}$ $2\cos^3 t + \cos^2 t - 2\cos t - 1 = 0 \Rightarrow \cos^2 t (2\cos t + 1) - 1(2\cos t + 1) = 0 \Rightarrow$

$(\cos^2 t - 1)(2\cos t + 1) = 0 \Rightarrow \cos t = \pm 1, -\frac{1}{2} \Rightarrow t = 0, \pi, \frac{2\pi}{3}, \frac{4\pi}{3}$

$\boxed{29}$ $\sin\beta + 2\cos^2\beta = 1 \Rightarrow \sin\beta + 2(1 - \sin^2\beta) = 1 \Rightarrow 2\sin^2\beta - \sin\beta - 1 = 0 \Rightarrow$

$(2\sin\beta + 1)(\sin\beta - 1) = 0 \Rightarrow \sin\beta = -\frac{1}{2}, 1 \Rightarrow \beta = \frac{7\pi}{6}, \frac{11\pi}{6}, \frac{\pi}{2}$

$\boxed{32}$ $\tan 2x \cos 2x = \sin 2x \Rightarrow \sin 2x = \sin 2x.$ This is an identity and is true for all values

of x in $[0, 2\pi)$ except those that make $\tan 2x$ undefined, or, equivalently, those that

make $\cos 2x$ equal to 0. $\cos 2x = 0 \Rightarrow 2x = \frac{\pi}{2} + \pi n \Rightarrow x = \frac{\pi}{4} + \frac{\pi}{2}n.$

Hence, the solutions are all x in $[0, 2\pi)$ except $\frac{\pi}{4}, \frac{3\pi}{4}, \frac{5\pi}{4}, \frac{7\pi}{4}.$

$\boxed{33}$ $2\cos 3x \cos 2x = 1 - 2\sin 3x \sin 2x \Rightarrow 2\cos 3x \cos 2x + 2\sin 3x \sin 2x = 1 \Rightarrow$

$2(\cos 3x \cos 2x + \sin 3x \sin 2x) = 1 \Rightarrow \cos(3x - 2x) = \frac{1}{2} \Rightarrow \cos x = \frac{1}{2} \Rightarrow x = \frac{\pi}{3}, \frac{5\pi}{3}$

$\boxed{35}$ $\cos\pi x + \sin\pi x = 0 \Rightarrow \sin\pi x = -\cos\pi x \Rightarrow \frac{\sin\pi x}{\cos\pi x} = \frac{-\cos\pi x}{\cos\pi x} \{\text{divide by } \cos\pi x\} \Rightarrow$

$\tan\pi x = -1 \Rightarrow \pi x = \frac{3\pi}{4} + \pi n \Rightarrow x = \frac{3}{4} + n \Rightarrow x = \frac{3}{4}, \frac{7}{4}, \frac{11}{4}, \frac{15}{4}, \frac{19}{4}, \frac{23}{4}$

$\boxed{37}$ $2\cos^2\frac{1}{2}\theta - 3\cos\theta = 0 \Rightarrow 2\left(\frac{1 + \cos\theta}{2}\right) - 3\cos\theta = 0 \Rightarrow$

$(1 + \cos\theta) - 3\cos\theta = 0 \Rightarrow 1 - 2\cos\theta = 0 \Rightarrow \cos\theta = \frac{1}{2} \Rightarrow \theta = \frac{\pi}{3}, \frac{5\pi}{3}$

$\boxed{39}$ $\sin 5x = \sin 3x \Rightarrow \sin 5x - \sin 3x = 0 \Rightarrow [S2]\ 2\cos\frac{5x + 3x}{2}\sin\frac{5x - 3x}{2} = 0 \Rightarrow$

$\cos 4x \sin x = 0 \Rightarrow 4x = \frac{\pi}{2} + \pi n$ or $x = \pi n \Rightarrow x = \frac{\pi}{8} + \frac{\pi}{4}n$ or $x = 0, \pi \Rightarrow$

$x = 0, \frac{\pi}{8}, \frac{3\pi}{8}, \frac{5\pi}{8}, \frac{7\pi}{8}, \pi, \frac{9\pi}{8}, \frac{11\pi}{8}, \frac{13\pi}{8}, \frac{15\pi}{8}$

$\boxed{40}$ $\cos 3x = -\cos 2x \Rightarrow \cos 3x + \cos 2x = 0 \Rightarrow [S3]\ 2\cos\frac{3x + 2x}{2}\cos\frac{3x - 2x}{2} = 0 \Rightarrow$

$\cos\frac{5}{2}x \cos\frac{1}{2}x = 0 \Rightarrow \frac{5}{2}x = \frac{\pi}{2} + \pi n$ or $\frac{1}{2}x = \frac{\pi}{2} + \pi n \Rightarrow x = \frac{\pi}{5} + \frac{2\pi}{5}n$ or $x = \pi + 2\pi n \Rightarrow$

$x = \frac{\pi}{5}, \frac{3\pi}{5}, \pi, \frac{7\pi}{5}, \frac{9\pi}{5}$

$\boxed{42}$ $\tan 285° = \tan(225° + 60°) =$

$\frac{\tan 225° + \tan 60°}{1 - \tan 225° \tan 60°} = \frac{1 + \sqrt{3}}{1 - 1 \cdot \sqrt{3}} = \frac{1 + \sqrt{3}}{1 - \sqrt{3}} \cdot \frac{1 + \sqrt{3}}{1 + \sqrt{3}} = \frac{4 + 2\sqrt{3}}{-2} = -2 - \sqrt{3}$

$\boxed{44}$ $\csc\dfrac{\pi}{8} = \dfrac{1}{\csc\frac{\pi}{8}} = \dfrac{1}{\sin\left(\frac{1}{2}\cdot\frac{\pi}{4}\right)} = \dfrac{1}{\sqrt{\dfrac{1-\cos\frac{\pi}{4}}{2}}} = \dfrac{1}{\sqrt{\dfrac{1-\sqrt{2}/2}{2}}} = \dfrac{1}{\sqrt{\dfrac{2-\sqrt{2}}{4}}} = \dfrac{2}{\sqrt{2-\sqrt{2}}}$

$\boxed{45}$ $\csc\theta = \frac{5}{3}$ and $\cos\phi = \frac{8}{17} \Rightarrow \sin\theta = \frac{3}{5}$, $\cos\theta = \frac{4}{5}$, $\tan\theta = \frac{3}{4}$ and $\sin\phi = \frac{15}{17}$, $\tan\phi = \frac{15}{8}$.

$$\sin(\theta+\phi) = \sin\theta\,\cos\phi + \cos\theta\,\sin\phi = \tfrac{3}{5}\cdot\tfrac{8}{17} + \tfrac{4}{5}\cdot\tfrac{15}{17} = \tfrac{84}{85}$$

$\boxed{48}$ $\tan(\theta-\phi) = \dfrac{\tan\theta - \tan\phi}{1 + \tan\theta\,\tan\phi} = \dfrac{\frac{3}{4} - \frac{15}{8}}{1 + \frac{3}{4}\cdot\frac{15}{8}}\cdot\dfrac{32}{32} = \dfrac{24-60}{32+45} = -\dfrac{36}{77}$

$\boxed{49}$ $\sin(\phi-\theta) = \sin\phi\,\cos\theta - \cos\phi\,\sin\theta = \frac{15}{17}\cdot\frac{4}{5} - \frac{8}{17}\cdot\frac{3}{5} = \frac{36}{85}$

$\boxed{50}$ First recognize the relationship to Exercise 49.

$$\sin(\theta-\phi) = \sin\big[-(\phi-\theta)\big] = -\sin(\phi-\theta) = -\tfrac{36}{85}$$

$\boxed{52}$ $\cos 2\phi = \cos^2\phi - \sin^2\phi = \left(\frac{8}{17}\right)^2 - \left(\frac{15}{17}\right)^2 = -\frac{161}{289}$

$\boxed{54}$ $\sin\frac{1}{2}\theta = \sqrt{\dfrac{1-\cos\theta}{2}} = \sqrt{\dfrac{1-\frac{4}{5}}{2}} = \sqrt{\dfrac{\frac{1}{5}}{2}} = \sqrt{\dfrac{1}{10}\cdot\dfrac{10}{10}} = \dfrac{1}{10}\sqrt{10}$

$\boxed{55}$ $\tan\frac{1}{2}\theta = \dfrac{1-\cos\theta}{\sin\theta} = \dfrac{1-\frac{4}{5}}{\frac{3}{5}} = \dfrac{\frac{1}{5}}{\frac{3}{5}} = \dfrac{1}{3}$

$\boxed{57}$ (a) $\sin 7t\,\sin 4t = [\text{P4}]\ \frac{1}{2}[\cos(7t-4t) - \cos(7t+4t)] = \frac{1}{2}\cos 3t - \frac{1}{2}\cos 11t$

(b) $\cos\frac{1}{4}u\,\cos\left(-\frac{1}{6}u\right) = [\text{P3}]\ \frac{1}{2}\left\{\cos\left[\frac{1}{4}u + \left(-\frac{1}{6}u\right)\right] + \cos\left[\frac{1}{4}u - \left(-\frac{1}{6}u\right)\right]\right\} =$

$$\tfrac{1}{2}\left(\cos\tfrac{2}{24}u + \cos\tfrac{10}{24}u\right) = \tfrac{1}{2}\cos\tfrac{1}{12}u + \tfrac{1}{2}\cos\tfrac{5}{12}u$$

$\boxed{58}$ (b) $\cos 3\theta - \cos 8\theta = [\text{S4}]\ -2\sin\dfrac{3\theta+8\theta}{2}\,\sin\dfrac{3\theta-8\theta}{2} = -2\sin\tfrac{11}{2}\theta\,\sin\left(-\tfrac{5}{2}\theta\right) =$

$$2\sin\tfrac{11}{2}\theta\,\sin\tfrac{5}{2}\theta$$

(c) $\sin\frac{1}{4}t - \sin\frac{1}{5}t = [\text{S2}]\ 2\cos\dfrac{\frac{1}{4}t + \frac{1}{5}t}{2}\,\sin\dfrac{\frac{1}{4}t - \frac{1}{5}t}{2} = 2\cos\dfrac{\frac{5}{20}t + \frac{4}{20}t}{2}\,\sin\dfrac{\frac{5}{20}t - \frac{4}{20}t}{2} =$

$$2\cos\tfrac{9}{40}t\,\sin\tfrac{1}{40}t$$

$\boxed{62}$ $\arccos\left(\tan\frac{3\pi}{4}\right) = \arccos(-1) = \pi$

$\boxed{63}$ $\arcsin\left(\sin\frac{5\pi}{4}\right) = \arcsin\left(-\dfrac{\sqrt{2}}{2}\right) = -\dfrac{\pi}{4}$ $\boxed{64}$ $\cos^{-1}\left(\cos\frac{5\pi}{4}\right) = \cos^{-1}\left(-\dfrac{\sqrt{2}}{2}\right) = \dfrac{3\pi}{4}$

$\boxed{69}$ Let $\alpha = \sin^{-1}\frac{15}{17}$ and $\beta = \sin^{-1}\frac{8}{17}$.

$\cos\left(\sin^{-1}\frac{15}{17} - \sin^{-1}\frac{8}{17}\right) = \cos(\alpha-\beta) = \cos\alpha\,\cos\beta + \sin\alpha\,\sin\beta = \frac{8}{17}\cdot\frac{15}{17} + \frac{15}{17}\cdot\frac{8}{17} = \frac{240}{289}$.

$\boxed{70}$ Let $\alpha = \sin^{-1}\frac{4}{5}$. $\cos\left(2\sin^{-1}\frac{4}{5}\right) = \cos(2\alpha) = \cos^2\alpha - \sin^2\alpha = \left(\frac{3}{5}\right)^2 - \left(\frac{4}{5}\right)^2 = -\frac{7}{25}$.

73 $y = 1 - \sin^{-1} x = -\sin^{-1} x + 1$ •

Reflect $y = \sin^{-1} x$ through the x-axis and shift it up 1 unit.

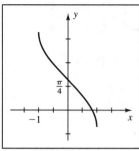

Figure 73

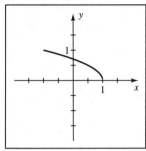

Figure 74

74 If $\alpha = \cos^{-1} x$, then $\cos \alpha = x$, where $0 \le \alpha \le \pi$.

Hence, $y = \sin\left(\frac{1}{2} \cos^{-1} x\right) = \sin \frac{1}{2}\alpha = \sqrt{\frac{1 - \cos \alpha}{2}} = \sqrt{\frac{1 - x}{2}}$.

Thus, we have the graph of the half-parabola $y = \sqrt{\frac{1}{2}(1 - x)}$ on the interval $[-1, 1]$.

75 $\cos(\alpha + \beta + \gamma)$

$$= \cos\big[(\alpha + \beta) + \gamma\big]$$
$$= \cos(\alpha + \beta) \cos \gamma - \sin(\alpha + \beta) \sin \gamma$$
$$= (\cos \alpha \cos \beta - \sin \alpha \sin \beta) \cos \gamma - (\sin \alpha \cos \beta + \cos \alpha \sin \beta) \sin \gamma$$
$$= \cos \alpha \cos \beta \cos \gamma - \sin \alpha \sin \beta \cos \gamma - \sin \alpha \cos \beta \sin \gamma - \cos \alpha \sin \beta \sin \gamma$$

76 (a) $t = -\frac{\pi}{2b} \Rightarrow F = A\left[\cos\left(-\frac{\pi}{2}\right) - a \cos\left(-\frac{3\pi}{2}\right)\right] = A(0 - a \cdot 0) = 0$

$$t = \frac{\pi}{2b} \Rightarrow F = A\left(\cos \frac{\pi}{2} - a \cos \frac{3\pi}{2}\right) = A(0 - a \cdot 0) = 0$$

(b) $a = \frac{1}{3} \Rightarrow \sin 3bt = \sin bt \Rightarrow \sin 3bt - \sin bt = 0 \Rightarrow$

[S2] $2 \cos \dfrac{3bt + bt}{2} \sin \dfrac{3bt - bt}{2} = 0 \Rightarrow \cos 2bt \sin bt = 0 \Rightarrow$

$\cos 2bt = 0$ or $\sin bt = 0 \Rightarrow 2bt = \frac{\pi}{2} + \pi n$ or $bt = \pi n \Rightarrow$

$$t = \frac{\pi}{4b} + \frac{\pi}{2b}n \text{ or } t = \frac{\pi}{b}n. \text{ Since } -\frac{\pi}{2b} < t < \frac{\pi}{2b}, \ t = \pm\frac{\pi}{4b}, \ 0.$$

(c) Using the values from part (b), $t = 0 \Rightarrow F = A\left(\cos 0 - \frac{1}{3}\cos 0\right) = A\left(1 - \frac{1}{3}\right) = \frac{2}{3}A$.

$$t = \pm\frac{\pi}{4b} \Rightarrow F = A\left[\cos\left(\pm\frac{\pi}{4}\right) - \frac{1}{3}\cos\left(\pm\frac{3\pi}{4}\right)\right] = A\left(\frac{\sqrt{2}}{2} + \frac{\sqrt{2}}{6}\right) = \frac{4\sqrt{2}}{6}A = \frac{2}{3}\sqrt{2}\,A.$$

The second value is $\sqrt{2}$ times the first, hence $\frac{2}{3}\sqrt{2}\,A$ is the maximum force.

78 (a) Bisect θ to form two right triangles. $\tan \frac{1}{2}\theta = \dfrac{\frac{1}{2}x}{d} \Rightarrow x = 2d \tan \frac{1}{2}\theta$.

(b) Using part (a) with $x = 0.5$ ft and $\theta = 0.0005$ radian,

$$\text{we have } d = \frac{x}{2 \tan \frac{1}{2}\theta} \approx 1000 \text{ ft, so } d \le 1000 \text{ ft.}$$

[79] (a) Bisect θ to form two right triangles.

$$\cos\tfrac{1}{2}\theta = \frac{r}{d+r} \Rightarrow d+r = \frac{r}{\cos\frac{1}{2}\theta} \Rightarrow d = r\,\sec\tfrac{1}{2}\theta - r = r\left(\sec\tfrac{1}{2}\theta - 1\right).$$

(b) $d = 300$ and $r = 4000 \Rightarrow \cos\tfrac{1}{2}\theta = \frac{r}{d+r} = \frac{4000}{4300} \Rightarrow \tfrac{1}{2}\theta \approx 21.5° \Rightarrow \theta \approx 43°.$

[80] $N = h/w$ and $\tan\theta = N \Rightarrow \tan\theta = \frac{h}{w}.$

(a) $\tan\theta = \frac{h}{w} = \frac{400}{80} = 5 \Rightarrow \theta = \tan^{-1}5 \approx 78.7°$

(b) $\tan\theta = \frac{h}{w} = \frac{55}{30} = \frac{11}{6} \Rightarrow \theta = \tan^{-1}\frac{11}{6} \approx 61.4°$

Chapter 8: Applications of Trigonometry

1 $\beta = 180° - \alpha - \gamma = 180° - 41° - 77° = 62°.$

$$\frac{b}{\sin \beta} = \frac{a}{\sin \alpha} \Rightarrow b = \frac{a \sin \beta}{\sin \alpha} = \frac{10.5 \sin 62°}{\sin 41°} \approx 14.1.$$

$$\frac{c}{\sin \gamma} = \frac{a}{\sin \alpha} \Rightarrow c = \frac{a \sin \gamma}{\sin \alpha} = \frac{10.5 \sin 77°}{\sin 41°} \approx 15.6.$$

3 $\gamma = 180° - \alpha - \beta = 180° - 27°40' - 52°10' = 100°10'.$

$$\frac{b}{\sin \beta} = \frac{a}{\sin \alpha} \Rightarrow b = \frac{a \sin \beta}{\sin \alpha} = \frac{32.4 \sin 52°10'}{\sin 27°40'} \approx 55.1.$$

$$\frac{c}{\sin \gamma} = \frac{a}{\sin \alpha} \Rightarrow c = \frac{a \sin \gamma}{\sin \alpha} = \frac{32.4 \sin 100°10'}{\sin 27°40'} \approx 68.7.$$

7 $\dfrac{\sin \beta}{b} = \dfrac{\sin \gamma}{c} \Rightarrow \beta = \sin^{-1}\left(\dfrac{b \sin \gamma}{c}\right) = \sin^{-1}\left(\dfrac{12 \sin 81°}{11}\right) \approx \sin^{-1}(1.0775).$ Since 1.0775 is

not in the domain of the inverse sine function, which is $[-1, 1]$, *no triangle exists.*

9 $\dfrac{\sin \alpha}{a} = \dfrac{\sin \gamma}{c} \Rightarrow \alpha = \sin^{-1}\left(\dfrac{a \sin \gamma}{c}\right) = \sin^{-1}\left(\dfrac{140 \sin 53°20'}{115}\right) \approx \sin^{-1}(0.9765) \approx$

77°30′ or 102°30′ {rounded to the nearest 10 minutes}. When using the inverse sine function to solve for an angle, remember that there are always *two* values if $\theta \neq 90°$. The first angle is the one you obtain using your calculator, and the second is the reference angle for the first angle in the second quadrant. After obtaining these values, we need to check the sum of the angles. If this sum is less than 180°, a triangle is formed. If this sum is greater than or equal to 180°, no triangle is formed. For this exercise, there are two triangles possible since in either case $\alpha + \gamma < 180°.$

$\beta = (180° - \gamma) - \alpha \approx (180° - 53°20') - (77°30' \text{ or } 102°30') = 49°10' \text{ or } 24°10'.$

$$\frac{b}{\sin \beta} = \frac{c}{\sin \gamma} \Rightarrow b = \frac{c \sin \beta}{\sin \gamma} \approx \frac{115 \sin (49°10' \text{ or } 24°10')}{\sin 53°20'} \approx 108 \text{ or } 58.7.$$

11 $\dfrac{\sin \alpha}{a} = \dfrac{\sin \gamma}{c} \Rightarrow \alpha = \sin^{-1}\left(\dfrac{a \sin \gamma}{c}\right) = \sin^{-1}\left(\dfrac{131.08 \sin 47.74°}{97.84}\right) \approx \sin^{-1}(0.9915) \approx$

82.54° or 97.46°. There are two triangles possible since in either case $\alpha + \gamma < 180°.$

$\beta = (180° - \gamma) - \alpha \approx (180° - 47.74°) - (82.54° \text{ or } 97.46°) = 49.72° \text{ or } 34.80°.$

$$\frac{b}{\sin \beta} = \frac{c}{\sin \gamma} \Rightarrow b = \frac{c \sin \beta}{\sin \gamma} \approx \frac{97.84 \sin (49.72° \text{ or } 34.80°)}{\sin 47.74°} \approx 100.85 \text{ or } 75.45.$$

13 $\dfrac{\sin \beta}{b} = \dfrac{\sin \alpha}{a} \Rightarrow \beta = \sin^{-1}\left(\dfrac{b \sin \alpha}{a}\right) = \sin^{-1}\left(\dfrac{18.9 \sin 65°10'}{21.3}\right) \approx \sin^{-1}(0.8053) \approx$

53°40′ or 126°20′ {rounded to the nearest 10 minutes}. Reject 126°20′ because then $\alpha + \beta \geq 180°.$ $\gamma = 180° - \alpha - \beta \approx 180° - 65°10' - 53°40' = 61°10'.$

$$\frac{c}{\sin \gamma} = \frac{a}{\sin \alpha} \Rightarrow c = \frac{a \sin \gamma}{\sin \alpha} \approx \frac{21.3 \sin 61°10'}{\sin 65°10'} \approx 20.6.$$

15　$\dfrac{\sin \gamma}{c} = \dfrac{\sin \beta}{b} \Rightarrow \gamma = \sin^{-1}\left(\dfrac{c \sin \beta}{b}\right) = \sin^{-1}\left(\dfrac{0.178 \sin 121.624°}{0.283}\right) \approx \sin^{-1}(0.5356) \approx$

$32.383°$ or $147.617°$. Reject $147.617°$ because then $\beta + \gamma \geq 180°$.

$\alpha = 180° - \beta - \gamma \approx 180° - 121.624° - 32.383° = 25.993°$.

$$\frac{a}{\sin \alpha} = \frac{b}{\sin \beta} \Rightarrow a = \frac{b \sin \alpha}{\sin \beta} \approx \frac{0.283 \sin 25.993°}{\sin 121.624°} \approx 0.146.$$

17　$\angle ABC = 180° - 54°10' - 63°20' = 62°30'. \quad \dfrac{\overline{AB}}{\sin 54°10'} = \dfrac{240}{\sin 62°30'} \Rightarrow \overline{AB} \approx 219.36$ yd

19　(a) $\angle ABP = 180° - 65° = 115°. \quad \angle APB = 180° - 21° - 115° = 44°.$

$$\frac{\overline{AP}}{\sin 115°} = \frac{1.2}{\sin 44°} \Rightarrow \overline{AP} \approx 1.57, \text{ or } 1.6 \text{ mi.}$$

(b) $\sin 21° = \dfrac{\text{height of } P}{\overline{AP}} \Rightarrow \text{height of } P = \dfrac{1.2 \sin 115° \sin 21°}{\sin 44°} \{\text{from part (a)}\} \approx 0.56,$

or 0.6 mi.

21　Let C denote the base of the balloon and P its projection on the ground.

$\angle ACB = 180° - 24°10' - 47°40' = 108°10'. \quad \dfrac{\overline{AC}}{\sin 47°40'} = \dfrac{8.4}{\sin 108°10'} \Rightarrow \overline{AC} \approx 6.5$ mi.

$$\sin 24°10' = \frac{\overline{PC}}{\overline{AC}} \Rightarrow \overline{PC} = \frac{8.4 \sin 47°40' \sin 24°10'}{\sin 108°10'} \approx 2.7 \text{ mi.}$$

23　$\angle APQ = 57° - 22° = 35°. \quad \angle AQP = 180° - (63° - 22°) = 139°.$

$$\angle PAQ = 180° - 139° - 35° = 6°. \quad \frac{\overline{AP}}{\sin 139°} = \frac{100}{\sin 6°} \Rightarrow \overline{AP} = \frac{100 \sin 139°}{\sin 6°} \approx 628 \text{ m.}$$

25　$\angle FAB = 90° - 27°10' = 62°50'.$

$\angle FBA = 90° - 52°40' = 37°20'.$

$\angle AFB = 180° - 62°50' - 37°20' = 79°50'.$

$\dfrac{\overline{AF}}{\sin 37°20'} = \dfrac{6}{\sin 79°50'} \Rightarrow \overline{AF} \approx 3.70$ mi.

$\dfrac{\overline{BF}}{\sin 62°50'} = \dfrac{6}{\sin 79°50'} \Rightarrow \overline{BF} \approx 5.42$ mi.

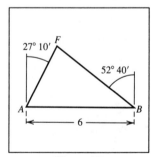

Figure 25

27　Let A denote the base of the hill, B the base of the cathedral, and C the top of the

spire. The angle at the base of the hill is $180° - 48° = 132°$. The angle at the top

of the spire is $180° - 132° - 41° = 7°. \quad \dfrac{\overline{AC}}{\sin 41°} = \dfrac{200}{\sin 7°} \Rightarrow \overline{AC} = \dfrac{200 \sin 41°}{\sin 7°} \approx 1077$ ft.

$\angle BAC = 48° - 32° = 16°. \quad \angle ACB = 90° - 48° = 42°. \quad \angle ABC = 180° - 42° - 16° = 122°.$

$$\frac{\overline{BC}}{\sin 16°} = \frac{\overline{AC}}{\sin 122°} \Rightarrow \overline{BC} = \frac{200 \sin 41° \sin 16°}{\sin 7° \sin 122°} \approx 350 \text{ ft.}$$

29 (a) In the triangle that forms the base, the third angle is $180° - 103° - 52° = 25°$.

Let l denote the length of the dashed line. $\dfrac{l}{\sin 103°} = \dfrac{12.0}{\sin 25°} \Rightarrow l \approx 27.7$ units.

Now $\tan 34° = \dfrac{h}{l} \Rightarrow h \approx 18.7$ units.

(b) Draw a line from the 103° angle that is perpendicular to l and call it d.

$\sin 52° = \dfrac{d}{12} \Rightarrow d \approx 9.5$ units. The area of the triangular base is $B = \frac{1}{2} l d$.

The volume V is $\frac{1}{3}(\frac{1}{2} l d)h = 288 \sin 52° \sin^2 103° \tan 34° \csc^2 25° \approx 814$ cubic units.

31 Draw a line through P perpendicular to the x-axis. Locate points A and B on this line so that $\angle PAQ = \angle PBR = 90°$. $\overline{AP} = 5127.5 - 3452.8 = 1674.7$,

$\overline{AQ} = 3145.8 - 1487.7 = 1658.1$, and $\tan \angle APQ = \frac{1658.1}{1674.7} \Rightarrow \angle APQ \approx 44°43'$.

Thus, $\angle BPR \approx 180° - 55°50' - 44°43' = 79°27'$.

By the distance formula, $\overline{PQ} \approx \sqrt{(1674.7)^2 + (1658.1)^2} \approx 2356.7$.

Now $\dfrac{\overline{PR}}{\sin 65°22'} = \dfrac{\overline{PQ}}{\sin (180° - 55°50' - 65°22')} \Rightarrow \overline{PR} = \dfrac{2356.7 \sin 65°22'}{\sin 58°48'} \approx 2504.5$.

$\sin \angle BPR = \dfrac{\overline{BR}}{\overline{PR}} \Rightarrow \overline{BR} \approx (2504.5)(\sin 79°27') \approx 2462.2$. $\cos \angle BPR = \dfrac{\overline{BP}}{\overline{PR}} \Rightarrow$

$\overline{BP} \approx (2504.5)(\cos 79°27') \approx 458.6$. Using the coordinates of P, we see that

$$R(x, y) \approx (1487.7 + 2462.2, \; 3452.8 - 458.6) = (3949.9, \; 2994.2).$$

8.2 Exercises

Note: These formulas will be used to solve problems that involve the law of cosines.

(1) $a^2 = b^2 + c^2 - 2bc \cos \alpha \Rightarrow a = \sqrt{b^2 + c^2 - 2bc \cos \alpha}$

(Similar formulas are used for b and c.)

(2) $a^2 = b^2 + c^2 - 2bc \cos \alpha \Rightarrow 2bc \cos \alpha = b^2 + c^2 - a^2 \Rightarrow$

$\cos \alpha = \left(\dfrac{b^2 + c^2 - a^2}{2bc} \right) \Rightarrow \alpha = \cos^{-1}\left(\dfrac{b^2 + c^2 - a^2}{2bc} \right)$

(Similar formulas are used for β and γ.)

1 $a = \sqrt{b^2 + c^2 - 2bc \cos \alpha} = \sqrt{20^2 + 30^2 - 2(20)(30) \cos 60°} = \sqrt{700} \approx 26$.

$\beta = \cos^{-1}\left(\dfrac{a^2 + c^2 - b^2}{2ac} \right) = \cos^{-1}\left(\dfrac{700 + 30^2 - 20^2}{2(\sqrt{700})(30)} \right) \approx \cos^{-1}(0.7559) \approx 41°$.

$\gamma = 180° - \alpha - \beta \approx 180° - 60° - 41° = 79°$.

$\boxed{3}$ $b = \sqrt{a^2 + c^2 - 2ac\cos\beta} = \sqrt{23{,}400 + 4500\sqrt{3}} \approx 177$, or 180.

$\alpha = \cos^{-1}\!\left(\dfrac{b^2 + c^2 - a^2}{2bc}\right) \approx \cos^{-1}(0.9054) \approx 25°10'$, or 25°.

Note: We do not have to worry about an "ambiguous" case of the law of cosines as we did with the law of sines. This is true because the inverse cosine function gives us a unique angle from 0° to 180° for all values in its domain.

$$\gamma = 180° - \alpha - \beta \approx 180° - 25°10' - 150° = 4°50', \text{ or } 5°.$$

$\boxed{7}$ $\alpha = \cos^{-1}\!\left(\dfrac{b^2 + c^2 - a^2}{2bc}\right) = \cos^{-1}(0.875) \approx 29°.$

$\beta = \cos^{-1}\!\left(\dfrac{a^2 + c^2 - b^2}{2ac}\right) = \cos^{-1}(0.6875) \approx 47°.$

$$\gamma = 180° - \alpha - \beta \approx 180° - 29° - 47° = 104°.$$

$\boxed{9}$ $\alpha = \cos^{-1}\!\left(\dfrac{b^2 + c^2 - a^2}{2bc}\right) \approx \cos^{-1}(0.9766) \approx 12°30'.$

$\beta = \cos^{-1}\!\left(\dfrac{a^2 + c^2 - b^2}{2ac}\right) = \cos^{-1}(-0.725) \approx 136°30'.$

$$\gamma = 180° - \alpha - \beta \approx 180° - 12°30' - 136°30' = 31°00'.$$

Note: When deciding whether to use the law of cosines or the law of sines, it may be helpful to remember that the law of cosines must be used when you have:

(1) 3 sides, or

(2) 2 sides and the angle between them.

$\boxed{11}$ We have two sides and the angle between them. Hence, we apply the law of cosines.

$$\text{Third side} = \sqrt{175^2 + 150^2 - 2(175)(150)\cos 73°40'} \approx 196 \text{ feet.}$$

$\boxed{13}$ 20 minutes $= \frac{1}{3}$ hour $\Rightarrow$ the cars have traveled $60(\frac{1}{3}) = 20$ miles and $45(\frac{1}{3}) = 15$ miles, respectively. The distance d apart is $d = \sqrt{20^2 + 15^2 - 2(20)(15)\cos 84°} \approx 24$ miles.

$\boxed{15}$ The first ship travels $(24)(2) = 48$ miles in two hours. The second ship travels $(18)(1\frac{1}{2}) = 27$ miles in $1\frac{1}{2}$ hours. The angle between the paths is $20° + 35° = 55°$.

$$\overline{AB} = \sqrt{27^2 + 48^2 - 2(27)(48)\cos 55°} \approx 39 \text{ miles.}$$

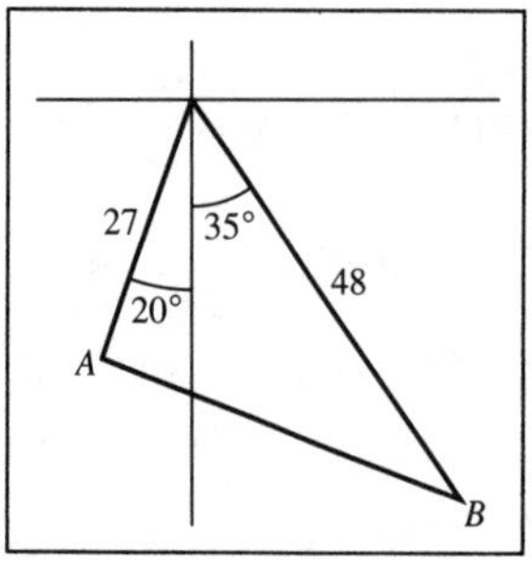

Figure 15

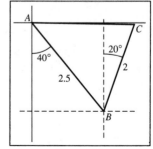

Figure 17

17 $\angle ABC = 40° + 20° = 60°$.

$$\overline{AB} = \left(\frac{1 \text{ mile}}{8 \text{ min}} \cdot 20 \text{ min}\right) = 2.5 \text{ miles and } \overline{BC} = \left(\frac{1 \text{ mile}}{8 \text{ min}} \cdot 16 \text{ min}\right) = 2 \text{ miles}.$$

$$\overline{AC} = \sqrt{2.5^2 + 2^2 - 2(2)(2.5) \cos 60°} = \sqrt{5.25} \approx 2.3 \text{ miles. See } \textit{Figure 17.}$$

19 $\gamma = \cos^{-1}\left(\dfrac{2^2 + 3^2 - 4^2}{2 \cdot 2 \cdot 3}\right) = \cos^{-1}(-0.25) \approx 104°29'. \quad \phi \approx 104°29' - 70° = 34°29'.$

The direction that the third side was traversed is approximately

$$\text{N}(90° - 34°29')\text{E} = \text{N}55°31'\text{E}.$$

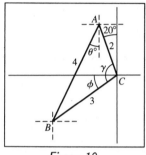

Figure 19

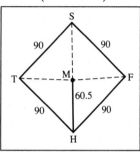

Figure 21

21 Let H denote home plate, M the mound, F first base, S second base, and T third base. $\overline{HS} = \sqrt{90^2 + 90^2} = 90\sqrt{2} \approx 127.3 \text{ ft.}$ $\overline{MS} = 90\sqrt{2} - 60.5 \approx 66.8 \text{ ft.}$

$$\angle MHF = 45° \text{ so } \overline{MF} = \sqrt{60.5^2 + 90^2 - 2(60.5)(90) \cos 45°} \approx 63.7 \text{ ft.}$$

$$\overline{MT} = \overline{MF} \text{ by the symmetry of the field.}$$

23 $\angle RTP = 21°$ and $\angle RSP = 37°$. $\sin \angle RSP = \dfrac{10,000}{\overline{SP}} \Rightarrow \overline{SP} = 10,000 \csc 37° \approx$

16,616 ft. $\sin \angle RTP = \dfrac{10,000}{\overline{TP}} \Rightarrow \overline{TP} = 10,000 \csc 21° \approx 27,904 \text{ ft.}$

$$\overline{ST} = \sqrt{\overline{SP}^2 + \overline{TP}^2 - 2(\overline{SP})(\overline{TP}) \cos 110°} \approx 37,039 \text{ ft} \approx 7 \text{ miles.}$$

25 Let $d = \overline{ES}$. $d^2 = R^2 + R^2 - 2RR \cos \theta \Rightarrow d^2 = 2R^2(1 - \cos \theta) \Rightarrow$

$d^2 = 4R^2\left(\dfrac{1 - \cos \theta}{2}\right) \Rightarrow d = 2R\sqrt{\dfrac{1 - \cos \theta}{2}} \Rightarrow d = 2R \sin \dfrac{\theta}{2}.$

Since $d = vt$, $t = \dfrac{d}{v} = \dfrac{2R}{v} \sin \dfrac{\theta}{2}.$

27 (a) $\angle BCP = \frac{1}{2}(\angle BCD) = \frac{1}{2}(72°) = 36°$. $\triangle BPC$ is isosceles so

$\angle BPC = \angle PBC$ and $2\angle BPC = 180° - 36° \Rightarrow \underline{\angle BPC = 72°}$.

$\angle APB = 180° - \angle BPC = 180° - 72° = \underline{108°}.$

$$\angle ABP = 180° - \angle APB - \angle BAP = 180° - 108° - 36° = \underline{36°}.$$

(b) $\overline{BP} = \sqrt{\overline{BC}^2 + \overline{PC}^2 - 2(\overline{BC})(\overline{PC}) \cos 36°} = \sqrt{1^2 + 1^2 - 2(1)(1) \cos 36°} \approx 0.62.$

(c) $\text{Area}_{\text{kite}} = 2(\text{Area of } \triangle BPC) = 2 \cdot \frac{1}{2}(\overline{CB})(\overline{CP}) \sin \angle BCP = \sin 36° \approx 0.59.$

$\text{Area}_{\text{dart}} = 2(\text{Area of } \triangle ABP) = 2 \cdot \frac{1}{2}(\overline{AB})(\overline{BP}) \sin \angle ABP = \overline{BP} \sin 36° \approx 0.36.$

$\{\overline{BP} \text{ was found in part (b)}\}$

Note: Exer. 29–36: $\mathcal{A}$ (the area) is measured in square units.

$\boxed{29}$ Since α is the angle between sides b and c, we may apply the area of a triangle

formula listed in this section. $\mathcal{A} = \frac{1}{2}bc \sin \alpha = \frac{1}{2}(20)(30) \sin 60° = 300(\sqrt{3}/2) \approx 260$.

$\boxed{31}$ $\gamma = 180° - \alpha - \beta = 180° - 40.3° - 62.9° = 76.8°$.

$$\frac{a}{\sin \alpha} = \frac{b}{\sin \beta} \Rightarrow a = \frac{b \sin \alpha}{\sin \beta} = \frac{5.63 \sin 40.3°}{\sin 62.9°}. \quad \mathcal{A} = \frac{1}{2}ab \sin \gamma \approx 11.21.$$

$\boxed{33}$ $\dfrac{\sin \beta}{b} = \dfrac{\sin \alpha}{a} \Rightarrow \sin \beta = \dfrac{b \sin \alpha}{a} = \dfrac{3.4 \sin 80.1°}{8.0} \approx 0.4187 \Rightarrow \beta \approx 24.8°$ or $155.2°$.

Reject $155.2°$ because then $\alpha + \beta = 235.3° \geq 180°$. $\gamma \approx 180° - 80.1° - 24.8° = 75.1°$.

$$\mathcal{A} = \frac{1}{2}ab \sin \gamma = \frac{1}{2}(8.0)(3.4) \sin 75.1° \approx 13.1.$$

$\boxed{35}$ Given the lengths of the 3 sides of a triangle, we compute the semiperimeter and then

use Heron's formula to find the area of the triangle.

$$s = \frac{1}{2}(a + b + c) = \frac{1}{2}(25.0 + 80.0 + 60.0) = 82.5.$$
$$\mathcal{A} = \sqrt{s(s - a)(s - b)(s - c)} = \sqrt{(82.5)(57.5)(2.5)(22.5)} \approx 516.56, \text{ or } 517.0.$$

$\boxed{37}$ $s = \frac{1}{2}(a + b + c) = \frac{1}{2}(115 + 140 + 200) = 227.5.$ $\mathcal{A} = \sqrt{s(s - a)(s - b)(s - c)} =$

$$\sqrt{(227.5)(112.5)(87.5)(27.5)} \approx 7847.6 \text{ yd}^2, \text{ or } \mathcal{A}/4840 \approx 1.62 \text{ acres}.$$

$\boxed{39}$ The area of the parallelogram is twice the area of the triangle formed by the two

sides and the included angle. $\mathcal{A} = 2(\frac{1}{2})(12.0)(16.0) \sin 40° \approx 123.4 \text{ ft}^2$.

8.3 Exercises

$\boxed{1}$ $|3 - 4i| = \sqrt{3^2 + (-4)^2} = \sqrt{25} = 5$ $\boxed{5}$ $|8i| = |0 + 8i| = \sqrt{0^2 + 8^2} = \sqrt{64} = 8$

$\boxed{7}$ From §2.4, $i^m = i, -1, -i,$ or 1. Since all of these are 1 unit from the origin,

$|i^m| = 1$ for any integer m. As an alternate solution,

$$|i^{500}| = |(i^4)^{125}| = |(1)^{125}| = |1| = |1 + 0i| = \sqrt{1^2 + 0^2} = \sqrt{1} = 1.$$

$\boxed{9}$ $|0| = |0 + 0i| = \sqrt{0^2 + 0^2} = \sqrt{0} = 0$

$\boxed{11}$ $4 + 2i$ $\boxed{13}$ $3 - 5i$ $\boxed{15}$ $-(3 - 6i) = -3 + 6i$

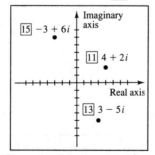

Figure for Exercises 11, 13, 15

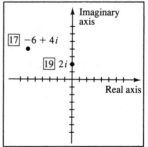

Figure for Exercises 17, 19

$\boxed{17}$ $2i(2 + 3i) = 4i + 6i^2 = 4i - 6 = -6 + 4i$

$\boxed{19}$ $(1 + i)^2 = 1 + 2(1)(i) + i^2 = 1 + 2i - 1 = 2i$

Note: For each of the following exercises, we need to find r and θ.

If $z = a + bi$, then $r = \sqrt{a^2 + b^2}$. To find θ, we will use the fact that

$\tan\theta = \frac{b}{a}$ and our knowledge of what quadrant the terminal side of θ is in.

$\boxed{21}$ $z = 1 - i \Rightarrow r = \sqrt{1 + (-1)^2} = \sqrt{2}$. $\tan\theta = \frac{-1}{1} = -1$ and θ in QIV $\Rightarrow \theta = \frac{7\pi}{4}$.

Thus, $z = 1 - i = \sqrt{2}\left(\cos\frac{7\pi}{4} + i\sin\frac{7\pi}{4}\right)$, or simply $\sqrt{2}\operatorname{cis}\frac{7\pi}{4}$.

$\boxed{23}$ $z = -4\sqrt{3} + 4i \Rightarrow r = \sqrt{(-4\sqrt{3})^2 + 4^2} = \sqrt{64} = 8$.

$\tan\theta = \frac{4}{-4\sqrt{3}} = -\frac{1}{\sqrt{3}}$ and θ in QII $\Rightarrow \theta = \frac{5\pi}{6}$. $z = 8\operatorname{cis}\frac{5\pi}{6}$.

$\boxed{27}$ $z = -4 - 4i \Rightarrow r = \sqrt{(-4)^2 + (-4)^2} = \sqrt{32} = 4\sqrt{2}$.

$\tan\theta = \frac{-4}{-4} = 1$ and θ in QIII $\Rightarrow \theta = \frac{5\pi}{4}$. $z = 4\sqrt{2}\operatorname{cis}\frac{5\pi}{4}$.

$\boxed{29}$ $z = -20i \Rightarrow r = 20$. θ on the negative y-axis $\Rightarrow \theta = \frac{3\pi}{2}$. $z = 20\operatorname{cis}\frac{3\pi}{2}$.

$\boxed{31}$ $z = 12 \Rightarrow r = 12$. θ on the positive x-axis $\Rightarrow \theta = 0$. $z = 12\operatorname{cis}0$.

$\boxed{33}$ $z = -7 \Rightarrow r = 7$. θ on the negative x-axis $\Rightarrow \theta = \pi$. $z = 7\operatorname{cis}\pi$.

$\boxed{35}$ $z = 6i \Rightarrow r = 6$. θ on the positive y-axis $\Rightarrow \theta = \frac{\pi}{2}$. $z = 6\operatorname{cis}\frac{\pi}{2}$.

$\boxed{39}$ $z = 2 + i \Rightarrow r = \sqrt{2^2 + 1^2} = \sqrt{5}$. $\tan\theta = \frac{1}{2}$ and θ in QI $\Rightarrow \theta = \tan^{-1}\frac{1}{2}$.

$z = \sqrt{5}\operatorname{cis}\left(\tan^{-1}\frac{1}{2}\right)$.

$\boxed{41}$ $z = -3 + i \Rightarrow r = \sqrt{(-3)^2 + 1^2} = \sqrt{10}$.

$\tan\theta = \frac{1}{-3}$ and θ in QII $\Rightarrow \theta = \tan^{-1}\left(-\frac{1}{3}\right) + \pi$. We must add π to $\tan^{-1}\left(-\frac{1}{3}\right)$

because $-\frac{\pi}{2} < \tan^{-1}\left(-\frac{1}{3}\right) < 0$ and we want θ to be in the interval $\left(\frac{\pi}{2}, \pi\right)$.

$z = \sqrt{10}\operatorname{cis}\left[\tan^{-1}\left(-\frac{1}{3}\right) + \pi\right]$.

$\boxed{43}$ $z = -5 - 3i \Rightarrow r = \sqrt{(-5)^2 + (-3)^2} = \sqrt{34}$. $\tan\theta = \frac{-3}{-5} = \frac{3}{5}$ and

θ in QIII $\Rightarrow \theta = \tan^{-1}\frac{3}{5} + \pi$. We must add π to $\tan^{-1}\frac{3}{5}$ because $0 < \tan^{-1}\frac{3}{5} < \frac{\pi}{2}$ and

we want θ to be in the interval $\left(\pi, \frac{3\pi}{2}\right)$. $z = \sqrt{34}\operatorname{cis}\left(\tan^{-1}\frac{3}{5} + \pi\right)$.

$\boxed{45}$ $z = 4 - 3i \Rightarrow r = \sqrt{4^2 + (-3)^2} = \sqrt{25} = 5$.

$\tan\theta = \frac{-3}{4}$ and θ in QIV $\Rightarrow \theta = \tan^{-1}\left(-\frac{3}{4}\right) + 2\pi$. We must add 2π to $\tan^{-1}\left(-\frac{3}{4}\right)$

because $-\frac{\pi}{2} < \tan^{-1}\left(-\frac{3}{4}\right) < 0$ and we want θ to be in the interval $\left(\frac{3\pi}{2}, 2\pi\right)$.

$z = 5\operatorname{cis}\left[\tan^{-1}\left(-\frac{3}{4}\right) + 2\pi\right]$.

$\boxed{47}$ $4\left(\cos\frac{\pi}{4} + i\sin\frac{\pi}{4}\right) = 4\left(\frac{\sqrt{2}}{2} + \frac{\sqrt{2}}{2}i\right) = 2\sqrt{2} + 2\sqrt{2}\,i$

$\boxed{51}$ $5\left(\cos\pi + i\sin\pi\right) = 5(-1 + 0i) = -5$

53 For any given angle θ, where $\theta = \tan^{-1}\frac{y}{x}$, we have $r = \sqrt{x^2 + y^2}$. In this case,

$\theta = \tan^{-1}\frac{3}{5}$, so $r = \sqrt{3^2 + 5^2} = \sqrt{34}$. You may want to draw a figure to represent x,

y, and r, as we did in previous chapters. Also, note that $\cos\theta = \frac{x}{r}$ and $\sin\theta = \frac{y}{r}$.

$$\sqrt{34}\,\text{cis}\,(\tan^{-1}\tfrac{3}{5}) = \sqrt{34}\Big[\cos(\tan^{-1}\tfrac{3}{5}) + i\sin(\tan^{-1}\tfrac{3}{5})\Big] = \sqrt{34}\left(\frac{5}{\sqrt{34}} + \frac{3}{\sqrt{34}}i\right) = 5 + 3i$$

55 $\sqrt{5}\,\text{cis}\Big[\tan^{-1}\big(-\tfrac{1}{2}\big)\Big] = \sqrt{5}\Big\{\cos\Big[\tan^{-1}\big(-\tfrac{1}{2}\big)\Big] + i\sin\Big[\tan^{-1}\big(-\tfrac{1}{2}\big)\Big]\Big\} =$

$$\sqrt{5}\left(\frac{2}{\sqrt{5}} - \frac{1}{\sqrt{5}}i\right) = 2 - i$$

Note: For Exercises 57–64, the trigonometric forms for z_1 and z_2 are listed and then used in the theorem in this section. The trigonometric forms can be found as in Exercises 21–46.

57 $z_1 = \sqrt{2}\,\text{cis}\,\frac{3\pi}{4}$ and $z_2 = \sqrt{2}\,\text{cis}\,\frac{\pi}{4}$. $z_1 z_2 = \sqrt{2}\cdot\sqrt{2}\,\text{cis}\,\big(\frac{3\pi}{4} + \frac{\pi}{4}\big) = 2\,\text{cis}\,\pi = -2 + 0i$.

$$\frac{z_1}{z_2} = \frac{\sqrt{2}}{\sqrt{2}}\,\text{cis}\,\big(\tfrac{3\pi}{4} - \tfrac{\pi}{4}\big) = 1\,\text{cis}\,\tfrac{\pi}{2} = 0 + i$$

59 $z_1 = 4\,\text{cis}\,\frac{4\pi}{3}$ and $z_2 = 5\,\text{cis}\,\frac{\pi}{2}$. $z_1 z_2 = 4\cdot 5\,\text{cis}\,\big(\frac{4\pi}{3} + \frac{\pi}{2}\big) = 20\,\text{cis}\,\frac{11\pi}{6} = 10\sqrt{3} - 10i$.

$$\frac{z_1}{z_2} = \tfrac{4}{5}\,\text{cis}\,\big(\tfrac{4\pi}{3} - \tfrac{\pi}{2}\big) = \tfrac{4}{5}\,\text{cis}\,\tfrac{5\pi}{6} = -\tfrac{2}{5}\sqrt{3} + \tfrac{2}{5}i.$$

61 $z_1 = 10\,\text{cis}\,\pi$ and $z_2 = 4\,\text{cis}\,\pi$. $z_1 z_2 = 10\cdot 4\,\text{cis}\,(\pi + \pi) = 40\,\text{cis}\,2\pi = 40 + 0i$.

$$\frac{z_1}{z_2} = \tfrac{10}{4}\,\text{cis}\,(\pi - \pi) = \tfrac{5}{2}\,\text{cis}\,0 = \tfrac{5}{2} + 0i.$$

63 $z_1 = 4\,\text{cis}\,0$ and $z_2 = \sqrt{5}\,\text{cis}\Big[\tan^{-1}\big(-\tfrac{1}{2}\big)\Big]$. Let $\theta = \tan^{-1}\big(-\tfrac{1}{2}\big)$.

Thus $\cos\theta = \dfrac{2}{\sqrt{5}}$ and $\sin\theta = -\dfrac{1}{\sqrt{5}}$.

$$z_1 z_2 = 4\cdot\sqrt{5}\,\text{cis}\,(0 + \theta) = 4\sqrt{5}\,(\cos\theta + i\sin\theta) = 4\sqrt{5}\left(\frac{2}{\sqrt{5}} + \frac{-1}{\sqrt{5}}i\right) = 8 - 4i.$$

$$\frac{z_1}{z_2} = \frac{4}{\sqrt{5}}\,\text{cis}\,(0 - \theta) = \frac{4}{\sqrt{5}}\big[\cos(-\theta) + i\sin(-\theta)\big]$$

$$= \frac{4}{\sqrt{5}}\,(\cos\theta - i\sin\theta) = \frac{4}{\sqrt{5}}\left(\frac{2}{\sqrt{5}} + \frac{1}{\sqrt{5}}i\right) = \tfrac{8}{5} + \tfrac{4}{5}i.$$

65 Let $z_1 = r_1\,\text{cis}\,\theta_1$ and $z_2 = r_2\,\text{cis}\,\theta_2$.

$$\frac{z_1}{z_2} = \frac{r_1\,\text{cis}\,\theta_1}{r_2\,\text{cis}\,\theta_2} = \frac{r_1\,(\cos\theta_1 + i\sin\theta_1)(\cos\theta_2 - i\sin\theta_2)}{r_2\,(\cos\theta_2 + i\sin\theta_2)(\cos\theta_2 - i\sin\theta_2)}$$

$$\{\text{multiplying by the conjugate of the denominator}\}$$

$$= \frac{r_1\big[(\cos\theta_1\,\cos\theta_2 + \sin\theta_1\,\sin\theta_2) + i\,(\sin\theta_1\,\cos\theta_2 - \sin\theta_2\,\cos\theta_1)\big]}{r_2\big[(\cos^2\theta_2 + \sin^2\theta_2) + i\,(\sin\theta_2\,\cos\theta_2 - \cos\theta_2\,\sin\theta_2)\big]}$$

$$= \frac{r_1\big[\cos(\theta_1 - \theta_2) + i\sin(\theta_1 - \theta_2)\big]}{r_2\,(1 + 0i)} = \frac{r_1}{r_2}\,\text{cis}\,(\theta_1 - \theta_2).$$

8.4 Exercises

Note: In this section, it is assumed that the reader can transform complex numbers to their trigonometric from. If this is not true, see Exercises 8.3.

$\boxed{1}$ $(3+3i)^5 = (3\sqrt{2}\operatorname{cis}\frac{\pi}{4})^5 = (3\sqrt{2})^5\operatorname{cis}(5\cdot\frac{\pi}{4}) = (3\sqrt{2})^5\operatorname{cis}\frac{5\pi}{4} =$

$$972\sqrt{2}\left(-\frac{\sqrt{2}}{2}-\frac{\sqrt{2}}{2}i\right) = -972-972i$$

$\boxed{3}$ $(1-i)^{10} = (\sqrt{2}\operatorname{cis}\frac{7\pi}{4})^{10} = (\sqrt{2})^{10}\operatorname{cis}(10\cdot\frac{7\pi}{4}) = (\sqrt{2})^{10}\operatorname{cis}\frac{35\pi}{2} =$

$$2^5\operatorname{cis}(16\pi+\frac{3\pi}{2}) = 32\operatorname{cis}\frac{3\pi}{2} = 32(0-i) = -32i$$

$\boxed{7}$ $\left(-\frac{\sqrt{2}}{2}+\frac{\sqrt{2}}{2}i\right)^{15} = (1\operatorname{cis}\frac{3\pi}{4})^{15} = 1^{15}\operatorname{cis}\frac{45\pi}{4} = \operatorname{cis}\frac{5\pi}{4} = -\frac{\sqrt{2}}{2}-\frac{\sqrt{2}}{2}i$

$\boxed{9}$ $\left(-\frac{\sqrt{3}}{2}-\frac{1}{2}i\right)^{20} = (1\operatorname{cis}\frac{7\pi}{6})^{20} = 1^{20}\operatorname{cis}\frac{70\pi}{3} = \operatorname{cis}\frac{4\pi}{3} = -\frac{1}{2}-\frac{\sqrt{3}}{2}i$

$\boxed{13}$ $1+\sqrt{3}i = 2\operatorname{cis}60°$. $w_k = \sqrt{2}\operatorname{cis}\left(\frac{60°+360°k}{2}\right)$ for $k=0,1$.

$$w_0 = \sqrt{2}\operatorname{cis}30° = \sqrt{2}\left(\frac{\sqrt{3}}{2}+\frac{1}{2}i\right) = \frac{\sqrt{6}}{2}+\frac{\sqrt{2}}{2}i.$$

$$w_1 = \sqrt{2}\operatorname{cis}210° = \sqrt{2}\left(-\frac{\sqrt{3}}{2}-\frac{1}{2}i\right) = -\frac{\sqrt{6}}{2}-\frac{\sqrt{2}}{2}i.$$

$\boxed{15}$ $-1-\sqrt{3}i = 2\operatorname{cis}240°$. $w_k = \sqrt[4]{2}\operatorname{cis}\left(\frac{240°+360°k}{4}\right)$ for $k=0,1,2,3$.

$$w_0 = \sqrt[4]{2}\operatorname{cis}60° = \sqrt[4]{2}\left(\frac{1}{2}+\frac{\sqrt{3}}{2}i\right) = \frac{\sqrt[4]{2}}{2}+\frac{\sqrt[4]{18}}{2}i.$$

$$\{\text{since }\sqrt[4]{2}\cdot\sqrt{3} = \sqrt[4]{2}\cdot\sqrt[4]{9} = \sqrt[4]{18}\}$$

$$w_1 = \sqrt[4]{2}\operatorname{cis}150° = \sqrt[4]{2}\left(-\frac{\sqrt{3}}{2}+\frac{1}{2}i\right) = -\frac{\sqrt[4]{18}}{2}+\frac{\sqrt[4]{2}}{2}i.$$

$$w_2 = \sqrt[4]{2}\operatorname{cis}240° = \sqrt[4]{2}\left(-\frac{1}{2}-\frac{\sqrt{3}}{2}i\right) = -\frac{\sqrt[4]{2}}{2}-\frac{\sqrt[4]{18}}{2}i.$$

$$w_3 = \sqrt[4]{2}\operatorname{cis}330° = \sqrt[4]{2}\left(\frac{\sqrt{3}}{2}-\frac{1}{2}i\right) = \frac{\sqrt[4]{18}}{2}-\frac{\sqrt[4]{2}}{2}i.$$

$\boxed{17}$ $-27i = 27\operatorname{cis}270°$. $w_k = \sqrt[3]{27}\operatorname{cis}\left(\frac{270°+360°k}{3}\right)$ for $k=0,1,2$.

$$w_0 = 3\operatorname{cis}90° = 3(0+i) = 3i.$$

$$w_1 = 3\operatorname{cis}210° = 3\left(-\frac{\sqrt{3}}{2}-\frac{1}{2}i\right) = -\frac{3\sqrt{3}}{2}-\frac{3}{2}i.$$

$$w_2 = 3\operatorname{cis}330° = 3\left(\frac{\sqrt{3}}{2}-\frac{1}{2}i\right) = \frac{3\sqrt{3}}{2}-\frac{3}{2}i.$$

$\boxed{19}$ $1 = 1\operatorname{cis}0°$. $w_k = \sqrt[6]{1}\operatorname{cis}\left(\dfrac{0° + 360°k}{6}\right)$ for $k = 0, 1, 2, 3, 4, 5$. See *Figure 19*.

$w_0 = 1\operatorname{cis}0° = 1 + 0i$. $w_1 = 1\operatorname{cis}60° = \dfrac{1}{2} + \dfrac{\sqrt{3}}{2}i$.

$w_2 = 1\operatorname{cis}120° = -\dfrac{1}{2} + \dfrac{\sqrt{3}}{2}i$. $w_3 = 1\operatorname{cis}180° = -1 + 0i$.

$w_4 = 1\operatorname{cis}240° = -\dfrac{1}{2} - \dfrac{\sqrt{3}}{2}i$. $w_5 = 1\operatorname{cis}300° = \dfrac{1}{2} - \dfrac{\sqrt{3}}{2}i$.

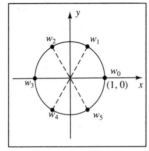
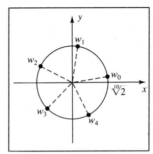

Figure 19   Figure 21

$\boxed{21}$ $1 + i = \sqrt{2}\operatorname{cis}45°$. $w_k = \sqrt{\sqrt[5]{2}}\operatorname{cis}\left(\dfrac{45° + 360°k}{5}\right)$ for $k = 0, 1, 2, 3, 4$.

$$w_k = \sqrt[10]{2}\operatorname{cis}\theta \text{ with } \theta = 9°, 81°, 153°, 225°, 297°.$$

$\boxed{23}$ $x^4 - 16 = 0 \Rightarrow x^4 = 16$. The problem is now to find the 4 fourth roots of 16.

$16 = 16 + 0i = 16\operatorname{cis}0°$. $w_k = \sqrt[4]{16}\operatorname{cis}\left(\dfrac{0° + 360°k}{4}\right)$ for $k = 0, 1, 2, 3$.

$w_0 = 2\operatorname{cis}0° = 2(1 + 0i) = 2$. $w_1 = 2\operatorname{cis}90° = 2(0 + i) = 2i$.

$w_2 = 2\operatorname{cis}180° = 2(-1 + 0i) = -2$. $w_3 = 2\operatorname{cis}270° = 2(0 - i) = -2i$.

$\boxed{25}$ $x^6 + 64 = 0 \Rightarrow x^6 = -64$. The problem is now to find the 6 sixth roots of -64.

$-64 = -64 + 0i = 64\operatorname{cis}180°$. $w_k = \sqrt[6]{64}\operatorname{cis}\left(\dfrac{180° + 360°k}{6}\right)$ for $k = 0, 1, \ldots, 5$.

$w_0 = 2\operatorname{cis}30° = 2\left(\dfrac{\sqrt{3}}{2} + \dfrac{1}{2}i\right) = \sqrt{3} + i$.

$w_1 = 2\operatorname{cis}90° = 2(0 + i) = 2i$.

$w_2 = 2\operatorname{cis}150° = 2\left(-\dfrac{\sqrt{3}}{2} + \dfrac{1}{2}i\right) = -\sqrt{3} + i$.

$w_3 = 2\operatorname{cis}210° = 2\left(-\dfrac{\sqrt{3}}{2} - \dfrac{1}{2}i\right) = -\sqrt{3} - i$.

$w_4 = 2\operatorname{cis}270° = 2(0 - i) = -2i$.

$w_5 = 2\operatorname{cis}330° = 2\left(\dfrac{\sqrt{3}}{2} - \dfrac{1}{2}i\right) = \sqrt{3} - i$.

27 $x^3 + 8i = 0 \Rightarrow x^3 = -8i$. The problem is now to find the 3 cube roots of $-8i$.

$$-8i = 0 - 8i = 8\operatorname{cis}270°. \quad w_k = \sqrt[3]{8}\operatorname{cis}\left(\frac{270° + 360°k}{3}\right) \text{ for } k = 0, 1, 2.$$

$$w_0 = 2\operatorname{cis}90° = 2(0 + i) = 2i.$$

$$w_1 = 2\operatorname{cis}210° = 2\left(-\frac{\sqrt{3}}{2} - \frac{1}{2}i\right) = -\sqrt{3} - i.$$

$$w_2 = 2\operatorname{cis}330° = 2\left(\frac{\sqrt{3}}{2} - \frac{1}{2}i\right) = \sqrt{3} - i.$$

29 $x^5 - 243 = 0 \Rightarrow x^5 = 243$. The problem is now to find the 5 fifth roots of 243.

$$243 = 243 + 0i = 243\operatorname{cis}0°. \quad w_k = \sqrt[5]{243}\operatorname{cis}\left(\frac{0° + 360°k}{5}\right) \text{ for } k = 0, 1, 2, 3, 4.$$

$$w_k = 3\operatorname{cis}\theta \text{ with } \theta = 0°, 72°, 144°, 216°, 288°.$$

8.5 Exercises

1   $\mathbf{a} + \mathbf{b} = \langle 2, -3\rangle + \langle 1, 4\rangle = \langle 2 + 1, -3 + 4\rangle = \langle 3, 1\rangle.$

$\mathbf{a} - \mathbf{b} = \langle 2, -3\rangle - \langle 1, 4\rangle = \langle 2 - 1, -3 - 4\rangle = \langle 1, -7\rangle.$

$$4\mathbf{a} + 5\mathbf{b} = 4\langle 2, -3\rangle + 5\langle 1, 4\rangle = \langle 4(2), 4(-3)\rangle + \langle 5(1), 5(4)\rangle$$
$$= \langle 8, -12\rangle + \langle 5, 20\rangle = \langle 8 + 5, -12 + 20\rangle = \langle 13, 8\rangle.$$

From above, $4\mathbf{a} - 5\mathbf{b} = \langle 8, -12\rangle - \langle 5, 20\rangle = \langle 8 - 5, -12 - 20\rangle = \langle 3, -32\rangle.$

3   Simplifying $\mathbf{a}$ and $\mathbf{b}$ first, we have $\mathbf{a} = -\langle 7, -2\rangle = \langle -(7), -(-2)\rangle = \langle -7, 2\rangle$ and

$$\mathbf{b} = 4\langle -2, 1\rangle = \langle 4(-2), 4(1)\rangle = \langle -8, 4\rangle.$$

$\mathbf{a} + \mathbf{b} = \langle -7, 2\rangle + \langle -8, 4\rangle = \langle -7 + (-8), 2 + 4\rangle = \langle -15, 6\rangle.$

$\mathbf{a} - \mathbf{b} = \langle -7, 2\rangle - \langle -8, 4\rangle = \langle -7 - (-8), 2 - 4\rangle = \langle 1, -2\rangle.$

$4\mathbf{a} + 5\mathbf{b} = 4\langle -7, 2\rangle + 5\langle -8, 4\rangle = \langle -28, 8\rangle + \langle -40, 20\rangle = \langle -68, 28\rangle.$

From above, $4\mathbf{a} - 5\mathbf{b} = \langle -28, 8\rangle - \langle -40, 20\rangle = \langle 12, -12\rangle.$

5   $\mathbf{a} + \mathbf{b} = (\mathbf{i} + 2\mathbf{j}) + (3\mathbf{i} - 5\mathbf{j}) = (1 + 3)\mathbf{i} + (2 - 5)\mathbf{j} = 4\mathbf{i} - 3\mathbf{j}.$

$\mathbf{a} - \mathbf{b} = (\mathbf{i} + 2\mathbf{j}) - (3\mathbf{i} - 5\mathbf{j}) = (1 - 3)\mathbf{i} + (2 - (-5))\mathbf{j} = -2\mathbf{i} + 7\mathbf{j}.$

$4\mathbf{a} + 5\mathbf{b} = 4(\mathbf{i} + 2\mathbf{j}) + 5(3\mathbf{i} - 5\mathbf{j}) = (4\mathbf{i} + 8\mathbf{j}) + (15\mathbf{i} - 25\mathbf{j}) = 19\mathbf{i} - 17\mathbf{j}.$

From above, $4\mathbf{a} - 5\mathbf{b} = (4\mathbf{i} + 8\mathbf{j}) - (15\mathbf{i} - 25\mathbf{j}) = -11\mathbf{i} + 33\mathbf{j}.$

$\boxed{7}$ $\mathbf{a} = 3\mathbf{i} + 2\mathbf{j}$ and $\mathbf{b} = -\mathbf{i} + 5\mathbf{j} \Rightarrow \mathbf{a} + \mathbf{b} = 2\mathbf{i} + 7\mathbf{j}$, $2\mathbf{a} = 6\mathbf{i} + 4\mathbf{j}$, and $-3\mathbf{b} = 3\mathbf{i} - 15\mathbf{j}$.

Terminal points of the vectors are $(3, 2)$, $(-1, 5)$, $(2, 7)$, $(6, 4)$, and $(3, -15)$.

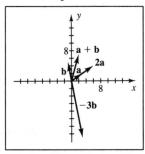

Figure 7   Figure 9

$\boxed{9}$ $\mathbf{a} = \langle -4, 6 \rangle$ and $\mathbf{b} = \langle -2, 3 \rangle \Rightarrow \mathbf{a} + \mathbf{b} = \langle -6, 9 \rangle$, $2\mathbf{a} = \langle -8, 12 \rangle$, and $-3\mathbf{b} = \langle 6, -9 \rangle$.

Terminal points of the vectors are $(-4, 6)$, $(-2, 3)$, $(-6, 9)$, $(-8, 12)$, and $(6, -9)$.

$\boxed{11}$ $\mathbf{a} + \mathbf{b} = \langle 2, 0 \rangle + \langle -1, 0 \rangle = \langle 1, 0 \rangle = -\langle -1, 0 \rangle = -\mathbf{b}$

$\boxed{13}$ $\mathbf{b} + \mathbf{e} = \langle -1, 0 \rangle + \langle 2, 2 \rangle = \langle 1, 2 \rangle = \mathbf{f}$

$\boxed{15}$ $\mathbf{b} + \mathbf{d} = \langle -1, 0 \rangle + \langle 0, -1 \rangle = \langle -1, -1 \rangle = -\frac{1}{2}\langle 2, 2 \rangle = -\frac{1}{2}\mathbf{e}$

$\boxed{17}$ $\mathbf{a} + (\mathbf{b} + \mathbf{c}) = \langle a_1, a_2 \rangle + (\langle b_1, b_2 \rangle + \langle c_1, c_2 \rangle)$

$\qquad\qquad = \langle a_1, a_2 \rangle + \langle b_1 + c_1, b_2 + c_2 \rangle$

$\qquad\qquad = \langle a_1 + b_1 + c_1, a_2 + b_2 + c_2 \rangle$

$\qquad\qquad = \langle a_1 + b_1, a_2 + b_2 \rangle + \langle c_1, c_2 \rangle$

$\qquad\qquad = (\langle a_1, a_2 \rangle + \langle b_1, b_2 \rangle) + \langle c_1, c_2 \rangle = (\mathbf{a} + \mathbf{b}) + \mathbf{c}$

$\boxed{21}$ $(mn)\mathbf{a} = (mn)\langle a_1, a_2 \rangle$

$\qquad\quad = \langle (mn)\,a_1, (mn)\,a_2 \rangle$

$\qquad\quad = \langle mna_1, mna_2 \rangle$

$\qquad\quad = m\langle na_1, na_2 \rangle \qquad$ or $n\langle ma_1, ma_2 \rangle$

$\qquad\quad = m(n\langle a_1, a_2 \rangle) \qquad$ or $n(m\langle a_1, a_2 \rangle)$

$\qquad\quad = m(n\mathbf{a}) \qquad\qquad$ or $n(m\mathbf{a})$

$\boxed{25}$ $-(\mathbf{a} + \mathbf{b}) = -(\langle a_1, a_2 \rangle + \langle b_1, b_2 \rangle)$

$\qquad\qquad = -(\langle a_1 + b_1, a_2 + b_2 \rangle)$

$\qquad\qquad = \langle -(a_1 + b_1), -(a_2 + b_2) \rangle$

$\qquad\qquad = \langle -a_1 - b_1, -a_2 - b_2 \rangle$

$\qquad\qquad = \langle -a_1, -a_2 \rangle + \langle -b_1, -b_2 \rangle$

$\qquad\qquad = -\mathbf{a} + (-\mathbf{b}) = -\mathbf{a} - \mathbf{b}$

$\boxed{27}$ $\|2\mathbf{v}\| = \|2\langle a, b \rangle\| = \|\langle 2a, 2b \rangle\| = \sqrt{(2a)^2 + (2b)^2} = \sqrt{4a^2 + 4b^2} =$

$\qquad\qquad\qquad\qquad\qquad 2\sqrt{a^2 + b^2} = 2\|\langle a, b \rangle\| = 2\|\mathbf{v}\|$

$\boxed{29}$ $\|\mathbf{a}\| = \sqrt{3^2 + (-3)^2} = \sqrt{18} = 3\sqrt{2}$. $\tan\theta = \frac{-3}{3} = -1$ and θ in QIV $\Rightarrow \theta = \frac{7\pi}{4}$.

$\boxed{31}$ $\|\mathbf{a}\| = 5$. The terminal side of θ is on the negative x-axis $\Rightarrow \theta = \pi$.

$\boxed{33}$ $\|\mathbf{a}\| = \sqrt{41}$. $\tan\theta = \frac{5}{-4}$ and θ in QII $\Rightarrow \theta = \tan^{-1}\left(-\frac{5}{4}\right) + \pi$.

$\boxed{35}$ $\|\mathbf{a}\| = 18$. The terminal side of θ is on the negative y-axis $\Rightarrow \theta = \frac{3\pi}{2}$.

Note: Exercises 37–42: Each resultant force is found by completing the parallelogram and then applying the law of cosines.

$\boxed{37}$ $\|\mathbf{r}\| = \sqrt{40^2 + 70^2 - 2(40)(70)\cos 135°} = \sqrt{6500 + 2800\sqrt{2}} \approx 102.3$, or 102 lb.

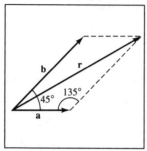

Figure 37

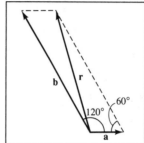

Figure 39

$\boxed{39}$ $\|\mathbf{r}\| = \sqrt{2^2 + 8^2 - 2(2)(8)\cos 60°} = \sqrt{68 - 16} = \sqrt{52} \approx 7.2$ kg.

$\boxed{41}$ $\|\mathbf{r}\| = \sqrt{90^2 + 60^2 - 2(90)(60)\cos 70°} \approx 89.48$, or 89 kg.

Using the law of cosines, $\alpha = \cos^{-1}\left(\dfrac{90^2 + \|\mathbf{r}\|^2 - 60^2}{2(90)(\|\mathbf{r}\|)}\right) \approx$

$\cos^{-1}(0.7765) \approx 39°$, which is 24° under the negative x-axis.

This angle is 204°, or S66°W.

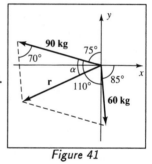

Figure 41

$\boxed{43}$ We will use a component approach for this exercise.

(a) $= \langle 6\cos 110°, \, 6\sin 110° \rangle \approx \langle -2.05, \, 5.64 \rangle$.

(b) $= \langle 2\cos 215°, \, 2\sin 215° \rangle \approx \langle -1.64, \, -1.15 \rangle$.

$\mathbf{a} + \mathbf{b} \approx \langle -3.69, \, 4.49 \rangle$ and $\|\mathbf{a} + \mathbf{b}\| \approx 5.8$ lb. $\tan\theta \approx \frac{4.49}{-3.69} \Rightarrow \theta \approx 129°$ since θ is in QII.

$\boxed{45}$ Horizontal $= 50\cos 35° \approx 40.96$. Vertical $= 50\sin 35° \approx 28.68$.

$\boxed{47}$ Horizontal $= 20\cos 108° \approx -6.18$. Vertical $= 20\sin 108° \approx 19.02$.

$\boxed{49}$ (a) $\mathbf{F} = \mathbf{F_1} + \mathbf{F_2} + \mathbf{F_3} = \langle 4, \, 3 \rangle + \langle -2, \, -3 \rangle + \langle 5, \, 2 \rangle = \langle 7, \, 2 \rangle$.

(b) $\mathbf{F} + \mathbf{G} = \mathbf{0} \Rightarrow \mathbf{G} = -\mathbf{F} = \langle -7, \, -2 \rangle$.

$\boxed{51}$ (a) $\mathbf{F} = \mathbf{F_1} + \mathbf{F_2} = \langle 6\cos 130°, \, 6\sin 130° \rangle + \langle 4\cos(-120°), \, 4\sin(-120°) \rangle \approx \langle -5.86, \, 1.13 \rangle$.

(b) $\mathbf{F} + \mathbf{G} = \mathbf{0} \Rightarrow \mathbf{G} = -\mathbf{F} \approx \langle 5.86, \, -1.13 \rangle$.

[53] The vertical components of the forces must add up to zero for the large ship to move along the line segment AB. The vertical component of the smaller tug is $3200 \sin(-30°) = -1600$. The vertical component of the larger tug is $4000 \sin \theta$.

$$4000 \sin \theta = 1600 \Rightarrow \theta = \sin^{-1}\left(\frac{1600}{4000}\right) = \sin^{-1}(0.4) \approx 23.6°.$$

Note: Exercises 55–60: Measure angles from the positive x-axis.

[55] $\mathbf{p} = \langle 200 \cos 40°, \ 200 \sin 40° \rangle \approx \langle 153.21, \ 128.56 \rangle$. $\mathbf{w} = \langle 40 \cos 0°, \ 40 \sin 0° \rangle = \langle 40, \ 0 \rangle$.

$\mathbf{p} + \mathbf{w} \approx \langle 193.21, \ 128.56 \rangle$ and $\|\mathbf{p} + \mathbf{w}\| \approx 232.07$, or 232 mi/hr.

$$\tan \theta \approx \frac{128.56}{193.21} \Rightarrow \theta \approx 34°. \ \text{The true course is then N}(90° - 34°)\text{E, or N56°E.}$$

[57] $\mathbf{w} = \langle 50 \cos 90°, \ 50 \sin 90° \rangle = \langle 0, \ 50 \rangle$.

$\mathbf{r} = \langle 400 \cos 200°, \ 400 \sin 200° \rangle \approx \langle -375.88, \ -136.81 \rangle$, where $\mathbf{r}$ is the desired resultant of $\mathbf{p} + \mathbf{w}$. Since $\mathbf{r} = \mathbf{p} + \mathbf{w}$, $\mathbf{p} = \mathbf{r} - \mathbf{w} \approx \langle -375.88, \ -186.81 \rangle$. $\|\mathbf{p}\| \approx 419.74$, or 420 mi/hr.

$\tan \theta \approx \frac{-186.81}{-375.88}$ and θ is in QIII $\Rightarrow \theta \approx 206°$ from the positive x-axis,

or 244° using the directional form.

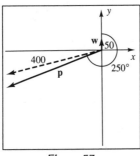

Figure 57

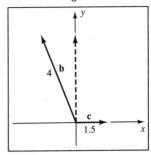

Figure 59

[59] Let the vectors $\mathbf{c}$, $\mathbf{b}$, and $\mathbf{r}$ denote the current, the boat, and the resultant, respectively. $\mathbf{c} = \langle 1.5 \cos 0°, \ 1.5 \sin 0° \rangle = \langle 1.5, \ 0 \rangle$. $\mathbf{r} = \langle s \cos 90°, \ s \sin 90° \rangle = \langle 0, \ s \rangle$, where s is the resulting speed. $\mathbf{b} = \langle 4 \cos \theta, \ 4 \sin \theta \rangle$. Also, $\mathbf{b} = \mathbf{r} - \mathbf{c} = \langle -1.5, \ s \rangle$.

$$4 \cos \theta = -1.5 \Rightarrow \theta \approx 112°, \ \text{or N22°W.}$$

[61] In the figure in the text, suppose $\mathbf{v}_1$ was pointing upwards instead of downwards. If we then write $\mathbf{v}_1$ in terms of its vertical and horizontal components, we have $\mathbf{v}_1 = \|\mathbf{v}_1\| \cos \theta_1 \mathbf{j} - \|\mathbf{v}_1\| \sin \theta_1 \mathbf{i}$. We want the negative of this vector since $\mathbf{v}_1$ is actually pointing downward, so using $\|\mathbf{v}_1\| = 8.2$ and $\theta_1 = 30°$, we obtain $\mathbf{v}_1 = \|\mathbf{v}_1\| \sin \theta_1 \mathbf{i} - \|\mathbf{v}_1\| \cos \theta_1 \mathbf{j} = 8.2(\frac{1}{2}) \mathbf{i} - 8.2(\sqrt{3}/2) \mathbf{j} = 4.1 \mathbf{i} - 4.1\sqrt{3} \mathbf{j} \approx 4.1 \mathbf{i} - 7.10 \mathbf{j}$. The angle θ_2 can now be computed using the given relationship in the text.

$$\frac{\|\mathbf{v}_1\|}{\|\mathbf{v}_2\|} = \frac{\tan \theta_1}{\tan \theta_2} \Rightarrow \tan \theta_2 = \frac{\|\mathbf{v}_2\|}{\|\mathbf{v}_1\|} \tan \theta_1 = \frac{3.8}{8.2} \times \frac{1}{\sqrt{3}} \Rightarrow \tan \theta_2 \approx 0.2676 \Rightarrow \theta_2 \approx 14.98°.$$

We now write $\mathbf{v}_2$ in terms of its horizontal and vertical components, and using $\|\mathbf{v}_2\| = 3.8$, it follows that $\mathbf{v}_2 = \|\mathbf{v}_2\| \sin \theta_2 \mathbf{i} - \|\mathbf{v}_2\| \cos \theta_2 \mathbf{j} \approx 0.98 \mathbf{i} - 3.67 \mathbf{j}$.

63 (a) $\mathbf{a} = 15\cos 40°\mathbf{i} + 15\sin 40°\mathbf{j} \approx 11.49\mathbf{i} + 9.64\mathbf{j}.$

  $\mathbf{b} = 17\cos 40°\mathbf{i} + 17\sin 40°\mathbf{j} \approx 13.02\mathbf{i} + 10.93\mathbf{j}.$

  $\overrightarrow{PR} = \mathbf{a} + \mathbf{b} \approx 24.51\mathbf{i} + 20.57\mathbf{j} \Rightarrow R \approx (24.51,\ 20.57).$

(b) $\mathbf{c} = 15\cos(40° + 85°)\mathbf{i} + 15\sin 125°\mathbf{j} \approx -8.60\mathbf{i} + 12.29\mathbf{j}.$

  $\mathbf{d} = 17\cos(40° + 85° + 35°)\mathbf{i} + 17\sin 160°\mathbf{j} \approx -15.97\mathbf{i} + 5.81\mathbf{j}.$

  $\overrightarrow{PR} = \mathbf{c} + \mathbf{d} \approx -24.57\mathbf{i} + 18.10\mathbf{j} \Rightarrow R \approx (-24.57,\ 18.10).$

8.6 Exercises

1 (a) The dot product of the two vectors is

$$\langle -2,\ 5 \rangle \cdot \langle 3,\ 6 \rangle = (-2)(3) + (5)(6) = -6 + 30 = 24.$$

(b) The angle between the two vectors is

$$\theta = \cos^{-1}\left(\frac{\langle -2,\ 5 \rangle \cdot \langle 3,\ 6 \rangle}{\|\langle -2,\ 5 \rangle\|\,\|\langle 3,\ 6 \rangle\|}\right) = \cos^{-1}\left(\frac{24}{\sqrt{29}\,\sqrt{45}}\right) \approx 48°22'.$$

3 (a) $(4\mathbf{i} - \mathbf{j}) \cdot (-3\mathbf{i} + 2\mathbf{j}) = (4)(-3) + (-1)(2) = -12 - 2 = -14$

(b) $\theta = \cos^{-1}\left(\dfrac{(4\mathbf{i} - \mathbf{j}) \cdot (-3\mathbf{i} + 2\mathbf{j})}{\|4\mathbf{i} - \mathbf{j}\|\,\|-3\mathbf{i} + 2\mathbf{j}\|}\right) = \cos^{-1}\left(\dfrac{-14}{\sqrt{17}\,\sqrt{13}}\right) \approx 160°21'$

7 (a) $\langle 10,\ 7 \rangle \cdot \langle -2,\ -\frac{7}{5} \rangle = (10)(-2) + (7)(-\frac{7}{5}) = -\frac{149}{5}$

(b) $\theta = \cos^{-1}\left(\dfrac{\langle 10,\ 7 \rangle \cdot \langle -2,\ -\frac{7}{5} \rangle}{\|\langle 10,\ 7 \rangle\|\,\|\langle -2,\ -\frac{7}{5} \rangle\|}\right) = \cos^{-1}\left(\dfrac{-149/5}{\sqrt{149}\,\sqrt{149/25}}\right) = \cos^{-1}(-1) = 180°$

This result $\{\theta = 180°\}$ indicates that the vectors have the opposite direction.

9 $\langle 4,\ -1 \rangle \cdot \langle 2,\ 8 \rangle = 8 - 8 = 0 \Rightarrow$ vectors are orthogonal since their dot product is zero.

11 $(-4\mathbf{j}) \cdot (-7\mathbf{i}) = 0 + 0 = 0 \Rightarrow$ vectors are orthogonal.

13 We first find the angle between the two vectors.

If this angle is 0 or π, then the vectors are parallel.

$$\cos\theta = \frac{\mathbf{a} \cdot \mathbf{b}}{\|\mathbf{a}\|\,\|\mathbf{b}\|} = \frac{(3)(-\frac{12}{7}) + (-5)(\frac{20}{7})}{\sqrt{9 + 25}\,\sqrt{\frac{144}{49} + \frac{400}{49}}} = \frac{-\frac{136}{7}}{\sqrt{\frac{18,496}{49}}} = \frac{-\frac{136}{7}}{\frac{136}{7}} = -1 \Rightarrow$$

$\theta = \cos^{-1}(-1) = \pi.$ $\mathbf{b} = c\mathbf{a} \Rightarrow -\frac{12}{7}\mathbf{i} + \frac{20}{7}\mathbf{j} = 3c\mathbf{i} - 5c\mathbf{j} \Rightarrow$

$3c = -\frac{12}{7}$ and $-5c = \frac{20}{7} \Rightarrow c = -\frac{4}{7} < 0 \Rightarrow \mathbf{a}$ and $\mathbf{b}$ have the opposite direction.

15 $\cos\theta = \dfrac{\mathbf{a} \cdot \mathbf{b}}{\|\mathbf{a}\|\,\|\mathbf{b}\|} = \dfrac{(\frac{2}{3})(8) + (\frac{1}{2})(6)}{\sqrt{\frac{4}{9} + \frac{1}{4}}\,\sqrt{64 + 36}} = \dfrac{\frac{25}{3}}{\sqrt{\frac{25}{36}} \cdot 10} = 1 \Rightarrow \theta = \cos^{-1} 1 = 0.$

$\mathbf{b} = c\mathbf{a} \Rightarrow 8\mathbf{i} + 6\mathbf{j} = \frac{2}{3}c\mathbf{i} + \frac{1}{2}c\mathbf{j} \Rightarrow 8 = \frac{2}{3}c$ and $6 = \frac{1}{2}c \Rightarrow c = 12 > 0 \Rightarrow$

$\mathbf{a}$ and $\mathbf{b}$ have the same direction.

17 We need to have the dot product of the two vectors equal 0.

$$(3\mathbf{i} - 2\mathbf{j}) \cdot (4\mathbf{i} + 5c\mathbf{j}) = 0 \Rightarrow 12 - 10c = 0 \Rightarrow c = \tfrac{6}{5}.$$

19 $(9\mathbf{i} - 16c\mathbf{j}) \cdot (\mathbf{i} + 4c\mathbf{j}) = 0 \Rightarrow 9 - 64c^2 = 0 \Rightarrow c^2 = \frac{9}{64} \Rightarrow c = \pm\frac{3}{8}.$

21 (a) $\mathbf{a} \cdot (\mathbf{b} + \mathbf{c}) = \langle 2, -3 \rangle \cdot (\langle 3, 4 \rangle + \langle -1, 5 \rangle) = \langle 2, -3 \rangle \cdot \langle 2, 9 \rangle = 4 - 27 = -23$

   (b) $\mathbf{a} \cdot \mathbf{b} + \mathbf{a} \cdot \mathbf{c} = \langle 2, -3 \rangle \cdot \langle 3, 4 \rangle + \langle 2, -3 \rangle \cdot \langle -1, 5 \rangle = (6 - 12) + (-2 - 15) = -23$

23 $(2\mathbf{a} + \mathbf{b}) \cdot (3\mathbf{c}) = (2\langle 2, -3 \rangle + \langle 3, 4 \rangle) \cdot (3\langle -1, 5 \rangle)$

$$= (\langle 4, -6 \rangle + \langle 3, 4 \rangle) \cdot \langle -3, 15 \rangle$$

$$= \langle 7, -2 \rangle \cdot \langle -3, 15 \rangle = -21 - 30 = -51$$

25 $\text{comp}_{\mathbf{c}}\mathbf{b} = \dfrac{\mathbf{b} \cdot \mathbf{c}}{\|\mathbf{c}\|} = \dfrac{\langle 3, 4 \rangle \cdot \langle -1, 5 \rangle}{\|\langle -1, 5 \rangle\|} = \dfrac{17}{\sqrt{26}} \approx 3.33$

27 $\text{comp}_{\mathbf{b}}(\mathbf{a} + \mathbf{c}) = \dfrac{(\mathbf{a} + \mathbf{c}) \cdot \mathbf{b}}{\|\mathbf{b}\|} = \dfrac{(\langle 2, -3 \rangle + \langle -1, 5 \rangle) \cdot \langle 3, 4 \rangle}{\|\langle 3, 4 \rangle\|} = \dfrac{\langle 1, 2 \rangle \cdot \langle 3, 4 \rangle}{5} = \dfrac{11}{5} = 2.2$

29 $\mathbf{c} \cdot \overrightarrow{PQ} = \langle 3, 4 \rangle \cdot \langle 5, -2 \rangle = 15 - 8 = 7.$

31 The following hint was missing in the first printing of the text:

             (*Hint:* Find a vector $\mathbf{b} = \langle b_1, b_2 \rangle$ such that $\mathbf{b} = \overrightarrow{PQ}$.)

We want a vector with initial point at the origin and terminal point located so that this vector has the same magnitude and direction as $\overrightarrow{PQ}$. Following the hint in the text, $\mathbf{b} = \overrightarrow{PQ} \Rightarrow \langle b_1, b_2 \rangle = \langle 4 - 2, 3 - (-1) \rangle \Rightarrow \langle b_1, b_2 \rangle = \langle 2, 4 \rangle.$

$$\mathbf{c} \cdot \mathbf{b} = \langle 6, 4 \rangle \cdot \langle 2, 4 \rangle = 12 + 16 = 28.$$

33 The force is described by the vector $\langle 0, 4 \rangle$.

$$\text{The work done is } \langle 0, 4 \rangle \cdot \langle 8, 3 \rangle = 0 + 12 = 12.$$

35 $\mathbf{a} \cdot \mathbf{a} = \langle a_1, a_2 \rangle \cdot \langle a_1, a_2 \rangle = a_1^2 + a_2^2 = \left(\sqrt{a_1^2 + a_2^2}\right)^2 = \|\mathbf{a}\|^2$

37 $(c\mathbf{a}) \cdot \mathbf{b} = (c\langle a_1, a_2 \rangle) \cdot \langle b_1, b_2 \rangle$

$$= \langle ca_1, ca_2 \rangle \cdot \langle b_1, b_2 \rangle$$

$$= ca_1 b_1 + ca_2 b_2$$

$$= c(a_1 b_1 + a_2 b_2) = c(\mathbf{a} \cdot \mathbf{b})$$

41 Using the horizontal and vertical components of a vector from Section 8.5, we have the force vector as $\langle 20\cos 30°, 20\sin 30° \rangle = \langle 10\sqrt{3}, 10 \rangle.$

The distance (direction vector) can be described by the vector $\langle 100, 0 \rangle$.

$$\text{The work done is } \langle 10\sqrt{3}, 10 \rangle \cdot \langle 100, 0 \rangle = 1000\sqrt{3} \approx 1732 \text{ ft-lb.}$$

43 (a) The horizontal component has magnitude 93×10^6 and the vertical component has magnitude 0.432×10^6.

       Thus, $\mathbf{v} = (93 \times 10^6)\mathbf{i} + (0.432 \times 10^6)\mathbf{j}$ and $\mathbf{w} = (93 \times 10^6)\mathbf{i} - (0.432 \times 10^6)\mathbf{j}.$

   (b) $\cos\theta = \dfrac{\mathbf{v} \cdot \mathbf{w}}{\|\mathbf{v}\|\|\mathbf{w}\|} = \dfrac{(93 \times 10^6)^2 - (0.432 \times 10^6)^2}{\sqrt{(93 \times 10^6)^2 + (0.432 \times 10^6)^2}\,\sqrt{(93 \times 10^6)^2 + (0.432 \times 10^6)^2}} \approx$

$$0.99995684 \Rightarrow \theta \approx 0.53°.$$

45 From the "Theorem on the Dot Product," we know that $\mathbf{F} \cdot \mathbf{v} = \|\mathbf{F}\| \|\mathbf{v}\| \cos\theta$, so

$P = \frac{1}{550}(\mathbf{F} \cdot \mathbf{v}) = \frac{1}{550}\|\mathbf{F}\| \|\mathbf{v}\| \cos\theta = \frac{1}{550}(2200)(8) \cos 30° = 16\sqrt{3} \approx 27.7$ horsepower.

Chapter 8 Review Exercises

1 We are given 2 sides of a triangle and the angle between them—so we use the law of cosines to find the third side.

$$a = \sqrt{b^2 + c^2 - 2bc \cos\alpha} = \sqrt{6^2 + 7^2 - 2(6)(7)\cos 60°} = \sqrt{43}.$$

$$\beta = \cos^{-1}\left(\frac{a^2 + c^2 - b^2}{2ac}\right) = \cos^{-1}\left(\frac{43 + 49 - 36}{2\sqrt{43}\,(7)}\right) = \cos^{-1}\left(\frac{4}{\sqrt{43}}\right).$$

$$\gamma = \cos^{-1}\left(\frac{a^2 + b^2 - c^2}{2ab}\right) = \cos^{-1}\left(\frac{43 + 36 - 49}{2\sqrt{43}\,(6)}\right) = \cos^{-1}\left(\frac{5}{2\sqrt{43}}\right).$$

2 $\dfrac{\sin\alpha}{a} = \dfrac{\sin\gamma}{c} \Rightarrow \alpha = \sin^{-1}\left(\dfrac{a \sin\gamma}{c}\right) = \sin^{-1}\left(\dfrac{2\sqrt{3} \cdot \frac{1}{2}}{2}\right) = \sin^{-1}\left(\dfrac{\sqrt{3}}{2}\right) = 60°$ or $120°$.

There are two triangles possible since in either case $\alpha + \gamma < 180°$.

$\beta = (180° - \gamma) - \alpha = (180° - 30°) - (60°$ or $120°) = 90°$ or $30°$.

$$\frac{b}{\sin\beta} = \frac{c}{\sin\gamma} \Rightarrow b = \frac{c \sin\beta}{\sin\gamma} = \frac{2\sin(90°\text{ or }30°)}{\sin 30°} = 4 \text{ or } 2.$$

3 $\gamma = 180° - \alpha - \beta = 180° - 60° - 45° = 75°$.

$$\frac{a}{\sin\alpha} = \frac{b}{\sin\beta} \Rightarrow a = \frac{b \sin\alpha}{\sin\beta} = \frac{100\sin 60°}{\sin 45°} = \frac{100 \cdot (\sqrt{3}/2)}{\sqrt{2}/2} \cdot \frac{\sqrt{2}}{\sqrt{2}} = 50\sqrt{6}.$$

$$\frac{c}{\sin\gamma} = \frac{b}{\sin\beta} \Rightarrow c = \frac{b \sin\gamma}{\sin\beta} = \frac{100\sin(45° + 30°)}{\sqrt{2}/2} =$$

$$100\sqrt{2}\,(\sin 45° \cos 30° + \cos 45° \sin 30°) = 100\sqrt{2}\left(\frac{\sqrt{2}}{2} \cdot \frac{\sqrt{3}}{2} + \frac{\sqrt{2}}{2} \cdot \frac{1}{2}\right) =$$

$$\tfrac{100}{4}\sqrt{2}\,(\sqrt{6} + \sqrt{2}) = 25(2\sqrt{3} + 2) = 50(1 + \sqrt{3}).$$

4 $\alpha = \cos^{-1}\left(\dfrac{b^2 + c^2 - a^2}{2bc}\right) = \cos^{-1}\left(\dfrac{9 + 16 - 4}{2(3)(4)}\right) = \cos^{-1}\left(\dfrac{7}{8}\right).$

$\beta = \cos^{-1}\left(\dfrac{a^2 + c^2 - b^2}{2ac}\right) = \cos^{-1}\left(\dfrac{4 + 16 - 9}{2(2)(4)}\right) = \cos^{-1}\left(\dfrac{11}{16}\right).$

$\gamma = \cos^{-1}\left(\dfrac{a^2 + b^2 - c^2}{2ab}\right) = \cos^{-1}\left(\dfrac{4 + 9 - 16}{2(2)(3)}\right) = \cos^{-1}\left(-\dfrac{1}{4}\right).$

6 $\dfrac{\sin\gamma}{c} = \dfrac{\sin\alpha}{a} \Rightarrow \gamma = \sin^{-1}\left(\dfrac{c \sin\alpha}{a}\right) = \sin^{-1}\left(\dfrac{125\sin 23°30'}{152}\right) \approx \sin^{-1}(0.3279) \approx$

$19°10'$ or $160°50'$ {rounded to the nearest 10 minutes}. Reject $160°50'$ because then $\alpha + \gamma \geq 180°$. $\beta = 180° - \alpha - \gamma \approx 180° - 23°30' - 19°10' = 137°20'$.

$$\frac{b}{\sin\beta} = \frac{a}{\sin\alpha} \Rightarrow b = \frac{a \sin\beta}{\sin\alpha} = \frac{152\sin 137°20'}{\sin 23°30'} \approx 258.3, \text{ or } 258.$$

$\boxed{9}$ Since we are given two sides and the included angle of a triangle,

we use the formula for the area of a triangle from Section 8.2.

$$\mathcal{A} = \tfrac{1}{2}bc \sin \alpha = \tfrac{1}{2}(20)(30) \sin 75° \approx 289.8, \text{ or } 290 \text{ square units.}$$

$\boxed{10}$ Given the three sides of a triangle, we apply Heron's formula to find the area.

$$s = \tfrac{1}{2}(a+b+c) = \tfrac{1}{2}(4+7+10) = 10.5.$$

$$\mathcal{A} = \sqrt{s(s-a)(s-b)(s-c)} = \sqrt{(10.5)(6.5)(3.5)(0.5)} \approx 10.9 \text{ square units.}$$

$\boxed{11}$ $z = -10 + 10i \Rightarrow r = \sqrt{(-10)^2 + 10^2} = \sqrt{200} = 10\sqrt{2}.$

$$\tan \theta = \tfrac{10}{-10} = -1 \text{ and } \theta \text{ in QII} \Rightarrow \theta = \tfrac{3\pi}{4}. \quad z = 10\sqrt{2} \operatorname{cis} \tfrac{3\pi}{4}.$$

$\boxed{13}$ $z = -17 \Rightarrow r = 17.$ θ on the negative x-axis $\Rightarrow \theta = \pi.$ $z = 17 \operatorname{cis} \pi.$

$\boxed{16}$ $z = 4 + 5i \Rightarrow r = \sqrt{4^2 + 5^2} = \sqrt{41}.$ $\tan \theta = \tfrac{5}{4}$ and θ in QI $\Rightarrow \theta = \tan^{-1} \tfrac{5}{4}.$

$$z = \sqrt{41} \operatorname{cis} (\tan^{-1} \tfrac{5}{4}).$$

$\boxed{17}$ $20 \left(\cos \tfrac{11\pi}{6} + i \sin \tfrac{11\pi}{6} \right) = 20 \left(\tfrac{\sqrt{3}}{2} - \tfrac{1}{2}i \right) = 10\sqrt{3} - 10i$

$\boxed{19}$ $z_1 = -3\sqrt{3} - 3i = 6 \operatorname{cis} \tfrac{7\pi}{6}$ and $z_2 = 2\sqrt{3} + 2i = 4 \operatorname{cis} \tfrac{\pi}{6}.$

$$z_1 z_2 = 6 \cdot 4 \operatorname{cis} \left(\tfrac{7\pi}{6} + \tfrac{\pi}{6} \right) = 24 \operatorname{cis} \tfrac{4\pi}{3} = 24 \left(-\tfrac{1}{2} - \tfrac{\sqrt{3}}{2}i \right) = -12 - 12\sqrt{3}\,i.$$

$$\tfrac{z_1}{z_2} = \tfrac{6}{4} \operatorname{cis} \left(\tfrac{7\pi}{6} - \tfrac{\pi}{6} \right) = \tfrac{3}{2} \operatorname{cis} \pi = \tfrac{3}{2}(-1 + 0i) = -\tfrac{3}{2}.$$

$\boxed{21}$ $(-\sqrt{3} + i)^9 = (2 \operatorname{cis} \tfrac{5\pi}{6})^9 = 2^9 \operatorname{cis} (9 \cdot \tfrac{5\pi}{6}) = 2^9 \operatorname{cis} \tfrac{15\pi}{2} = 512 \operatorname{cis} \tfrac{3\pi}{2} = 512(0 - i) = -512i$

$\boxed{23}$ $(3 - 3i)^5 = (3\sqrt{2} \operatorname{cis} \tfrac{7\pi}{4})^5 = (3\sqrt{2})^5 \operatorname{cis} \tfrac{35\pi}{4} = 972\sqrt{2} \operatorname{cis} \tfrac{3\pi}{4} = 972\sqrt{2} \left(-\tfrac{\sqrt{2}}{2} + \tfrac{\sqrt{2}}{2}i \right) =$

$$-972 + 972i$$

$\boxed{25}$ $-27 + 0i = 27 \operatorname{cis} 180°.$ $w_k = \sqrt[3]{27} \operatorname{cis} \left(\dfrac{180° + 360°k}{3} \right)$ for $k = 0, 1, 2.$

$$w_0 = 3 \operatorname{cis} 60° = 3 \left(\tfrac{1}{2} + \tfrac{\sqrt{3}}{2}i \right) = \tfrac{3}{2} + \tfrac{3\sqrt{3}}{2}i.$$

$$w_1 = 3 \operatorname{cis} 180° = 3(-1 + 0i) = -3.$$

$$w_2 = 3 \operatorname{cis} 300° = 3 \left(\tfrac{1}{2} - \tfrac{\sqrt{3}}{2}i \right) = \tfrac{3}{2} - \tfrac{3\sqrt{3}}{2}i.$$

$\boxed{27}$ $x^5 - 32 = 0 \Rightarrow x^5 = 32.$ The problem is now to find the 5 fifth roots of 32.

$$32 = 32 + 0i = 32 \operatorname{cis} 0°. \quad w_k = \sqrt[5]{32} \operatorname{cis} \left(\dfrac{0° + 360°k}{5} \right) \text{ for } k = 0, 1, 2, 3, 4.$$

$$w_k = 2 \operatorname{cis} \theta \text{ with } \theta = 0°, 72°, 144°, 216°, 288°.$$

$\boxed{29}$ (a) $4\mathbf{a} + \mathbf{b} = 4(2\mathbf{i} + 5\mathbf{j}) + (4\mathbf{i} - \mathbf{j}) = 8\mathbf{i} + 20\mathbf{j} + 4\mathbf{i} - \mathbf{j} = 12\mathbf{i} + 19\mathbf{j}.$

(b) $2\mathbf{a} - 3\mathbf{b} = 2(2\mathbf{i} + 5\mathbf{j}) - 3(4\mathbf{i} - \mathbf{j}) = 4\mathbf{i} + 10\mathbf{j} - 12\mathbf{i} + 3\mathbf{j} = -8\mathbf{i} + 13\mathbf{j}.$

(c) $\| \mathbf{a} - \mathbf{b} \| = \| (2\mathbf{i} + 5\mathbf{j}) - (4\mathbf{i} - \mathbf{j}) \| = \| -2\mathbf{i} + 6\mathbf{j} \| = \sqrt{(-2)^2 + 6^2} = \sqrt{40} = 2\sqrt{10} \approx$

$$6.32.$$

(d) $\| \mathbf{a} \| - \| \mathbf{b} \| = \| 2\mathbf{i} + 5\mathbf{j} \| - \| 4\mathbf{i} - \mathbf{j} \| = \sqrt{2^2 + 5^2} - \sqrt{4^2 + (-1)^2} = \sqrt{29} - \sqrt{17} \approx 1.26.$

30 $\|\mathbf{r} - \mathbf{a}\| = c \Rightarrow \| \langle x, y \rangle - \langle a_1, a_2 \rangle \| = c \Rightarrow \| \langle x - a_1, y - a_2 \rangle \| = c \Rightarrow$

$\sqrt{(x - a_1)^2 + (y - a_2)^2} = c \Rightarrow (x - a_1)^2 + (y - a_2)^2 = c^2.$

This is a circle with center (a_1, a_2) and radius c.

33 S60°E is equivalent to 330° and N74°E is equivalent to 16°.

$\langle 72 \cos 330°, 72 \sin 330° \rangle + \langle 46 \cos 16°, 46 \sin 16° \rangle = \mathbf{r} \approx \langle 106.57, -23.32 \rangle.$

$\|\mathbf{r}\| \approx 109$ kg. $\tan \theta \approx \frac{-23.32}{106.57} \Rightarrow \theta \approx -12°$, or equivalently, S78°E.

34 $\mathbf{p} = \langle 400 \cos 10°, 400 \sin 10° \rangle \approx \langle 393.92, 69.46 \rangle.$

$\mathbf{r} = \langle 390 \cos 0°, 390 \sin 0° \rangle = \langle 390, 0 \rangle.$

$\mathbf{w} = \mathbf{r} - \mathbf{p} \approx \langle -3.92, -69.46 \rangle$ and $\|\mathbf{w}\| \approx 69.57$, or 70 mi/hr.

$\tan \theta \approx \frac{-69.46}{-3.92} \Rightarrow \theta \approx 267°$, or in the direction of 183°.

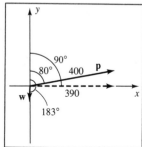

Figure 34

36 (a) $(2\mathbf{a} - 3\mathbf{b}) \cdot \mathbf{a} = [2(6\mathbf{i} - 2\mathbf{j}) - 3(\mathbf{i} + 3\mathbf{j})] \cdot (6\mathbf{i} - 2\mathbf{j})$

$= (9\mathbf{i} - 13\mathbf{j}) \cdot (6\mathbf{i} - 2\mathbf{j}) = 54 + 26 = 80.$

(b) $\mathbf{c} = \mathbf{a} + \mathbf{b} = (6\mathbf{i} - 2\mathbf{j}) + (\mathbf{i} + 3\mathbf{j}) = 7\mathbf{i} + \mathbf{j}.$ The angle between $\mathbf{a}$ and $\mathbf{c}$ is

$\theta = \cos^{-1}\left(\frac{\mathbf{a} \cdot \mathbf{c}}{\|\mathbf{a}\|\|\mathbf{c}\|}\right) = \cos^{-1}\left(\frac{(6\mathbf{i} - 2\mathbf{j}) \cdot (7\mathbf{i} + \mathbf{j})}{\|6\mathbf{i} - 2\mathbf{j}\|\|7\mathbf{i} + \mathbf{j}\|}\right) = \cos^{-1}\left(\frac{40}{\sqrt{40}\sqrt{50}}\right) \approx 26°34'.$

(c) $\text{comp}_{\mathbf{a}}(\mathbf{a} + \mathbf{b}) = \text{comp}_{\mathbf{a}}\mathbf{c} = \frac{\mathbf{c} \cdot \mathbf{a}}{\|\mathbf{a}\|} = \frac{40}{\sqrt{40}} = \sqrt{40} = 2\sqrt{10} \approx 6.32.$

38 $\frac{\sin \gamma}{150} = \frac{\sin 27.4°}{200} \Rightarrow$

$\gamma = \sin^{-1}\left(\frac{150 \sin 27.4°}{200}\right) \approx \sin^{-1}(0.3451) \approx 20.2°.$

$\beta = 180° - \alpha - \beta \approx 180° - 27.4° - 20.2° = 132.4°.$

The angle between the hill and the horizontal is then

$180° - 132.4° = 47.6°.$

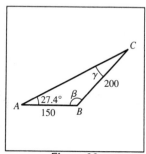

Figure 38

39 Let P denote the point at the base of the shorter building, S the point at the top of the shorter building, T the point at the top of the skyscraper, Q the point 50 feet up the side of the skyscraper, and h the height of the skyscraper.

(a) $\angle SPT = 90° - 62° = 28°.$ $\angle PST = 90° + 59° = 149°.$

Thus, $\angle STP = 180° - 28° - 149° = 3°.$ $\frac{\overline{ST}}{\sin 28°} = \frac{50}{\sin 3°} \Rightarrow \overline{ST} \approx 448.52$, or 449 ft.

(b) $h = \overline{QT} + 50 = \overline{ST} \sin 59° + 50 \approx 434.45$, or 434 ft.

41 Let E denote the middle point. $\angle CDA = \angle BDC - \angle BDA = 125° - 100° = 25°$.

In $\triangle CAD$, $\angle CAD = 180° - \angle ACD - \angle CDA = 180° - 115° - 25° = 40°$.

$\dfrac{\overline{AD}}{\sin 115°} = \dfrac{120}{\sin 40°} \Rightarrow \overline{AD} \approx 169.20$. $\angle DCB = \angle ACD - \angle ACB = 115° - 92° = 23°$.

In $\triangle DBC$, $\angle DBC = 180° - \angle BDC - \angle DCB = 180° - 125° - 23° = 32°$.

$\dfrac{\overline{BD}}{\sin 23°} = \dfrac{120}{\sin 32°} \Rightarrow \overline{BD} \approx 88.48$.

In $\triangle ADB$, $\overline{AB}^2 = \overline{AD}^2 + \overline{BD}^2 - 2(\overline{AD})(\overline{BD})\cos\angle BDA \Rightarrow$

$$\overline{AB} \approx \sqrt{(169.20)^2 + (88.48)^2 - 2(169.20)(88.48)\cos 100°} \approx 204.1, \text{ or } 204 \text{ ft.}$$

42 If d denotes the distance each girl walks before losing contact with each other,

then $d = 5t$, where t is in hours. Using the law of cosines,

$$10^2 = d^2 + d^2 - 2(d)(d)\cos 105° \Rightarrow 100 = 2d^2(1 - \cos 105°) \Rightarrow d \approx 6.30 \Rightarrow$$

$$t = d/5 \approx 1.26 \text{ hours, or } 1 \text{ hour and } 16 \text{ minutes.}$$

43 (a) Draw a vertical line l through C and label its x-intercept D. Since we have

alternate interior angles, $\angle ACD = \theta_1$. $\angle DCP = 180° - \theta_2$.

$$\text{Thus } \angle ACP = \angle ACD + \angle DCP = \theta_1 + (180° - \theta_2) = 180° - (\theta_2 - \theta_1).$$

(b) Let $k = d(A, P)$. $k^2 = 17^2 + 17^2 - 2(17)(17)\cos\big[180° - (\theta_2 - \theta_1)\big]$.

Since $\cos(180° - \alpha) = \cos 180° \cos\alpha + \sin 180° \sin\alpha = -\cos\alpha$, we have

$k^2 = 578 + 578\cos(\theta_2 - \theta_1) = 578\big[1 + \cos(\theta_2 - \theta_1)\big]$. Using the distance

formula with the points $A(0, 26)$ and $P(x, y)$, we also have $k^2 = x^2 + (y - 26)^2$.

Hence, $578\big[1 + \cos(\theta_2 - \theta_1)\big] = x^2 + (y - 26)^2 \Rightarrow 1 + \cos(\theta_2 - \theta_1) = \dfrac{x^2 + (y - 26)^2}{578}$.

(c) If $x = 25$, $y = 4$, and $\theta_1 = 135°$, then $1 + \cos(\theta_2 - 135°) = \dfrac{25^2 + (-22)^2}{578} = \dfrac{1109}{578} \Rightarrow$

$$\cos(\theta_2 - 135°) = \tfrac{531}{578} \Rightarrow \theta_2 - 135° \approx 23.3° \Rightarrow \theta_2 \approx 158.3°, \text{ or } 158°.$$

44 (a) Let d denote the length of the rescue tunnel. Using the law of cosines,

$d^2 = 45^2 + 50^2 - 2(45)(50)\cos 78° \Rightarrow d \approx 59.91$ ft. Now using the law of sines,

$$\dfrac{\sin\theta}{45} = \dfrac{\sin 78°}{d} \Rightarrow \theta = \sin^{-1}\Big(\dfrac{45\sin 78°}{d}\Big) \approx 47.28°, \text{ or } 47°.$$

(b) If x denotes the number of hours needed, then

$$d \text{ ft} = (3 \text{ ft/hr})(x \text{ hr}) \Rightarrow x = \tfrac{1}{3}d = \tfrac{1}{3}(59.91) \approx 20 \text{ hr.}$$

[45] (a) $\angle CBA = 180° - 136° = 44°$ and

$$d = \overline{AC} = \sqrt{22.9^2 + 17.2^2 - 2(22.9)(17.2)\cos 44°} \approx 15.9. \quad \text{Let } \alpha = \angle BAC.$$

Using the law of sines, $\dfrac{\sin \alpha}{22.9} = \dfrac{\sin 44°}{d} \Rightarrow \alpha = \sin^{-1}\left(\dfrac{22.9 \sin 44°}{d}\right) \approx 87.4°.$

Let $\beta = \angle CAD$. Using the law of cosines, $5.7^2 = d^2 + 16^2 - 2(d)(16)\cos \beta \Rightarrow$

$$\beta = \cos^{-1}\left(\frac{d^2 + 16^2 - 5.7^2}{2(d)(16)}\right) \approx 20.6°. \quad \phi \approx 180° - 87.4° - 20.6° = 72°.$$

(b) The area of $ABCD$ is the sum of the areas of $\triangle CBA$ and $\triangle ADC$.

Area $= \frac{1}{2}(\text{base } \overline{BC})(\text{height to } A) + \frac{1}{2}(\text{base } \overline{AC})(\text{height to } D)$

$= \frac{1}{2}(\overline{BC})(\overline{BA})\sin \angle CBA + \frac{1}{2}(\overline{AC})(\overline{AD})\sin \angle CAD$

$= \frac{1}{2}(22.9)(17.2)\sin 44° + \frac{1}{2}(15.9)(16)\sin 20.6° \approx 136.8 + 44.8 = 181.6 \text{ ft}^2.$

(c) Let h denote the perpendicular distance from $\overline{BA}$ to C.

$$\sin 44° = \frac{h}{22.9} \Rightarrow h \approx 15.9. \quad \text{The wing span } \overline{CC'} \text{ is } 2h + 5.8 \approx 37.6 \text{ ft}.$$

Chapter 9: Systems of Equations and Inequalities

Note: The notation E_1 and E_2 refers to the first equation and the second equation.

$\boxed{1}$ Substituting y in E_2 into E_1 yields $2x - 1 = x^2 - 4 \Rightarrow x^2 - 2x - 3 = 0 \Rightarrow$

$(x - 3)(x + 1) = 0 \Rightarrow x = 3,\ -1$. Substituting $x = 3$ into E_2 gives us $y = 5$,

so $(3,\ 5)$ is a solution of the system. Similarly, $(-1,\ -3)$ is a solution.

$\boxed{3}$ Solving E_2 for x, $x = 1 - 2y$, and substituting into E_1 yields $y^2 = 1 - (1 - 2y) \Rightarrow$

$y^2 - 2y = 0 \Rightarrow y(y - 2) = 0 \Rightarrow y = 0,\ 2;\ x = 1,\ -3$. $\qquad \bigstar\ (1,\ 0),\ (-3,\ 2)$

$\boxed{7}$ Solving E_1 for x, $x = -2y - 1$, and substituting into E_2 yields

$2(-2y - 1) - 3y = 12 \Rightarrow -7y = 14 \Rightarrow y = -2;\ x = 3$. $\qquad \bigstar\ (3,\ -2)$

$\boxed{9}$ Solving E_1 for x, $x = \frac{1}{2} + \frac{3}{2}y$, and substituting into E_2 yields

$-6(\frac{1}{2} + \frac{3}{2}y) + 9y = 4 \Rightarrow -3 = 4$, a contradiction. There are *no solutions.*

$\boxed{11}$ Solving E_1 for x, $x = 5 - 3y$, and substituting into E_2 yields $(5 - 3y)^2 + y^2 = 25 \Rightarrow$

$10y^2 - 30y = 0 \Rightarrow 10y(y - 3) = 0 \Rightarrow y = 0,\ 3;\ x = 5,\ -4$. $\qquad \bigstar\ (-4,\ 3),\ (5,\ 0)$

$\boxed{15}$ Solving E_2 for y, $y = 3x + 2$, and substituting into E_1 yields

$x^2 + (3x + 2)^2 = 9 \Rightarrow 10x^2 + 12x - 5 = 0 \Rightarrow x = \dfrac{-6 \pm \sqrt{86}}{10} = -\dfrac{3}{5} \pm \dfrac{1}{10}\sqrt{86}$.

$y = 3\left(\dfrac{-6 \pm \sqrt{86}}{10}\right) + 2 = \dfrac{-18 \pm 3\sqrt{86}}{10} + \dfrac{20}{10} = \dfrac{2 \pm 3\sqrt{86}}{10} = \dfrac{1}{5} \pm \dfrac{3}{10}\sqrt{86}$.

$\bigstar\ (-\frac{3}{5} + \frac{1}{10}\sqrt{86},\ \frac{1}{5} + \frac{3}{10}\sqrt{86}),\ (-\frac{3}{5} - \frac{1}{10}\sqrt{86},\ \frac{1}{5} - \frac{3}{10}\sqrt{86})$

$\boxed{21}$ Substituting y in E_1 into E_2 yields $20/x^2 = 9 - x^2$ {multiply by x^2} $\Rightarrow$

$x^4 - 9x^2 + 20 = 0 \Rightarrow (x^2 - 4)(x^2 - 5) = 0 \Rightarrow x^2 = 4,\ 5 \Rightarrow x = \pm 2,\ \pm\sqrt{5};\ y = 5,\ 4$.

$\bigstar\ (\pm 2,\ 5),\ (\pm\sqrt{5},\ 4)$

$\boxed{23}$ Solving E_1 for y^2, $y^2 = 4x^2 + 4$, and substituting into E_2 yields

$9(4x^2 + 4) + 16x^2 = 140 \Rightarrow 52x^2 = 104 \Rightarrow x^2 = 2 \Rightarrow x = \pm\sqrt{2};\ y = \pm 2\sqrt{3}$.

There are four solutions. $\qquad \bigstar\ (\sqrt{2},\ \pm 2\sqrt{3}),\ (-\sqrt{2},\ \pm 2\sqrt{3})$

27 Solving E_2 for y, $y = 2x + z - 9$, and substituting into E_1 and E_3 yields

$$\begin{cases} x + 2(2x + z - 9) - z & = -1 \\ x + 3(2x + z - 9) + 3z & = 6 \end{cases} \Rightarrow \begin{cases} 5x + z & = 17 \quad (E_4) \\ 7x + 6z & = 33 \quad (E_5) \end{cases}$$

Solving E_4 for z, $z = 17 - 5x$, and substituting into E_5 yields $7x + 6(17 - 5x) = 33 \Rightarrow$

$-23x = -69 \Rightarrow x = 3$. Now $z = 17 - 5x = 17 - 5(3) = 2$ and

$y = 2x + z - 9 = 2(3) + 2 - 9 = -1$.   ★ (3, −1, 2)

31 Using $P = 40 = 2l + 2w$ and $A = 96 = lw$, we have $\begin{cases} 2l + 2w & = 40 \quad (E_1) \\ lw & = 96 \quad (E_2) \end{cases}$

Solving E_1 for l, $l = 20 - w$, and substituting into E_2 yields $(20 - w)w = 96 \Rightarrow$

$20w - w^2 = 96 \Rightarrow w^2 - 20w + 96 = 0 \Rightarrow (w - 8)(w - 12) = 0 \Rightarrow w = 8, 12; l = 12, 8.$

In either case, the rectangle is 12 in. × 8 in.

37 (a) Using $R = aS/(S + b)$, let $S = 40,000$ and $R = 60,000$ for the data for 1987 and

1988. Then let $S = 60,000$ and $R = 72,000$ for the data for 1988 and 1989.

$$\begin{cases} 60,000 = 40,000a/(40,000 + b) \\ 72,000 = 60,000a/(60,000 + b) \end{cases} \Rightarrow \begin{cases} 3 = 2a/(40,000 + b) \\ 6 = 5a/(60,000 + b) \end{cases} \Rightarrow \begin{cases} 120,000 + 3b = 2a \quad (E_1) \\ 360,000 + 6b = 5a \quad (E_2) \end{cases}$$

Solving E_1 for a, $a = 60,000 + \frac{3}{2}b$, and substituting into E_2 yields

$$360,000 + 6b = 5(60,000 + \tfrac{3}{2}b) \Rightarrow 60,000 = \tfrac{3}{2}b \Rightarrow b = 40,000; a = 120,000.$$

(b) Now let $S = 72,000$ and thus $R = \dfrac{(120,000)(72,000)}{72,000 + 40,000} = \dfrac{540,000}{7} \approx 77,143.$

39 Let R_1 and R_2 equal 0. The system is then $\begin{cases} 0 = 0.01x(50 - x - y) \\ 0 = 0.02y(100 - y - 0.5x) \end{cases}$

The first equation is zero if $x = 0$ or if $50 - x - y = 0$. The second equation is zero if $y = 0$ or if $100 - y - 0.5x = 0$. Hence, there are 4 possible solutions.

(1) A solution is $x = 0$ and $y = 0$, or (0, 0).

(2) A second solution is $x = 0$ and $100 - y - 0.5x = 0$ $\{y = 100\}$, or (0, 100).

(3) A third solution is $y = 0$ and $50 - x - y = 0$ $\{x = 50\}$, or (50, 0).

(4) A fourth solution occurs if $50 - x - y = 0$ and $100 - y - 0.5x = 0$.

Solve the first equation for y $\{y = 50 - x\}$ and substitute into the second

equation: $100 - (50 - x) - 0.5x = 0 \Rightarrow 50 = -\frac{1}{2}x \Rightarrow x = -100$; $y = 150$.

This solution is meaningless for this problem since x and y are nonnegative.

41 Since we have an *open* top instead of a closed top, we use $3xy$ instead of $4xy$ for the

surface area formula. $\begin{cases} x^2y = 2 & \textit{Volume} \\ 2x^2 + 3xy = 8 & \textit{Surface Area} \end{cases}$

Solving E_1 for y and substituting into E_2 yields $2x^2 + 3x(2/x^2) = 8 \Rightarrow$

$2x^2 + (6/x) = 8$ { multiply by x } $\Rightarrow 2x^3 - 8x + 6 = 0 \Rightarrow 2(x^3 - 4x + 3) = 0$.

We know that $x = 1$ is a solution of $x^3 - 4x + 3 = 0$ since the sum of the coefficients

is zero. Continuing, $2(x - 1)(x^2 + x - 3) = 0 \Rightarrow \{x > 0\}$ $x = 1, \dfrac{-1 + \sqrt{13}}{2}$.

There are two solutions: 1 ft $\times$ 1 ft $\times$ 2 ft or

$$\frac{\sqrt{13} - 1}{2} \text{ ft} \times \frac{\sqrt{13} - 1}{2} \text{ ft} \times \frac{8}{(\sqrt{13} - 1)^2} \text{ ft} \approx 1.30 \text{ ft} \times 1.30 \text{ ft} \times 1.18 \text{ ft.}$$

45 (a) The slope of the line from $(-4, -3)$ to the origin is $\frac{3}{4}$ so the slope of the tangent

line (which is perpendicular to the line to the origin) is $-\frac{4}{3}$. An equation of the

line through $(-4, -3)$ with slope $-\frac{4}{3}$ is $y + 3 = -\frac{4}{3}(x + 4)$, or, equivalently,

$4x + 3y = -25$. Letting $y = -50$, we find that $x = 31.25$.

(b) The slope of the line from an arbitrary point (x, y) on the circle to the origin is

$\frac{y}{x}$, so the slope of the tangent line is $-\frac{x}{y}$. The line through $(0, -50)$ is

$y + 50 = \left(-\frac{x}{y}\right)(x - 0)$, or, equivalently, $y^2 + 50y = -x^2$. The equation of the

circle is $x^2 + y^2 = 25$. Substituting $x^2 = 25 - y^2$ into $y^2 + 50y = -x^2$ gives us

$50y = -25$, or $y = -\frac{1}{2}$. Substituting $y = -\frac{1}{2}$ into $x^2 = 25 - y^2$ gives us $x^2 = \frac{99}{4}$

and hence $x = \pm\frac{3}{2}\sqrt{11}$. The positive solution corresponds to releasing the

hammer from a clockwise spin, whereas the negative solution corresponds to

releasing the hammer from a counterclockwise spin as depicted in the figures.

Hence, the hammer should be released at $(-\frac{3}{2}\sqrt{11}, -\frac{1}{2}) \approx (-4.975, -0.5)$.

Note: The notation E_1 and E_2 refers to equation 1 and equation 2. $3E_2$ symbolizes "3 times equation 2". After the value of one variable is found, the value(s) of the other variable(s) will be stated and can be found by substituting the known value(s) back into the original equation(s).

[1] The easiest choice for multipliers is to take -2 times the second equation and eliminate x. If we wanted to eliminate y, we could choose 2 times the first equation and 3 times the second equation. $-2E_2 + E_1 \Rightarrow 7y = -14 \Rightarrow y = -2;\ x = 4$

★ $(4, -2)$

[3] $3E_1 - 2E_2 \Rightarrow 29y = 0 \Rightarrow y = 0;\ x = 8$ ★ $(8, 0)$

[7] $3E_1 - 5E_2 \Rightarrow -53y = -28 \Rightarrow y = \frac{28}{53}$. Instead of substituting into one of the equations to find the value of the other variable, it is usually easier to pick different multipliers and re-solve the system for the other variable.

$7E_1 + 6E_2 \Rightarrow 53x = 76 \Rightarrow x = \frac{76}{53}$.   ★ $\left(\frac{76}{53}, \frac{28}{53}\right)$

[9] We will first eliminate all fractions by multiplying both sides of each equation by its

lcd. $\begin{cases} 6E_1 \\ 3E_2 \end{cases} \Rightarrow \begin{cases} 2c + 3d = 30 & (E_3) \\ 3c - 2d = -3 & (E_4) \end{cases}$ $3E_3 - 2E_4 \Rightarrow 13d = 96 \Rightarrow d = \frac{96}{13}$.

$2E_3 + 3E_4 \Rightarrow 13c = 51 \Rightarrow c = \frac{51}{13}$ ★ $\left(\frac{51}{13}, \frac{96}{13}\right)$

[11] The least common multiple of 2 and 3 is 6, so we will multiply the first equation by $\sqrt{3}$ and the second equation by $\sqrt{2}$ so that the coefficients of y are $\sqrt{6}$ and $-\sqrt{6}$.
$\sqrt{3}E_1 + \sqrt{2}E_2 \Rightarrow 7x = 8 \Rightarrow x = \frac{8}{7}$. We will now re-solve the system to obtain a value for y. $2\sqrt{2}E_1 - \sqrt{3}E_2 \Rightarrow -7y = 3\sqrt{6} \Rightarrow y = -\frac{3}{7}\sqrt{6}$.

★ $\left(\frac{8}{7}, -\frac{3}{7}\sqrt{6}\right)$

[19] Using the hint, we let $u = 1/x$ and $v = 1/y$, and the system is $\begin{cases} 2u + 3v = -2 & (E_3) \\ 4u - 5v = 1 & (E_4) \end{cases}$

$-2E_3 + E_4 \Rightarrow -11v = 5 \Rightarrow v = -\frac{5}{11}$. $5E_3 + 3E_4 \Rightarrow 22u = -7 \Rightarrow u = -\frac{7}{22}$.
Resubstituting, we have $x = 1/u = -\frac{22}{7}$ and $y = 1/v = -\frac{11}{5}$.   ★ $\left(-\frac{22}{7}, -\frac{11}{5}\right)$

[23] The volume for a cylinder is $\pi r^2 h$ and the volume for a cone is $\frac{1}{3}\pi r^2 h$.

The radius of the cylinder is $\frac{1}{2}$ cm. $\begin{cases} x + y = 8 & \text{\textit{length}} \\ \pi\left(\frac{1}{2}\right)^2 x + \frac{1}{3}\pi\left(\frac{1}{2}\right)^2 y = 5 & \text{\textit{volume}} \end{cases}$

Solving E_1 for y and substituting into E_2 yields $\frac{\pi}{4}x + \frac{\pi}{12}(8 - x) = 5 \Rightarrow$

$$\frac{\pi}{6}x = \frac{15 - 2\pi}{3} \Rightarrow x = \frac{30 - 4\pi}{\pi} = \frac{30}{\pi} - 4 \approx 5.55 \text{ cm.}$$

$$y = 8 - \left(\frac{30 - 4\pi}{\pi}\right) = \frac{12\pi - 30}{\pi} = 12 - \frac{30}{\pi} \approx 2.45 \text{ cm.}$$

25 The perimeter is composed of 2 sides of the rectangular portion of the table, $2l$, and 2

edges of the semicircular regions, $2 \cdot \frac{1}{2}(2\pi r) = 2\pi r$. Since the radius is $\frac{1}{2}w$, the system

can be represented by $\begin{cases} 2l + 2\pi(\frac{1}{2}w) = 40 & \textit{perimeter} \\ lw = 2\left[\pi(\frac{1}{2}w)^2\right] & \textit{area} \end{cases}$

Solving E_1 for l, $l = \frac{40 - \pi w}{2}$, and substituting into E_2 yields $\left(\frac{40 - \pi w}{2}\right)w = \frac{\pi w^2}{2} \Rightarrow$

$(40 - \pi w)w = \pi w^2 \Rightarrow 40w = 2\pi w^2 \Rightarrow 2\pi w^2 - 40w = 0 \Rightarrow 2w(\pi w - 20) = 0 \Rightarrow$

$w = 0, \frac{20}{\pi}$. We discard $w = 0$ and use $w = 20/\pi \approx 6.37$ ft.

$$\text{Thus, } l = \frac{40 - \pi(20/\pi)}{2} = \frac{20}{2} = 10 \text{ ft.}$$

27 Let x denote the number of adults and y the number of kittens. Thus,

$(\frac{1}{2}x)$ is the number of adult females. $\begin{cases} x + y = 6000 & \textit{total} \\ y = 3(\frac{1}{2}x) & \textit{3 kittens per adult female} \end{cases}$

Substituting y from E_2 into E_1 yields $x + \frac{3}{2}x = 6000 \Rightarrow x = 2400; y = 3600$.

31 Let x denote the speed of the plane and y the speed of the wind. Use $d = rt$.

$\begin{cases} 1200 = (x + y)(2) & \textit{with the wind} \\ 1200 = (x - y)(2\frac{1}{2}) & \textit{against the wind} \end{cases} \Rightarrow \begin{cases} 600 = x + y \\ 480 = x - y \end{cases}$

$$E_1 + E_2 \Rightarrow 2x = 1080 \Rightarrow x = 540 \text{ mi/hr}; y = 60 \text{ mi/hr}$$

37 (a) The expression $\underline{6x + 5y}$ represents the total bill for the plumber's business. This

should equal the plumber's income, which is the plumber's number of hours times

his/her hourly wage—that is, $(6 + 4)(x) = 10x$. The expression $\underline{4x + 6y}$

represents the total bill for the electrician's business. This should equal the

electrician's income, that is, $(5 + 6)(y) = 11y$.

$\begin{cases} 6x + 5y = 10x \\ 4x + 6y = 11y \end{cases} \Rightarrow \begin{cases} 5y = 4x \\ 4x = 5y \end{cases} \Rightarrow y = \frac{4}{5}x$, or, equivalently, $y = 0.80x$

(b) The electrician should charge 80% of what the plumber charges—80% of $20 per

hour is $16 per hour.

39 Let $t = 0$ correspond to the year 1891. The average daily maximum can then be

approximated by the linear equation $y_1 = 0.011t + 15.1$ and the average daily

minimum by the linear equation $y_2 = 0.019t + 5.8$. We must determine t when y_1

and y_2 differ by 9. $y_1 - y_2 = 9 \Rightarrow (0.011t + 15.1) - (0.019t + 5.8) = 9 \Rightarrow$

$-0.008t + 9.3 = 9 \Rightarrow -0.008t = -0.3 \Rightarrow t = 37.5$. $1891 + 37.5 = 1928.5$, or during the

year 1928. Also, $t = 37.5 \Rightarrow y_1 = 0.011(37.5) + 15.1 = 15.5125 \approx 15.5\,°C$.

[41] (a) Since consumers will buy 1,000,000 T-shirts if the selling price is $10 and they will buy 900,000 if the price is $11, we can model this information with a line through the two points $(10, 1,000,000)$ and $(11, 900,000)$. Using the point-slope formula of a line, we have $Q - 1,000,000 = \frac{900,000 - 1,000,000}{11 - 10}(p - 10) \Rightarrow$

$$Q - 1,000,000 = -100,000(p - 10) \Rightarrow Q = -100,000p + 2,000,000.$$

(b) As in part (a), use the two points $(15, 2,000,000)$ and $(16, 2,150,000)$.

$$K - 2,000,000 = \frac{150,000}{1}(p - 15) \Rightarrow K = 150,000p - 250,000.$$

(c) $Q = K \Rightarrow -100,000p + 2,000,000 = 150,000p - 250,000 \Rightarrow$

$$2,250,000 = 250,000p \Rightarrow p = \$9.00$$

9.3 Exercises

Note: Most systems are solved using the back substitution method. Solutions for Exercises 7 and 17 use the reduced echelon method. To avoid fractions, some solutions include linear combinations of rows—that is, $m\mathrm{R}_i + n\mathrm{R}_j$.

[1]
$$\begin{bmatrix} 1 & -2 & -3 & -1 \\ 2 & 1 & 1 & 6 \\ 1 & 3 & -2 & 13 \end{bmatrix} \begin{array}{l} \mathrm{R}_2 - 2\mathrm{R}_1 \to \mathrm{R}_2 \\ \mathrm{R}_3 - \mathrm{R}_1 \to \mathrm{R}_3 \end{array}$$

$$\begin{bmatrix} 1 & -2 & -3 & -1 \\ 0 & 5 & 7 & 8 \\ 0 & 5 & 1 & 14 \end{bmatrix} \mathrm{R}_3 - \mathrm{R}_2 \to \mathrm{R}_3$$

$$\begin{bmatrix} 1 & -2 & -3 & -1 \\ 0 & 5 & 7 & 8 \\ 0 & 0 & -6 & 6 \end{bmatrix} -\tfrac{1}{6}\mathrm{R}_3 \to \mathrm{R}_3$$

R_3: $z = -1$

R_2: $5y + 7z = 8 \Rightarrow y = 3$ $\star$ $(2, 3, -1)$

R_1: $x - 2y - 3z = -1 \Rightarrow x = 2$

[5]
$$\begin{bmatrix} 2 & 6 & -4 & 1 \\ 1 & 3 & -2 & 4 \\ 2 & 1 & -3 & -7 \end{bmatrix} \mathrm{R}_1 \leftrightarrow \mathrm{R}_2$$

$$\begin{bmatrix} 1 & 3 & -2 & 4 \\ 2 & 6 & -4 & 1 \\ 2 & 1 & -3 & -7 \end{bmatrix} \begin{array}{l} \mathrm{R}_2 - 2\mathrm{R}_1 \to \mathrm{R}_2 \\ \mathrm{R}_3 - 2\mathrm{R}_1 \to \mathrm{R}_3 \end{array}$$

$$\begin{bmatrix} 1 & 3 & -2 & 4 \\ 0 & 0 & 0 & -7 \\ 0 & -5 & 1 & -15 \end{bmatrix}$$

The second row, $0x + 0y + 0z = -7$, has *no solution*.

$\boxed{7}$ $\begin{bmatrix} 2 & -3 & 2 & -3 \\ -3 & 2 & 1 & 1 \\ 4 & 1 & -3 & 4 \end{bmatrix}$ $R_1 \leftrightarrow R_2$ and then $R_2 \leftrightarrow R_3$

$\begin{bmatrix} -3 & 2 & 1 & 1 \\ 4 & 1 & -3 & 4 \\ 2 & -3 & 2 & -3 \end{bmatrix}$ $R_1 + R_2 \rightarrow R_1$

$\begin{bmatrix} 1 & 3 & -2 & 5 \\ 4 & 1 & -3 & 4 \\ 2 & -3 & 2 & -3 \end{bmatrix}$ $R_2 - 4 R_1 \rightarrow R_2$
$R_3 - 2 R_1 \rightarrow R_3$

$\begin{bmatrix} 1 & 3 & -2 & 5 \\ 0 & -11 & 5 & -16 \\ 0 & -9 & 6 & -13 \end{bmatrix}$ $4 R_2 - 5 R_3 \rightarrow R_2$ ($*$ see note)

$\begin{bmatrix} 1 & 3 & -2 & 5 \\ 0 & 1 & -10 & 1 \\ 0 & -9 & 6 & -13 \end{bmatrix}$ $R_3 + 9 R_2 \rightarrow R_3$

$\begin{bmatrix} 1 & 3 & -2 & 5 \\ 0 & 1 & -10 & 1 \\ 0 & 0 & -84 & -4 \end{bmatrix}$ $-\frac{1}{84} R_3 \rightarrow R_3$

R_3: $z = \frac{1}{21}$

R_2: $y - 10z = 1 \Rightarrow y = \frac{31}{21}$ $\qquad\qquad$ $\bigstar$ $\left(\frac{2}{3}, \frac{31}{21}, \frac{1}{21}\right)$

R_1: $x + 3y - 2z = 5 \Rightarrow x = \frac{14}{21} = \frac{2}{3}$

($*$) We could have just used $-\frac{1}{11}R_2 \rightarrow R_2$, but we would then have to work with many fractions and increase our chance of making a mistake. We will solve this system again, but this time we will obtain the reduced echelon form. The first 4 matrices are exactly the same. Starting with the fourth matrix we have:

$\begin{bmatrix} 1 & 3 & -2 & 5 \\ 0 & 1 & -10 & 1 \\ 0 & -9 & 6 & -13 \end{bmatrix}$ $R_1 - 3 R_2 \rightarrow R_1$
$R_3 + 9 R_2 \rightarrow R_3$

$\begin{bmatrix} 1 & 0 & 28 & 2 \\ 0 & 1 & -10 & 1 \\ 0 & 0 & -84 & -4 \end{bmatrix}$ $-\frac{1}{84} R_3 \rightarrow R_3$

$\begin{bmatrix} 1 & 0 & 28 & 2 \\ 0 & 1 & -10 & 1 \\ 0 & 0 & 1 & \frac{1}{21} \end{bmatrix}$ $R_1 - 28 R_3 \rightarrow R_1$
$R_2 + 10 R_3 \rightarrow R_2$

$\begin{bmatrix} 1 & 0 & 0 & \frac{14}{21} \\ 0 & 1 & 0 & \frac{31}{21} \\ 0 & 0 & 1 & \frac{1}{21} \end{bmatrix}$

R_1: $x = \frac{14}{21} = \frac{2}{3}$; R_2: $y = \frac{31}{21}$; R_3: $z = \frac{1}{21}$

Note: Exer. 9–16: There are other forms for the answers; c is any real number.

$\boxed{9}$ $\begin{bmatrix} 1 & 3 & 1 & 0 \\ 1 & 1 & -1 & 0 \\ 1 & -2 & -4 & 0 \end{bmatrix}$ $\begin{array}{l} R_2 - R_1 \to R_2 \\ R_3 - R_1 \to R_3 \end{array}$

$\begin{bmatrix} 1 & 3 & 1 & 0 \\ 0 & -2 & -2 & 0 \\ 0 & -5 & -5 & 0 \end{bmatrix}$ $\begin{array}{l} -\frac{1}{2}R_2 \to R_2 \\ \\ -\frac{1}{5}R_3 \to R_3 \end{array}$

$\begin{bmatrix} 1 & 3 & 1 & 0 \\ 0 & 1 & 1 & 0 \\ 0 & 1 & 1 & 0 \end{bmatrix}$ $\begin{array}{l} R_1 - 3R_2 \to R_1 \\ \\ R_3 - R_2 \to R_3 \end{array}$

$\begin{bmatrix} 1 & 0 & -2 & 0 \\ 0 & 1 & 1 & 0 \\ 0 & 0 & 0 & 0 \end{bmatrix}$

R_1: $x - 2z = 0 \Rightarrow x = 2z$
R_2: $y + z = 0 \Rightarrow y = -z$ $\qquad\qquad\qquad$ ★ $(2c, -c, c)$

$\boxed{11}$ $\begin{bmatrix} 2 & 1 & 1 & 0 \\ 1 & -2 & -2 & 0 \\ 1 & 1 & 1 & 0 \end{bmatrix}$ $R_1 \leftrightarrow R_2$

$\begin{bmatrix} 1 & -2 & -2 & 0 \\ 2 & 1 & 1 & 0 \\ 1 & 1 & 1 & 0 \end{bmatrix}$ $\begin{array}{l} R_2 - 2R_1 \to R_2 \\ R_3 - R_1 \to R_3 \end{array}$

$\begin{bmatrix} 1 & -2 & -2 & 0 \\ 0 & 5 & 5 & 0 \\ 0 & 3 & 3 & 0 \end{bmatrix}$ $\begin{array}{l} \frac{1}{5}R_2 \to R_2 \\ \\ \frac{1}{3}R_3 \to R_3 \end{array}$

$\begin{bmatrix} 1 & -2 & -2 & 0 \\ 0 & 1 & 1 & 0 \\ 0 & 1 & 1 & 0 \end{bmatrix}$ $\begin{array}{l} R_1 + 2R_2 \to R_1 \\ \\ R_3 - R_2 \to R_3 \end{array}$

$\begin{bmatrix} 1 & 0 & 0 & 0 \\ 0 & 1 & 1 & 0 \\ 0 & 0 & 0 & 0 \end{bmatrix}$

R_1: $x = 0$

R_2: $y + z = 0 \Rightarrow y = -z$ $\qquad\qquad\qquad$ ★ $(0, -c, c)$

$\boxed{15}$ $\begin{bmatrix} 4 & -2 & 1 & 5 \\ 3 & 1 & -4 & 0 \end{bmatrix}$ $R_1 - R_2 \to R_1$

$\begin{bmatrix} 1 & -3 & 5 & 5 \\ 3 & 1 & -4 & 0 \end{bmatrix}$ $R_2 - 3R_1 \to R_2$

$\begin{bmatrix} 1 & -3 & 5 & 5 \\ 0 & 10 & -19 & -15 \end{bmatrix}$

R_2: $10y - 19z = -15 \Rightarrow y = \frac{19}{10}z - \frac{3}{2}$

R_1: $x - 3y + 5z = 5 \Rightarrow x = 3\left(\frac{19}{10}z - \frac{3}{2}\right) - 5z + 5 = \frac{7}{10}z + \frac{1}{2}$ $\qquad$ ★ $\left(\frac{7}{10}c + \frac{1}{2}, \frac{19}{10}c - \frac{3}{2}, c\right)$

17

$$\begin{bmatrix} 1 & 2 & -1 & -3 & 2 \\ 3 & 1 & -2 & -1 & 6 \\ 1 & 1 & 3 & -2 & -3 \\ -2 & -2 & 3 & 1 & -9 \end{bmatrix} \begin{array}{l} R_2 - 3\,R_1 \to R_2 \\ R_3 - R_1 \to R_3 \\ R_4 + 2\,R_1 \to R_4 \end{array}$$

$$\begin{bmatrix} 1 & 2 & -1 & -3 & 2 \\ 0 & -5 & 1 & 8 & 0 \\ 0 & -1 & 4 & 1 & -5 \\ 0 & 2 & 1 & -5 & -5 \end{bmatrix} \quad -R_3 \longleftrightarrow R_2$$

$$\begin{bmatrix} 1 & 2 & -1 & -3 & 2 \\ 0 & 1 & -4 & -1 & 5 \\ 0 & -5 & 1 & 8 & 0 \\ 0 & 2 & 1 & -5 & -5 \end{bmatrix} \begin{array}{l} R_1 - 2\,R_2 \to R_1 \\ \\ R_3 + 5\,R_2 \to R_3 \\ R_4 - 2\,R_2 \to R_4 \end{array}$$

$$\begin{bmatrix} 1 & 0 & 7 & -1 & -8 \\ 0 & 1 & -4 & -1 & 5 \\ 0 & 0 & -19 & 3 & 25 \\ 0 & 0 & 9 & -3 & -15 \end{bmatrix} \quad -R_3 - 2\,R_4 \to R_3$$

$$\begin{bmatrix} 1 & 0 & 7 & -1 & -8 \\ 0 & 1 & -4 & -1 & 5 \\ 0 & 0 & 1 & 3 & 5 \\ 0 & 0 & 9 & -3 & -15 \end{bmatrix} \begin{array}{l} R_1 - 7\,R_3 \to R_1 \\ R_2 + 4\,R_3 \to R_2 \\ \\ \frac{1}{3}R_4 - 3\,R_3 \to R_4 \end{array}$$

$$\begin{bmatrix} 1 & 0 & 0 & -22 & -43 \\ 0 & 1 & 0 & 11 & 25 \\ 0 & 0 & 1 & 3 & 5 \\ 0 & 0 & 0 & -10 & -20 \end{bmatrix} \quad -\frac{1}{10}R_4 \to R_4$$

$$\begin{bmatrix} 1 & 0 & 0 & -22 & -43 \\ 0 & 1 & 0 & 11 & 25 \\ 0 & 0 & 1 & 3 & 5 \\ 0 & 0 & 0 & 1 & 2 \end{bmatrix} \begin{array}{l} R_1 + 22\,R_4 \to R_1 \\ R_2 - 11\,R_4 \to R_2 \\ R_3 - 3\,R_4 \to R_3 \end{array}$$

$$\begin{bmatrix} 1 & 0 & 0 & 0 & 1 \\ 0 & 1 & 0 & 0 & 3 \\ 0 & 0 & 1 & 0 & -1 \\ 0 & 0 & 0 & 1 & 2 \end{bmatrix} \qquad \bigstar\,(1,\,3,\,-1,\,2)$$

19

$$\begin{bmatrix} 2 & -1 & -2 & 2 & -5 & 2 \\ 1 & 3 & -2 & 1 & -2 & -5 \\ -1 & 4 & 2 & -3 & 8 & -4 \\ 3 & -2 & -4 & 1 & -3 & -3 \\ 4 & -6 & 1 & -2 & 1 & 10 \end{bmatrix} \quad R_1 \leftrightarrow R_2$$

$$\begin{bmatrix} 1 & 3 & -2 & 1 & -2 & -5 \\ 2 & -1 & -2 & 2 & -5 & 2 \\ -1 & 4 & 2 & -3 & 8 & -4 \\ 3 & -2 & -4 & 1 & -3 & -3 \\ 4 & -6 & 1 & -2 & 1 & 10 \end{bmatrix} \begin{array}{l} \\ R_2 - 2\,R_1 \to R_2 \\ R_3 + R_1 \to R_3 \\ R_4 - 3\,R_1 \to R_4 \\ R_5 - 4\,R_1 \to R_5 \end{array}$$

$$
\left[\begin{array}{rrrrrr}
1 & 3 & -2 & 1 & -2 & -5 \\
0 & -7 & 2 & 0 & -1 & 12 \\
0 & 7 & 0 & -2 & 6 & -9 \\
0 & -11 & 2 & -2 & 3 & 12 \\
0 & -18 & 9 & -6 & 9 & 30
\end{array}\right]
\qquad
\begin{array}{l}
3\,R_2 - 2\,R_4 \to R_2 \\[1.5em]
\\[1em]
-\tfrac{1}{3}R_5 \to R_5
\end{array}
$$

$$
\left[\begin{array}{rrrrrr}
1 & 3 & -2 & 1 & -2 & -5 \\
0 & 1 & 2 & 4 & -9 & 12 \\
0 & 7 & 0 & -2 & 6 & -9 \\
0 & -11 & 2 & -2 & 3 & 12 \\
0 & 6 & -3 & 2 & -3 & -10
\end{array}\right]
\qquad
\begin{array}{l}
R_1 - 3\,R_2 \to R_1 \\[1em]
R_3 - 7\,R_2 \to R_3 \\
R_4 + 11\,R_2 \to R_4 \\
R_5 - 6\,R_2 \to R_5
\end{array}
$$

$$
\left[\begin{array}{rrrrrr}
1 & 0 & -8 & -11 & 25 & -41 \\
0 & 1 & 2 & 4 & -9 & 12 \\
0 & 0 & -14 & -30 & 69 & -93 \\
0 & 0 & 24 & 42 & -96 & 144 \\
0 & 0 & -15 & -22 & 51 & -82
\end{array}\right]
\qquad
\begin{array}{l}
\\[2em]
R_3 - R_5 \to R_3 \\
\tfrac{1}{2}R_4 \to R_4
\end{array}
$$

$$
\left[\begin{array}{rrrrrr}
1 & 0 & -8 & -11 & 25 & -41 \\
0 & 1 & 2 & 4 & -9 & 12 \\
0 & 0 & 1 & -8 & 18 & -11 \\
0 & 0 & 12 & 21 & -48 & 72 \\
0 & 0 & -15 & -22 & 51 & -82
\end{array}\right]
\qquad
\begin{array}{l}
R_1 + 8\,R_3 \to R_1 \\
R_2 - 2\,R_3 \to R_2 \\[1em]
R_4 - 12\,R_3 \to R_4 \\
R_5 + 15\,R_3 \to R_5
\end{array}
$$

$$
\left[\begin{array}{rrrrrr}
1 & 0 & 0 & -75 & 169 & -129 \\
0 & 1 & 0 & 20 & -45 & 34 \\
0 & 0 & 1 & -8 & 18 & -11 \\
0 & 0 & 0 & 117 & -264 & 204 \\
0 & 0 & 0 & -142 & 321 & -247
\end{array}\right]
\qquad
\begin{array}{l}
\\[3em]
17\,R_4 + 14\,R_5 \to R_4
\end{array}
$$

$$
\left[\begin{array}{rrrrrr}
1 & 0 & 0 & -75 & 169 & -129 \\
0 & 1 & 0 & 20 & -45 & 34 \\
0 & 0 & 1 & -8 & 18 & -11 \\
0 & 0 & 0 & 1 & 6 & 10 \\
0 & 0 & 0 & -142 & 321 & -247
\end{array}\right]
\qquad
\begin{array}{l}
R_1 + 75\,R_4 \to R_1 \\
R_2 - 20\,R_4 \to R_2 \\
R_3 + 8\,R_4 \to R_3 \\[1em]
R_5 + 142\,R_4 \to R_5
\end{array}
$$

$$
\left[\begin{array}{rrrrrr}
1 & 0 & 0 & 0 & 619 & 621 \\
0 & 1 & 0 & 0 & -165 & -166 \\
0 & 0 & 1 & 0 & 66 & 69 \\
0 & 0 & 0 & 1 & 6 & 10 \\
0 & 0 & 0 & 0 & 1173 & 1173
\end{array}\right]
\qquad
\begin{array}{l}
\\[3em]
\tfrac{1}{1173}R_5 \to R_5
\end{array}
$$

$$
\left[\begin{array}{rrrrrr}
1 & 0 & 0 & 0 & 619 & 621 \\
0 & 1 & 0 & 0 & -165 & -166 \\
0 & 0 & 1 & 0 & 66 & 69 \\
0 & 0 & 0 & 1 & 6 & 10 \\
0 & 0 & 0 & 0 & 1 & 1
\end{array}\right]
\qquad
\begin{array}{l}
R_1 - 619\,R_5 \to R_1 \\
R_2 + 165\,R_5 \to R_2 \\
R_3 - 66\,R_5 \to R_3 \\
R_4 - 6\,R_5 \to R_4
\end{array}
$$

$$
\left[\begin{array}{rrrrrr}
1 & 0 & 0 & 0 & 0 & 2 \\
0 & 1 & 0 & 0 & 0 & -1 \\
0 & 0 & 1 & 0 & 0 & 3 \\
0 & 0 & 0 & 1 & 0 & 4 \\
0 & 0 & 0 & 0 & 1 & 1
\end{array}\right]
\qquad
\bigstar\ (2,\,-1,\,3,\,4,\,1)
$$

23 $\begin{bmatrix} 4 & -3 & 1 \\ 2 & 1 & -7 \\ -1 & 1 & -1 \end{bmatrix}$ $R_1 + 3\,R_3 \rightarrow R_1$

$\begin{bmatrix} 1 & 0 & -2 \\ 2 & 1 & -7 \\ -1 & 1 & -1 \end{bmatrix}$ $R_2 - 2\,R_1 \rightarrow R_2$
$R_3 + R_1 \rightarrow R_3$

$\begin{bmatrix} 1 & 0 & -2 \\ 0 & 1 & -3 \\ 0 & 1 & -3 \end{bmatrix}$ $R_3 - R_2 \rightarrow R_3$

$\begin{bmatrix} 1 & 0 & -2 \\ 0 & 1 & -3 \\ 0 & 0 & 0 \end{bmatrix}$ $R_2:\ y = -3;\ R_1:\ x = -2$ ★ $(-2, -3)$

25 $\begin{bmatrix} 2 & 3 & 5 \\ 1 & -3 & 4 \\ 1 & 1 & -2 \end{bmatrix}$ $R_1 \leftrightarrow R_2$

$\begin{bmatrix} 1 & -3 & 4 \\ 2 & 3 & 5 \\ 1 & 1 & -2 \end{bmatrix}$ $R_2 - 2\,R_1 \rightarrow R_2$
$R_3 - R_1 \rightarrow R_3$

$\begin{bmatrix} 1 & -3 & 4 \\ 0 & 9 & -3 \\ 0 & 4 & -6 \end{bmatrix}$ $R_2 - 2\,R_3 \rightarrow R_2$

$\begin{bmatrix} 1 & -3 & 4 \\ 0 & 1 & 9 \\ 0 & 4 & -6 \end{bmatrix}$ $R_1 + 3\,R_2 \rightarrow R_1$
$R_3 - 4\,R_2 \rightarrow R_3$

$\begin{bmatrix} 1 & 0 & 31 \\ 0 & 1 & 9 \\ 0 & 0 & -42 \end{bmatrix}$

There is a contradiction in row 3, therefore there is *no solution.*

29 Let x, y, and z denote the number of hours needed for A, B, and C, respectively, to produce 1000 items. In one hour, A, B, and C produce $\frac{1000}{x}$, $\frac{1000}{y}$, and $\frac{1000}{z}$ items, respectively. In 6 hours, A and B produce $\frac{6000}{x}$ and $\frac{6000}{y}$ items. From the table, this sum must equal 4500. The system of equations is then:

$$\begin{cases} \frac{6000}{x} + \frac{6000}{y} & = 4500 \\ \frac{8000}{x} + \frac{8000}{z} & = 3600 \\ \frac{7000}{y} + \frac{7000}{z} & = 4900 \end{cases}$$

To simplify, let $a = 1/x$, $b = 1/y$, $c = 1/z$ and divide each equation by its greatest common factor $\{1500, 400, \text{and } 700\}$.

$$\begin{cases} 4a + \ 4b \ \ \ \ = 3 & (\text{E}_1) \\ 20a + \ \ \ \ + 20c = 9 & (\text{E}_2) \\ 10b + 10c = 7 & (\text{E}_3) \end{cases}$$

$\text{E}_2 - 5\,\text{E}_1 \Rightarrow 20c - 20b = -6$ (E_4)

$\text{E}_4 + 2\,\text{E}_3 \Rightarrow 40c = 8 \Rightarrow c = \frac{1}{5}$; $b = \frac{1}{2}$; $a = \frac{1}{4}$. Resubstituting, $x = 4$, $y = 2$, and $z = 5$.

31 Let x, y, and z denote the amounts of G_1, G_2, and G_3, respectively.

$$\begin{cases} x + y + z = 600 & quantity & (\text{E}_1) \\ 0.30x + 0.20y + 0.15z = (0.25)(600) & quality & (\text{E}_2) \\ z = 100 + y & constraint & (\text{E}_3) \end{cases}$$

Substitute $z = 100 + y$ into E_1 and $100\,\text{E}_2$ to obtain

$$\begin{cases} x + \ 2y = \ \ \ \ \ 500 & (\text{E}_4) \\ 30x + 35y = \ 13{,}500 & (\text{E}_5) \end{cases}$$

$$\text{E}_5 - 30\,\text{E}_4 \Rightarrow -25y = -1500 \Rightarrow y = 60; \ z = 160; \ x = 380.$$

33 (a) $I_1 - I_2 + I_3 = 0$ $\qquad I_1 = I_2 - I_3 \quad (\text{E}_1)$

$ 3I_1 + 3I_2 = 6 \qquad \Rightarrow \qquad I_1 + I_2 = 2 \quad (\text{E}_2)$

$ 3I_2 + 3I_3 = 12 \qquad\qquad I_2 + I_3 = 4 \quad (\text{E}_3)$

Substitute I_1 in E_1 into E_2 to obtain $2I_2 - I_3 = 2$ (E_4).

$$\text{E}_4 + \text{E}_3 \Rightarrow 3I_2 = 6 \Rightarrow I_2 = 2; \ I_3 = 2; \ I_1 = 0.$$

(b) $I_1 = I_2 - I_3 \qquad (\text{E}_1)$

$ 4I_1 + I_2 = 6 \qquad (\text{E}_2)$

$ I_2 + 4I_3 = 12 \qquad (\text{E}_3)$

Substitute I_1 in E_1 into E_2 to obtain $5I_2 - 4I_3 = 6$ (E_4).

$$\text{E}_4 + \text{E}_3 \Rightarrow 6I_2 = 18 \Rightarrow I_2 = 3; \ I_3 = \frac{9}{4}; \ I_1 = \frac{3}{4}.$$

35 Let x, y, and z denote the amount of Columbian, Brazilian, and Kenyan coffee used, respectively.

$$\begin{cases} x + y + z = 1 & \text{\textit{quantity}} & (\text{E}_1) \\ 10x + 6y + 8z = (8.50)(1) & \text{\textit{quality}} & (\text{E}_2) \\ x = 3y & \text{\textit{constraint}} & (\text{E}_3) \end{cases}$$

Substitute $x = 3y$ into E_1 and E_2 to obtain

$$\begin{cases} 4y + \quad z = \quad 1 & (\text{E}_4) \\ 36y + 8z = 8.5 & (\text{E}_5) \end{cases} \quad \text{E}_5 - 8\,\text{E}_4 \Rightarrow 4y = \tfrac{1}{2} \Rightarrow y = \tfrac{1}{8};\ z = \tfrac{1}{2};\ x = \tfrac{3}{8}.$$

37 (a) A: $x_1 + x_4 = 50 + 25 = 75$, B: $x_1 + x_2 = 100 + 50 = 150$,

   C: $x_2 + x_3 = 150 + 75 = 225$, D: $x_3 + x_4 = 100 + 50 = 150$

(b) From C, $x_3 = 100 \Rightarrow x_2 = 225 - 100 = 125$.

   From D, $x_3 = 100 \Rightarrow x_4 = 150 - 100 = 50$.

   From A, $x_4 = 50 \Rightarrow x_1 = 75 - 50 = 25$.

(c) From D in part (a), $x_3 = 150 - x_4 \Rightarrow x_3 \le 150$ since $x_4 \ge 0$. From C,

   $x_3 = 225 - x_2 = 225 - (150 - x_1)\ \{\text{from B}\} = 75 + x_1 \Rightarrow x_3 \ge 75$ since $x_1 \ge 0$.

39 $t = 2070 - 1990 = 80$. $rt = (0.025)(80) = 2$ and $A = 800$ for E_1,

$(0.015)(80) = 1.2$ and $A = 560$ for E_2, and $(0.01)(0) = 0$ and $A = 340$ for E_3.

Substituting into $A = a + ct + ke^{rt}$ and summarizing as a system, we have:

$$\begin{cases} a + 80c + e^2 k = 800 & (\text{E}_1) \\ a + 80c + e^{1.2}k = 560 & (\text{E}_2) \\ a \quad + \quad k = 340 & (\text{E}_3) \end{cases}$$

We want to find t when A has doubled—that is, $A = 2(340)$. First we find c and k.

$\text{E}_1 - \text{E}_2 \Rightarrow e^2 k - e^{1.2}k = 240 \Rightarrow k = \dfrac{240}{e^2 - e^{1.2}} \approx 58.98$.

Substituting into E_3 gives $a = 340 - k \approx 281.02$.

Substituting into E_1 gives $c = \dfrac{800 - a - e^2 k}{80} \approx \dfrac{800 - 281.02 - (58.98)e^2}{80} \approx 1.04$.

Thus, $A = 281.02 + 1.04t + 58.98e^{rt}$. If $A = 680$ and $r = 0.01$, then

$680 = 281.02 + 1.04t + 58.98e^{0.01t} \Rightarrow 1.04t + 58.98e^{0.01t} - 398.98 = 0$. Graphing

$y = 1.04t + 58.98e^{0.01t} - 398.98$, we see there is an x-intercept at $x \approx 144.08$.

$1990 + 144.08 = 2134.08$, or during the year 2134.

$[0, 1050]$ by $[-350, 350]$

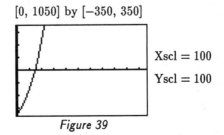

Xscl $= 100$

Yscl $= 100$

Figure 39

41 The parabola has an equation of the form $y = ax^2 + bx + c$.

Substituting the x and y values of P, Q, and R into this equation yields:

$$\begin{cases} 4a + 2b + c = 5 & P \quad (E_1) \\ 4a - 2b + c = -3 & Q \quad (E_2) \\ a + b + c = 6 & R \quad (E_3) \end{cases}$$

Solving E_3 for c $\{c = 6 - a - b\}$ and substituting into E_1 and E_2 yields:

$$\begin{cases} 3a + b = -1 & (E_4) \\ 3a - 3b = -9 & (E_5) \end{cases}$$

$E_4 - E_5 \Rightarrow 4b = 8 \Rightarrow b = 2$; $a = -1$; $c = 5$. The equation is $y = -x^2 + 2x + 5$.

45 The circle has an equation of the form $x^2 + y^2 + ax + by + c = 0$.

Substituting the x and y values of P, Q, and R into this equation yields:

$$\begin{cases} 2a + b + c = -5 & P \quad (E_1) \\ -a - 4b + c = -17 & Q \quad (E_2) \\ 3a + c = -9 & R \quad (E_3) \end{cases}$$

Solving E_3 for c $\{c = -9 - 3a\}$ and substituting into E_1 and E_2 yields:

$$\begin{cases} -a + b = 4 & (E_4) \\ -4a - 4b = -8 & (E_5) \end{cases} \Rightarrow \begin{cases} -a + b = 4 & (E_6) \\ a + b = 2 & (E_7) \end{cases}$$

$E_6 + E_7 \Rightarrow 2b = 6 \Rightarrow b = 3$; $a = -1$; $c = -6$. The equation is $x^2 + y^2 - x + 3y - 6 = 0$.

47
$$\begin{cases} -a + b - c + d = 2 & (-1, 2) \\ 0.125a + 0.25b + 0.5c + d = 2 & (0.5, 2) \\ a + b + c + d = 3 & (1, 3) \\ 8a + 4b + 2c + d = 4.5 & (2, 4.5) \end{cases}$$

Solving the system yields $a = -\frac{4}{9}$, $b = \frac{11}{9}$, $c = \frac{17}{18}$, and $d = \frac{23}{18}$.

9.4 Exercises

Note: The general outline for the solutions in this section is as follows:

1st line)   The expression is shown on the left side of the equation and its decomposition is on the right side.

2nd line)   The equation in the first line is multiplied by its least common denominator and left in factored form.

3rd line and beyond)   Values are substituted into the equation in the second line and the coefficients are found by solving the resulting equations. It will be stated when the method of equating coefficients is used.

3 $\dfrac{x + 34}{(x - 6)(x + 2)} = \dfrac{A}{x - 6} + \dfrac{B}{x + 2}$

$x + 34 = A(x + 2) + B(x - 6)$

$x = -2$: $32 = -8B \Rightarrow B = -4$

$x = 6$: $40 = 8A \Rightarrow A = 5$

$\bigstar \ \dfrac{5}{x - 6} - \dfrac{4}{x + 2}$

$\boxed{7}$ $\dfrac{4x^2 - 5x - 15}{x(x-5)(x+1)} = \dfrac{A}{x} + \dfrac{B}{x-5} + \dfrac{C}{x+1}$

$4x^2 - 5x - 15 = A(x-5)(x+1) + Bx(x+1) + Cx(x-5)$

$x = -1: -6 = 6C \Rightarrow C = -1$

$x = 0: -15 = -5A \Rightarrow A = 3$ $\qquad\qquad\qquad \bigstar\; \dfrac{3}{x} + \dfrac{2}{x-5} - \dfrac{1}{x+1}$

$x = 5: 60 = 30B \Rightarrow B = 2$

$\boxed{11}$ $\dfrac{19x^2 + 50x - 25}{x^2(3x-5)} = \dfrac{A}{x} + \dfrac{B}{x^2} + \dfrac{C}{3x-5}$

$19x^2 + 50x - 25 = Ax(3x-5) + B(3x-5) + Cx^2$

$x = \frac{5}{3}: \dfrac{1000}{9} = \dfrac{25}{9}C \Rightarrow C = 40$

$x = 0: -25 = -5B \Rightarrow B = 5$ $\qquad\qquad\qquad \bigstar\; -\dfrac{7}{x} + \dfrac{5}{x^2} + \dfrac{40}{3x-5}$

$x = 1: 44 = -2A - 2B + C \Rightarrow A = -7$

$\boxed{15}$ $\dfrac{3x^3 + 11x^2 + 16x + 5}{x(x+1)^3} = \dfrac{A}{x} + \dfrac{B}{x+1} + \dfrac{C}{(x+1)^2} + \dfrac{D}{(x+1)^3}$

$3x^3 + 11x^2 + 16x + 5 = A(x+1)^3 + Bx(x+1)^2 + Cx(x+1) + Dx$

$x = -1: -3 = -D \Rightarrow D = 3$

$x = 0: 5 = A$

$x = 1: 35 = 8A + 4B + 2C + D \quad (\text{E}_1)$

$x = -2: -7 = -A - 2B + 2C - 2D \quad (\text{E}_2)$

Substituting the values for A and D into E_1 and E_2 yields

$\begin{cases} 4B + 2C &= -8 \\ -2B + 2C &= 4 \end{cases} \Rightarrow \begin{cases} 2B + C &= -4 \quad (\text{E}_3) \\ -B + C &= 2 \quad (\text{E}_4) \end{cases}$ $\qquad \bigstar\; \dfrac{5}{x} - \dfrac{2}{x+1} + \dfrac{3}{(x+1)^3}$

$\text{E}_3 + 2\,\text{E}_4 \Rightarrow 3C = 0 \Rightarrow C = 0;\; B = -2$

$\boxed{19}$ $\dfrac{9x^2 - 3x + 8}{x(x^2+2)} = \dfrac{A}{x} + \dfrac{Bx + C}{x^2+2}$

$9x^2 - 3x + 8 = A(x^2+2) + (Bx+C)x$

$x = 0: 8 = 2A \Rightarrow A = 4$

$x = 1: 14 = 3A + B + C \quad (\text{E}_1)$

$x = -1: 20 = 3A + B - C \quad (\text{E}_2)$ $\qquad\qquad\qquad \bigstar\; \dfrac{4}{x} + \dfrac{5x - 3}{x^2+2}$

$\text{E}_1 - \text{E}_2 \Rightarrow -6 = 2C \Rightarrow C = -3;\; B = 5$

[21] $\dfrac{4x^3 - x^2 + 4x + 2}{(x^2+1)^2} = \dfrac{Ax+B}{x^2+1} + \dfrac{Cx+D}{(x^2+1)^2}$

$$4x^3 - x^2 + 4x + 2 = (Ax+B)(x^2+1) + Cx + D$$
$$= Ax^3 + Bx^2 + (A+C)x + (B+D)$$

Equating coefficients, i.e., the coefficient of x^3 on the left side must equal the

coefficient of x^3 on the right side, we have the following:

x^3 $: A = 4$

x^2 $: B = -1$

x $: A + C = 4 \Rightarrow C = 0$ $\star\ \dfrac{4x-1}{x^2+1} + \dfrac{3}{(x^2+1)^2}$

constant $: B + D = 2 \Rightarrow D = 3$

[23] The degree of the numerator is not lower than the degree of the denominator.

Thus, we must first use long division and then decompose the remaining expression.

Hence, by first dividing and then factoring, we have the following:

$$2x + \dfrac{4x^2 - 3x + 1}{(x^2+1)(x-1)} = 2x + \dfrac{Ax+B}{x^2+1} + \dfrac{C}{x-1}$$

$$4x^2 - 3x + 1 = (Ax+B)(x-1) + C(x^2+1)$$

$x = 1$: $2 = 2C \Rightarrow C = 1$

$x = 0$: $1 = -B + C \Rightarrow B = 0$ $\star\ 2x + \dfrac{1}{x-1} + \dfrac{3x}{x^2+1}$

$x = -1$: $8 = 2A - 2B + 2C \Rightarrow A = 3$

[25] By first dividing and then factoring, we have the following:

$$3 + \dfrac{12x - 16}{x(x-4)} = 3 + \dfrac{A}{x} + \dfrac{B}{x-4}$$

$$12x - 16 = A(x-4) + Bx$$

$x = 0$: $-16 = -4A \Rightarrow A = 4$

      $\star\ 3 + \dfrac{4}{x} + \dfrac{8}{x-4}$

$x = 4$: $32 = 4B \Rightarrow B = 8$

[27] By first dividing and then factoring, we have the following:

$$2x + 3 + \dfrac{x+5}{(2x+1)(x-1)} = 2x + 3 + \dfrac{A}{2x+1} + \dfrac{B}{x-1}$$

$$x + 5 = A(x-1) + B(2x+1)$$

$x = 1$: $6 = 3B \Rightarrow B = 2$

$x = -\frac{1}{2}$: $\frac{9}{2} = -\frac{3}{2}A \Rightarrow A = -3$ $\star\ 2x + 3 + \dfrac{2}{x-1} - \dfrac{3}{2x+1}$

9.5 Exercises

1 $3x - 2y < 6 \Leftrightarrow y > \frac{3}{2}x - 3$. Sketch the graph of $y = \frac{3}{2}x - 3$ with dashes. The point $(0, 0)$ is clearly on one side of the line, so we will substitute $x = 0$ and $y = 0$ into the inequality $y > \frac{3}{2}x - 3$. Checking, we have $0 > -3$, a true statement. Hence, we shade *all* points that are on the same side of the line as $(0, 0)$.

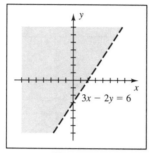

Figure 1

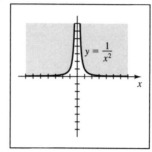

Figure 9

9 $yx^2 \geq 1 \Leftrightarrow y \geq 1/x^2 \ \{x \neq 0\}$. Sketch the graph of $y = 1/x^2$ with a solid curve. The point $(1, 0)$ is clearly not on the graph. Substituting $x = 1$ and $y = 0$ into $y \geq 1/x^2$ yields $0 \geq 1$, a false statement. Hence, we do not shade the region containing $(1, 0)$, but we do shade the regions in the first and second quadrants that are above the graph.

Note: The will use the notation $V @ (a, b), (c, d), \ldots$ to denote the intersection point(s) of the solution region of the graph. These can be found by solving the system of *equalities* that correspond to the given system of inequalities.

11 $\begin{cases} 3x + y < 3 \\ 4 - y < 2x \end{cases} \Leftrightarrow \begin{cases} y < -3x + 3 \\ y > -2x + 4 \end{cases}$ $V @ (-1, 6)$

Solving the system $y = -3x + 3$ and $y = -2x + 4$ gives us the solution $(-1, 6)$. Testing the point $(0, 0)$ in both inequalities, we see that we need to shade under $y = -3x + 3$ and above $y = -2x + 4$, as shown in *Figure 11*.

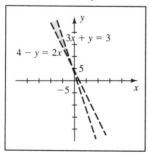

Figure 11

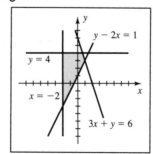

Figure 15

$\boxed{15}$ $\begin{cases} 3x + y \le 6 \\ y - 2x \ge 1 \\ x \ge -2 \\ y \le 4 \end{cases}$ $\Leftrightarrow$ $\begin{cases} y \le -3x + 6 \\ y \ge 2x + 1 \\ x \ge -2 \\ y \le 4 \end{cases}$ $V @ (-2, -3), (-2, 4), (\frac{2}{3}, 4), (1, 3)$

$\boxed{21}$ $|x + 2| \le 1 \Leftrightarrow -1 \le x + 2 \le 1 \Leftrightarrow -3 \le x \le -1$.

This region is bounded by the vertical lines $x = -3$ and $x = -1$, including the lines.

$|y - 3| < 5 \Leftrightarrow -5 < y - 3 < 5 \Leftrightarrow -2 < y < 8$.

This region is bounded by the horizontal lines $y = -2$ and $y = 8$, excluding the lines.

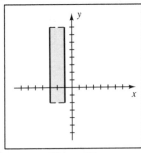

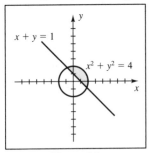

Figure 21   Figure 23

$\boxed{23}$ $\begin{cases} x^2 + y^2 \le 4 \\ x + y \ge 1 \end{cases}$ $\Leftrightarrow$ $\begin{cases} x^2 + y^2 \le 2^2 \\ y \ge -x + 1 \end{cases}$ $V @ (\frac{1}{2} \mp \frac{1}{2}\sqrt{7}, \frac{1}{2} \pm \frac{1}{2}\sqrt{7})$

$\boxed{27}$ The shaded region lies between $x = 0$ and $x = 3$, including $x = 0$ and excluding $x = 3$. This region is described by $0 \le x < 3$.

The dashed line goes through $(0, 4)$ and $(4, 0)$. Hence, its slope is -1, its y-intercept is 4, and an equation for it is $y = -x + 4$. Substituting $x = 0$ and $y = 0$ into $y = -x + 4$ gives us $0 = 4$. We replace $=$ by $<$ {the line is dashed} to make this statement a true statement. Hence, we have the inequality $y < -x + 4$.

The solid line goes through $(0, -4)$ and $(4, 0)$. Hence, its slope is 1, its y-intercept is -4, and an equation for it is $y = x - 4$. Substituting $x = 0$ and $y = 0$ into $y = x - 4$ gives us $0 = -4$. We replace $=$ by $\ge$ {the line is solid} to make this statement a true statement. Hence, we have the inequality $y \ge x - 4$. Thus, the graph may be described by the system $\begin{cases} 0 \le x < 3 \\ y < -x + 4 \\ y \ge x - 4 \end{cases}$

$\boxed{31}$ The center of the circle is $(2, 2)$ and it passes through $(0, 0)$. The radius is the distance from $(0, 0)$ to $(2, 2)$, which is $\sqrt{8}$. An equation of the circle is $(x - 2)^2 + (y - 2)^2 = 8$. The center is clearly inside the circle, and if we substitute $x = 2$ and $y = 2$ in the equation of the circle, we get $0 = 8$. Since we want to make this a true statement and include the boundary, we use $(x - 2)^2 + (y - 2)^2 \le 8$ to describe the shaded circular region and its boundary.

For the dashed line $y = x$, test the point $(1, 0)$. This test gives us $0 = 1$, which we can make true by replacing $=$ by $<$. Hence $y < x$ describes the shaded region under the line $y = x$. The other inequality may be found as in Exercise 27. Thus,

the graph may be described by the system $\begin{cases} y < x \\ y \le -x + 4 \\ (x-2)^2 + (y-2)^2 \le 8 \end{cases}$

33 For the dashed line through $(-4, 0)$ and $(4, 1)$, we can use the point-slope form to find an equation of the line. $y - 0 = \frac{1-0}{4-(-4)}(x - (-4)) \Leftrightarrow y = \frac{1}{8}x + \frac{1}{2}$. Checking the point $(0, 1)$ {which is in the shaded region} gives us $1 = \frac{1}{2}$, so we change $=$ to $>$ to obtain $y > \frac{1}{8}x + \frac{1}{2}$.

Similarly, for the solid line with x-intercept -4 and y-intercept 4, the shaded region can be described by $y \le x + 4$.

Lastly, the solid line passing through $(4, 1)$ has slope $-\frac{3}{4}$, y-intercept 4, and its shaded region can be described by $y \le -\frac{3}{4}x + 4$. Thus, the system is $\begin{cases} y > \frac{1}{8}x + \frac{1}{2} \\ y \le x + 4 \\ y \le -\frac{3}{4}x + 4 \end{cases}$

35 Let x and y denote the number of sets of brand A and brand B, respectively. "Necessary to stock at least twice as many sets of brand A as of brand B" may be symbolized as $x \ge 2y$. "Necessary to have on hand at least 10 sets of brand B" may be symbolized as $y \ge 10$. Consequently, $x \ge 20$ from the first inequality. "Room for not more than 100 sets in the store" may be symbolized as $x + y \le 100$. We now sketch the system of inequalities and shade the region that they have in common. The graph is the region bounded by the triangle with vertices $(20, 10)$, $(90, 10)$, and $\left(\frac{200}{3}, \frac{100}{3}\right)$.

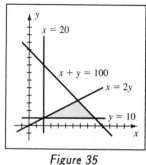

Figure 35

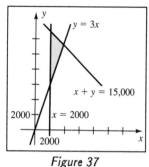

Figure 37

37 If x and y denote the amount placed in the high-risk and low-risk investment, respectively, then a system is $x \ge 2000$, $y \ge 3x$, $x + y \le 15{,}000$. The graph is the region bounded by the triangle with vertices $(2000, 6000)$, $(2000, 13{,}000)$, and $(3750, 11{,}250)$.

39 A system is $x + y \leq 9$, $y \geq x$, $x \geq 1$. To justify the condition $y \geq x$, start with

$$\frac{\text{cylinder volume}}{\text{total volume}} \geq 0.75 \Rightarrow \frac{\pi r^2 y}{\pi r^2 y + \frac{1}{3}\pi r^2 x} \geq \frac{3}{4} \Rightarrow 4\pi r^2 y \geq 3\pi r^2 y + \pi r^2 x \Rightarrow$$

$\pi r^2 y \geq \pi r^2 x \Rightarrow y \geq x$. The graph is the region bounded by the triangle with

vertices $(1, 1)$, $(1, 8)$, and $(\frac{9}{2}, \frac{9}{2})$.

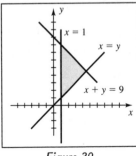

Figure 39

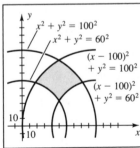

Figure 41

41   If the plant is located at (x, y), then a system is $(60)^2 \leq x^2 + y^2 \leq (100)^2$,

$(60)^2 \leq (x - 100)^2 + y^2 \leq (100)^2$, $y \geq 0$. The graph is the region in the first quadrant

that lies between the two concentric circles with center $(0, 0)$ and radii 60 and 100,

and also between the two concentric circles with center $(100, 0)$ and radii 60 and 100.

  Equating the different circle equations, we obtain the vertices of the solution

region $(50, 50\sqrt{3})$, $(50, 10\sqrt{11})$, $(18, 6\sqrt{91})$, and $(82, 6\sqrt{91})$. For example, to

determine where the circle $x^2 + y^2 = 100^2$ intersects the circle $(x - 100)^2 + y^2 = 60^2$,

we   subtract   the   first   equation   from   the   second   to   obtain

$[(x - 100)^2 + y^2] - (x^2 + y^2) = 60^2 - 100^2 \Rightarrow -200x + 10,000 = -6400 \Rightarrow$

$200x = 16,400 \Rightarrow x = 82$.   Substituting $x = 82$ into $x^2 + y^2 = 100^2$ gives   us

$y^2 = 100^2 - 82^2 = 3276 \Rightarrow y = \pm\sqrt{3276} = \pm6\sqrt{91} \approx \pm57.2$.

43 $64y^3 - x^3 \leq e^{1 - 2x} \Rightarrow y \leq \frac{1}{4}(e^{1 - 2x} + x^3)^{1/3}$ Graph $y = \frac{1}{4}(e^{1 - 2x} + x^3)^{1/3}$ $\{Y_1\}$.

The solution includes the graph and the region below the graph.

Shading was obtained by using the command Shade(Ymin, Y_1, Xres, Xmin, Xmax).

$[-3.5, 4]$ by $[-1, 4]$

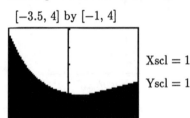

Figure 43

Xscl $= 1$

Yscl $= 1$

$[-1.5, 1.5]$ by $[-1, 1]$

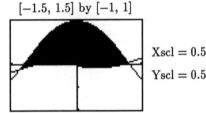

Figure 45

Xscl $= 0.5$

Yscl $= 0.5$

$\boxed{45}$ $5^{1-y} \geq x^4 + x^2 + 1 \Rightarrow 1 - y \geq \log_5(x^4 + x^2 + 1) \Rightarrow y \leq 1 - \log_5(x^4 + x^2 + 1)$.

$x + 3y \geq x^{5/3} \Rightarrow y \geq \frac{1}{3}(x^{5/3} - x)$.   Graph $y = 1 - \log_5(x^4 + x^2 + 1)$ $\{Y_1\}$ and

$y = \frac{1}{3}(x^{5/3} - x)$ $\{Y_2\}$. The graphs intersect at approximately $(1.21, 0.05)$ and

$(-1.32, -0.09)$. The solution is located between the points of intersection. It

includes the graphs and the region below the first graph and above the second graph.

Shading was obtained by using the command Shade(Y_2, Y_1, Xres, -1.32, 1.21). See

Figure 45.

9.6 Exercises

$\boxed{1}$ We substitute the x and y values of each vertex into $C = 3x + 2y + 5$ and summarize

the results in the following table. We see that there is a maximum of 27 at $(6, 2)$,

and a minimum of 9 at $(0, 2)$.

(x, y)	(0, 2)	(0, 4)	(3, 5)	(6, 2)	(5, 0)	(2, 0)
C	9 ■	13	24	27 ■	20	11

$\boxed{3}$ The region R is sketched in *Figure 3.* The vertices are found by obtaining the

intersection points of the system of equations corresponding to the given inequalities.

(x, y)	(0, 0)	(0, 3)	(4, 6)	(6, 3)	(5, 0)
C	0	3	18	21 ■	15

$C = 3x + y$;

maximum of 21 at $(6, 3)$

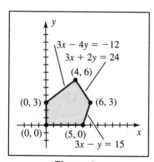

Figure 3

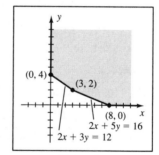

Figure 5

$\boxed{5}$

(x, y)	(8, 0)	(3, 2)	(0, 4)
C	24	21 ■	24

$C = 3x + 6y$;

minimum of 21 at $(3, 2)$

7

(x, y)	$(0, 0)$	$(0, 4)$	$(2, 5)$	$(6, 3)$	$(8, 0)$
C	0	16	24 ■	24 ■	16

As in Example 2, $C = 2x + 4y$ has the maximum value 24 for *any* point on the line segment from $(2, 5)$ to $(6, 3)$. To show that the last statement is true, we will substitute $6 - \frac{1}{2}x$ for y in C. $C = 2x + 4y = 2x + 4(6 - \frac{1}{2}x) = 2x + 24 - 2x = 24$.

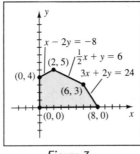

Figure 7

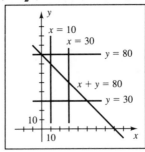

Figure 9

Note: For the linear programming application exercises, the solution outline is as follows:

(1) Define the variables used in the problem.

(2) Define the function to be maximized or minimized.

(3) List the system of inequalities that determine the solution region R.

(4) A table of intersection points and values of the function in (2) at those points is given. The maximum or minimum value is denoted by a ■.

(5) A summarizing statement is given.

(6) A figure is shown for the system of inequalities.

9 Let x and y denote the number of oversized and standard rackets, respectively.

Profit function: $P = 15x + 8y$

$$\begin{cases} 30 \le y \le 80 \\ 10 \le x \le 30 \\ x + y \le 80 \end{cases}$$

(x, y)	$(10, 30)$	$(30, 30)$	$(30, 50)$	$(10, 70)$
P	390	690	850 ■	710

The maximum profit of \$850 per day occurs when 30 oversized rackets and 50 standard rackets are manufactured.

11 Let x and y denote the number of pounds of S and T, respectively.

Cost function: $C = 3x + 4y$

(x, y)	$(0, 4.5)$	$(3.5, 1)$	$(5, 0)$
C	18	14.5 ■	15

$$\begin{cases} 2x + 2y \geq 9 & \text{amount of I} \\ 4x + 6y \geq 20 & \text{amount of G} \\ x \geq 0 \\ y \geq 0 \end{cases}$$

The minimum cost of $14.50 occurs when 3.5 pounds of S and 1 pound of T are used.

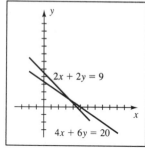

Figure 11

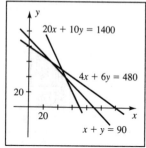

Figure 15

15 Let x and y denote the number of acres planted with alfalfa and corn, respectively.

Profit function: $P = 110x - 4x - 20x + 150y - 6y - 10y = \underline{86x + 134y}$

$$\begin{cases} 4x + 6y \leq 480 & \text{seed cost} \\ 20x + 10y \leq 1400 & \text{labor cost} \\ x + y \leq 90 & \text{area} \\ x, y \geq 0 \end{cases}$$

(x, y)	$(70, 0)$	$(50, 40)$	$(30, 60)$	$(0, 80)$	$(0, 0)$
P	6020	9660	10,620	10,720 ■	0

The maximum profit of $10,720 occurs when

0 acres of alfalfa are planted and 80 acres of corn are planted.

$\boxed{17}$ Let x, y, and z denote the number of ounces of X, Y, and Z, respectively.

Cost function: $C = 0.25x + 0.35y + 0.50z$

$$= 0.25x + 0.35y + 0.50(20 - x - y) = \underline{10 - 0.25x - 0.15y}$$

$$\left\{ \begin{array}{l} 0.20x + 0.20y + 0.10z \geq 0.14(20) \quad \text{amount of A} \\ 0.10x + 0.40y + 0.20z \geq 0.16(20) \quad \text{amount of B} \\ 0.25x + 0.15y + 0.25z \geq 0.20(20) \quad \text{amount of C} \end{array} \right. \Rightarrow \left\{ \begin{array}{r} x + y \geq 8 \\ x - 2y \leq 8 \\ y \leq 10 \\ x + y \leq 20 \\ 0 \leq x,\, y \leq 20 \end{array} \right.$$

The new restrictions are found by substituting $z = 20 - x - y$ into the 3 inequalities, simplifying, and adding the last 2 inequalities.

(x, y)	(8, 0)	(16, 4)	(10, 10)	(0, 10)	(0, 8)
C	8.00	5.40 ■	6.00	8.50	8.80 ■■

The minimum cost of $5.40 requires 16 oz of X, 4 oz of Y, and 0 oz of Z.

The maximum cost of $8.80 requires 0 oz of X, 8 oz of Y, and 12 oz of Z.

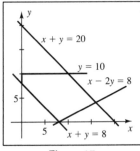

Figure 17

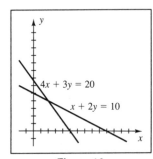

Figure 19

$\boxed{19}$ Let x and y denote the number of vans and buses purchased, respectively.

$$\left\{ \begin{array}{l} 10{,}000x + 20{,}000y \leq 100{,}000 \quad \text{purchase} \\ 100x + 75y \leq 500 \quad\quad\quad \text{maintenance} \\ x \geq 0 \\ y \geq 0 \end{array} \right. \Rightarrow \left\{ \begin{array}{r} x + 2y \leq 10 \\ 4x + 3y \leq 20 \\ x \geq 0 \\ y \geq 0 \end{array} \right.$$

(x, y)	(5, 0)	(2, 4)	(0, 5)	(0, 0)
P	75	130 ■	125	0

Passenger capacity function: $P = 15x + 25y$

The maximum passenger capacity of 130 would occur if the community purchases 2 vans and 4 buses.

21 Let x and y denote the number of trout and bass, respectively.

Pound function: $P = 3x + 4y$

$$\begin{cases} x + y \le 5000 & \textit{number of fish} \\ 0.50x + 0.75y \le 3000 & \textit{cost} \\ x \ge 0 \\ y \ge 0 \end{cases} \Rightarrow \begin{cases} x + y \le 5000 \\ 2x + 3y \le 12{,}000 \\ x \ge 0 \\ y \ge 0 \end{cases}$$

(x, y)	(5000, 0)	(3000, 2000)	(0, 4000)	(0, 0)
P	15,000	17,000 ■	16,000	0

The total number of pounds of fish will be a maximum of 17,000 if 3000 trout and

2000 bass are purchased.

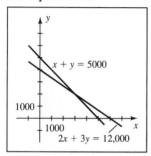

Figure 21

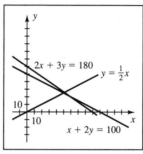

Figure 23

23 Let x and y denote the number of basic and deluxe units constructed, respectively.

$$\begin{cases} 300x + 600y \le 30{,}000 & \textit{cost} \\ x \ge 2y & \textit{ratio} \\ 80x + 120y \le 7200 & \textit{area} \\ x, y \ge 0 \end{cases} \Rightarrow \begin{cases} x + 2y \le 100 \\ y \le \frac{1}{2}x \\ 2x + 3y \le 180 \\ x, y \ge 0 \end{cases}$$

(x, y)	(90, 0)	(60, 20)	(50, 25)	(0, 0)
R	3600	3900 ■	3875	0

Revenue function: $R = 40x + 75y$

The maximum monthly revenue of $3900 occurs if 60 basic units and 20 deluxe units

are constructed.

9.7 Exercises

Note: For Exercises 1–8, $(A + B)$ and $(A - B)$ are possible only if A and B have the

same number of rows and columns. {#7 & #8 are not possible} $(2A)$ and $(-3B)$

are always possible by simply multiplying the elements of A and B by 2 and -3,

respectively.

1 $A + B = \begin{bmatrix} 5 & -2 \\ 1 & 3 \end{bmatrix} + \begin{bmatrix} 4 & 1 \\ -3 & 2 \end{bmatrix} = \begin{bmatrix} 5 + 4 & -2 + 1 \\ 1 + (-3) & 3 + 2 \end{bmatrix} = \begin{bmatrix} 9 & -1 \\ -2 & 5 \end{bmatrix}$

$A - B = \begin{bmatrix} 5 & -2 \\ 1 & 3 \end{bmatrix} - \begin{bmatrix} 4 & 1 \\ -3 & 2 \end{bmatrix} = \begin{bmatrix} 5 - 4 & -2 - 1 \\ 1 - (-3) & 3 - 2 \end{bmatrix} = \begin{bmatrix} 1 & -3 \\ 4 & 1 \end{bmatrix}$

$$2A = 2\begin{bmatrix} 5 & -2 \\ 1 & 3 \end{bmatrix} = \begin{bmatrix} 2(5) & 2(-2) \\ 2(1) & 2(3) \end{bmatrix} = \begin{bmatrix} 10 & -4 \\ 2 & 6 \end{bmatrix}$$

$$-3B = -3\begin{bmatrix} 4 & 1 \\ -3 & 2 \end{bmatrix} = \begin{bmatrix} -3(4) & -3(1) \\ -3(-3) & -3(2) \end{bmatrix} = \begin{bmatrix} -12 & -3 \\ 9 & -6 \end{bmatrix}$$

7 $A + B$ and $A - B$ are not possible since A and B are different sizes.

$$2A = 2\begin{bmatrix} 3 & -2 & 2 \\ 0 & 1 & -4 \\ -3 & 2 & -1 \end{bmatrix} = \begin{bmatrix} 6 & -4 & 4 \\ 0 & 2 & -8 \\ -6 & 4 & -2 \end{bmatrix}, -3B = -3\begin{bmatrix} 4 & 0 \\ 2 & -1 \\ -1 & 3 \end{bmatrix} = \begin{bmatrix} -12 & 0 \\ -6 & 3 \\ 3 & -9 \end{bmatrix}$$

Note: For Exercises 9–20, AB is possible only if the number of columns of A equal the number of rows of B. $\{BA$ is not possible in #19, AB is not possible in #20$\}$ Multiplying an $\boxed{m} \times n$ matrix with an $n \times \boxed{k}$ matrix will result in an $m \times k$ matrix.

9 $AB = \begin{bmatrix} 2 & 6 \\ 3 & -4 \end{bmatrix}\begin{bmatrix} 5 & -2 \\ 1 & 7 \end{bmatrix}$

$$= \begin{bmatrix} 2(5) + 6(1) & 2(-2) + 6(7) \\ 3(5) + (-4)(1) & 3(-2) + (-4)(7) \end{bmatrix} = \begin{bmatrix} 10 + 6 & -4 + 42 \\ 15 - 4 & -6 - 28 \end{bmatrix} = \begin{bmatrix} 16 & 38 \\ 11 & -34 \end{bmatrix}$$

$$BA = \begin{bmatrix} 5 & -2 \\ 1 & 7 \end{bmatrix}\begin{bmatrix} 2 & 6 \\ 3 & -4 \end{bmatrix}$$

$$= \begin{bmatrix} 5(2) + (-2)(3) & 5(6) + (-2)(-4) \\ 1(2) + 7(3) & 1(6) + 7(-4) \end{bmatrix} = \begin{bmatrix} 4 & 38 \\ 23 & -22 \end{bmatrix}$$

13 $AB = \begin{bmatrix} 4 & -3 & 1 \\ -5 & 2 & 2 \end{bmatrix}\begin{bmatrix} 2 & 1 \\ 0 & 1 \\ -4 & 7 \end{bmatrix}$

$$= \begin{bmatrix} 4(2) + (-3)(0) + 1(-4) & 4(1) + (-3)(1) + 1(7) \\ -5(2) + 2(0) + 2(-4) & -5(1) + 2(1) + 2(7) \end{bmatrix} = \begin{bmatrix} 4 & 8 \\ -18 & 11 \end{bmatrix}$$

$$BA = \begin{bmatrix} 2 & 1 \\ 0 & 1 \\ -4 & 7 \end{bmatrix}\begin{bmatrix} 4 & -3 & 1 \\ -5 & 2 & 2 \end{bmatrix}$$

$$= \begin{bmatrix} 2(4) + 1(-5) & 2(-3) + 1(2) & 2(1) + 1(2) \\ 0(4) + 1(-5) & 0(-3) + 1(2) & 0(1) + 1(2) \\ -4(4) + 7(-5) & -4(-3) + 7(2) & -4(1) + 7(2) \end{bmatrix} = \begin{bmatrix} 3 & -4 & 4 \\ -5 & 2 & 2 \\ -51 & 26 & 10 \end{bmatrix}$$

17 $AB = \begin{bmatrix} -3 & 7 & 2 \end{bmatrix}\begin{bmatrix} 1 \\ 4 \\ -5 \end{bmatrix} = \begin{bmatrix} -3(1) + 7(4) + 2(-5) \end{bmatrix} = \begin{bmatrix} 15 \end{bmatrix}$

$$BA = \begin{bmatrix} 1 \\ 4 \\ -5 \end{bmatrix} \begin{bmatrix} -3 & 7 & 2 \end{bmatrix} = \begin{bmatrix} 1(-3) & 1(7) & 1(2) \\ 4(-3) & 4(7) & 4(2) \\ -5(-3) & -5(7) & -5(2) \end{bmatrix} = \begin{bmatrix} -3 & 7 & 2 \\ -12 & 28 & 8 \\ 15 & -35 & -10 \end{bmatrix}$$

19 $AB = \begin{bmatrix} 2 & 0 & 1 \\ -1 & 2 & 0 \end{bmatrix} \begin{bmatrix} 1 & -1 & 2 \\ 3 & 1 & 0 \\ 0 & 2 & 1 \end{bmatrix}$

$$= \begin{bmatrix} 2(1) + 0(3) + 1(0) & 2(-1) + 0(1) + 1(2) & 2(2) + 0(0) + 1(1) \\ -1(1) + 2(3) + 0(0) & -1(-1) + 2(1) + 0(2) & -1(2) + 2(0) + 0(1) \end{bmatrix}$$

$$= \begin{bmatrix} 2 & 0 & 5 \\ 5 & 3 & -2 \end{bmatrix}. \quad BA \text{ is a } 3 \times 3 \text{ matrix times a } 2 \times 3 \text{ matrix. Since the number}$$

of columns of B, 3, is not equal to the number of rows of A, 2, BA is not possible.

23 $AB = \begin{bmatrix} 2 & 1 & 0 & -3 \\ -7 & 0 & -2 & 4 \end{bmatrix} \begin{bmatrix} 4 & -2 & 0 \\ 1 & 1 & -2 \\ 0 & 0 & 5 \\ -3 & -1 & 0 \end{bmatrix} = \begin{bmatrix} c_{11} & c_{12} & c_{13} \\ c_{21} & c_{22} & c_{23} \end{bmatrix}$, where

$c_{11} = 2(4) + 1(1) + 0(0) + (-3)(-3)$ $c_{12} = 2(-2) + 1(1) + 0(0) + (-3)(-1)$

$c_{13} = 2(0) + 1(-2) + 0(5) + (-3)(0)$ $c_{21} = -7(4) + 0(1) + (-2)(0) + 4(-3)$

$c_{22} = -7(-2) + 0(1) + (-2)(0) + 4(-1)$ $c_{23} = -7(0) + 0(-2) + (-2)(5) + 4(0)$

Hence, $AB = \begin{bmatrix} 18 & 0 & -2 \\ -40 & 10 & -10 \end{bmatrix}$.

25 $(A + B)(A - B) = \begin{bmatrix} 3 & 1 \\ 3 & -2 \end{bmatrix} \begin{bmatrix} -1 & 3 \\ -3 & -4 \end{bmatrix} = \begin{bmatrix} -6 & 5 \\ 3 & 17 \end{bmatrix};$

$A^2 - B^2 = \begin{bmatrix} 1 & -4 \\ 0 & 9 \end{bmatrix} - \begin{bmatrix} 1 & -3 \\ 9 & -2 \end{bmatrix} = \begin{bmatrix} 0 & -1 \\ -9 & 11 \end{bmatrix}; (A + B)(A - B) \neq A^2 - B^2$

27 $A(B + C) = \begin{bmatrix} 1 & 2 \\ 0 & -3 \end{bmatrix} \begin{bmatrix} 5 & 0 \\ 1 & 1 \end{bmatrix} = \begin{bmatrix} 7 & 2 \\ -3 & -3 \end{bmatrix};$

$AB + AC = \begin{bmatrix} 8 & 1 \\ -9 & -3 \end{bmatrix} + \begin{bmatrix} -1 & 1 \\ 6 & 0 \end{bmatrix} = \begin{bmatrix} 7 & 2 \\ -3 & -3 \end{bmatrix}$

29 $m(A + B) = m \begin{bmatrix} a+p & b+q \\ c+r & d+s \end{bmatrix} = \begin{bmatrix} m(a+p) & m(b+q) \\ m(c+r) & m(d+s) \end{bmatrix}$

$$= \begin{bmatrix} ma+mp & mb+mq \\ mc+mr & md+ms \end{bmatrix} = \begin{bmatrix} ma & mb \\ mc & md \end{bmatrix} + \begin{bmatrix} mp & mq \\ mr & ms \end{bmatrix}$$

$$= m \begin{bmatrix} a & b \\ c & d \end{bmatrix} + m \begin{bmatrix} p & q \\ r & s \end{bmatrix} = mA + mB$$

31 $\quad A(B+C) = \begin{bmatrix} a & b \\ c & d \end{bmatrix}\begin{bmatrix} p+w & q+x \\ r+y & s+z \end{bmatrix}$

$$= \begin{bmatrix} a(p+w)+b(r+y) & a(q+x)+b(s+z) \\ c(p+w)+d(r+y) & c(q+x)+d(s+z) \end{bmatrix}$$

$$= \begin{bmatrix} ap+aw+br+by & aq+ax+bs+bz \\ cp+cw+dr+dy & cq+cx+ds+dz \end{bmatrix}$$

$$= \begin{bmatrix} ap+br & aq+bs \\ cp+dr & cq+ds \end{bmatrix} + \begin{bmatrix} aw+by & ax+bz \\ cw+dy & cx+dz \end{bmatrix}$$

$$= \begin{bmatrix} a & b \\ c & d \end{bmatrix}\begin{bmatrix} p & q \\ r & s \end{bmatrix} + \begin{bmatrix} a & b \\ c & d \end{bmatrix}\begin{bmatrix} w & x \\ y & z \end{bmatrix} = AB + AC$$

9.8 Exercises

Note: Exer. 1–10: Let A denote the given matrix.

1 $\quad \begin{bmatrix} 2 & -4 & | & 1 & 0 \\ 1 & 3 & | & 0 & 1 \end{bmatrix} R_1 - R_2 \to R_1 \quad \Rightarrow \begin{bmatrix} 1 & -7 & | & 1 & -1 \\ 1 & 3 & | & 0 & 1 \end{bmatrix} R_2 - R_1 \to R_2$

$\begin{bmatrix} 1 & -7 & | & 1 & -1 \\ 0 & 10 & | & -1 & 2 \end{bmatrix} \frac{1}{10} R_2 \to R_2 \quad \Rightarrow \begin{bmatrix} 1 & -7 & | & 1 & -1 \\ 0 & 1 & | & -\frac{1}{10} & \frac{2}{10} \end{bmatrix} R_1 + 7 R_2 \to R_1$

$\begin{bmatrix} 1 & 0 & | & \frac{3}{10} & \frac{4}{10} \\ 0 & 1 & | & -\frac{1}{10} & \frac{2}{10} \end{bmatrix} \quad \Rightarrow A^{-1} = \frac{1}{10}\begin{bmatrix} 3 & 4 \\ -1 & 2 \end{bmatrix}$

3 $\quad \begin{bmatrix} 2 & 4 & | & 1 & 0 \\ 4 & 8 & | & 0 & 1 \end{bmatrix} \frac{1}{2} R_1 \to R_1 \quad \Rightarrow \begin{bmatrix} 1 & 2 & | & \frac{1}{2} & 0 \\ 4 & 8 & | & 0 & 1 \end{bmatrix} R_2 - 4 R_1 \to R_2$

$\begin{bmatrix} 1 & 2 & | & \frac{1}{2} & 0 \\ 0 & 0 & | & -2 & 1 \end{bmatrix}$

Since the identity matrix cannot be obtained on the left, *no inverse exists.*

7 $\quad \begin{bmatrix} -2 & 2 & 3 & | & 1 & 0 & 0 \\ 1 & -1 & 0 & | & 0 & 1 & 0 \\ 0 & 1 & 4 & | & 0 & 0 & 1 \end{bmatrix} R_1 + 2 R_2 \leftrightarrow R_2$

$\begin{bmatrix} 1 & -1 & 0 & | & 0 & 1 & 0 \\ 0 & 0 & 3 & | & 1 & 2 & 0 \\ 0 & 1 & 4 & | & 0 & 0 & 1 \end{bmatrix} R_1 + R_3 \to R_1$

$\begin{bmatrix} 1 & 0 & 4 & | & 0 & 1 & 1 \\ 0 & 0 & 3 & | & 1 & 2 & 0 \\ 0 & 1 & 4 & | & 0 & 0 & 1 \end{bmatrix} \frac{1}{3} R_2 \leftrightarrow R_3$

$\begin{bmatrix} 1 & 0 & 4 & | & 0 & 1 & 1 \\ 0 & 1 & 4 & | & 0 & 0 & 1 \\ 0 & 0 & 1 & | & \frac{1}{3} & \frac{2}{3} & 0 \end{bmatrix} \begin{matrix} R_1 - 4 R_3 \to R_1 \\ R_2 - 4 R_3 \to R_2 \end{matrix}$

$$
\begin{bmatrix}
1 & 0 & 0 & -\frac{4}{3} & -\frac{5}{3} & 1 \\
0 & 1 & 0 & -\frac{4}{3} & -\frac{8}{3} & 1 \\
0 & 0 & 1 & \frac{1}{3} & \frac{2}{3} & 0
\end{bmatrix}
\qquad
A^{-1} = \frac{1}{3}\begin{bmatrix}
-4 & -5 & 3 \\
-4 & -8 & 3 \\
1 & 2 & 0
\end{bmatrix}
$$

11. $\begin{bmatrix} a & 0 & 1 & 0 \\ 0 & b & 0 & 1 \end{bmatrix} \begin{matrix} (1/a)\,R_1 \to R_1 \\ (1/b)\,R_2 \to R_2 \end{matrix} \Rightarrow \begin{bmatrix} 1 & 0 & 1/a & 0 \\ 0 & 1 & 0 & 1/b \end{bmatrix}$

The inverse is the matrix with main diagonal elements $(1/a)$ and $(1/b)$.

The required conditions are that a and b are nonzero to avoid division by zero.

15. (a) $X = A^{-1}B = \frac{1}{10}\begin{bmatrix} 3 & 4 \\ -1 & 2 \end{bmatrix}\begin{bmatrix} 3 \\ 1 \end{bmatrix} = \frac{1}{10}\begin{bmatrix} 13 \\ -1 \end{bmatrix};\quad (\frac{13}{10}, -\frac{1}{10})$

(b) $X = A^{-1}B = \frac{1}{10}\begin{bmatrix} 3 & 4 \\ -1 & 2 \end{bmatrix}\begin{bmatrix} -2 \\ 5 \end{bmatrix} = \frac{1}{10}\begin{bmatrix} 14 \\ 12 \end{bmatrix};\quad (\frac{7}{5}, \frac{6}{5})$

17. (a) $X = A^{-1}B = \frac{1}{3}\begin{bmatrix} -4 & -5 & 3 \\ -4 & -8 & 3 \\ 1 & 2 & 0 \end{bmatrix}\begin{bmatrix} 1 \\ 3 \\ -2 \end{bmatrix} = \frac{1}{3}\begin{bmatrix} -25 \\ -34 \\ 7 \end{bmatrix};\quad (-\frac{25}{3}, -\frac{34}{3}, \frac{7}{3})$

(b) $X = A^{-1}B = \frac{1}{3}\begin{bmatrix} -4 & -5 & 3 \\ -4 & -8 & 3 \\ 1 & 2 & 0 \end{bmatrix}\begin{bmatrix} -1 \\ 0 \\ 4 \end{bmatrix} = \frac{1}{3}\begin{bmatrix} 16 \\ 16 \\ -1 \end{bmatrix};\quad (\frac{16}{3}, \frac{16}{3}, -\frac{1}{3})$

19. (a) $AX = B \Leftrightarrow \begin{bmatrix} 3.1 & 6.7 & -8.7 \\ 4.1 & -5.1 & 0.2 \\ 0.6 & 1.1 & -7.4 \end{bmatrix}\begin{bmatrix} x \\ y \\ z \end{bmatrix} = \begin{bmatrix} 1.5 \\ 2.1 \\ 3.9 \end{bmatrix}$

(b) The inverse should be found using some type of computational device. If you are using a TI-81, enter the 9 values into the matrix [A]. Change Float to 4 via MODE. Now find $[A]^{-1}$ { be sure to use the $\boxed{x^{-1}}$ key } and STOre this matrix into matrix [B] to use in part (c). Use the right arrow key to see the rightmost elements in [A].

$$
A^{-1} \approx \begin{bmatrix}
0.1474 & 0.1572 & -0.1691 \\
0.1197 & -0.0696 & -0.1426 \\
0.0297 & 0.0024 & -0.1700
\end{bmatrix}
$$

(c) Following the instructions in part (b), enter the 3 values into [C] { a 3×1 matrix }, and then evaluate [B]*[C].

$$
X = A^{-1}B \approx \begin{bmatrix}
0.1474 & 0.1572 & -0.1691 \\
0.1197 & -0.0696 & -0.1426 \\
0.0297 & 0.0024 & -0.1700
\end{bmatrix}\begin{bmatrix} 1.5 \\ 2.1 \\ 3.9 \end{bmatrix} \approx \begin{bmatrix} -0.1081 \\ -0.5227 \\ -0.6135 \end{bmatrix}
$$

9.9 Exercises

Note: The minor M_{ij} and the cofactor A_{ij} are equal if $(i+j)$ is even and of opposite sign if $(i+j)$ is odd.

1 The minor M_{11} is obtained by deleting the first row and first column from

$$A = \begin{bmatrix} 7 & -1 \\ 5 & 0 \end{bmatrix}. \quad \text{Thus,} \quad M_{11} = 0 = A_{11}. \quad \text{Similarly,} \quad M_{12} = 5 \quad \text{and} \quad A_{12} = -5;$$

$M_{21} = -1$ and $A_{21} = 1$; and $M_{22} = 7 = A_{22}$.

3 Let $A = \begin{bmatrix} 2 & 4 & -1 \\ 0 & 3 & 2 \\ -5 & 7 & 0 \end{bmatrix}.$

Be sure you understand the note preceding the solution for Exercise 1.

$$M_{11} = \begin{vmatrix} 3 & 2 \\ 7 & 0 \end{vmatrix} = (3)(0) - (7)(2) = 0 - 14 = -14 = A_{11};$$

$$M_{12} = \begin{vmatrix} 0 & 2 \\ -5 & 0 \end{vmatrix} = (0)(0) - (-5)(2) = 0 - (-10) = 10; \; A_{12} = -10;$$

$$M_{13} = \begin{vmatrix} 0 & 3 \\ -5 & 7 \end{vmatrix} = 15 = A_{13}; \qquad M_{21} = \begin{vmatrix} 4 & -1 \\ 7 & 0 \end{vmatrix} = 7; \; A_{21} = -7;$$

$$M_{22} = \begin{vmatrix} 2 & -1 \\ -5 & 0 \end{vmatrix} = -5 = A_{22}; \qquad M_{23} = \begin{vmatrix} 2 & 4 \\ -5 & 7 \end{vmatrix} = 34; \; A_{23} = -34;$$

$$M_{31} = \begin{vmatrix} 4 & -1 \\ 3 & 2 \end{vmatrix} = 11 = A_{31}; \qquad M_{32} = \begin{vmatrix} 2 & -1 \\ 0 & 2 \end{vmatrix} = 4; \; A_{32} = -4;$$

$$M_{33} = \begin{vmatrix} 2 & 4 \\ 0 & 3 \end{vmatrix} = 6 = A_{33}.$$

Note: Exercises 5–20: Let A denote the given matrix.

7 Expanding $|A|$ by the first column and using the cofactor values from Exercise 3,

we obtain $|A| = a_{11}A_{11} + a_{21}A_{21} + a_{31}A_{31} = 2(-14) + 0(A_{21}) - 5(11) = -83.$

13 Expand by the first row.

$$|A| = a_{11}A_{11} + a_{12}A_{12} + a_{13}A_{13}$$

$$= (3)(-1)^{1+1}\begin{vmatrix} 2 & 5 \\ 3 & -1 \end{vmatrix} + (1)(-1)^{1+2}\begin{vmatrix} 4 & 5 \\ -6 & -1 \end{vmatrix} + (-2)(-1)^{1+3}\begin{vmatrix} 4 & 2 \\ -6 & 3 \end{vmatrix}$$

$$= (3)(1)(-17) + (1)(-1)(26) + (-2)(1)(24)$$

$$= -51 - 26 - 48 = -125$$

17 Expand $|A|$ by the third row. $\quad |A| = 6A_{32} = -6M_{32} = -6\begin{vmatrix} 3 & 2 & 0 \\ 4 & -3 & 5 \\ 1 & -4 & 2 \end{vmatrix}.$

Expand M_{32} by the first row.

$$M_{32} = 3(14) + 2(-3) + 0(-13) = 36 \Rightarrow |A| = -6(36) = -216.$$

21 LS $= ad - bc$; $\qquad\qquad\qquad$ RS $= -(bc - ad) = ad - bc$

25 LS $= ad - bc$; $\qquad\qquad\qquad$ RS $= abk + ad - abk - bc = ad - bc$

$\boxed{27}$ $LS = ad - bc + af - ce;$ $\qquad$ $RS = ad + af - bc - ce$

$\boxed{29}$ Consider the matrix in Exercise 20. If we expanded along the first column, we would obtain a times its cofactor. Expanding along the first column again, we obtain ab times another cofactor. This exercise is similar since all elements in A *above* {rather than below} the main diagonal are zero. We can evaluate the determinant using n expansions by the first row, and obtain $|A| = a_{11}a_{22}\cdots a_{nn}$.

$\boxed{31}$ (a) $A - xI = \begin{bmatrix} 1 & 2 \\ 3 & 2 \end{bmatrix} - x\begin{bmatrix} 1 & 0 \\ 0 & 1 \end{bmatrix} = \begin{bmatrix} 1-x & 2 \\ 3 & 2-x \end{bmatrix}.$

$\qquad f(x) = |A - xI| = \begin{vmatrix} 1-x & 2 \\ 3 & 2-x \end{vmatrix}$

$\qquad\qquad = (1-x)(2-x) - (3)(2) = (2 - 3x + x^2) - 6 = x^2 - 3x - 4.$

(b) $x^2 - 3x - 4 = 0 \Rightarrow (x-4)(x+1) = 0 \Rightarrow x = -1, 4$

$\boxed{37}$ (a) $f(x) = \begin{vmatrix} 0-x & 2 & -2 \\ -1 & 3-x & 1 \\ -3 & 3 & 1-x \end{vmatrix}$ $\qquad$ {Expand by the first row.}

$\qquad\qquad = (-x)[(3-x)(1-x) - 3] - 2[(x-1) + 3] - 2[-3 + 3(3-x)]$

$\qquad\qquad = (-x)[(3 - 4x + x^2) - 3] - 2(x+2) - 2(-3x+6)$

$\qquad\qquad = (-x^3 + 4x^2) - 2x - 4 + 6x - 12$

$\qquad\qquad = -x^3 + 4x^2 + 4x - 16$

(b) By trying possible rational roots, we determine that 2 is a zero of f.

$\qquad$ Thus, $-x^3 + 4x^2 + 4x - 16 = 0 \Rightarrow (x-2)(-x^2 + 2x + 8) = 0 \Rightarrow$

$\qquad\qquad\qquad (x+2)(x-2)(-x+4) = 0 \Rightarrow x = -2, 2, 4.$

$\boxed{39}$ Expand the determinant by the first row.

$\begin{vmatrix} i & j & k \\ 2 & -1 & 6 \\ -3 & 5 & 1 \end{vmatrix} = i\begin{vmatrix} -1 & 6 \\ 5 & 1 \end{vmatrix} - j\begin{vmatrix} 2 & 6 \\ -3 & 1 \end{vmatrix} + k\begin{vmatrix} 2 & -1 \\ -3 & 5 \end{vmatrix} = -31i - 20j + 7k$

$\boxed{43}$ (a) $f(x) = |A - xI| = \begin{vmatrix} 1-x & 0 & 1 \\ 0 & 2-x & 1 \\ 1 & 1 & -2-x \end{vmatrix} = -x^3 + x^2 + 6x - 7$

(b) The characteristic values of A are equal to the zeros of f. From the graph,

$\qquad$ we see that the zeros are approximately -2.51, 1.22, and 2.29.

$[-10, 11]$ by $[-12, 2]$

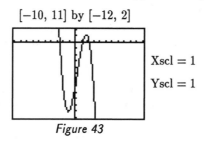

$Xscl = 1$

$Yscl = 1$

Figure 43

9.10 Exercises

1 R_2 and R_3 are interchanged. The determinant value is negated.

3 R_3 is replaced by $(R_3 - R_1)$. There is no change in the determinant value.

5 The number 2 can be factored out of R_1, yielding $2\begin{vmatrix} 1 & 2 & 1 \\ 1 & 2 & 4 \\ 2 & 6 & 4 \end{vmatrix}$.

Next, the number 2 can be factored out of R_3, yielding $4\begin{vmatrix} 1 & 2 & 1 \\ 1 & 2 & 4 \\ 1 & 3 & 2 \end{vmatrix}$.

7 R_1 and R_3 are identical. The determinant is 0.

9 The number -1 can be factored out of R_2, yielding the determinant on the right side.

11 Every number in C_2 is 0. The determinant is zero.

13 C_3 is replaced by $(2C_1 + C_3)$. There is no change in the determinant value.

Note: The notation $\{R_i\,(C_i)\}$ means expand the determinant by the ith row (column).

15 There are many possibilities for introducing zeros. In this case, obtaining a zero in the third row, second column would lead to an easy evaluation using the second column.

$$\begin{vmatrix} 3 & 1 & 0 \\ -2 & 0 & 1 \\ 1 & 3 & -1 \end{vmatrix} R_3 - 3R_1 \rightarrow R_3 = \begin{vmatrix} 3 & 1 & 0 \\ -2 & 0 & 1 \\ -8 & 0 & -1 \end{vmatrix} \{C_2\} = (-1)\begin{vmatrix} -2 & 1 \\ -8 & -1 \end{vmatrix} =$$

$$(-1)(2+8) = -10$$

19 $\begin{vmatrix} 2 & 2 & -3 \\ 3 & 6 & 9 \\ -2 & 5 & 4 \end{vmatrix} \{\,3 \text{ is a common factor of } R_2\,\} = (3)\begin{vmatrix} 2 & 2 & -3 \\ 1 & 2 & 3 \\ -2 & 5 & 4 \end{vmatrix}$

Since there is a "1" in the second row, first column, we will obtain zeros in the other two locations of the first column.

$$(3)\begin{vmatrix} 2 & 2 & -3 \\ 1 & 2 & 3 \\ -2 & 5 & 4 \end{vmatrix} \begin{matrix} R_1 - 2R_2 \rightarrow R_1 \\ \\ R_3 + 2R_2 \rightarrow R_3 \end{matrix} = (3)\begin{vmatrix} 0 & -2 & -9 \\ 1 & 2 & 3 \\ 0 & 9 & 10 \end{vmatrix} \{C_1\}$$

$$= (3)(-1)\begin{vmatrix} -2 & -9 \\ 9 & 10 \end{vmatrix} = (-3)(-20+81) = -183$$

21 $\begin{vmatrix} 3 & 1 & -2 & 2 \\ 2 & 0 & 1 & 4 \\ 0 & 1 & 3 & 5 \\ -1 & 2 & 0 & -3 \end{vmatrix} \begin{matrix} \\ \\ R_3 - R_1 \rightarrow R_3 \\ R_4 - 2R_1 \rightarrow R_4 \end{matrix} = \begin{vmatrix} 3 & 1 & -2 & 2 \\ 2 & 0 & 1 & 4 \\ -3 & 0 & 5 & 3 \\ -7 & 0 & 4 & -7 \end{vmatrix} \{C_2\}$

$$= (-1)\begin{vmatrix} 2 & 1 & 4 \\ -3 & 5 & 3 \\ -7 & 4 & -7 \end{vmatrix} \begin{matrix} R_2 - 5R_1 \rightarrow R_2 \\ R_3 - 4R_1 \rightarrow R_3 \end{matrix} = (-1)\begin{vmatrix} 2 & 1 & 4 \\ -13 & 0 & -17 \\ -15 & 0 & -23 \end{vmatrix} \{C_2\}$$

$$= (-1)(-1)\begin{vmatrix} -13 & -17 \\ -15 & -23 \end{vmatrix} = (1)(299-255) = 44$$

$$
\boxed{23} \quad
\begin{vmatrix}
2 & -2 & 0 & 0 & -3 \\
3 & 0 & 3 & 2 & -1 \\
0 & 1 & -2 & 0 & 2 \\
-1 & 2 & 0 & 3 & 0 \\
0 & 4 & 1 & 0 & 0
\end{vmatrix}
\begin{array}{l} C_2 - 4\,C_3 \to C_2 \\ = \end{array}
\begin{vmatrix}
2 & -2 & 0 & 0 & -3 \\
3 & -12 & 3 & 2 & -1 \\
0 & 9 & -2 & 0 & 2 \\
-1 & 2 & 0 & 3 & 0 \\
0 & 0 & 1 & 0 & 0
\end{vmatrix}
\{R_5\}
$$

$$
= (1)
\begin{vmatrix}
2 & -2 & 0 & -3 \\
3 & -12 & 2 & -1 \\
0 & 9 & 0 & 2 \\
-1 & 2 & 3 & 0
\end{vmatrix}
\begin{array}{l} R_1 + 2\,R_4 \to R_1 \\ R_2 + 3\,R_4 \to R_2 \\ = \end{array}
\begin{vmatrix}
0 & 2 & 6 & -3 \\
0 & -6 & 11 & -1 \\
0 & 9 & 0 & 2 \\
-1 & 2 & 3 & 0
\end{vmatrix}
\{C_1\}
$$

$$
= (1)
\begin{vmatrix}
2 & 6 & -3 \\
-6 & 11 & -1 \\
9 & 0 & 2
\end{vmatrix}
\begin{array}{l} R_1 - 3\,R_2 \to R_1 \\ \\ R_3 + 2\,R_2 \to R_3 \end{array}
=
\begin{vmatrix}
20 & -27 & 0 \\
-6 & 11 & -1 \\
-3 & 22 & 0
\end{vmatrix}
\{C_3\}
$$

$$
= (1)
\begin{vmatrix}
20 & -27 \\
-3 & 22
\end{vmatrix}
= (1)(440 - 81) = 359
$$

$$
\boxed{25} \quad
\begin{vmatrix}
1 & 1 & 1 \\
a & b & c \\
a^2 & b^2 & c^2
\end{vmatrix}
\begin{array}{l} C_1 - C_2 \to C_1 \\ \\ C_3 - C_2 \to C_3 \end{array}
$$

$$
=
\begin{vmatrix}
0 & 1 & 0 \\
a - b & b & c - b \\
a^2 - b^2 & b^2 & c^2 - b^2
\end{vmatrix}
\begin{array}{l} a - b \text{ is a common factor of } C_1 \\ \\ c - b \text{ is a common factor of } C_3 \end{array}
$$

$$
= (a - b)(c - b)
\begin{vmatrix}
0 & 1 & 0 \\
1 & b & 1 \\
a + b & b^2 & c + b
\end{vmatrix}
\{R_1\}
$$

$$
= (a - b)(c - b)(-1)
\begin{vmatrix}
1 & 1 \\
a + b & c + b
\end{vmatrix}
$$

$$
= (a - b)(b - c)(c + b - a - b) = \underline{(a - b)(b - c)(c - a)}
$$

$$
\boxed{27} \quad
\begin{vmatrix}
a_{11} & a_{12} & a_{13} & a_{14} \\
0 & a_{22} & a_{23} & a_{24} \\
0 & 0 & a_{33} & a_{34} \\
0 & 0 & 0 & a_{44}
\end{vmatrix}
\{C_1\} = (a_{11})
\begin{vmatrix}
a_{22} & a_{23} & a_{24} \\
0 & a_{33} & a_{34} \\
0 & 0 & a_{44}
\end{vmatrix}
\{C_1\}
$$

$$
= (a_{11})(a_{22})
\begin{vmatrix}
a_{33} & a_{34} \\
0 & a_{44}
\end{vmatrix}
= (a_{11}a_{22})(a_{33}a_{44} - 0) = a_{11}\,a_{22}\,a_{33}\,a_{44}
$$

$$
\boxed{29} \quad |AB| =
\begin{vmatrix}
a_{11}b_{11} + a_{12}b_{21} & a_{11}b_{12} + a_{12}b_{22} \\
a_{21}b_{11} + a_{22}b_{21} & a_{21}b_{12} + a_{22}b_{22}
\end{vmatrix}
$$

$$
= (a_{11}b_{11} + a_{12}b_{21})(a_{21}b_{12} + a_{22}b_{22}) - (a_{11}b_{12} + a_{12}b_{22})(a_{21}b_{11} + a_{22}b_{21})
$$

$$
= \quad a_{11}b_{11}a_{21}b_{12} + a_{11}b_{11}a_{22}b_{22} + a_{12}b_{21}a_{21}b_{12} + a_{12}b_{21}a_{22}b_{22}
$$

$$
- a_{11}b_{12}a_{21}b_{11} - a_{11}b_{12}a_{22}b_{21} - a_{12}b_{22}a_{21}b_{11} - a_{12}b_{22}a_{22}b_{21}
$$

$$
= a_{11}a_{22}b_{11}b_{22} - a_{11}a_{22}b_{21}b_{12} - a_{21}a_{12}b_{11}b_{22} + a_{21}a_{12}b_{21}b_{12}
$$

$$
= (a_{11}a_{22} - a_{21}a_{12})(b_{11}b_{22} - b_{21}b_{12}) = |A|\,|B|
$$

31 Expanding by the first row yields $Ax + By + C = 0$ {an equation of a line} where A, B, and C are constants. To show that the line contains (x_1, y_1) and (x_2, y_2), we must show that these points are solutions of the equation. Substituting x_1 for x and y_1 for y, we obtain two identical rows and the determinant is zero. Hence, (x_1, y_1) is a solution of the equation and a similar argument can be made for (x_2, y_2).

33 For the system $\begin{cases} 2x + 3y = 2 \\ x - 2y = 8 \end{cases}$, $|D| = \begin{vmatrix} 2 & 3 \\ 1 & -2 \end{vmatrix} = -4 - 3 = -7$.

Since $|D| = -7 \neq 0$, we may solve the system using Cramer's rule.

$|D_x| = \begin{vmatrix} 2 & 3 \\ 8 & -2 \end{vmatrix} = -4 - 24 = -28.$ $|D_y| = \begin{vmatrix} 2 & 2 \\ 1 & 8 \end{vmatrix} = 16 - 2 = 14.$

$x = \dfrac{|D_x|}{|D|} = \dfrac{-28}{-7} = 4.$ $y = \dfrac{|D_y|}{|D|} = \dfrac{14}{-7} = -2.$ ★ $(4, -2)$

37 $|D| = \begin{vmatrix} 2 & -3 \\ -6 & 9 \end{vmatrix} = 18 - 18 = 0$, so Cramer's rule cannot be used.

41 $|D| = \begin{vmatrix} 5 & 2 & -1 \\ 1 & -2 & 2 \\ 0 & 3 & 1 \end{vmatrix}$ {R_3} $= -3(11) + 1(-12) = -45$

$|D_x| = \begin{vmatrix} -7 & 2 & -1 \\ 0 & -2 & 2 \\ 17 & 3 & 1 \end{vmatrix}$ {C_1} $= -7(-8) + 17(2) = 90$

$|D_y| = \begin{vmatrix} 5 & -7 & -1 \\ 1 & 0 & 2 \\ 0 & 17 & 1 \end{vmatrix}$ {R_2} $= -1(10) - 2(85) = -180$

$|D_z| = \begin{vmatrix} 5 & 2 & -7 \\ 1 & -2 & 0 \\ 0 & 3 & 17 \end{vmatrix}$ {C_1} $= 5(-34) - 1(55) = -225$

$x = \dfrac{|D_x|}{|D|} = \dfrac{90}{-45} = -2;$ $y = \dfrac{|D_y|}{|D|} = \dfrac{-180}{-45} = 4;$ $z = \dfrac{|D_z|}{|D|} = \dfrac{-225}{-45} = 5.$

★ $(-2, 4, 5)$

Chapter 9 Review Exercises

3 Solve E_2 for y, $y = -2x - 1$, and substitute into E_1 to yield $x^2 + 2x - 3 = 0 \Rightarrow$ $(x + 3)(x - 1) = 0 \Rightarrow x = -3, 1$ and $y = 5, -3$.   ★ $(-3, 5)$, $(1, -3)$

6 From E_3, $x^2 = xz \Rightarrow x^2 - xz = 0 \Rightarrow x(x - z) = 0 \Rightarrow x = 0$ or $x = z$.

If $x = 0$, E_1 is $0 = y^2 + 3z$ and E_2 is $z = 1 - y^2$. Substituting z into E_1 yields
$$2y^2 = 3 \text{ or } y = \pm\tfrac{1}{2}\sqrt{6} \text{ and } z \text{ is } -\tfrac{1}{2} \text{ for both values of } y.$$

If $x = z$, E_1 is $0 = y^2 + z$ and E_2 is $y^2 = 1$.

Thus, $y = \pm 1$ and in either case, $x = z = -1$.

There are four solutions: $(-1, \pm 1, -1)$, $(0, \pm\tfrac{1}{2}\sqrt{6}, -\tfrac{1}{2})$.

$\boxed{8}$ Treat 3^{y+1} as $3 \cdot 3^y$ and 2^{x+1} as $2 \cdot 2^x$.

Thus, E_1 is $1(2^x) + 3(3^y) = 10$ and E_2 is $2(2^x) - 1(3^y) = 5$.

Now $E_1 + 3 E_2 \Rightarrow 7 \cdot 2^x = 25 \Rightarrow 2^x = \frac{25}{7} \Rightarrow x = \log_2 \frac{25}{7} = \dfrac{\log \frac{25}{7}}{\log 2} \approx 1.84$.

Resolving the original system for y, we have

$$-2 E_1 + E_2 \Rightarrow -7 \cdot 3^y = -15 \Rightarrow 3^y = \frac{15}{7} \Rightarrow y = \log_3 \frac{15}{7} = \dfrac{\log \frac{15}{7}}{\log 3} \approx 0.69.$$

$\boxed{9}$ Solve E_3 for z, $z = 4x + 5y + 2$, and substitute into E_1 and E_2 to yield

$$\begin{cases} 3x + y - 2(4x + 5y + 2) = -1 \\ 2x - 3y + (4x + 5y + 2) = 4 \end{cases} \Rightarrow \begin{cases} -5x - 9y = 3 & (E_4) \\ 6x + 2y = 2 & (E_5) \end{cases}$$

$6 E_4 + 5 E_5 \Rightarrow -44y = 28 \Rightarrow y = -\frac{7}{11}$;

$2 E_4 + 9 E_5 \Rightarrow 44x = 24 \Rightarrow x = \frac{6}{11}$; $z = 1$ $\qquad\qquad$ $\bigstar$ $(\frac{6}{11}, -\frac{7}{11}, 1)$

$\boxed{11}$ Solve E_2 for x, $x = y + z$, and substitute into E_1 and E_3 to yield

$$\begin{cases} 4(y + z) - 3y - z = 0 \\ 3(y + z) - y + 3z = 0 \end{cases} \Rightarrow \begin{cases} y + 3z = 0 & (E_4) \\ 2y + 6z = 0 & (E_5) \end{cases}$$

Now E_5 is $2 E_4$, hence $y = -3z$ and $x = y + z = -3z + z = -2z$.

$\qquad\qquad$ The general solution is $(-2c, -3c, c)$ for any real number c.

$\boxed{15}$ Let $a = 1/x$, $b = 1/y$, and $c = 1/z$ to obtain the system

$$\begin{cases} 4a + b + 2c = 4 & (E_1) \\ 2a + 3b - c = 1 & (E_2) \\ a + b + c = 4 & (E_3) \end{cases}$$

Solving E_2 for c and substituting into E_1 and E_3 yields

$$\begin{cases} 4a + b + 2(2a + 3b - 1) = 4 \\ a + b + (2a + 3b - 1) = 4 \end{cases} \Rightarrow \begin{cases} 8a + 7b = 6 & (E_4) \\ 3a + 4b = 5 & (E_5) \end{cases}$$

$3 E_4 - 8 E_5 \Rightarrow -11b = -22 \Rightarrow b = 2$

$\qquad\qquad 4 E_4 - 7 E_5 \Rightarrow 11a = -11 \Rightarrow a = -1$, $c = 3$; $(x, y, z) = (-1, \frac{1}{2}, \frac{1}{3})$

$\boxed{19}$ $\dfrac{x^2 + 14x - 13}{x^3 + 5x^2 + 4x + 20} = \dfrac{A}{x + 5} + \dfrac{Bx + C}{x^2 + 4}$

$x^2 + 14x - 13 = A(x^2 + 4) + (Bx + C)(x + 5)$

$x = -5$: $-58 = 29A \Rightarrow A = -2$

$x = 0$: $-13 = 4A + 5C \Rightarrow C = -1$ $\qquad\qquad$ $\bigstar$ $-\dfrac{2}{x + 5} + \dfrac{3x - 1}{x^2 + 4}$

$x = 1$: $2 = 5A + 6B + 6C \Rightarrow B = 3$

[21] $\begin{cases} x^2 + y^2 < 16 \\ y - x^2 > 0 \end{cases} \Leftrightarrow \begin{cases} x^2 + y^2 < 4^2 \\ y > x^2 \end{cases}$ $V @ \left(\pm \sqrt{-\frac{1}{2} + \frac{1}{2}\sqrt{65}}, \; -\frac{1}{2} + \frac{1}{2}\sqrt{65} \right) \approx (\pm 1.88, \; 3.53)$

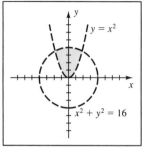

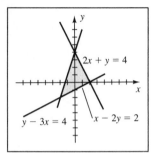

Figure 21   Figure 23

[23] $\begin{cases} x - 2y \le 2 \\ y - 3x \le 4 \\ 2x + y \le 4 \end{cases} \Leftrightarrow \begin{cases} y \ge \frac{1}{2}x - 1 \\ y \le 3x + 4 \\ y \le -2x + 4 \end{cases}$ $V @ (-2, -2), \; (0, 4), \; (2, 0)$

[27] $\begin{bmatrix} 2 & 0 \\ 1 & 4 \\ -2 & 3 \end{bmatrix} \begin{bmatrix} 0 & 2 & -3 \\ 4 & 5 & 1 \end{bmatrix} =$

$$\begin{bmatrix} 2(0) + 0(4) & 2(2) + 0(5) & 2(-3) + 0(1) \\ 1(0) + 4(4) & 1(2) + 4(5) & 1(-3) + 4(1) \\ -2(0) + 3(4) & -2(2) + 3(5) & -2(-3) + 3(1) \end{bmatrix} = \begin{bmatrix} 0 & 4 & -6 \\ 16 & 22 & 1 \\ 12 & 11 & 9 \end{bmatrix}$$

[33] $\begin{bmatrix} 1 & 2 \\ 3 & 4 \end{bmatrix} \left\{ \begin{bmatrix} 2 & -4 \\ 3 & 7 \end{bmatrix} + \begin{bmatrix} 1 & 5 \\ -2 & -3 \end{bmatrix} \right\} = \begin{bmatrix} 1 & 2 \\ 3 & 4 \end{bmatrix} \begin{bmatrix} 3 & 1 \\ 1 & 4 \end{bmatrix} = \begin{bmatrix} 5 & 9 \\ 13 & 19 \end{bmatrix}$

Note: Let A denote each of the matrices in Exercises 35–50.

[37] $\left[\begin{array}{ccc|ccc} 1 & 0 & 0 & 1 & 0 & 0 \\ 0 & 4 & 7 & 0 & 1 & 0 \\ 0 & 1 & 2 & 0 & 0 & 1 \end{array} \right] \; R_3 \leftrightarrow R_2$

$\left[\begin{array}{ccc|ccc} 1 & 0 & 0 & 1 & 0 & 0 \\ 0 & 1 & 2 & 0 & 0 & 1 \\ 0 & 4 & 7 & 0 & 1 & 0 \end{array} \right] \; R_3 - 4\,R_2 \rightarrow R_3$

$\left[\begin{array}{ccc|ccc} 1 & 0 & 0 & 1 & 0 & 0 \\ 0 & 1 & 2 & 0 & 0 & 1 \\ 0 & 0 & -1 & 0 & 1 & -4 \end{array} \right] \; -R_3 \rightarrow R_3$

$\left[\begin{array}{ccc|ccc} 1 & 0 & 0 & 1 & 0 & 0 \\ 0 & 1 & 2 & 0 & 0 & 1 \\ 0 & 0 & 1 & 0 & -1 & 4 \end{array} \right] \; R_2 - 2\,R_3 \rightarrow R_2$

$\left[\begin{array}{ccc|ccc} 1 & 0 & 0 & 1 & 0 & 0 \\ 0 & 1 & 0 & 0 & 2 & -7 \\ 0 & 0 & 1 & 0 & -1 & 4 \end{array} \right]$ $A^{-1} = \begin{bmatrix} 1 & 0 & 0 \\ 0 & 2 & -7 \\ 0 & -1 & 4 \end{bmatrix}$

[39] $X = A^{-1}B = -\frac{1}{2}\begin{bmatrix} 2 & 4 \\ 3 & 5 \end{bmatrix}\begin{bmatrix} 30 \\ -16 \end{bmatrix} = -\frac{1}{2}\begin{bmatrix} -4 \\ 10 \end{bmatrix} = \begin{bmatrix} 2 \\ -5 \end{bmatrix}$; $(x, y) = (2, -5)$

[41] $A = \begin{bmatrix} -6 \end{bmatrix} \Rightarrow |A| = -6.$

[45] $\{R_1\}$ $|A| = 2(-7) + 3(-5) + 5(-11) = -84$

[47] From Exercise 29 of §9.9, the determinant of A is the product of the main diagonal

elements of A—that is, $|A| = (5)(-3)(-4)(2) = 120.$

[49] C_2 and C_4 are identical, so $|A| = 0.$

[51] $\begin{vmatrix} 2-x & 3 \\ 1 & -4-x \end{vmatrix} = 0 \Rightarrow (2-x)(-4-x) - 3 = 0 \Rightarrow$

$$x^2 + 2x - 11 = 0 \Rightarrow x = \frac{-2 \pm \sqrt{4+44}}{2} = \frac{-2 \pm 4\sqrt{3}}{2} = -1 \pm 2\sqrt{3}$$

[55] This is an extension of Exercise 29 of §9.9. Expanding by C_1, only a_{11} is not 0.

Expanding by the new C_1 again, only a_{22} is not 0. Repeating this process yields

$$|A| = a_{11}a_{22}a_{33}\cdots a_{nn}, \text{ the product of the main diagonal elements.}$$

[59] Let x and y denote the length and width, respectively, of the rectangle. A diagonal

of the field is 100 ft. We can use the Pythagorean theorem to formulate an equation

that relates the sides and a diagonal.

$$\begin{cases} xy = 4000 & \text{area} & (E_1) \\ x^2 + y^2 = 100^2 & \text{diagonal} & (E_2) \end{cases}$$

Solve E_1 for y, $y = 4000/x$, and substitute into E_2.

$x^2 + \dfrac{4000^2}{x^2} = 100^2 \Rightarrow x^4 - 10{,}000x^2 + 16{,}000{,}000 = 0 \Rightarrow$

$(x^2 - 2000)(x^2 - 8000) = 0 \Rightarrow x = 20\sqrt{5}, 40\sqrt{5}$ and $y = 40\sqrt{5}, 20\sqrt{5}.$

The dimensions are $20\sqrt{5}$ ft $\times\, 40\sqrt{5}$ ft.

[60] Following the hint, we let $y = mx + 3$ and substitute into $x^2 + y^2 = 1$, obtaining

$x^2 + (mx + 3)^2 = 1 \Rightarrow x^2 + (m^2x^2 + 6mx + 9) = 1 \Rightarrow (m^2 + 1)x^2 + (6m)x + (8) = 0 \Rightarrow$

$$x = \frac{-6m \pm \sqrt{(6m)^2 - 4(m^2+1)(8)}}{2(m^2+1)} = \frac{-6m \pm \sqrt{36m^2 - 32m^2 - 32}}{2(m^2+1)}.$$

If there is to be only one solution to the system, i.e., one point of intersection

between the circle and the line, then the discriminant must equal 0.

$4m^2 - 32 = 0 \Rightarrow m = \pm 2\sqrt{2}$ and the equations of the lines are $y = \pm 2\sqrt{2}\,x + 3.$

[62] Let r_1 and r_2 denote the inside radius and the outside radius, respectively.

Inside distance $= 90\%$ (outside distance) $\Rightarrow 2\pi r_1 = 0.90\,(2\pi r_2) \Rightarrow$

$r_1 = 0.90\,(r_1 + 10)$ {since $r_2 = r_1 + 10$} $\Rightarrow 0.1r_1 = 9 \Rightarrow r_1 = 90$ ft and $r_2 = 100$ ft

$\boxed{64}$ Let x and y denote the number of desks shipped from the western warehouse and the eastern warehouse, respectively.

$$\begin{cases} x + y = 150 & \text{\textit{quantity}} \\ 24x + 35y = 4205 & \text{\textit{price}} \end{cases} \qquad E_2 - 24\,E_1 \Rightarrow 11y = 605 \Rightarrow y = 55;\ x = 95$$

$\boxed{65}$ If x and y denote the length and the width, respectively, then a system is $x \le 12$, $y \le 8$, $y \ge \frac{1}{2}x$. The graph is the region bounded by the quadrilateral with vertices $(0, 0)$, $(0, 8)$, $(12, 8)$, and $(12, 6)$.

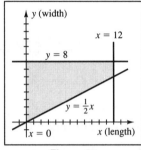

Figure 65

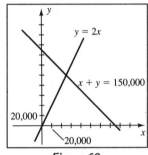

Figure 68

$\boxed{68}$ Let x and y denote the amount in the high- and low-risk investments, respectively. The amount in bonds is $150,000 - x - y$.

Profit function: $P = 0.15x + 0.10y + 0.08(150,000 - x - y) = 12,000 + 0.07x + 0.02y$

(x, y)	$(0, 0)$	$(0, 150,000)$	$(50,000, 100,000)$
P	12,000	15,000	17,500 ■

$$\begin{cases} x + y \le 150,000 \\ y \ge 2x \\ x,\ y \ge 0 \end{cases}$$

The maximum return of $17,500 occurs when

$50,000 is invested in the high-risk investment,

$100,000 is invested in the low-risk investment, and $0 is invested in bonds.

Chapter 10: Sequences, Series, and Probability

Note: For Exercises 1–16, the answers are listed in the order a_1, a_2, a_3, a_4; and a_8.

Simply substitute 1, 2, 3, 4, and 8 for n in the formula for a_n to obtain the results.

$\boxed{3}$ $a_n = \dfrac{3n-2}{n^2+1}$ • $a_1 = \dfrac{3(1)-2}{(1)^2+1} = \dfrac{1}{2}$, $a_2 = \dfrac{3(2)-2}{(2)^2+1} = \dfrac{4}{5}$, $a_3 = \dfrac{3(3)-2}{(3)^2+1} = \dfrac{7}{10}$,

$a_4 = \dfrac{3(4)-2}{(4)^2+1} = \dfrac{10}{17}$; $a_8 = \dfrac{3(8)-2}{(8)^2+1} = \dfrac{22}{65}$ ★ $\dfrac{1}{2}$, $\dfrac{4}{5}$, $\dfrac{7}{10}$, $\dfrac{10}{17}$; $\dfrac{22}{65}$

$\boxed{7}$ $a_n = 2 + (-0.1)^n$ • $a_1 = 2 + (-0.1)^1 = 2 - 0.1 = 1.9$,

$a_2 = 2 + (-0.1)^2 = 2 + 0.01 = 2.01$,

$a_3 = 2 + (-0.1)^3 = 2 - 0.001 = 1.999$,

$a_4 = 2 + (-0.1)^4 = 2 + 0.0001 = 2.0001$;

$a_8 = 2 + (-0.1)^8 = 2 + 0.00000001 = 2.00000001$

★ 1.9, 2.01, 1.999, 2.0001; 2.00000001

$\boxed{9}$ $a_n = (-1)^{n-1}\left(\dfrac{n+7}{2n}\right)$ •

$a_1 = (-1)^{1-1}\left(\dfrac{1+7}{2(1)}\right) = (-1)^0 \left(\tfrac{8}{2}\right) = (1)(4) = 4$,

$a_2 = (-1)^{2-1}\left(\dfrac{2+7}{2(2)}\right) = (-1)^1 \left(\tfrac{9}{4}\right) = -\dfrac{9}{4}$,

$a_3 = (-1)^{3-1}\left(\dfrac{3+7}{2(3)}\right) = (-1)^2 \left(\tfrac{10}{6}\right) = \dfrac{5}{3}$,

$a_4 = (-1)^{4-1}\left(\dfrac{4+7}{2(4)}\right) = (-1)^3 \left(\tfrac{11}{8}\right) = -\dfrac{11}{8}$;

$a_8 = (-1)^{8-1}\left(\dfrac{8+7}{2(8)}\right) = (-1)^7 \left(\tfrac{15}{16}\right) = -\dfrac{15}{16}$ ★ 4, $-\dfrac{9}{4}$, $\dfrac{5}{3}$, $-\dfrac{11}{8}$; $-\dfrac{15}{16}$

$\boxed{13}$ $a_n = \dfrac{2^n}{n^2+2}$ • $a_1 = \dfrac{2^1}{1^2+2} = \dfrac{2}{3}$, $a_2 = \dfrac{2^2}{2^2+2} = \dfrac{4}{6} = \dfrac{2}{3}$, $a_3 = \dfrac{2^3}{3^2+2} = \dfrac{8}{11}$,

$a_4 = \dfrac{2^4}{4^2+2} = \dfrac{16}{18} = \dfrac{8}{9}$; $a_8 = \dfrac{2^8}{8^2+2} = \dfrac{256}{66} = \dfrac{128}{33}$ ★ $\dfrac{2}{3}$, $\dfrac{2}{3}$, $\dfrac{8}{11}$, $\dfrac{8}{9}$; $\dfrac{128}{33}$

$\boxed{15}$ a_n is the number of decimal places in $(0.1)^n$. •

$(0.1)^1 = 0.1$ { 1 decimal place } $\Rightarrow a_1 = 1$,

$(0.1)^2 = 0.01$ { 2 decimal places } $\Rightarrow a_2 = 2$,

$(0.1)^3 = 0.001$ { 3 decimal places } $\Rightarrow a_3 = 3$,

$(0.1)^4 = 0.0001$ { 4 decimal places } $\Rightarrow a_4 = 4$;

$(0.1)^8 = 0.00000001$ { 8 decimal places } $\Rightarrow a_8 = 8$ ★ 1, 2, 3, 4; 8

$\boxed{17}$ $a_1 = 2,$ $\qquad a_{k+1} = 3a_k - 5$ •

$a_2 = a_{1+1} \{k=1\} = 3a_1 - 5 = 3(2) - 5 = 1,$ $\quad a_3 = 3a_2 - 5 = 3(1) - 5 = -2,$

$a_4 = 3a_3 - 5 = 3(-2) - 5 = -11,$ $\qquad\qquad a_5 = 3a_4 - 5 = 3(-11) - 5 = -38$

$\boxed{21}$ $a_1 = 5,$ $\qquad a_{k+1} = ka_k$ •

$a_2 = a_{1+1} \{k=1\} = 1a_1 = 1(5) = 5,$ $\qquad a_3 = 2a_2 = 2(5) = 10,$

$a_4 = 3a_3 = 3(10) = 30,$ $\qquad\qquad a_5 = 4a_4 = 4(30) = 120$

$\boxed{23}$ $a_1 = 2,$ $\qquad a_{k+1} = (a_k)^k$ •

$a_2 = a_{1+1} \{k=1\} = (a_1)^1 = (2)^1 = 2,$ $\qquad a_3 = (a_2)^2 = (2)^2 = 4,$

$a_4 = (a_3)^3 = (4)^3 = 4^3$ or $64,$ $\qquad\qquad a_5 = (a_4)^4 = (4^3)^4 = 4^{12}$ or $16{,}777{,}216$

$\boxed{25}$ $a_n = 3 + \frac{1}{2}n$ • $S_1 = a_1 = 3 + \frac{1}{2} = \frac{7}{2}.$ $S_2 = S_1 + a_2 = \frac{7}{2} + 4 = \frac{15}{2}.$ Alternatively, to

find S_2 we could use $a_1 + a_2.$ However, in most cases, it is easier to compute S_{n+1}

by using $S_n + a_{n+1}.$ $S_3 = S_2 + a_3 = \frac{15}{2} + \frac{9}{2} = 12.$ $S_4 = S_3 + a_4 = 12 + 5 = 17.$

$\boxed{27}$ $a_n = (-1)^n n^{-1/2}$ • $S_1 = a_1 = -1.$

$S_2 = S_1 + a_2 = -1 + 1/\sqrt{2}.$

$S_3 = S_2 + a_3 = -1 + 1/\sqrt{2} - 1/\sqrt{3}.$

$S_4 = S_3 + a_4 = -1 + 1/\sqrt{2} - 1/\sqrt{3} + \frac{1}{2} = -\frac{1}{2} + 1/\sqrt{2} - 1/\sqrt{3}.$

$\boxed{31}$ $\displaystyle\sum_{k=1}^{4} (k^2 - 5) = (-4) + (-1) + 4 + 11 = 10$

$\boxed{33}$ $\displaystyle\sum_{k=0}^{5} k(k-2) = 0 + (-1) + 0 + 3 + 8 + 15 = 25$

$\boxed{37}$ $\displaystyle\sum_{k=1}^{5} (-3)^{k-1} = 1 + (-3) + 9 + (-27) + 81 = 61$

$\boxed{39}$ $\displaystyle\sum_{k=1}^{100} 100 = 100(100) = 10{,}000$

$\boxed{41}$ $\displaystyle\sum_{k=253}^{571} \frac{1}{3} = (571 - 253 + 1)(\frac{1}{3}) = 319(\frac{1}{3}) = \frac{319}{3}$

$\boxed{43}$ $\displaystyle\sum_{k=1}^{n} (a_k - b_k) = (a_1 - b_1) + (a_2 - b_2) + \cdots + (a_n - b_n)$

$\qquad\qquad = (a_1 + a_2 + \cdots + a_n) + (-b_1 - b_2 - \cdots - b_n)$

$\qquad\qquad = (a_1 + a_2 + \cdots + a_n) - (b_1 + b_2 + \cdots + b_n)$

$\qquad\qquad = \displaystyle\sum_{k=1}^{n} a_k - \sum_{k=1}^{n} b_k$

47 Probably the easiest way to describe this sequence is by using a piecewise-defined function. If $1 \leq n \leq 4$, then $2n$ describes the sequence 2, 4, 6, 8. For $n = 5$, we need to obtain a as the general term—one way to symbolize this is $(n-4)a$. Hence, one possibility is $a_n = \begin{cases} 2n & \text{if } 1 \leq n \leq 4 \\ (n-4)a & \text{if } n \geq 5 \end{cases}$

A more sophisticated (and complicated) approach is to add a term to $2n$—call this term k—so that k is 0 for $n = 1, 2, 3, 4$, but equal to $(a - 10)$ for $n = 5$ {the *minus* 10 is needed to compensate for the *plus* 10 from the $2n$ term}. We can do this by starting with the expression $(n-1)(n-2)(n-3)(n-4)$, which is zero for $n = 1, 2, 3, 4$. For $n = 5$, this expression is $4 \cdot 3 \cdot 2 \cdot 1 = 24$, so we will divide it by 24 and multiply it by $(a - 10)$ to obtain $(a - 10)$. Thus, another possibility is $a_n = 2n + \dfrac{(n-1)(n-2)(n-3)(n-4)(a-10)}{24}$.

51 (a) $a_1 = 1$, $a_2 = 1$, $a_3 = a_2 + a_1 = 1 + 1 = 2$,

$a_4 = a_3 + a_2 = 2 + 1 = 3$,

$a_5 = a_4 + a_3 = 3 + 2 = 5$,

$a_6 = a_5 + a_4 = 5 + 3 = 8$,

$a_7 = a_6 + a_5 = 8 + 5 = 13$,

$a_8 = a_7 + a_6 = 13 + 8 = 21$,

$a_9 = a_8 + a_7 = 21 + 13 = 34$,

$a_{10} = a_9 + a_8 = 34 + 21 = 55$

(b) $r_1 = \frac{1}{1} = 1$, $\qquad\qquad r_2 = \frac{2}{1} = 2$, $\qquad\qquad r_3 = \frac{3}{2} = 1.5$,

$r_4 = \frac{5}{3} = 1.\overline{6}$, $\qquad\qquad r_5 = \frac{8}{5} = 1.6$, $\qquad\qquad r_6 = \frac{13}{8} = 1.625$,

$r_7 = \frac{21}{13} \approx 1.6153846$, $\qquad r_8 = \frac{34}{21} \approx 1.6190476$, $\qquad r_9 = \frac{55}{34} \approx 1.6176471$,

and $r_{10} = \frac{89}{55} \approx 1.6181818$.

53 Graph $y = \left(1 + \dfrac{1}{x} + \dfrac{1}{2x^2}\right)^x$ on the interval $[1, 100]$.

The graph approaches the horizontal asymptote $y \approx 2.718 \approx e$.

For increasing values of n, the terms of the sequence appear to approximate e.

$[1, 100]$ by $[0, 3]$

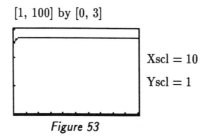

$\mathrm{Xscl} = 10$

$\mathrm{Yscl} = 1$

Figure 53

1 To show that the given sequence, $-6, -2, 2, \ldots, 4n-10, \ldots$, is arithmetic, we must show that $a_{k+1} - a_k$ is equal to some constant, which is the common difference.

$$a_n = 4n - 10 \Rightarrow a_{k+1} - a_k = \left[4(k+1) - 10\right] - \left[4(k) - 10\right] = 4k + 4 - 10 - 4k + 10 = 4$$

Note: For Exercises 3–10, we will find the nth term first, and then use that term to find a_5 and a_{10}.

5 The common difference can be found by *subtracting* any term from its successor.

$$d = 2.7 - 3 = -0.3; \; a_n = a_1 + (n-1)d = 3 + (n-1)(-0.3) = -0.3n + 3.3;$$

$$a_5 = -0.3(5) + 3.3 = 1.8; \; a_{10} = -0.3(10) + 3.3 = 0.3$$

7 $d = -3.9 - (-7) = 3.1; \; a_n = -7 + (n-1)(3.1) = 3.1n - 10.1; \; a_5 = 5.4; \; a_{10} = 20.9$

9 An equivalent sequence is $\ln 3, \ln 3^2, \ln 3^3, \ln 3^4, \ldots$, which is also equivalent to the sequence $\ln 3, 2\ln 3, 3\ln 3, 4\ln 3, \ldots; \; d = 2\ln 3 - \ln 3 = \ln 3;$

$$a_n = \ln 3 + (n-1)(\ln 3) = n\ln 3 \text{ or } \ln 3^n; \; a_5 = 5\ln 3 \text{ or } \ln 3^5; \; a_{10} = 10\ln 3 \text{ or } \ln 3^{10}$$

11 $a_6 = a_1 + 5d$ and $a_2 = a_1 + d \Rightarrow a_6 - a_2 = (a_1 + 5d) - (a_1 + d) = 4d.$

But $a_6 - a_2 = -11 - 21 = -32.$ Hence, $4d = -32$ and $d = -8.$

15 $d = a_7 - a_6 = 5.2 - 2.7 = 2.5; \; a_6 = a_1 + 5d \Rightarrow 2.7 = a_1 + 5(2.5) \Rightarrow a_1 = -9.8$

17 As in the solution for Exercise 11, $a_3 = 7$ and $a_{20} = 43 \Rightarrow 17d = 36 \Rightarrow d = \frac{36}{17}.$

$$a_{15} = a_3 + 12d = 7 + 12\left(\tfrac{36}{17}\right) = \tfrac{551}{17}.$$

Note: To find the sum in Exercises 19–26, we use the formulas

$$S_n = \tfrac{n}{2}\left[2a_1 + (n-1)(d)\right] \quad \text{and} \quad S_n = \tfrac{n}{2}(a_1 + a_n).$$

19 $a_1 = 40, \; d = -3, \; n = 30 \Rightarrow S_{30} = \tfrac{30}{2}\left[2(40) + (29)(-3)\right] = -105.$

21 $a_1 = -9, \; a_{10} = 15, \; n = 10 \Rightarrow S_{10} = \tfrac{10}{2}(-9 + 15) = 30.$

25 $\displaystyle\sum_{k=1}^{18} \left(\tfrac{1}{2}k + 7\right)$ • $a_1 = \tfrac{15}{2}, \; a_{18} = 16, \; n = 18 \Rightarrow S_{18} = \tfrac{18}{2}\left(\tfrac{15}{2} + 16\right) = \tfrac{423}{2}.$

27 $1 + 3 + 5 + 7.$ Since the difference in terms is 2, the general term is

$$a_1 + (n-1)d = 1 + (n-1)2 = 2n - 1. \quad \sum_{n=1}^{4} (2n - 1)$$

29 $1 + 3 + 5 + \cdots + 73.$ From Exercise 27, the general term is $2n - 1$ with

n starting at 1. $2n - 1 = 73 \Rightarrow n = 37$, the largest value. $\displaystyle\sum_{n=1}^{37} (2n - 1)$

31 $\tfrac{3}{7} + \tfrac{6}{11} + \tfrac{9}{15} + \tfrac{12}{19} + \tfrac{15}{23} + \tfrac{18}{27}.$ The numerators increase by 3, the denominators increase

by 4. The general terms are $3 + (n-1)3 = 3n$ and $7 + (n-1)4 = 4n + 3.$ $\displaystyle\sum_{n=1}^{6} \frac{3n}{4n+3}$

33 $a_1 = -2$, $d = \frac{1}{4}$, $S = 21$, and $S_n = \frac{n}{2}[2a_1 + (n-1)(d)] \Rightarrow$

$21 = \frac{n}{2}\left[2(-2) + (n-1)(\frac{1}{4})\right] \Rightarrow 42 = n(\frac{1}{4}n - \frac{17}{4}) \Rightarrow 168 = n^2 - 17n \Rightarrow$

$n^2 - 17n - 168 = 0 \Rightarrow (n-24)(n+7) = 0 \Rightarrow n = 24, -7$.

Since n can not be negative, $n = 24$.

35 If we insert five arithmetic means between 2 and 10, there will be 6 differences that span the distance from 2 to 10. Hence, $6d = 10 - 2 \Rightarrow d = \frac{4}{3}$;

The terms are 2, $\frac{10}{3}$, $\frac{14}{3}$, 6, $\frac{22}{3}$, $\frac{26}{3}$, 10.

37 (a) The first integer greater than 32 that is divisible by 6 is 36 $\{6 \cdot 6\}$

and the last integer less than 395 that is divisible by 6 is 390 $\{65 \cdot 6\}$.

The number of terms is then $65 - 6 + 1 = 60$.

(b) The sum is $S_{60} = \frac{60}{2}(36 + 390) = 12{,}780$.

39 There are $(24 - 10 + 1) = 15$ layers. Model this problem as an arithmetic sequence

with $a_1 = 10$ and $a_{15} = 24$. $S_{15} = \frac{15}{2}(10 + 24) = 255$.

41 This is similar to inserting 9 arithmetic means between 4 and 24.

$10d = 20 \Rightarrow d = 2$. The circumference of each ring is πD with $D = 4, 6, 8, \ldots, 24$.

$a_1 = 4\pi$ and $a_{11} = 24\pi \Rightarrow S_{11} = \frac{11}{2}(4\pi + 24\pi) = 154\pi$ ft.

43 $n = 5$, $S_5 = 5000$, $d = -100 \Rightarrow$

$5000 = \frac{5}{2}\left[2a_1 + 4(-100)\right] \Rightarrow 2000 = 2a_1 - 400 \Rightarrow a_1 = \$1200$

47 If the nth term is $\frac{1}{x_n}$ and $x_{n+1} = \frac{x_n}{1 + x_n}$, then the $(n+1)$st term is

$$\frac{1}{x_{n+1}} = \frac{1}{\frac{x_n}{1 + x_n}} = \frac{1 + x_n}{x_n} = \frac{1}{x_n} + \frac{x_n}{x_n} = 1 + \frac{1}{x_n},$$

which is 1 greater than the nth term and therefore the sequence is arithmetic.

49 (a) $T_8 = 1 + 2 + \cdots + 8 = 36$. $A_k = \frac{n - k + 1}{T_n} \Rightarrow A_1 = \frac{8 - 1 + 1}{36} = \frac{8}{36}$.

$A_2 = \frac{7}{36}$, $A_3 = \frac{6}{36}$, $A_4 = \frac{5}{36}$, $A_5 = \frac{4}{36}$, $A_6 = \frac{3}{36}$, $A_7 = \frac{2}{36}$, $A_8 = \frac{1}{36}$.

(b) $d = A_{k+1} - A_k = -\frac{1}{36}$ for $k = 1, 2, \ldots, 7$. $S_8 = \sum_{k=1}^{8} A_k = \frac{8}{36} + \frac{7}{36} + \cdots + \frac{1}{36} = 1$.

(c) $\$1000\left(\frac{8}{36} + \frac{7}{36} + \frac{6}{36} + \frac{5}{36}\right) \approx \$722.22$

1 To show that the given sequence, $5, -\frac{5}{4}, \frac{5}{16}, \ldots, 5(-\frac{1}{4})^{n-1}, \ldots$, is geometric, we must show that $\frac{a_{k+1}}{a_k}$ is equal to some constant, which is the common ratio.

$$a_n = a_1 r^{n-1} = 5(-\tfrac{1}{4})^{n-1} \Rightarrow \frac{a_{k+1}}{a_k} = \frac{5(-\frac{1}{4})^{(k+1)-1}}{5(-\frac{1}{4})^{k-1}} = -\frac{1}{4}$$

Note: For Exercises 3–14, we will find the nth term first and then use that term to find a_5 and a_8.

3 The common ratio can be found by *dividing* any term by its successor. $r = \frac{4}{8} = \frac{1}{2}$;

$$a_n = a_1 r^{n-1} = 8(\tfrac{1}{2})^{n-1} = 2^3(2^{-1})^{n-1} = 2^3 2^{1-n} = 2^{4-n};$$

$$a_5 = 2^{4-5} = 2^{-1} = \tfrac{1}{2}; \qquad\qquad a_8 = 2^{4-8} = 2^{-4} = \tfrac{1}{16}$$

5 $r = \frac{-30}{300} = -0.1$; $a_n = a_1 r^{n-1} = 300(-0.1)^{n-1}$;

$$a_5 = 300(-0.1)^4 = 0.03; \qquad\qquad a_8 = 300(-0.1)^7 = -0.00003$$

11 $r = \frac{-x^2}{1} = -x^2$; $a_n = a_1 r^{n-1} = 1(-x^2)^{n-1} = (-1)^{n-1}x^{2n-2}$;

$$a_5 = (-1)^4 x^{10-2} = x^8; \qquad\qquad a_8 = (-1)^7 x^{16-2} = -x^{14}$$

13 $r = \frac{2^{x+1}}{2} = 2^x$; $a_n = a_1 r^{n-1} = 2(2^x)^{n-1} = 2^{(n-1)x+1}$; $a_5 = 2^{4x+1}$; $a_8 = 2^{7x+1}$

15 $\frac{a_6}{a_4} = \frac{9}{3} = 3$ and $\frac{a_6}{a_4} = \frac{a_1 r^5}{a_1 r^3} = r^2$. Hence, $r^2 = 3$ and $r = \pm\sqrt{3}$.

17 $r = \frac{6}{4} = \frac{3}{2}$; $a_6 = a_1 r^{n-1} = 4(\frac{3}{2})^5 = \frac{243}{8}$

19 $\frac{a_7}{a_4} = \frac{12}{4} = 3$ and $\frac{a_7}{a_4} = \frac{a_1 r^6}{a_1 r^3} = r^3$. Hence, $r^3 = 3$ and $r = \sqrt[3]{3}$. Since the tenth term

can be obtained by multiplying the seventh term by the common ratio three times, $a_{10} = a_7 r^3 = 12(3) = 36$.

23 $\displaystyle\sum_{k=0}^{9} (-\tfrac{1}{2})^{k+1} = \sum_{k=1}^{10} (-\tfrac{1}{2})^k = -\frac{1}{2} \cdot \frac{1-(-\frac{1}{2})^{10}}{1-(-\frac{1}{2})} = -\frac{1}{2} \cdot \frac{\frac{1023}{1024}}{\frac{3}{2}} = -\frac{1023}{3072}$

27 $\frac{1}{4} - \frac{1}{12} + \frac{1}{36} - \frac{1}{108} = \frac{1}{4} - \frac{1}{4}\cdot\frac{1}{3^1} + \frac{1}{4}\cdot\frac{1}{3^2} - \frac{1}{4}\cdot\frac{1}{3^3} = \displaystyle\sum_{n=1}^{4} (-1)^{n+1}\frac{1}{4}(\frac{1}{3})^{n-1}$

29 $1 - \frac{1}{2} + \frac{1}{4} - \frac{1}{8} + \cdots$ • $a_1 = 1$, $r = -\frac{1}{2}$, $S = \frac{a}{1-r} = \frac{1}{1-(-\frac{1}{2})} = \frac{1}{\frac{3}{2}} = \frac{2}{3}$

33 $\sqrt{2} - 2 + \sqrt{8} - 4 + \cdots$ •

The ratio $r = \dfrac{-2}{\sqrt{2}} = -\sqrt{2}$, but since $|r| = \sqrt{2} > 1$, the sum does not exist.

35 $256 + 192 + 144 + 108 + \cdots$ • $a_1 = 256$, $r = \dfrac{192}{256} = \dfrac{3}{4}$, $S = \dfrac{a}{1-r} = \dfrac{256}{1 - \frac{3}{4}} = 1024$

37 $0.\overline{23}$ • If we write $0.\overline{23}$ as $\dfrac{23}{100} + \dfrac{23}{10,000} + \cdots$, we see that the first term is $\dfrac{23}{100}$ and

the common ratio is $\dfrac{1}{100}$. $a_1 = 0.23$, $r = 0.01$, $S = \dfrac{0.23}{1 - 0.01} = \dfrac{23}{99}$

41 $5.\overline{146}$ • In this problem, we work only with $0.\overline{146}$ and add its rational number

representation to 5. $a_1 = 0.146$, $r = 0.001$,

$$S = \frac{0.146}{1 - 0.001} = \frac{146}{999}; \ 5.\overline{146} = 5 + \frac{146}{999} = \frac{5141}{999}$$

43 $1.\overline{6124}$ • $a_1 = 0.6124$, $r = 0.0001$, $S = \dfrac{0.6124}{1 - 0.0001} = \dfrac{6124}{9999}$;

$$1.\overline{6124} = 1 + \frac{6124}{9999} = \frac{16,123}{9999}$$

45 The geometric mean of 12 and 48 is $\sqrt{12 \cdot 48} = \sqrt{576} = 24$.

47 Inserting 2 geometric means results in a sequence that looks like 4, x, y, 500.

Now $4r = x$, $4r^2 = y$, and $4r^3 = 500$. Thus, $4r^3 = 500 \Rightarrow r^3 = \dfrac{500}{4} = 125 \Rightarrow r = 5$.

The terms are 4, 20, 100, and 500.

49 Let $a_1 = x$ { the original amount of air in the container } and $r = \dfrac{1}{2}$.

$a_{11} = x(\frac{1}{2})^{10} = \dfrac{1}{1024}x$. This is $(\dfrac{1}{1024} \cdot 100)\%$ or $\dfrac{25}{256}\%$ or approximately 0.1% of x.

55 Total amount of local spending $= 2{,}000{,}000(0.60) + 2{,}000{,}000(0.60)^2 + \cdots$. We have

$a_1 = 2{,}000{,}000(0.60) = 1{,}200{,}000$ and $r = 0.60$, so $S = \ = \dfrac{1{,}200{,}000}{1 - 0.60} = \$3{,}000{,}000$.

57 (a) A half-life of 2 hours means there will be $(\frac{1}{2})(\frac{1}{2}D) = \frac{1}{4}D$ after 4 hours. The

amount remaining after n doses { not hours } for a given dose is $a_n = D(\frac{1}{4})^{n-1}$.

Since D mg are administered every 4 hours, the amount of the drug in the

bloodstream after n doses is $\displaystyle\sum_{k=1}^{n} a_k = \sum_{k=1}^{n} D(\frac{1}{4})^{k-1} = D + \frac{1}{4}D + \cdots + (\frac{1}{4})^{n-1}D$.

Since $r = \frac{1}{4} < 1$, S_n may be approximated by $S = \dfrac{a_1}{1-r}$ for large n.

$S = \dfrac{D}{1 - \frac{1}{4}} = \frac{4}{3}D$.

(b) Since the amount of the drug in the bloodstream is given by $\frac{4}{3}D$, and this amount

must be less than or equal to 500 mg, we have $\frac{4}{3}D \le 500$ mg, or, equivalently,

$D \le 375$ mg.

[59] (a) From the figure on the right, we see that

$$(\tfrac{1}{4}a_k)^2 + (\tfrac{3}{4}a_k)^2 = (a_{k+1})^2 \Rightarrow \tfrac{10}{16}a_k^2 = a_{k+1}^2 \Rightarrow a_{k+1} = \tfrac{1}{4}\sqrt{10}\,a_k.$$

(b) From part (a), the common ratio $r = \dfrac{a_{k+1}}{a_k} = \tfrac{1}{4}\sqrt{10}$, so

$a_n = a_1 r^{n-1} = a_1(\tfrac{1}{4}\sqrt{10})^{n-1}$. For the square S_{k+1}, the area $A_{k+1} = a_{k+1}^2$.

But from part (a), $a_{k+1}^2 = \tfrac{10}{16}a_k^2$. Since a_k^2 is the area A_k of square S_k, we have

$A_{k+1} = \tfrac{5}{8}A_k$, and hence $A_n = (\tfrac{5}{8})^{n-1}A_1$. Similarly, for the perimeter of the

square S_k, $P_{k+1} = 4a_{k+1} = 4 \cdot \tfrac{1}{4}\sqrt{10}\,a_k = \sqrt{10}\,a_k = \sqrt{10}(\tfrac{1}{4}P_k)$, and hence

$P_n = (\tfrac{1}{4}\sqrt{10})^{n-1}P_1$.

(c) $\displaystyle\sum_{n=1}^{\infty} P_n$ is an infinite geometric series with first term P_1 and $r = \tfrac{1}{4}\sqrt{10}$.

$$S = \frac{P_1}{1-\tfrac{1}{4}\sqrt{10}} \cdot \frac{4}{4} = \frac{4P_1}{4-\sqrt{10}} = \frac{16a_1}{4-\sqrt{10}}.$$

[61] Let $a_k = 100\left(1 + \dfrac{0.06}{12}\right)^k = 100(1.005)^k$, where k represents the number of compounding

periods for each deposit. For the first deposit, $k = 18 \cdot 12 = 216$. For the last

deposit, $k = 1$. $\quad S_{216} = a_1 + a_2 + \cdots + a_{216}$

$$= 100(1.005)^1 + 100(1.005)^2 + \cdots + 100(1.005)^{216}$$

$$= 100(1.005)\left(\frac{1-(1.005)^{216}}{1-(1.005)}\right) \approx \$38{,}929.00$$

[65] (a) $A_1 = \tfrac{2}{5}\left(1-\tfrac{2}{5}\right)^{1-1} = \tfrac{2}{5}\left(\tfrac{3}{5}\right)^0 = \tfrac{2}{5}$.

$$A_2 = \tfrac{2}{5}\left(\tfrac{3}{5}\right)^1 = \tfrac{6}{25}, \; A_3 = \tfrac{2}{5}\left(\tfrac{3}{5}\right)^2 = \tfrac{18}{125}, \; A_4 = \tfrac{2}{5}\left(\tfrac{3}{5}\right)^3 = \tfrac{54}{625}, \; A_5 = \tfrac{2}{5}\left(\tfrac{3}{5}\right)^4 = \tfrac{162}{3125}.$$

(b) $r = A_{k+1}/A_k = \tfrac{3}{5}$ for $k = 1, 2, 3, 4$.

$$S_5 = \sum_{k=1}^{5} A_k = A_1 + A_2 + A_3 + A_4 + A_5 = \tfrac{2}{5} \cdot \frac{1-(\tfrac{3}{5})^5}{1-\tfrac{3}{5}} = 1 - (\tfrac{3}{5})^5 = \tfrac{2882}{3125} = 0.92224.$$

(c) $\$25{,}000\left(\tfrac{2}{5} + \tfrac{6}{25}\right) = \$25{,}000(\tfrac{16}{25}) = \$16{,}000$

10.4 Exercises

[1] (1) P_1 is true, since $2(1) = 1(1+1) = 2$.

(2) Assume P_k is true:

$2 + 4 + 6 + \cdots + 2k = k(k+1)$. Hence,

$$2 + 4 + 6 + \cdots + 2k + 2(k+1) = k(k+1) + 2(k+1)$$

$$= (k+1)(k+2)$$

$$= (k+1)(k+1+1).$$

Thus, P_{k+1} is true, and the proof is complete.

5 (1) P_1 is true, since $5(1) - 3 = \frac{1}{2}(1)[5(1) - 1] = 2$.

(2) Assume P_k is true:

$2 + 7 + 12 + \cdots + (5k - 3) = \frac{1}{2}k(5k - 1)$. Hence,

$2 + 7 + 12 + \cdots + (5k - 3) + 5(k + 1) - 3 = \frac{1}{2}k(5k - 1) + 5(k + 1) - 3$

$$= \tfrac{5}{2}k^2 + \tfrac{9}{2}k + 2$$
$$= \tfrac{1}{2}(5k^2 + 9k + 4)$$
$$= \tfrac{1}{2}(k + 1)(5k + 4)$$
$$= \tfrac{1}{2}(k + 1)[5(k + 1) - 1].$$

Thus, P_{k+1} is true, and the proof is complete.

7 (1) P_1 is true, since $1 \cdot 2^{1-1} = 1 + (1 - 1) \cdot 2^1 = 1$.

(2) Assume P_k is true:

$1 + 2 \cdot 2 + 3 \cdot 2^2 + \cdots + k \cdot 2^{k-1} = 1 + (k - 1) \cdot 2^k$. Hence,

$1 + 2 \cdot 2 + 3 \cdot 2^2 + \cdots + k \cdot 2^{k-1} + (k + 1) \cdot 2^k = 1 + (k - 1) \cdot 2^k + (k + 1) \cdot 2^k$

$$= 1 + k \cdot 2^k - 2^k + k \cdot 2^k + 2^k$$
$$= 1 + k \cdot 2^1 \cdot 2^k$$
$$= 1 + [(k + 1) - 1] \cdot 2^{k+1}.$$

Thus, P_{k+1} is true, and the proof is complete.

9 (1) P_1 is true, since $(1)^1 = \dfrac{1(1 + 1)[2(1) + 1]}{6} = 1$.

(2) Assume P_k is true:

$1^2 + 2^2 + 3^2 + \cdots + k^2 = \dfrac{k(k + 1)(2k + 1)}{6}$. Hence,

$1^2 + 2^2 + 3^2 + \cdots + k^2 + (k + 1)^2 = \dfrac{k(k + 1)(2k + 1)}{6} + (k + 1)^2$

$$= (k + 1)\left[\frac{k(2k + 1)}{6} + \frac{6(k + 1)}{6}\right]$$
$$= \frac{(k + 1)(2k^2 + 7k + 6)}{6}$$
$$= \frac{(k + 1)(k + 2)(2k + 3)}{6}.$$

Thus, P_{k+1} is true, and the proof is complete.

11 (1) P_1 is true, since $\frac{1}{1(1+1)} = \frac{1}{1+1} = \frac{1}{2}$.

(2) Assume P_k is true:

$$\frac{1}{1\cdot 2}+\frac{1}{2\cdot 3}+\frac{1}{3\cdot 4}+\cdots+\frac{1}{k(k+1)}=\frac{k}{k+1}. \text{ Hence,}$$

$$\frac{1}{1\cdot 2}+\frac{1}{2\cdot 3}+\frac{1}{3\cdot 4}+\cdots+\frac{1}{k(k+1)}+\frac{1}{(k+1)(k+2)}=\frac{k}{k+1}+\frac{1}{(k+1)(k+2)}$$

$$=\frac{k}{k+1}+\frac{1}{(k+1)(k+2)}$$

$$=\frac{k(k+2)+1}{(k+1)(k+2)}$$

$$=\frac{k^2+2k+1}{(k+1)(k+2)}$$

$$=\frac{k+1}{(k+1)+1}.$$

Thus, P_{k+1} is true, and the proof is complete.

13 (1) P_1 is true, since $3^1 = \frac{3}{2}(3^1-1) = 3$.

(2) Assume P_k is true:

$$3+3^2+3^3+\cdots+3^k=\frac{3}{2}(3^k-1). \text{ Hence,}$$

$$3+3^2+3^3+\cdots+3^k+3^{k+1}=\frac{3}{2}(3^k-1)+3^{k+1}$$

$$=\frac{3}{2}\cdot 3^k-\frac{3}{2}+3\cdot 3^k$$

$$=\frac{9}{2}\cdot 3^k-\frac{3}{2}$$

$$=\frac{3}{2}(3\cdot 3^k-1)$$

$$=\frac{3}{2}(3^{k+1}-1).$$

Thus, P_{k+1} is true, and the proof is complete.

15 (1) P_1 is true, since $1 < 2^1$.

(2) Assume P_k is true: $k < 2^k$. Now $k+1 < k+k = 2(k)$ for $k > 1$.

From P_k, we see that $2(k) < 2(2^k) = 2^{k+1}$ and conclude that $k+1 < 2^{k+1}$.

Thus, P_{k+1} is true, and the proof is complete.

17 (1) P_1 is true, since $1 < \frac{1}{8}[2(1)+1]^2 = \frac{9}{8}$.

(2) Assume P_k is true: $1+2+3+\cdots+k < \frac{1}{8}(2k+1)^2$. Hence,

$$1+2+3+\cdots+k+(k+1) < \frac{1}{8}(2k+1)^2+(k+1)$$

$$=\frac{1}{2}k^2+\frac{3}{2}k+\frac{9}{8}$$

$$=\frac{1}{8}(4k^2+12k+9)$$

$$=\frac{1}{8}(2k+3)^2$$

$$=\frac{1}{8}[2(k+1)+1]^2.$$

Thus, P_{k+1} is true, and the proof is complete.

19 (1) For $n = 1$, $n^3 - n + 3 = 3$ and 3 is a factor of 3.

   (2) Assume 3 is a factor of $k^3 - k + 3$. The $(k+1)$st term is

$$(k+1)^3 - (k+1) + 3 = k^3 + 3k^2 + 2k + 3$$
$$= (k^3 - k + 3) + 3k^2 + 3k$$
$$= (k^3 - k + 3) + 3(k^2 + k).$$

By the induction hypothesis, 3 is a factor of $k^3 - k + 3$ and 3 is a factor of $3(k^2 + k)$, so 3 is a factor of the $(k+1)$st term. Thus, P_{k+1} is true, and the proof is complete.

21 (1) For $n = 1$, $5^n - 1 = 4$ and 4 is a factor of 4.

   (2) Assume 4 is a factor of $5^k - 1$. The $(k+1)$st term is

$$5^{k+1} - 1 = 5 \cdot 5^k - 1$$
$$= 5 \cdot 5^k - 5 + 4$$
$$= 5(5^k - 1) + 4.$$

By the induction hypothesis, 4 is a factor of $5^k - 1$ and 4 is a factor of 4, so 4 is a factor of the $(k+1)$st term. Thus, P_{k+1} is true, and the proof is complete.

23 (1) If $a > 1$, then $a^1 = a > 1$, so P_1 is true.

   (2) Assume P_k is true: $a^k > 1$.

Multiply both sides by a to obtain $a^{k+1} > a$, but since $a > 1$, we have $a^{k+1} > 1$.

Thus, P_{k+1} is true, and the proof is complete.

25 (1) For $n = 1$, $a - b$ is a factor of $a^1 - b^1$.

   (2) Assume $a - b$ is a factor of $a^k - b^k$. Following the hint for the $(k+1)$st term,
$a^{k+1} - b^{k+1} = a^k \cdot a - b \cdot a^k + b \cdot a^k - b^k \cdot b = a^k(a - b) + (a^k - b^k)b$. Since $(a - b)$
is a factor of $a^k(a - b)$ and since by the induction hypothesis $a - b$ is a factor of
$(a^k - b^k)$, it follows that $a - b$ is a factor of the $(k+1)$st term. Thus,
P_{k+1} is true, and the proof is complete.

Note: For Exercises 27–32 in this section and Exercises 47–48 in the Chapter Review, there are several ways to find j. Possibilities include: solve the inequality, sketch the graphs of functions representing each side, and trial and error. Trial and error may be the easiest to use.

27 For j: $n^2 \geq n + 12 \Rightarrow n^2 - n - 12 \geq 0 \Rightarrow (n - 4)(n + 3) \geq 0 \Rightarrow n \geq 4 \ \{n > 0\}$

   (1) P_4 is true, since $4 + 12 \leq 4^2$.

   (2) Assume P_k is true: $k + 12 \leq k^2$. Hence,

$$(k+1) + 12 = (k+12) + 1 \leq (k^2) + 1 < k^2 + 2k + 1 = (k+1)^2.$$

Thus, P_{k+1} is true, and the proof is complete.

29 For j: By sketching $y = 5 + \log_2 x$ and $y = x$, we see that the solution for $x > 1$

must be larger than 5. By trial and error, $j = 8$.

(1) P_8 is true, since $5 + \log_2 8 \le 8$.

(2) Assume P_k is true: $5 + \log_2 k \le k$. Hence,

$$5 + \log_2 (k+1) < 5 + \log_2 (k+k)$$
$$= 5 + \log_2 2k$$
$$= 5 + \log_2 2 + \log_2 k$$
$$= (5 + \log_2 k) + 1$$
$$\le k + 1.$$

Thus, P_{k+1} is true, and the proof is complete.

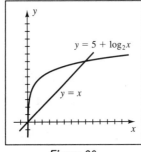

Figure 29

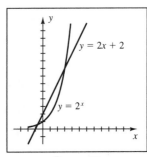

Figure 31

31 For j: By sketching $y = 2x + 2$ and $y = 2^x$, we see there is one positive solution.

By trial and error, $j = 3$.

(1) P_3 is true, since $2(3) + 2 \le 2^3$.

(2) Assume P_k is true: $2k + 2 \le 2^k$. Hence,

$$2(k+1) + 2 = (2k + 2) + 2 \le 2^k + 2^k = 2 \cdot 2^k = 2^{k+1}.$$

Thus, P_{k+1} is true, and the proof is complete.

33 Following the hint in the text:

$$\sum_{k=1}^{n} (k^2 + 3k + 5) = \sum_{k=1}^{n} k^2 + 3 \sum_{k=1}^{n} k + \sum_{k=1}^{n} 5$$
$$= \frac{n(n+1)(2n+1)}{6} + 3 \left[\frac{n(n+1)}{2} \right] + 5n$$
$$= \frac{n(n+1)(2n+1) + 9n(n+1) + 30n}{6}$$
$$= \frac{2n^3 + 12n^2 + 40n}{6}$$
$$= \frac{n^3 + 6n^2 + 20n}{3}$$

37 (a) $n = 1 \Rightarrow a(1)^3 + b(1)^2 + c(1) = 1^2 \Rightarrow a + b + c = 1$

$n = 2 \Rightarrow a(2)^3 + b(2)^2 + c(2) = 1^2 + 2^2 \Rightarrow 8a + 4b + 2c = 5$

$n = 3 \Rightarrow a(3)^3 + b(3)^2 + c(3) = 1^2 + 2^2 + 3^2 \Rightarrow 27a + 9b + 3c = 14$

$$AX = B \Rightarrow \begin{bmatrix} 1 & 1 & 1 \\ 8 & 4 & 2 \\ 27 & 9 & 3 \end{bmatrix} \begin{bmatrix} a \\ b \\ c \end{bmatrix} = \begin{bmatrix} 1 \\ 5 \\ 14 \end{bmatrix} \Rightarrow X = A^{-1}B = \begin{bmatrix} 1/3 \\ 1/2 \\ 1/6 \end{bmatrix}.$$

(b) $a = \frac{1}{3}, \; b = \frac{1}{2}, \; c = \frac{1}{6} \Rightarrow 1^2 + 2^2 + 3^2 + \cdots + n^2 = \frac{1}{3}n^3 + \frac{1}{2}n^2 + \frac{1}{6}n = \dfrac{n(n+1)(2n+1)}{6}$,

which is the formula found in Exercise 9. This method does not verify the formula for all n but only for $n = 1, 2, 3$. Mathematical induction should be used to verify the formula for all n as in Exercise 9.

39 (a) Following the pattern from Exercises 37–38, write

$1^4 + 2^4 + 3^4 + \cdots + n^4 = an^5 + bn^4 + cn^3 + dn^2 + en$. Then, it follows that:

$n = 1 \Rightarrow a(1)^5 + b(1)^4 + c(1)^3 + d(1)^2 + e(1) = 1^4 \Rightarrow a + b + c + d + e = 1$,

$n = 2 \Rightarrow a(2)^5 + b(2)^4 + c(2)^3 + d(2)^2 + e(2) = 1^4 + 2^4 \Rightarrow$
$$32a + 16b + 8c + 4d + 2e = 17,$$

$n = 3 \Rightarrow a(3)^5 + b(3)^4 + c(3)^3 + d(3)^2 + e(3) = 1^4 + 2^4 + 3^4 \Rightarrow$
$$243a + 81b + 27c + 9d + 3e = 98,$$

$n = 4 \Rightarrow a(4)^5 + b(4)^4 + c(4)^3 + d(4)^2 + e(4) = 1^4 + 2^4 + 3^4 + 4^4 \Rightarrow$
$$1024a + 256b + 64c + 16d + 4e = 354,$$

$n = 5 \Rightarrow a(5)^5 + b(5)^4 + c(5)^3 + d(5)^2 + e(5) = 1^4 + 2^4 + 3^4 + 4^4 + 5^4 \Rightarrow$
$$3125a + 625b + 125c + 25d + 5e = 979.$$

$$AX = B \Rightarrow \begin{bmatrix} 1 & 1 & 1 & 1 & 1 \\ 32 & 16 & 8 & 4 & 2 \\ 243 & 81 & 27 & 9 & 3 \\ 1024 & 256 & 64 & 16 & 4 \\ 3125 & 625 & 125 & 25 & 5 \end{bmatrix} \begin{bmatrix} a \\ b \\ c \\ d \\ e \end{bmatrix} = \begin{bmatrix} 1 \\ 17 \\ 98 \\ 354 \\ 979 \end{bmatrix} \Rightarrow X = A^{-1}B = \begin{bmatrix} 1/5 \\ 1/2 \\ 1/3 \\ 0 \\ -1/30 \end{bmatrix}$$

Thus, $1^4 + 2^4 + 3^4 + \cdots + n^4 = \frac{1}{5}n^5 + \frac{1}{2}n^4 + \frac{1}{3}n^3 - \frac{1}{30}n$.

This formula must be verified.

(b) Let P_n be the statement that $1^4 + 2^4 + 3^4 + \cdots + n^4 = \frac{1}{5}n^5 + \frac{1}{2}n^4 + \frac{1}{3}n^3 - \frac{1}{30}n$.

(1) $1^4 = \frac{1}{5} + \frac{1}{2} + \frac{1}{3} - \frac{1}{30} = 1$ and P_1 is true.

(2) Assume that P_k is true. We must show that P_{k+1} is true.

$P_k \Rightarrow 1^4 + 2^4 + 3^4 + \cdots + k^4 = \frac{1}{5}k^5 + \frac{1}{2}k^4 + \frac{1}{3}k^3 - \frac{1}{30}k \Rightarrow$

$1^4 + 2^4 + 3^4 + \cdots + k^4 + (k+1)^4 = \frac{1}{5}k^5 + \frac{1}{2}k^4 + \frac{1}{3}k^3 - \frac{1}{30}k + (k+1)^4 \Rightarrow$

$1^4 + 2^4 + 3^4 + \cdots + k^4 + (k+1)^4 =$

$$\frac{1}{5}k^5 + \frac{1}{2}k^4 + \frac{1}{3}k^3 - \frac{1}{30}k + (k^4 + 4k^3 + 6k^2 + 4k + 1) \Rightarrow$$

$$1^4 + 2^4 + 3^4 + \cdots + k^4 + (k+1)^4 = \tfrac{1}{5}k^5 + \tfrac{3}{2}k^4 + \tfrac{13}{3}k^3 + 6k^2 + \tfrac{119}{30}k + 1.$$

$$P_{k+1} \Rightarrow 1^4 + 2^4 + 3^4 + \cdots + k^4 + (k+1)^4$$

$$= \tfrac{1}{5}(k+1)^5 + \tfrac{1}{2}(k+1)^4 + \tfrac{1}{3}(k+1)^3 - \tfrac{1}{30}(k+1)$$

$$= \tfrac{1}{5}(k^5 + 5k^4 + 10k^3 + 10k^2 + 5k + 1) + \tfrac{1}{2}(k^4 + 4k^3 + 6k^2 + 4k + 1) +$$

$$\tfrac{1}{3}(k^3 + 3k^2 + 3k + 1) - \tfrac{1}{30}(k+1)$$

$$= \tfrac{1}{5}k^5 + \tfrac{3}{2}k^4 + \tfrac{13}{3}k^3 + 6k^2 + \tfrac{119}{30}k + 1. \text{ Thus, the formula is true for all } n.$$

43 (1) For $n = 1$, $\left[r\left(\cos\theta + i\sin\theta\right)\right]^1 = r^1\left[\cos\left(1\theta\right) + i\sin\left(1\theta\right)\right]$.

(2) Assume P_k is true: $\left[r\left(\cos\theta + i\sin\theta\right)\right]^k = r^k\left(\cos k\theta + i\sin k\theta\right)$. Hence,

$$\left[r\left(\cos\theta + i\sin\theta\right)\right]^{k+1} = \left[r\left(\cos\theta + i\sin\theta\right)\right]^k\left[r\left(\cos\theta + i\sin\theta\right)\right]$$

$$= r^k\left[\cos k\theta + i\sin k\theta\right]\left[r\left(\cos\theta + i\sin\theta\right)\right]$$

$$= r^{k+1}\left[\left(\cos k\theta \cos\theta - \sin k\theta \sin\theta\right)\right.$$

$$\left. + i\left(\sin k\theta \cos\theta + \cos k\theta \sin\theta\right)\right]$$

$$\{ \text{ Use the addition formulas for the sine and cosine. } \}$$

$$= r^{k+1}\left[\cos\left(k+1\right)\theta + i\sin\left(k+1\right)\theta\right].$$

Thus, P_{k+1} is true, and the proof is complete.

10.5 Exercises

1 $2!6! = (2 \cdot 1) \cdot (6 \cdot 5 \cdot 4 \cdot 3 \cdot 2 \cdot 1) = 2 \cdot 720 = 1440$

3 $7!0! = (7 \cdot 6 \cdot 5 \cdot 4 \cdot 3 \cdot 2 \cdot 1) \cdot (1) \{ \text{ Remember that } 0! = 1. \} = 5040$

5 $\dfrac{8!}{5!} = \dfrac{8 \cdot 7 \cdot 6 \cdot 5!}{5!} \{ \text{ cancel } 5! \} = 8 \cdot 7 \cdot 6 = 336$

9 $\dbinom{7}{5} = \dfrac{7!}{5!\,2!} \{ \text{ cancel } 5! \} = \dfrac{7 \cdot 6}{2} = 21$

11 $\dbinom{13}{4} = \dfrac{13!}{4!\,9!} \{ \text{ cancel } 9! \} = \dfrac{13 \cdot 12 \cdot 11 \cdot 10}{4 \cdot 3 \cdot 2} = 715$

13 $\dfrac{(2n+2)!}{(2n)!} = \dfrac{(2n+2)(2n+1)(2n)!}{(2n)!} \{ \text{ cancel } (2n)! \} = (2n+2)(2n+1)$

15 We use the binomial theorem formula with $a = 4x$, $b = -y$, and $n = 3$.

$$(4x - y)^3 = \dbinom{3}{0}(4x)^3(-y)^0 + \dbinom{3}{1}(4x)^2(-y)^1 + \dbinom{3}{2}(4x)^1(-y)^2 + \dbinom{3}{3}(4x)^0(-y)^3$$

$$= (1)(64x^3)(1) + (3)(16x^2)(-y) + (3)(4x)(y^2) + (1)(1)(-y^3)$$

$$= 64x^3 - 48x^2y + 12xy^2 - y^3$$

17 $(a+b)^6 = a^6 + \dbinom{6}{1}a^5b^1 + \dbinom{6}{2}a^4b^2 + \dbinom{6}{3}a^3b^3 + \dbinom{6}{4}a^2b^4 + \dbinom{6}{5}a^1b^5 + b^6$

$$= a^6 + 6a^5b + 15a^4b^2 + 20a^3b^3 + 15a^2b^4 + 6ab^5 + b^6$$

$\boxed{21}$ We use the binomial theorem formula with $a = 3x$, $b = -5y$, and $n = 4$.

$$(3x - 5y)^4 = \binom{4}{0}(3x)^4(-5y)^0 + \binom{4}{1}(3x)^3(-5y)^1 + \binom{4}{2}(3x)^2(-5y)^2 +$$
$$\binom{4}{3}(3x)^1(-5y)^3 + \binom{4}{4}(3x)^0(-5y)^4$$

$$= (1)(81x^4)(1) + (4)(27x^3)(-5y) + (6)(9x^2)(25y^2) +$$
$$(4)(3x)(-125y^3) + (1)(1)(625y^4)$$

$$= 81x^4 - 540x^3y + 1350x^2y^2 - 1500xy^3 + 625y^4$$

$\boxed{25}$ We use the binomial theorem formula with $a = x^{-2}$, $b = 3x$, and $n = 6$.

$$\left(\frac{1}{x^2} + 3x\right)^6 = (x^{-2} + 3x)^6$$

$$= \binom{6}{0}(x^{-2})^6(3x)^0 + \binom{6}{1}(x^{-2})^5(3x)^1 + \binom{6}{2}(x^{-2})^4(3x)^2 + \binom{6}{3}(x^{-2})^3(3x)^3 +$$
$$\binom{6}{4}(x^{-2})^2(3x)^4 + \binom{6}{5}(x^{-2})^1(3x)^5 + \binom{6}{6}(x^{-2})^0(3x)^6$$

$$= (1)(x^{-12})(1) + (6)(x^{-10})(3x^1) + (15)(x^{-8})(9x^2) + (20)(x^{-6})(27x^3) +$$
$$(15)(x^{-4})(81x^4) + (6)(x^{-2})(243x^5) + (1)(1)(729x^6)$$

$$= x^{-12} + 18x^{-9} + 135x^{-6} + 540x^{-3} + 1215 + 1458x^3 + 729x^6$$

$\boxed{27}$ We use the binomial theorem formula with $a = x^{1/2}$, $b = -x^{-1/2}$, and $n = 5$.

$$\left(\sqrt{x} - \frac{1}{\sqrt{x}}\right)^5 = (x^{1/2} - x^{-1/2})^5$$

$$= \binom{5}{0}(x^{1/2})^5(-x^{-1/2})^0 + \binom{5}{1}(x^{1/2})^4(-x^{-1/2})^1 + \binom{5}{2}(x^{1/2})^3(-x^{-1/2})^2 +$$
$$\binom{5}{3}(x^{1/2})^2(-x^{-1/2})^3 + \binom{5}{4}(x^{1/2})^1(-x^{-1/2})^4 + \binom{5}{5}(x^{1/2})^0(-x^{-1/2})^5$$

$$= (1)(x^{5/2})(1) + (5)(x^2)(-x^{-1/2}) + (10)(x^{3/2})(x^{-1}) +$$
$$(10)(x^1)(-x^{-3/2}) + (5)(x^{1/2})(x^{-2}) + (1)(1)(-x^{-5/2})$$

$$= x^{5/2} - 5x^{3/2} + 10x^{1/2} - 10x^{-1/2} + 5x^{-3/2} - x^{-5/2}$$

$\boxed{29}$ For the binomial expression $(3c^{2/5} + c^{4/5})^{25}$, the first three terms are

$$\sum_{k=0}^{2}\binom{25}{k}(3c^{2/5})^{25-k}(c^{4/5})^k$$

$$= \binom{25}{0}(3c^{2/5})^{25}(c^{4/5})^0 + \binom{25}{1}(3c^{2/5})^{24}(c^{4/5})^1 + \binom{25}{2}(3c^{2/5})^{23}(c^{4/5})^2$$

$$= (1)(3^{25}c^{10})(1) + (25)(3^{24}c^{48/5})(c^{4/5}) + (300)(3^{23}c^{46/5})(c^{8/5})$$

$$= 3^{25}c^{10} + 25 \cdot 3^{24}c^{52/5} + 300 \cdot 3^{23}c^{54/5}$$

31 For the binomial expression $(4b^{-1} - 3b)^{15}$, the last three terms are

$$\sum_{k=13}^{15} \binom{15}{k}(4b^{-1})^{15-k}(-3b)^k$$

$$= \binom{15}{13}(4b^{-1})^2(-3b)^{13} + \binom{15}{14}(4b^{-1})^1(-3b)^{14} + \binom{15}{15}(4b^{-1})^0(-3b)^{15}$$

$$= (105)(16b^{-2})(-3^{13}b^{13}) + (15)(4b^{-1})(3^{14}b^{14}) + (1)(1)(-3^{15}b^{15})$$

$$= -1680 \cdot 3^{13}b^{11} + 60 \cdot 3^{14}b^{13} - 3^{15}b^{15}$$

Note: For the following exercises, the general formula for the

kth term of $(a+b)^n$ is $\binom{n}{k-1}(a)^{n-(k-1)}(b)^{k-1} = \boxed{\binom{n}{k-1}(a)^{n-k+1}(b)^{k-1}}$.

33 $\left(\dfrac{3}{c} + \dfrac{c^2}{4}\right)^7$; sixth term $= \binom{7}{5}\left(\dfrac{3}{c}\right)^2\left(\dfrac{c^2}{4}\right)^5 = 21\left(\dfrac{9}{c^2}\right)\left(\dfrac{c^{10}}{1024}\right) = \dfrac{189}{1024}c^8$

37 Since there are 9 terms, the middle term is the fifth term $\left\{\dfrac{9+1}{2} = 5\right\}$.

Using the formula in the *Note* above, we obtain the 5th term of $(x^{1/2} + y^{1/2})^8$,

$$\binom{8}{4}(x^{1/2})^4(y^{1/2})^4 = 70x^2y^2.$$

39 $(2y + x^2)^8$; term that contains x^{10} •

Consider only the variable x in the expansion: $(x^2)^{k-1} = x^{10} \Rightarrow 2k - 2 = 10 \Rightarrow k = 6$;

$$6\text{th term} = \binom{8}{5}(2y)^3(x^2)^5 = 448y^3x^{10}$$

41 $(3b^3 - 2a^2)^4$; term that contains b^9 •

Consider only the variable b in the expansion:

$$(b^3)^{4-k+1} = b^9 \Rightarrow 15 - 3k = 9 \Rightarrow k = 2; \text{2nd term} = \binom{4}{1}(3b^3)^3(-2a^2)^1 = -216b^9a^2$$

43 $\left(3x - \dfrac{1}{4x}\right)^6$; term that does not contain x •

Consider only the variable x in the expansion:

$$x^{6-k+1}(x^{-1})^{k-1} = x^0 \Rightarrow x^{8-2k} = x^0 \Rightarrow k = 4; \text{4th term} = \binom{6}{3}(3x)^3\left(-\dfrac{1}{4x}\right)^3 = -\dfrac{135}{16}$$

45 The first three terms in the expansion of $(1 + 0.2)^{10}$ are

$$\sum_{k=0}^{2} \binom{10}{k}(1)^{10-k}(0.2)^k = \binom{10}{0}(1)^{10}(0.2)^0 + \binom{10}{1}(1)^9(0.2)^1 + \binom{10}{2}(1)^8(0.2)^2$$

$$= (1)(1)(1) + (10)(1)(0.2) + (45)(1)(0.04) = 1 + 2 + 1.8 = 4.8.$$

The calculator result for $(1.2)^{10}$ is approximately 6.19.

47 $\dfrac{(x+h)^4 - x^4}{h} = \dfrac{(x^4 + 4x^3h + 6x^2h^2 + 4xh^3 + h^4) - x^4}{h} = \dfrac{h(4x^3 + 6x^2h + 4xh^2 + h^3)}{h} =$

$$4x^3 + 6x^2h + 4xh^2 + h^3$$

49 $\binom{n}{1} = \dfrac{n!}{(n-1)!\,1!} = n$ and $\binom{n}{n-1} = \dfrac{n!}{[n-(n-1)]!\,(n-1)!} = \dfrac{n!}{1!\,(n-1)!} = n$

1 $P(7, 3) = \frac{7!}{4!} = \frac{7 \cdot 6 \cdot 5 \cdot 4!}{4!} = 7 \cdot 6 \cdot 5 = 210$

5 $P(5, 5) = \frac{5!}{0!} = \frac{5 \cdot 4 \cdot 3 \cdot 2 \cdot 1}{1} = 120$

7 $P(6, 1) = \frac{6!}{5!} = \frac{6 \cdot 5!}{5!} = 6$

9 (a) We can think of the three-digit numbers as filling 3 slots. There are 5 digits to pick from for filling the first slot. There are 4 remaining digits to pick from to fill the second slot. There are 3 remaining digits to pick from to fill the third slot. By the fundamental counting principle, there are a total of $5 \cdot 4 \cdot 3 = 60$ three-digit numbers.

(b) The difference between parts (a) and (b) is that we can use any of the 5 digits for all 3 slots. Hence, there are a total of $5 \cdot 5 \cdot 5 = 125$ three-digit numbers if repetitions are allowed.

11 There are 4 one digit numbers; $4 \cdot 3 = 12$ two digit numbers; $4 \cdot 3 \cdot 2 = 24$ three digit numbers; $4 \cdot 3 \cdot 2 \cdot 1 = 24$ four digit numbers.

Total is $4 + 12 + 24 + 24 = 64$.

17 (a) By the fundamental counting principle, $26 \cdot 9 \cdot 10^4 = 2{,}340{,}000$.

(b) By the fundamental counting principle, $24 \cdot 9 \cdot 10^4 = 2{,}160{,}000$.

19 (a) $P(10, 6) = \frac{10!}{4!} = 10 \cdot 9 \cdot 8 \cdot 7 \cdot 6 \cdot 5 = 151{,}200$

(b) Boy-girl: $6 \cdot 4 \cdot 5 \cdot 3 \cdot 4 \cdot 2 = 2880$. Girl-boy: $4 \cdot 6 \cdot 3 \cdot 5 \cdot 2 \cdot 4 = 2880$.

Total $= 2880 + 2880 = 5760$

27 (a) The number of choices for each letter are: $\underline{2} \cdot \underline{25} \cdot \underline{24} \cdot \underline{23} = 27{,}600$

(b) The number of choices for each letter are: $\underline{2} \cdot \underline{26} \cdot \underline{26} \cdot \underline{26} = 35{,}152$

33 There are 3! ways to choose the couples and 2 ways for each couple to sit. $3! \cdot 2^3 = 48$

35 $P(10, 10) = \frac{10!}{0!} = 10! = 3{,}628{,}800$

37 (a) There are 9 choices for the first digit, 10 for the second, 10 for the third,

and 1 for the fourth and fifth. $9 \cdot 10 \cdot 10 \cdot 1 \cdot 1 = 900$

(b) If n is even, we need to select the first $\frac{n}{2}$ digits. $9 \cdot 10^{(n/2) - 1}$

If n is odd, we need to select the first $\frac{n+1}{2}$ digits. $9 \cdot 10^{(n-1)/2}$

39 (a) There is a horizontal asymptote of $y = 1$.

(b) $\dfrac{n! \, e^n}{n^n \sqrt{2\pi n}} \approx 1 \Rightarrow n! \approx \dfrac{n^n \sqrt{2\pi n}}{e^n}.$

Example: $50! \approx \dfrac{50^{50} \sqrt{2\pi(50)}}{e^{50}} \approx 3.0363 \times 10^{64}.$

The actual value is closer to $3.0414 \times 10^{64}.$

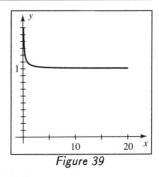

Figure 39

10.7 Exercises

1. $C(7, 3) = \dfrac{7!}{4! \, 3!} = \dfrac{7 \cdot 6 \cdot 5 \cdot 4!}{4! \, 3!} = \dfrac{7 \cdot 6 \cdot 5}{3!} = \dfrac{7 \cdot 6 \cdot 5}{6} = 7 \cdot 5 = 35$

5 $C(n, n-1) = \dfrac{n!}{\left[n - (n-1)\right]! \, (n-1)!} = \dfrac{n!}{1! \, (n-1)!} = n$

7 $C(7, 0) = \dfrac{7!}{7! \, 0!} = \dfrac{1}{0!} = \dfrac{1}{1} = 1$

9 There are 12! total permutations. We want the number of *distinguishable* permutations of the 12 disks. Using the theorem on distinguishable permutations,

$$\text{we have } \dfrac{(5 + 3 + 2 + 2)!}{5! \, 3! \, 2! \, 2!} = \dfrac{12!}{5! \, 3! \, 2! \, 2!} = 166{,}320.$$

13 There are $C(10, 5)$ ways to pick the first team.

The second team is determined once the first team is selected. $C(10, 5) = 252$

15 Two points determine a unique line. $C(8, 2) = 28$

19 Pick the center, $C(3, 1)$; two guards, $C(10, 2)$;

two tackles from the 8 remaining linemen, $C(8, 2)$; two ends, $C(4, 2)$;

two halfbacks, $C(6, 2)$; the quarterback, $C(3, 1)$; and the fullback, $C(4, 1)$.

$$3 \cdot C(10, 2) \cdot C(8, 2) \cdot C(4, 2) \cdot C(6, 2) \cdot 3 \cdot 4 = 4{,}082{,}400$$

21 There are $C(12, 3) = 220$ ways to pick the men and $C(8, 2) = 28$ ways to pick the women. By the fundamental counting principle, the total number of ways to pick the committee is $220 \cdot 28 = 6160$.

23 We need 3 U's out of 8 moves. $C(8, 3) = 56$

27 Let n denote the number of players. $C(n, 2) = 45 \Rightarrow \dfrac{n!}{(n-2)! \, 2!} = 45 \Rightarrow$

$$n(n-1) = 90 \Rightarrow (n-10)(n+9) = 0 \Rightarrow \{n > 0\} \; n = 10.$$

29 Each team must win 3 of the first 6 games for the series to be extended to a

7th game. $C(6, 3) = 20$

31 They may have computed $C(31, 3)$, which is 4495.

33 (a) $S_1 = \binom{1}{1} + \binom{1}{3} + \binom{1}{5} + \cdots = 1 + 0 + 0 + \cdots = 1.$

$S_2 = \binom{2}{1} + \binom{2}{3} + \binom{2}{5} + \cdots = 2 + 0 + 0 + \cdots = 2.$

$S_3 = 3 + 1 + 0 + \cdots = 4. \quad S_4 = 4 + 4 + 0 + \cdots = 8.$

$$S_5 = 16, \ S_6 = 32, \ S_7 = 64, \ S_8 = 128, \ S_9 = 256, \ S_{10} = 512.$$

(b) It appears that $S_n = 2^{n-1}$.

10.8 Exercises

1 (a) We may use ordered pairs to represent the outcomes of the sample space S of the experiment. A representative outcome is (nine of clubs, 3). The number of outcomes in the sample space S is $n(S) = 52 \cdot 6 = 312$.

(b) For each integer k, where $2 \le k \le 6$, there are 4 ways to obtain an outcome of the form $(k, \ k)$ since there are 4 suits. Because there are 5 values of k, $n(E_1) = 5 \cdot 4 = 20.$ $n(E_1') = n(S) - n(E_1) = 312 - 20 = 292.$

$$P(E_1) = \frac{n(E_1)}{n(S)} = \frac{20}{312} = \frac{5}{78}.$$

(c) No, if E_2 or E_3 occurs, then the other event may occur.

Yes, the occurrence of either E_2 or E_3 has no effect on the other event.

$P(E_2) = \dfrac{n(E_2)}{n(S)} = \dfrac{12 \cdot 6}{312} = \dfrac{72}{312} = \dfrac{3}{13}.$ $\qquad P(E_3) = \dfrac{n(E_3)}{n(S)} = \dfrac{52 \cdot 3}{312} = \dfrac{156}{312} = \dfrac{1}{2}.$

Since E_2 and E_3 are indep., $P(E_2 \cap E_3) = P(E_2) \cdot P(E_3) = \frac{3}{13} \cdot \frac{1}{2} = \frac{3}{26} = \frac{36}{312}.$

$P(E_2 \cup E_3) = P(E_2) + P(E_3) - P(E_2 \cap E_3) = \frac{72}{312} + \frac{156}{312} - \frac{36}{312} = \frac{192}{312} = \frac{8}{13}.$

(d) Yes, if E_1 or E_2 occurs, then the other event cannot occur. No, the occurrence of either E_1 or E_2 influences the occurrence of the other event. Remember, (non-empty) *mutually exclusive events* **cannot** *be independent events.*

Since E_1 and E_2 are mutually exclusive, $P(E_1 \cap E_2) = 0$ and

$$P(E_1 \cup E_2) = P(E_1) + P(E_2) = \tfrac{20}{312} + \tfrac{72}{312} = \tfrac{92}{312} = \tfrac{23}{78}.$$

3 The sample space is the 52-card deck, thus, $n(S) = 52$. Using $p(E) = \dfrac{n(E)}{n(S)}$,

we have (a) $\frac{4}{52} = \frac{1}{13}$ (b) $\frac{4}{52} + \frac{4}{52} = \frac{8}{52} = \frac{2}{13}$ (c) $\frac{4}{52} + \frac{4}{52} + \frac{4}{52} = \frac{12}{52} = \frac{3}{13}.$

5 $S = \{1, 2, 3, 4, 5, 6\}$ and $n(S) = 6$. (a) $\frac{1}{6}$ (b) $\frac{1}{6}$ (c) $\frac{1}{6} + \frac{1}{6} = \frac{2}{6} = \frac{1}{3}$

7 $n(S) = 5 + 6 + 4 = 15$ (a) $\frac{5}{15} = \frac{1}{3}$ (b) $\frac{6}{15} = \frac{2}{5}$ (c) $\frac{5}{15} + \frac{4}{15} = \frac{9}{15} = \frac{3}{5}$

9 *Note:* The table gives the results for the sum of two dice being tossed.

Sum of two dice	2	3	4	5	6	7	8	9	10	11	12
# of ways to obtain	1	2	3	4	5	6	5	4	3	2	1

(a) $\frac{2}{36} = \frac{1}{18}$ (b) $\frac{5}{36}$ (c) $\frac{2}{36} + \frac{5}{36} = \frac{7}{36}$

11 $n(S) = 6 \times 6 \times 6 = 216.$

There are 6 ways to make a sum of 5 (3 with 1, 1, 3 and 3 with 1, 2, 2). $\frac{6}{216} = \frac{1}{36}$

Note: For Exercises 15–20, there are $C(52, 5) = 2,598,960$ ways to draw 5 cards.

15 There are 13 denominations to pick from and any one of them could be combined

with any one of the remaining 48 cards. $\frac{48 \cdot 13}{C(52, 5)} = \frac{1}{4165} \approx 0.00024$

17 Pick 4 of the 13 diamonds and 1 of the 13 spades.

$$\frac{C(13, 4) \cdot C(13, 1)}{C(52, 5)} = \frac{143}{39,984} \approx 0.00358$$

21 Let E_1 be the event that the number is odd, E_2 that the number is prime.

$E_1 = \{1, 3, 5\}$ and $E_2 = \{2, 3, 5\}.$

$$P(E_1 \cup E_2) = P(E_1) + P(E_2) - P(E_1 \cap E_2) = \tfrac{3}{6} + \tfrac{3}{6} - \tfrac{2}{6} = \tfrac{4}{6} = \tfrac{2}{3}.$$

23 The probability that the player does *not* get a hit is $1 - 0.326 = 0.674$. The probability that the player does not get a hit four times in a row is $(0.674) \cdot (0.674) \cdot (0.674) \cdot (0.674) = (0.674)^4 \approx 0.2064.$

25 (a) $P(E_2) = P(2) + P(3) + P(4) = 0.10 + 0.15 + 0.20 = 0.45$

(b) $P(E_1 \cap E_2) = P(2) = 0.10$

(c) $P(E_1 \cup E_2) = P(E_1) + P(E_2) - P(E_1 \cap E_2) = 0.35 + 0.45 - 0.10 = 0.70$

(d) $P(E_2 \cup E_3') = P(E_2) + P(E_3') - P(E_2 \cap E_3')$. $E_3' = \{1, 2, 3, 5\}$ and

$$E_2 \cap E_3' = \{2, 3\} \Rightarrow P(E_2 \cup E_3') = 0.45 + 0.75 - 0.25 = 0.95.$$

Note: For Exercises 27–28, there are $C(60, 5) = 5,461,512$ ways to draw 5 chips.

27 (a) We want 5 blue and 0 non-blue. There are 20 blue and 40 non-blue chips in the

box. $\dfrac{C(20, 5) \cdot C(40, 0)}{C(60, 5)} = \dfrac{34}{11,977} \approx 0.0028.$

(b) $P(\text{at least 1 green}) = 1 - P(\text{no green}) =$

$$1 - \frac{C(30, 0) \cdot C(30, 5)}{C(60, 5)} = 1 - \frac{117}{4484} = \frac{4367}{4484} \approx 0.9739.$$

(c) $P(\text{at most 1 red}) = P(0 \text{ red}) + P(1 \text{ red}) =$

$$\frac{C(10, 0) \cdot C(50, 5)}{C(60, 5)} + \frac{C(10, 1) \cdot C(50, 4)}{C(60, 5)} = \frac{26,320}{32,509} \approx 0.8096.$$

29 (a) $\dfrac{C(8, 8)}{2^8} = \dfrac{1}{256} \approx 0.00391$ (b) $\dfrac{C(8, 7)}{2^8} = \dfrac{1}{32} = 0.03125$

(c) $\dfrac{C(8, 6)}{2^8} = \dfrac{7}{64} = 0.109375$ (d) $\dfrac{C(8, 6) + C(8, 7) + C(8, 8)}{2^8} = \dfrac{37}{256} \approx 0.14453$

31 $P(\text{obtaining at least one ace}) = 1 - P(\text{no aces}) =$

$$1 - \frac{C(48, 5)}{C(52, 5)} = 1 - \frac{35,673}{54,145} = \frac{18,472}{54,145} \approx 0.34116$$

33 Let k denote the sum.
$$P(k > 5) = 1 - P(k \le 5) = 1 - \left(\tfrac{1}{36} + \tfrac{2}{36} + \tfrac{3}{36} + \tfrac{4}{36}\right) = 1 - \tfrac{10}{36} = \tfrac{26}{36} = \tfrac{13}{18}.$$

37 (a) $\dfrac{C(4, 4)}{4!} = \tfrac{1}{24} \approx 0.04167$ (b) $\dfrac{C(4, 2)}{4!} = \tfrac{1}{4} = 0.25$

39 (a) The 3, 4, or 5 on the left die would need to combine with a 4, 3, or 2 on the right die to sum to 7, but the right die only has 1, 5, or 6. The probability is 0.

(b) To obtain 8, we would need a 3 on the left die and a 5 on the right.
$$\tfrac{1}{3} \cdot \tfrac{1}{3} = \tfrac{1}{9} = 0.\overline{1}$$

41 (a) The ball must take 4 "lefts". $\tfrac{1}{2} \cdot \tfrac{1}{2} \cdot \tfrac{1}{2} \cdot \tfrac{1}{2} = \tfrac{1}{16} = 0.0625$

(b) We need two "lefts". $\dfrac{C(4, 2)}{2^4} = \tfrac{6}{16} = \tfrac{3}{8} = 0.375$

43 For one ticket, $P(E) = \dfrac{n(E)}{n(S)} = \dfrac{C(6, 6)}{C(54, 6)} = \dfrac{1}{25{,}827{,}165}.$

For two tickets, $P(E) = \dfrac{2 \times 1}{25{,}827{,}165},$ or about 1 chance in 13 million.

45 (a) $P(7 \text{ or } 11) = P(7) + P(11) = \tfrac{6}{36} + \tfrac{2}{36} = \tfrac{8}{36}$

(b) To win with a 4 on the first roll, we must first get a 4, and then get another 4 before a 7. The probability of getting a 4 is $\tfrac{3}{36}$. The probability of getting another 4 before a 7 is $\tfrac{3}{3+6}$ since there are 3 ways to get a 4, 6 ways to get a 7, and numbers other than 4 and 7 are immaterial.

Thus, $P(\text{winning with 4}) = \tfrac{3}{36} \cdot \tfrac{3}{3+6} = \tfrac{1}{36}.$

(c) Let $P(k)$ denote the probability of winning a pass line bet with the number k.

We first note that $P(4) = P(10)$, $P(5) = P(9)$, and $P(6) = P(8)$.

$P(\text{winning}) = 2 \cdot P(4) + 2 \cdot P(5) + 2 \cdot P(6) + P(7) + P(11)$

$\phantom{P(\text{winning})} = 2 \cdot \tfrac{3}{36} \cdot \tfrac{3}{3+6} + 2 \cdot \tfrac{4}{36} \cdot \tfrac{4}{4+6} + 2 \cdot \tfrac{5}{36} \cdot \tfrac{5}{5+6} + \tfrac{6}{36} + \tfrac{2}{36}$

$\phantom{P(\text{winning})} = 2 \cdot \tfrac{1}{36} + 2 \cdot \tfrac{2}{45} + 2 \cdot \tfrac{25}{396} + \tfrac{1}{6} + \tfrac{1}{18} = \tfrac{488}{990} = \tfrac{244}{495} \approx 0.4929$

47 (a) $p = P((S_1 \cap S_2) \cup (S_3 \cap S_4))$

$\quad = P(S_1 \cap S_2) + P(S_3 \cap S_4) - P((S_1 \cap S_2) \cap (S_3 \cap S_4))$

$\quad = P(S_1) \cdot P(S_2) + P(S_3) \cdot P(S_4) - P(S_1 \cap S_2) \cdot P(S_3 \cap S_4)$

$\quad = P(S_1) \cdot P(S_2) + P(S_3) \cdot P(S_4) - P(S_1) \cdot P(S_2) \cdot P(S_3) \cdot P(S_4)$

Let $P(S_k) = x$. Then $p = x \cdot x + x \cdot x - x \cdot x \cdot x \cdot x = -x^4 + 2x^2$.

$$x = 0.9 \Rightarrow p = 0.9639.$$

[−2.25, 2.25] by [−2, 1]

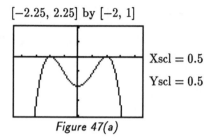

Xscl = 0.5

Yscl = 0.5

Figure 47(a)

[0.96, 1.05] by [−0.03, 0.04]

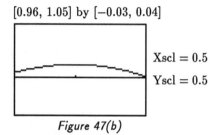

Xscl = 0.5

Yscl = 0.5

Figure 47(b)

(b) $p = 0.99 \Rightarrow -x^4 + 2x^2 = 0.99$. The graph of $y = -x^4 + 2x^2 - 0.99$ is shown in *Figure 47(a)*. The region near $x = 1$ is enlarged in *Figure 47(b)* to show that the graph is above the x-axis for some values of x. The approximate x-intercepts are ± 0.95, ± 1.05. Since $0 \le P(S_k) \le 1$, $P(S_k) = 0.95$.

49 (a) The number of ways that n people can all have a different birthday is $P(365, n)$.

The number of outcomes in the sample space is 365^n.

$$\text{Thus, } p = \frac{P(365, n)}{365^n} = \frac{365!}{365^n (365 - n)!}.$$

(b) $n = 32 \Rightarrow p = \dfrac{365!}{365^{32}\,333!} \Rightarrow \ln p = \ln\dfrac{365!}{365^{32}\,333!} =$

$\ln 365! - \ln 365^{32} - \ln 333! \approx (365 \ln 365 - 365) - (32 \ln 365) - (333 \ln 333 - 333) \approx$

-1.45. Thus, $p \approx e^{-1.45} \approx 0.24$.

The probability that two or more people have the same birthday is $1 - p \approx 0.76$.

Chapter 10 Review Exercises

3   $a_n = 1 + \left(-\frac{1}{2}\right)^{n-1}$ •

$\quad a_1 = 1 + \left(-\frac{1}{2}\right)^0 = 1 + 1 = 2$, $a_2 = 1 + \left(-\frac{1}{2}\right)^1 = 1 - \frac{1}{2} = \frac{1}{2}$

$\quad a_3 = 1 + \left(-\frac{1}{2}\right)^2 = 1 + \frac{1}{4} = \frac{5}{4}$, $a_4 = 1 + \left(-\frac{1}{2}\right)^3 = 1 - \frac{1}{8} = \frac{7}{8}$

$\quad a_7 = 1 + \left(-\frac{1}{2}\right)^6 = 1 + \frac{1}{64} = \frac{65}{64}$ ★ $2, \frac{1}{2}, \frac{5}{4}, \frac{7}{8}, \frac{65}{64}$

8   $a_1 = 1$, $a_{k+1} = (1 + a_k)^{-1}$ •

$\quad a_2 = a_{1+1}\ \{k = 1\} = (1 + a_1)^{-1} = (1 + 1)^{-1} = (2)^{-1} = \frac{1}{2}$

$\quad a_3 = a_{2+1} = (1 + a_2)^{-1} = \left(1 + \frac{1}{2}\right)^{-1} = \left(\frac{3}{2}\right)^{-1} = \frac{2}{3}$

$\quad a_4 = a_{3+1} = (1 + a_3)^{-1} = \left(1 + \frac{2}{3}\right)^{-1} = \left(\frac{5}{3}\right)^{-1} = \frac{3}{5}$

$\quad a_5 = a_{4+1} = (1 + a_4)^{-1} = \left(1 + \frac{3}{5}\right)^{-1} = \left(\frac{8}{5}\right)^{-1} = \frac{5}{8}$ ★ $1, \frac{1}{2}, \frac{2}{3}, \frac{3}{5}, \frac{5}{8}$

$\boxed{10}$ $\displaystyle\sum_{k=2}^{6} \frac{2k-8}{k-1} = (-4) + (-1) + 0 + \frac{1}{2} + \frac{4}{5} = -\frac{37}{10}$

$\boxed{14}$ We first note that $4 + 2 + 1 + \frac{1}{2} + \frac{1}{4} + \frac{1}{8}$ is the sum of six terms of a geometric

sequence with first term 4 and ratio $\frac{1}{2}$—that is, $a_n = a_1 r^{n-1} = 4(\frac{1}{2})^{n-1}$.

$$4 + 2 + 1 + \frac{1}{2} + \frac{1}{4} + \frac{1}{8} = \sum_{n=1}^{6} 4(\tfrac{1}{2})^{n-1} = \sum_{n=1}^{6} 2^2(2^{-1})^{n-1} = \sum_{n=1}^{6} 2^2 2^{1-n} = \sum_{n=1}^{6} 2^{3-n}$$

$\boxed{20}$ $1 - \frac{1}{2} + \frac{1}{3} - \frac{1}{4} + \frac{1}{5} - \frac{1}{6} + \frac{1}{7}$. The terms have alternating signs,

the numerator is 1, and the denominators increase by 1. $\displaystyle\sum_{n=1}^{7} (-1)^{n-1} \frac{1}{n}$

$\boxed{21}$ $a_0 + a_1 x^4 + a_2 x^8 + \cdots + a_{25} x^{100}$ { the exponents are multiples of 4 } $= \displaystyle\sum_{n=0}^{25} a_n x^{4n}$

$\boxed{24}$ $1 + x + \frac{x^2}{2} + \frac{x^3}{3} + \cdots + \frac{x^n}{n}$.

The pattern begins with the second term and the general term is listed. $1 + \displaystyle\sum_{k=1}^{n} \frac{x^k}{k}$

$\boxed{25}$ $d = 3 - (4 + \sqrt{3}) = -1 - \sqrt{3}$; $a_{10} = (4 + \sqrt{3}) + (9)(-1 - \sqrt{3}) = -5 - 8\sqrt{3}$;

$$S_{10} = \frac{10}{2}\left[(4 + \sqrt{3}) + (-5 - 8\sqrt{3}) \right] = -5 - 35\sqrt{3}.$$

$\boxed{27}$ $a_5 = a_1 + 4d = 5$ and $a_{13} = a_1 + 12d = 77$. $a_{13} - a_5 = (a_1 + 12d) - (a_1 + 4d) = 8d$ and

$a_{13} - a_5 = 77 - 5 = 72 \Rightarrow 8d = 72 \Rightarrow d = 9$. $a_5 = a_1 + 4d \Rightarrow 5 = a_1 + 36 \Rightarrow a_1 = -31$.

$$a_{10} = -31 + 9(9) = 50.$$

$\boxed{30}$ We can divide the fourth term by the third term to obtain the common ratio,

$r = \frac{-0.3}{3} = -0.1$. If we multiply the third term by r five times,

we will get the eighth term. $a_8 = a_3 r^5 = 3(-0.1)^5 = -0.00003$.

$\boxed{34}$ $a_5 = a_1 r^4 \Rightarrow \frac{1}{16} = a_1 \left(\frac{3}{2}\right)^4 \Rightarrow a_1 = \frac{1}{16} \cdot \frac{16}{81} = \frac{1}{81}$.

$$S_5 = \frac{1}{81} \cdot \frac{1 - (\frac{3}{2})^5}{1 - \frac{3}{2}} = \frac{1}{81} \cdot \frac{1 - \frac{243}{32}}{-\frac{1}{2}} = \frac{1}{81} \cdot \frac{-\frac{211}{32}}{-\frac{1}{2}} = \frac{211}{1296}.$$

$\boxed{37}$ This sum can be written as a sum of a geometric sequence and an arithmetic

sequence. $\displaystyle\sum_{k=1}^{10} (2^k - \tfrac{1}{2}) = \sum_{k=1}^{10} 2^k - \sum_{k=1}^{10} \tfrac{1}{2} = 2 \cdot \frac{1 - 2^{10}}{1 - 2} - 10(\tfrac{1}{2}) = 2046 - 5 = 2041$

$\boxed{39}$ $a_1 = 1$, $r = -\frac{2}{5} \Rightarrow S = \dfrac{1}{1 - (-\frac{2}{5})} = \dfrac{1}{\frac{7}{5}} = \dfrac{5}{7}$.

41 (1) P_1 is true, since $3(1) - 1 = \dfrac{1[3(1)+1]}{2} = 2$.

(2) Assume P_k is true: $2 + 5 + 8 + \cdots + (3k - 1) = \dfrac{k(3k+1)}{2}$. Hence,

$$
\begin{aligned}
2 + 5 + 8 + \cdots + (3k - 1) + 3(k + 1) - 1 &= \frac{k(3k+1)}{2} + 3(k+1) - 1 \\
&= \frac{3k^2 + k + 6k + 4}{2} \\
&= \frac{3k^2 + 7k + 4}{2} \\
&= \frac{(k+1)(3k+4)}{2} \\
&= \frac{(k+1)[3(k+1)+1]}{2}.
\end{aligned}
$$

Thus, P_{k+1} is true, and the proof is complete.

43 (1) P_1 is true, since $\dfrac{1}{[2(1)-1][2(1)+1]} = \dfrac{1}{2(1)+1} = \dfrac{1}{3}$.

(2) Assume P_k is true:

$$
\frac{1}{1 \cdot 3} + \frac{1}{3 \cdot 5} + \frac{1}{5 \cdot 7} + \cdots + \frac{1}{(2k-1)(2k+1)} = \frac{k}{2k+1}. \quad \text{Hence,}
$$

$$
\begin{aligned}
\frac{1}{1 \cdot 3} + \frac{1}{3 \cdot 5} + \frac{1}{5 \cdot 7} + \cdots &+ \frac{1}{(2k-1)(2k+1)} + \frac{1}{(2k+1)(2k+3)} \\
&= \frac{k}{2k+1} + \frac{1}{(2k+1)(2k+3)} \\
&= \frac{k(2k+3)+1}{(2k+1)(2k+3)} \\
&= \frac{2k^2 + 3k + 1}{(2k+1)(2k+3)} \\
&= \frac{(2k+1)(k+1)}{(2k+1)(2k+3)} \\
&= \frac{k+1}{2(k+1)+1}.
\end{aligned}
$$

Thus, P_{k+1} is true, and the proof is complete.

45 (1) For $n = 1$, $n^3 + 2n = 3$ and 3 is a factor of 3.

(2) Assume 3 is a factor of $k^3 + 2k$. The $(k+1)$st term is

$$
\begin{aligned}
(k+1)^3 + 2(k+1) &= k^3 + 3k^2 + 5k + 3 \\
&= (k^3 + 2k) + (3k^2 + 3k + 3) \\
&= (k^3 + 2k) + 3(k^2 + k + 1).
\end{aligned}
$$

By the induction hypothesis, 3 is a factor of $k^3 + 2k$ and 3 is a factor of $3(k^2 + k + 1)$, so 3 is a factor of the $(k+1)$st term. Thus, P_{k+1} is true, and the proof is complete.

46 (1) P_5 is true, since $5^2 + 3 < 2^5$.

(2) Assume P_k is true: $k^2 + 3 < 2^k$. Hence, $(k+1)^2 + 3 = k^2 + 2k + 4 =$

$$(k^2 + 3) + (k+1) < 2^k + (k+1) < 2^k + 2^k = 2 \cdot 2^k = 2^{k+1}.$$

Thus, P_{k+1} is true, and the proof is complete.

48 For j: $10^n \le n^n \Rightarrow \left(\frac{n}{10}\right)^n \ge 1$. This is true if $\frac{n}{10} \ge 1$ or $n \ge 10$. Thus, $j = 10$.

(1) P_{10} is true, since $10^{10} \le 10^{10}$.

(2) Assume P_k is true: $10^k \le k^k$. Hence,

$$10^{k+1} = 10 \cdot 10^k \le 10 \cdot k^k < (k+1) \cdot k^k < (k+1) \cdot (k+1)^k = (k+1)^{k+1}.$$

Thus, P_{k+1} is true, and the proof is complete.

50 We use the binomial theorem formula with $a = 2a$, $b = b^3$, and $n = 4$.

$$(2a + b^3)^4 = \binom{4}{0}(2a)^4(b^3)^0 + \binom{4}{1}(2a)^3(b^3)^1 + \binom{4}{2}(2a)^2(b^3)^2 + \binom{4}{3}(2a)^1(b^3)^3 +$$

$$\binom{4}{4}(2a)^0(b^3)^4$$

$$= (1)(16a^4)(1) + (4)(8a^3)(b^3) + (6)(4a^2)(b^6) + (4)(2a)(b^9) + (1)(1)(b^{12})$$

$$= 16a^4 + 32a^3b^3 + 24a^2b^6 + 8ab^9 + b^{12}$$

53 $(4a^2 - b)^7$; term that contains a^{10} •

Consider only the variable a in the expansion:

$$(a^2)^{7-k+1} = a^{10} \Rightarrow 16 - 2k = 10 \Rightarrow k = 3; \text{ 3rd term} = \binom{7}{2}(4a^2)^5(-b)^2 = 21{,}504a^{10}b^2$$

55 (a) $S_5 = 10 \Rightarrow 10 = \frac{5}{2}(2a_1 + 4d) \Rightarrow 4 = 2a_1 + 4d \Rightarrow 4d = 4 - 2a_1 \Rightarrow$

$$d = 1 - \tfrac{1}{2}a_1. \text{ Since } a_1 \text{ is positive, } 1 - \tfrac{1}{2}a_1 \text{ is less than 1 ft.}$$

(b) $a_1 = \frac{1}{2} \Rightarrow d = 1 - \frac{1}{2}(\frac{1}{2}) = \frac{3}{4}.$

The lengths of the other four pieces are $1\frac{1}{4}$, 2, $2\frac{3}{4}$, and $3\frac{1}{2}$ ft.

57 If $s_1 = 1$, then $s_2 = f$, $s_3 = f^2$, ... There are two of each of the s_k's.

Since $0 < f < 1$, and f is the common ratio, we can sum the infinite sequence.

The sum of the s_k's is $2(1 + f + f^2 + \cdots) = 2\left(\frac{1}{1-f}\right) = \frac{2}{1-f}.$

58 $\text{Time}_{\text{total}} = \text{Time}_{\text{down}} + \text{Time}_{\text{up}}$

$$= \left[\frac{\sqrt{10}}{4} + \frac{\sqrt{10 \cdot \frac{3}{4}}}{4} + \frac{\sqrt{10 \cdot (\frac{3}{4})^2}}{4} + \cdots\right] + \left[\frac{\sqrt{10 \cdot \frac{3}{4}}}{4} + \frac{\sqrt{10 \cdot (\frac{3}{4})^2}}{4} + \cdots\right]$$

$$= \frac{\sqrt{10}}{4} + 2\left[\frac{\sqrt{10 \cdot \frac{3}{4}}}{4} + \frac{\sqrt{10 \cdot (\frac{3}{4})^2}}{4} + \cdots\right] = \frac{\sqrt{10}}{4} + 2 \cdot \frac{\sqrt{\frac{30}{4}}/4}{1 - \sqrt{\frac{3}{4}}} = \tfrac{1}{4}\sqrt{10} + 2 \cdot \frac{\sqrt{30}/8}{1 - \frac{1}{2}\sqrt{3}}$$

$$= \tfrac{1}{4}\sqrt{10} + \frac{\sqrt{30}}{2(2 - \sqrt{3})} \cdot \frac{2 + \sqrt{3}}{2 + \sqrt{3}} = \tfrac{1}{4}\sqrt{10} + (\sqrt{30} + \tfrac{3}{2}\sqrt{10}) = \tfrac{7}{4}\sqrt{10} + \sqrt{30} \approx 11.01 \text{ sec.}$$

59 (a) $P(52, 13) \approx 3.954 \times 10^{21}$

(b) $P(13, 5) \cdot P(13, 3) \cdot P(13, 3) \cdot P(13, 2) \approx 7.094 \times 10^{13}$

64 (a) We need 4 of the 26 cards of one color. There are 2 colors.

$$\frac{P(26, 4) \cdot 2}{P(52, 4)} = \frac{92}{833} \approx 0.1104$$

(b) We need R-B-R-B. $\dfrac{26^2 \cdot 25^2}{P(52, 4)}$ or $\dfrac{26}{52} \cdot \dfrac{26}{51} \cdot \dfrac{25}{50} \cdot \dfrac{25}{49} = \dfrac{325}{4998} \approx 0.0650$

67 (a) $P(\text{passing}) = P(4, 5, \text{ or } 6 \text{ correct}) =$

$$\frac{C(6, 4) + C(6, 5) + C(6, 6)}{2^6} = \frac{15 + 6 + 1}{64} = \frac{22}{64} = \frac{11}{32}$$

(b) $P(\text{failing}) = 1 - P(\text{passing}) = 1 - \dfrac{22}{64} = \dfrac{42}{64}$

69 Let O denote the event that the individual is over 60 and F denote the event that the individual is female.

$$P(O \cup F) = P(O) + P(F) - P(O \cap F) = \frac{1000}{5000} + \frac{2000}{5000} - \frac{0.40(2000)}{5000} = \frac{2200}{5000} = 0.44$$

70 There are 2 ways to obtain 10 (6, 4 and 4, 6), 2 ways for 11 (6, 5 and 5, 6), 1 way for

12, 16, 20, and 24 (double 3's, 4's, 5's, and 6's). $\dfrac{2+2+1+1+1+1}{36} = \dfrac{8}{36} = \dfrac{2}{9} = 0.\overline{2}$

Chapter 11: Topics From Analytic Geometry

Note: For Exercises 1–12, we will put each parabola equation in one of the forms listed on page 705—either

$$(x - h)^2 = 4p(y - k) \quad \text{or} \quad (y - k)^2 = 4p(x - h).$$

Once in one of those forms, the information concerning the vertex, focus, and directrix is easily obtainable and illustrated in the chart in the text. Let V, F, and l denote the vertex, focus, and directrix, respectively.

3 $2y^2 = -3x \Rightarrow (y - 0)^2 = -\frac{3}{2}(x - 0) \Rightarrow 4p = -\frac{3}{2} \Rightarrow p = -\frac{3}{8}$. We know that the parabola opens either right or left since the variable "y" is squared. Since p is negative, we know that the parabola opens left and that the focus is $\frac{3}{8}$ unit to the left of the vertex. The directrix is $\frac{3}{8}$ unit to the right of the vertex.

$$V(0, 0); \ F(-\tfrac{3}{8}, 0); \ l: x = \tfrac{3}{8}$$

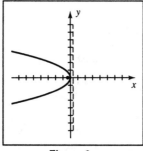

Figure 3

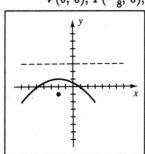

Figure 5

5 $(x + 2)^2 = -8(y - 1) \Rightarrow 4p = -8 \Rightarrow p = -2$. The $(x + 2)$ and $(y - 1)$ factors indicate that we need to shift the vertex, of the parabola having equation $x^2 = -8y$, 2 units left and 1 unit up—that is, move it from $(0, 0)$ to $(-2, 1)$.

$$V(-2, 1); \ F(-2, -1); \ l: y = 3$$

$\boxed{7}$ $(y-2)^2 = \frac{1}{4}(x-3) \Rightarrow 4p = \frac{1}{4} \Rightarrow p = \frac{1}{16}$. "$y$" squared and p positive imply that the parabola opens to the right and the focus is to the right of the vertex. The $(x-3)$ and $(y-2)$ factors indicate that we need to shift the vertex 3 units right and 2 units up from $(0, 0)$ to $(3, 2)$.

$$V(3, 2);\ F(\tfrac{49}{16}, 2);\ l\colon x = \tfrac{47}{16}$$

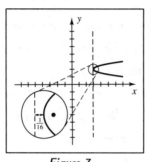

Figure 7

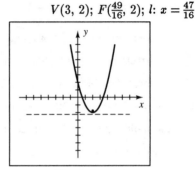

Figure 9

$\boxed{9}$ For this exercise, we need to "complete the square" in order to get the equation in proper form. The term we need to add is $\left[\frac{1}{2}(\text{coefficient of } x)\right]^2$. In this case, that value is $\left[\frac{1}{2}(-4)\right]^2 = 4$. Notice that we add and subtract the value 4 from the same side of the equation as opposed to adding 4 to both sides of the equation.

$y = x^2 - 4x + 2 = (x^2 - 4x + \underline{4\,}) + 2 - \underline{4} = (x-2)^2 - 2 \Rightarrow (y+2) = 1(x-2)^2 \Rightarrow$
$4p = 1 \Rightarrow p = \frac{1}{4}$.

$$V(2, -2);\ F(2, -\tfrac{7}{4});\ l\colon y = -\tfrac{9}{4}$$

$\boxed{13}$ Since the vertex is at $(1, 0)$ and the parabola has a horizontal axis, the standard equation has the form $y^2 = 4p(x-1)$. The distance from the focus $F(6, 0)$ to the vertex $V(1, 0)$ is $6 - 1 = 5$, which is the value of p. Hence, an equation of the parabola is $y^2 = 20(x-1)$.

$\boxed{15}$ $V(-2, 3) \Rightarrow (x+2)^2 = 4p(y-3)$.

$$x = 2,\ y = 2 \Rightarrow 16 = 4p(-1) \Rightarrow p = -4.\ (x+2)^2 = -16(y-3).$$

$\boxed{17}$ The distance from the directrix to the focus is $2 - (-2) = 4$ units. The vertex $V(0, 0)$ is halfway between the directrix and the focus—that is, 2 units from either one. Since the focus is 2 units to the right of the vertex, p is 2. Using one of the forms of an equation of a parabola with vertex at (h, k), we have $(y-0)^2 = 4p(x-0)$, or, equivalently, $y^2 = 8x$.

$\boxed{21}$ $V(3, -5)$ and $l\colon x = 2 \Rightarrow p = 1.$ $(y+5)^2 = 4p(x-3) \Rightarrow (y+5)^2 = 4(x-3).$

$\boxed{23}$ $V(-1, 0)$ and $F(-4, 0) \Rightarrow p = -3.$ $(y-0)^2 = 4p(x+1) \Rightarrow y^2 = -12(x+1).$

[25] The vertex at the origin and symmetric to the y-axis imply that the equation is of the form $y = ax^2$. Substituting $x = 2$ and $y = -3$ into that equation yields

$$-3 = a \cdot 4 \Rightarrow a = -\tfrac{3}{4}. \text{ Thus, an equation is } y = -\tfrac{3}{4}x^2, \text{ or } 3x^2 = -4y.$$

[27] The vertex at $(-3, 5)$ and axis parallel to the x-axis imply that the equation is of the form $(y - 5)^2 = 4p(x + 3)$. Substituting $x = 5$ and $y = 9$ into that equation yields $16 = 4p \cdot 8 \Rightarrow p = \tfrac{1}{2}$. Thus, an equation is $(y - 5)^2 = 2(x + 3)$.

[31] Refer to the definition of a parabola. The point $P(-6, 3)$ is the fixed point (focus) and the line $l: x = -2$ is the fixed line (directrix). The vertex is halfway between the focus and the directrix—that is, at $V(-4, 3)$. An equation is of the form $(y - 3)^2 = 4p(x + 4)$. The distance from the vertex to the focus is $p = -6 - (-4) = -2$. Thus, an equation is $(y - 3)^2 = -8(x + 4)$.

Note: To find an equation for a lower or upper half, we need to solve for y (use $-$ or $+$ respectively). For the left or right half, solve for x (use $-$ or $+$ respectively).

[33] $(y + 1)^2 = x + 3 \Rightarrow y + 1 = \pm\sqrt{x + 3} \Rightarrow y = -\sqrt{x + 3} - 1$

[35] $(x + 1)^2 = y - 4 \Rightarrow x + 1 = \pm\sqrt{y - 4} \Rightarrow x = \sqrt{y - 4} - 1$

[37] The parabola has an equation of the form $y = ax^2 + bx + c$.

Substituting the x and y values of P, Q, and R into this equation yields:

$$\begin{cases} 4a + 2b + c = 5 & P \quad (E_1) \\ 4a - 2b + c = -3 & Q \quad (E_2) \\ a + b + c = 6 & R \quad (E_3) \end{cases}$$

Solving E_3 for c $\{c = 6 - a - b\}$ and substituting into E_1 and E_2 yields:

$$\begin{cases} 3a + b = -1 & (E_4) \\ 3a - 3b = -9 & (E_5) \end{cases}$$

$E_4 - E_5 \Rightarrow 4b = 8 \Rightarrow b = 2; a = -1; c = 5$. The equation is $y = -x^2 + 2x + 5$.

[39] The parabola has an equation of the form $x = ay^2 + by + c$.

Substituting the x and y values of P, Q, and R into this equation yields:

$$\begin{cases} a + b + c = -1 & P \quad (E_1) \\ 4a - 2b + c = 11 & Q \quad (E_2) \\ a - b + c = 5 & R \quad (E_3) \end{cases}$$

Solving E_3 for c $\{c = 5 - a + b\}$ and substituting into E_1 and E_2 yields:

$$\begin{cases} 2b = -6 & (E_4) \\ 3a - b = 6 & (E_5) \end{cases}$$

$E_4 \Rightarrow b = -3; a = 1; c = 1$. The equation is $x = y^2 - 3y + 1$.

[41] A cross section of the mirror is a parabola with $V(0, 0)$ and passing through $P(4, 1)$. The incoming light will collect at the focus F. A general equation of this form of a parabola is $y = ax^2$. Substituting $x = 4$ and $y = 1$ gives us $1 = a(4)^2 \Rightarrow a = \tfrac{1}{16}$. $p = 1/(4a) = 1/(\tfrac{1}{4}) = 4$. The light will collect 4 inches from the center of the mirror.

45 $p = 5 \Rightarrow a = 1/(4p) = 1/(4 \cdot 5) = \frac{1}{20}.$

$y = ax^2 \; \{ y = 2 \; \underline{\text{ft}} = 24 \text{ inches} \} \Rightarrow 24 = \frac{1}{20}x^2 \Rightarrow x^2 = 480 \Rightarrow x = \sqrt{480}.$

The width is twice the value of x. Width $= 2\sqrt{480} \approx 43.82$ in.

47 (a) Let the parabola have the equation $x^2 = 4py$. Since the point (r, h) is on the

parabola, we can substitute r for x and h for y, giving us $r^2 = 4ph$. Solving for p

we have $p = \dfrac{r^2}{4h}.$

(b) $p = 10$ and $h = 5 \Rightarrow r^2 = 4(10)(5) \Rightarrow r = 10\sqrt{2}.$

49 With $a = 125$ and $p = 50$, $S = \dfrac{8\pi p^2}{3}\left[\left(1 + \dfrac{a^2}{4p^2}\right)^{3/2} - 1\right] \approx 64{,}968 \text{ ft}^2.$

51 Depending on the type of calculator or software used, we may need to solve for y in

terms of x. $x = -y^2 + 2y + 5 \Rightarrow y^2 - 2y + (x - 5) = 0$. This is a quadratic equation

in y. Using the quadratic formula to solve for y yields

$$y = \frac{-(-2) \pm \sqrt{(-2)^2 - 4(1)(x - 5)}}{2(1)} = 1 \pm \sqrt{6 - x}.$$

$$[-11, 10] \text{ by } [-7, 7]$$

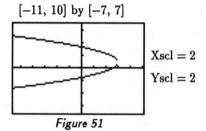

Xscl $= 2$

Yscl $= 2$

Figure 51

11.2 Exercises

Note: Let C, V, F, and M denote the center, the vertices, the foci, and the end points of

the minor axis, respectively. Let c denote the distance from the center of the

ellipse to a focus.

1 $\dfrac{x^2}{9} + \dfrac{y^2}{4} = 1$ • The x-intercepts are at $(\pm\sqrt{9}, 0)$, or, equivalently, $(\pm 3, 0)$. The

y-intercepts are at $(\pm\sqrt{4}, 0) = (\pm 2, 0)$. The major axis $\{$the longer of the two

axes$\}$ is the horizontal axis and has length $2(3) = 6$. The minor axis $\{$the shorter$\}$ is

the vertical axis and has length $2(2) = 4$. To find the foci, it is helpful to remember

the relationship

$$\left[\tfrac{1}{2}(\text{minor axis})\right]^2 + \left[c\right]^2 = \left[\tfrac{1}{2}(\text{major axis})\right]^2.$$

Using the values from above we have $\left[\tfrac{1}{2}(4)\right]^2 + c^2 = \left[\tfrac{1}{2}(6)\right]^2 \Rightarrow 4 + c^2 = 9 \Rightarrow$

$c = \pm\sqrt{5}.$

$V(\pm 3, 0); \; F(\pm\sqrt{5}, 0); \; M(0, \pm 2)$ See *Figure 1*.

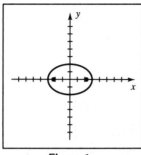

Figure 1

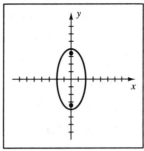

Figure 5

5 We first divide by 16 to obtain the "1" on the right side of the equation.

$4x^2 + y^2 = 16 \Rightarrow \frac{x^2}{4} + \frac{y^2}{16} = 1.$ Since the 16 in the denominator of the term with the

variable y is larger than the 4 in the denominator of the term with the variable x, the

vertices and foci are on the y-axis and the major axis is the vertical axis.

$4 + c^2 = 16 \Rightarrow c^2 = 16 - 4 \Rightarrow c = \pm 2\sqrt{3}.$

$$V(0, \pm 4); \; F(0, \pm 2\sqrt{3}); \; M(\pm 2, 0)$$

7 $4x^2 + 25y^2 = 1 \Rightarrow \frac{x^2}{\frac{1}{4}} + \frac{y^2}{\frac{1}{25}} = 1.$ $\frac{1}{25} + c^2 = \frac{1}{4} \Rightarrow c^2 = \frac{1}{4} - \frac{1}{25} = \frac{21}{100} \Rightarrow c = \pm \frac{1}{10}\sqrt{21}.$

$$V(\pm \tfrac{1}{2}, 0); \; F(\pm \tfrac{1}{10}\sqrt{21}, 0); \; M(0, \pm \tfrac{1}{5})$$

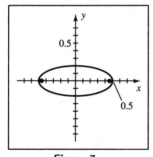

Figure 7

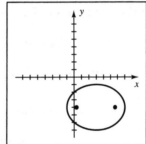

Figure 9

9 $\frac{(x-3)^2}{16} + \frac{(y+4)^2}{9} = 1$ • The effect of the factors $(x-3)$ and $(y+4)$ is to shift

the center of the ellipse from $(0, 0)$ to $(3, -4)$. Since the larger denominator, 16, is in

the term with x, the major axis will be horizontal. The end points are 4 units in

either direction of the center. Their coordinates are the points $(3 \pm 4, -4)$, or,

equivalently, $(7, -4)$ and $(-1, -4)$. The minor axis will be vertical with end points

$(3, -4 \pm 3)$, or, equivalently, $(3, -1)$ and $(3, -7)$. $9 + c^2 = 16 \Rightarrow c^2 = 16 - 9 \Rightarrow$

$c = \pm \sqrt{7}.$ Remember that c is the distance *from the center* to a focus. Hence, the

coordinates of the foci are $(3 \pm \sqrt{7}, -4)$.

$$C(3, -4); \; V(3 \pm 4, -4); \; F(3 \pm \sqrt{7}, -4); \; M(3, -4 \pm 3)$$

11 $4x^2 + 9y^2 - 32x - 36y + 64 = 0 \Rightarrow$

$\quad (4x^2 - 32x) + (9y^2 - 36y) = -64 \Rightarrow$

$\quad\quad$ { first group the x terms and the y terms }

$\quad 4(x^2 - 8x) + 9(y^2 - 4y) = -64 \Rightarrow$

$\quad\quad$ { factor out the coefficients of x^2 and y^2 }

$\quad 4(x^2 - 8x + \underline{16}) + 9(y^2 - 4y + \underline{4}) = -64 + \underline{64} + \underline{36} \Rightarrow$

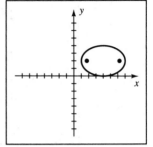

$\quad\quad$ { Complete the squares—remember that we have   *Figure 11*

$\quad\quad$ added $\underline{4}$ (16) and $\underline{9}$ (4), <u>not</u> 16 and 4. Thus, we add 64 and 36 to the right side

$\quad\quad$ of the equation.} $4(x-4)^2 + 9(y-2)^2 = 36 \Rightarrow \dfrac{(x-4)^2}{9} + \dfrac{(y-2)^2}{4} = 1.$

$c^2 = 9 - 4 \Rightarrow c = \pm\sqrt{5}.$

$$C(4, 2); \; V(4 \pm 3, 2); \; F(4 \pm \sqrt{5}, 2); \; M(4, 2 \pm 2)$$

Note: Let b denote the distance equal to $\frac{1}{2}$(minor axis length) and

$\quad\quad\quad\quad\quad\quad$ let a denote the distance equal to $\frac{1}{2}$(major axis length).

15 $a = 2$ and $b = 6 \Rightarrow \dfrac{x^2}{a^2} + \dfrac{y^2}{b^2} = 1$ is $\dfrac{x^2}{4} + \dfrac{y^2}{36} = 1.$

17 The center of the ellipse is $(-2, 1)$. $a = 5$ and $b = 2$ give us $\dfrac{(x+2)^2}{25} + \dfrac{(y-1)^2}{4} = 1.$

19 Since the vertices are at $(\pm 8, 0)$, $\frac{1}{2}$(major axis) $= 8$. Since the foci are at $(\pm 5, 0)$,

$\quad c = 5.$ Using the relationship $\left[\frac{1}{2}(\text{minor axis}) \right]^2 + \left[c \right]^2 = \left[\frac{1}{2}(\text{major axis}) \right]^2$, we have

$\quad \left[\frac{1}{2}(\text{minor axis}) \right]^2 + 5^2 = 8^2 \Rightarrow \left[\frac{1}{2}(\text{minor axis}) \right]^2 = 64 - 25 = 39.$

$\quad\quad\quad\quad\quad\quad\quad\quad\quad\quad\quad\quad\quad\quad\quad$ An equation is $\dfrac{x^2}{64} + \dfrac{y^2}{39} = 1.$

21 If the length of the minor axis is 3, then $b = \frac{3}{2}.$

$\quad\quad\quad\quad\quad\quad$ An equation is $\dfrac{x^2}{(\frac{3}{2})^2} + \dfrac{y^2}{5^2} = 1$, or, equivalently, $\dfrac{4x^2}{9} + \dfrac{y^2}{25} = 1.$

23 With the vertices at $(0, \pm 6)$, an equation of the ellipse is $\dfrac{x^2}{b^2} + \dfrac{y^2}{6^2} = 1.$ Substituting

$\quad x = 3$ and $y = 2$ and solving for b^2 yields $\dfrac{9}{b^2} + \dfrac{4}{36} = 1 \Rightarrow \dfrac{9}{b^2} = \dfrac{8}{9} \Rightarrow b^2 = \dfrac{81}{8}.$

$\quad\quad\quad\quad\quad\quad$ An equation is $\dfrac{x^2}{\frac{81}{8}} + \dfrac{y^2}{36} = 1$, or, equivalently, $\dfrac{8x^2}{81} + \dfrac{y^2}{36} = 1.$

25 With vertices $V(0, \pm 4)$, an equation of the ellipse is $\dfrac{x^2}{b^2} + \dfrac{y^2}{16} = 1.$ Think of the

$\quad$ formula for the eccentricity as

$$\boxed{e = \text{eccentricity} = \dfrac{\text{distance from center to a focus}}{\text{distance from center to a vertex}}}$$

$\quad$ Hence, $e = \frac{c}{a} = \frac{3}{4}$ and $a = 4 \Rightarrow c = 3.$ Thus, $b^2 + c^2 = a^2 \Rightarrow$

$\quad\quad\quad b^2 = a^2 - c^2 = 4^2 - 3^2 = 16 - 9 = 7.$ An equation is $\dfrac{x^2}{7} + \dfrac{y^2}{16} = 1.$

29 Remember to divide the lengths of the major and minor axes by 2.

$$\frac{x^2}{(\frac{1}{2}\cdot 8)^2}+\frac{y^2}{(\frac{1}{2}\cdot 5)^2}=1 \Rightarrow \frac{x^2}{16}+\frac{y^2}{\frac{25}{4}}=1 \Rightarrow \frac{x^2}{16}+\frac{4y^2}{25}=1.$$

31 The graph of $x^2+4y^2=20$, or, equivalently, $\frac{x^2}{20}+\frac{y^2}{5}=1$,

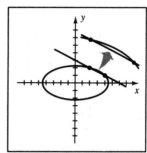

is that of an ellipse with x-intercepts at $(\pm\sqrt{20},\,0)$ and y-intercepts at $(0,\,\pm\sqrt{5})$. The graph of $x+2y=6$, or, equivalently, $y=-\frac{1}{2}x+3$, is that of a line with y-intercept 3 and slope $-\frac{1}{2}$. Substituting $x=6-2y$ into $x^2+4y^2=20$ yields $(6-2y)^2+4y^2=20 \Rightarrow$

$8y^2-24y+16=0 \Rightarrow 8(y^2-3y+2)=0 \Rightarrow$

Figure 31

$8(y-1)(y-2)=0 \Rightarrow y=1,\,2;\,x=4,\,2.$

The two points of intersection are $(2,\,2)$ and $(4,\,1)$.

33 Refer to the definition of an ellipse. As in the discussion on page 711, we will let the positive constant, k, equal $2a$. $k=2a=10 \Rightarrow a=5$. $F(3,\,0)$ and $F'(-3,\,0) \Rightarrow c=3$.

$$b^2=a^2-c^2=25-9=16. \quad \text{An equation is } \frac{x^2}{25}+\frac{y^2}{16}=1.$$

37 $y=11\sqrt{1-\frac{x^2}{49}} \Rightarrow \frac{y}{11}=\sqrt{1-\frac{x^2}{49}} \Rightarrow \frac{x^2}{49}+\frac{y^2}{121}=1.$

Since $y\geq 0$ in the original equation, its graph is the upper half of the ellipse.

39 $x=-\frac{1}{3}\sqrt{9-y^2} \Rightarrow -3x=\sqrt{9-y^2} \Rightarrow 9x^2=9-y^2 \Rightarrow x^2+\frac{y^2}{9}=1.$

Since $x\leq 0$ in the original equation, its graph is the left half of the ellipse.

41 $x=1+2\sqrt{1-\frac{(y+2)^2}{9}} \Rightarrow \frac{x-1}{2}=\sqrt{1-\frac{(y+2)^2}{9}} \Rightarrow \frac{(x-1)^2}{4}+\frac{(y+2)^2}{9}=1.$

Since $x\geq 1$ in the original equation, its graph is the right half of the ellipse.

43 $y=2-7\sqrt{1-\frac{(x+1)^2}{9}} \Rightarrow \frac{y-2}{-7}=\sqrt{1-\frac{(x+1)^2}{9}} \Rightarrow \frac{(x+1)^2}{9}+\frac{(y-2)^2}{49}=1.$

Since $y\leq 2$ in the original equation, its graph is the lower half of the ellipse.

45 Model this problem as an ellipse with $V(\pm 15,\,0)$ and $M(0,\,\pm 10)$.

Substituting $x=6$ into $\frac{x^2}{15^2}+\frac{y^2}{10^2}=1$ yields $\frac{y^2}{100}=\frac{189}{225} \Rightarrow y^2=84.$

The desired height is $\sqrt{84}=2\sqrt{21}\approx 9.165$ ft.

47 $e=\frac{c}{a}=0.017 \Rightarrow c=0.017a=0.017(93{,}000{,}000)=1{,}581{,}000.$ The maximum and minimum distances are $a+c=94{,}581{,}000$ miles and $a-c=91{,}419{,}000$ miles.

49 (a) Let c denote the distance from the center of the hemi-ellipsoid to F. Hence,

$$(\tfrac{1}{2}k)^2 + c^2 = h^2 \Rightarrow c^2 = h^2 - \tfrac{1}{4}k^2 \Rightarrow c = \sqrt{h^2 - \tfrac{1}{4}k^2}. \quad d = d(V,\,F) = h - c \Rightarrow$$

$$d = h - \sqrt{h^2 - \tfrac{1}{4}k^2} \text{ and } d' = d(V,\,F') = h + c \Rightarrow d' = h + \sqrt{h^2 - \tfrac{1}{4}k^2}.$$

(b) From part (a), $d' = h + c \Rightarrow c = d' - h = 32 - 17 = 15.$ $c = \sqrt{h^2 - \tfrac{1}{4}k^2} \Rightarrow$

$$15 = \sqrt{17^2 - \tfrac{1}{4}k^2} \Rightarrow 225 = 289 - \tfrac{1}{4}k^2 \Rightarrow \tfrac{1}{4}k^2 = 64 \Rightarrow k^2 = 256 \Rightarrow k = 16 \text{ cm.}$$

$$d = h - c = 17 - 15 = 2 \Rightarrow F \text{ should be located 2 cm from } V.$$

51 $c^2 = (\tfrac{1}{2} \cdot 50)^2 - 15^2 = 625 - 225 = 400 \Rightarrow c = 20.$

Their feet should be $25 - 20 = 5$ ft from the vertices.

55 $\dfrac{(x + 0.1)^2}{1.7} + \dfrac{y^2}{0.9} = 1 \Rightarrow y = \pm\sqrt{0.9[1 - (x + 0.1)^2/1.7]}.$

$$\dfrac{x^2}{0.9} + \dfrac{(y - 0.25)^2}{1.8} = 1 \Rightarrow y = 0.25 \pm \sqrt{1.8(1 - x^2/0.9)}.$$

From the graph, the points of intersection are approximately

$$(-0.88,\ 0.76),\ (-0.48, -0.91),\ (0.58, -0.81),\text{ and } (0.92,\ 0.59).$$

$[-3,\ 3]$ by $[-2,\ 2]$

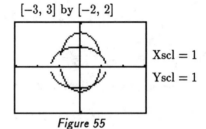

Xscl $= 1$

Yscl $= 1$

Figure 55

11.3 Exercises

Note: Let C, V, F, and W denote the center, the vertices, the foci, and the end points of the conjugate axis, respectively. Let c denote the distance from the center of the hyperbola to a focus.

⚊1⚊ $\frac{x^2}{9} - \frac{y^2}{4} = 1$ • The hyperbola will have a right branch and a left branch since the term containing x is positive. The vertices will be on the horizontal transverse axis, $\pm\sqrt{9} = \pm 3$ units from the center. The end points of the vertical conjugate axis are $(0, \pm\sqrt{4}) = (0, \pm 2)$. The asymptotes have equations

$$y = \pm\left[\frac{\frac{1}{2}(\text{vertical axis length})}{\frac{1}{2}(\text{horizontal axis length})}\right](x).$$

Note that the terms are "vertical" and "horizontal" and not transverse and conjugate since the latter can be either vertical or horizontal. In this case, we have $y = \pm\frac{\frac{1}{2}(4)}{\frac{1}{2}(6)}(x) = \pm\frac{2}{3}x$. The positive sign corresponds to the asymptote with positive slope and the negative sign corresponds to the asymptote with negative slope. To find the foci, it is helpful to remember the relationship

$$\left[\tfrac{1}{2}(\text{transverse axis})\right]^2 + \left[\tfrac{1}{2}(\text{conjugate axis})\right]^2 = \left[c\right]^2.$$

Using the values from above we have $\left[\frac{1}{2}(6)\right]^2 + \left[\frac{1}{2}(4)\right]^2 = c^2 \Rightarrow 9 + 4 = c^2 \Rightarrow c = \pm\sqrt{13}$.

$$V(\pm 3, 0); \ F(\pm\sqrt{13}, 0); \ W(0, \pm 2); \ y = \pm\tfrac{2}{3}x$$

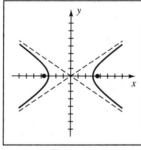

Figure 1

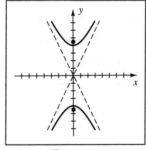

Figure 7

⚊7⚊ $y^2 - 4x^2 = 16 \Rightarrow \frac{y^2}{16} - \frac{x^2}{4} = 1$ • The hyperbola will have an upper branch and a lower branch since the term containing y is positive. The vertices will be $\pm\sqrt{16} = \pm 4$ units from the center on the vertical transverse axis. The end points of the horizontal conjugate axis are $(\pm\sqrt{4}, 0) = (\pm 2, 0)$. The asymptotes have equations $y = \pm\frac{\frac{1}{2}(8)}{\frac{1}{2}(4)}(x) = \pm 2x$. Using the foci relationship given in Exercise 1, we have $\left[\frac{1}{2}(8)\right]^2 + \left[\frac{1}{2}(4)\right]^2 = c^2 \Rightarrow 16 + 4 = c^2 \Rightarrow c = \pm 2\sqrt{5}$.

$$V(0, \pm 4); \ F(0, \pm 2\sqrt{5}); \ W(\pm 2, 0); \ y = \pm 2x$$

9 $16x^2 - 36y^2 = 1 \Rightarrow \dfrac{x^2}{\frac{1}{16}} - \dfrac{y^2}{\frac{1}{36}} = 1.$ $c^2 = \frac{1}{16} + \frac{1}{36} \Rightarrow c = \pm \frac{1}{12}\sqrt{13}.$

$$V(\pm \tfrac{1}{4}, 0); \; F(\pm \tfrac{1}{12}\sqrt{13}, 0); \; W(0, \pm \tfrac{1}{6}); \; y = \pm \tfrac{2}{3}x$$

Note that the branches of the hyperbola almost coincide with the asymptotes.

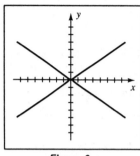
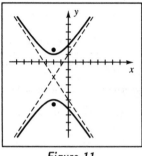
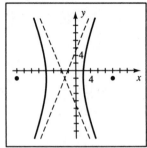

Figure 9   Figure 11   Figure 13

11 $\dfrac{(y+2)^2}{9} - \dfrac{(x+2)^2}{4} = 1$ • The effect of the factors $(x+2)$ and $(y+2)$ is to shift

the center of the hyperbola from $(0, 0)$ to $(-2, -2)$. Since the term involving y is positive, the transverse axis will be vertical. The vertices are 3 units in either direction of the center. Their coordinates are $(-2, -2 \pm 3)$ or equivalently, $(-2, 1)$ and $(-2, -5)$. The conjugate axis will be horizontal with end points $(-2 \pm 2, -2)$, or, equivalently, $(0, -2)$ and $(-4, -2)$. Remember that c is the distance *from the center* to a focus. $c^2 = 9 + 4 \Rightarrow c = \pm\sqrt{13}$. Hence, the coordinates of the foci are $(-2, -2 \pm \sqrt{13})$. If the center of the hyperbola was at the origin, we would have asymptote equations $y = \pm \frac{3}{2}x$. Since the center of the hyperbola has been shifted by the factors $(y+2)$ and $(x+2)$, we can shift the asymptote equations using the same factors. Hence, these equations are $(y+2) = \pm \frac{3}{2}(x+2)$.

$$C(-2, -2); \; V(-2, -2 \pm 3); \; F(-2, -2 \pm \sqrt{13}); \; W(-2 \pm 2, -2)$$

13 $144x^2 - 25y^2 + 864x - 100y - 2404 = 0 \Rightarrow$

$144(x^2 + 6x + \underline{\;9\;}) - 25(y^2 + 4y + \underline{\;4\;}) = 2404 + \underline{\;1296\;} - \underline{\;100\;} \Rightarrow$

$144(x+3)^2 - 25(y+2)^2 = 3600 \Rightarrow \dfrac{(x+3)^2}{25} - \dfrac{(y+2)^2}{144} = 1.$

$c^2 = 25 + 144 \Rightarrow c = \pm 13.$ See *Figure 13*.

$C(-3, -2); \; V(-3 \pm 5, -2); \; F(-3 \pm 13, -2); \; W(-3, -2 \pm 12); \; (y+2) = \pm \frac{12}{5}(x+3)$

Note: Let b denote the distance equal to $\frac{1}{2}$(conjugate axis length) and

let a denote the distance equal to $\frac{1}{2}$(transverse axis length).

17 $a = 3$ and $c = 5 \Rightarrow b^2 = c^2 - a^2 = 16.$ $\dfrac{x^2}{a^2} - \dfrac{y^2}{b^2} = 1$ is then $\dfrac{x^2}{9} - \dfrac{y^2}{16} = 1.$

19 The center of the hyperbola is $(-2, -3)$. $a = 1$ and $c = 2 \Rightarrow b^2 = 2^2 - 1^2 = 3$ and an

equation is $\dfrac{(y+3)^2}{1^2} - \dfrac{(x+2)^2}{3} = 1$, or, equivalently, $(y+3)^2 - \dfrac{(x+2)^2}{3} = 1$.

21 $F(0, \pm 4) \Rightarrow c = 4$. $V(0, \pm 1) \Rightarrow a = 1$. $a^2 + b^2 = c^2 \Rightarrow b^2 = 4^2 - 1^2 = 15$.

Since the vertices are on the y-axis, the "1^2" is associated with the y^2 term.

An equation is $\dfrac{y^2}{1} - \dfrac{x^2}{15} = 1$.

25 Conjugate axis of length 4 implies that $b = 2$. $F(0, \pm 5) \Rightarrow c = 5$.

$a^2 + b^2 = c^2 \Rightarrow a^2 = 5^2 - 2^2 = 21$. An equation is $\dfrac{y^2}{21} - \dfrac{x^2}{4} = 1$.

27 Since the asymptote equations are $y = \pm 2x$ and we know that the point $(3, 0)$ is on

the hyperbola, we conclude that the upper right corner of the rectangle formed by the

transverse and conjugate axes has coordinates $(3, 6)$ { substitute $x = 3$ in $y = 2x$ to

obtain the 6 }. Thus we have end points of the conjugate axis at $(0, \pm 6)$.

An equation is $\dfrac{x^2}{9} - \dfrac{y^2}{36} = 1$.

31 Since the transverse axis is vertical, the y^2 term will be positive.

An equation is $\dfrac{y^2}{(\frac{1}{2} \cdot 10)^2} - \dfrac{x^2}{(\frac{1}{2} \cdot 14)^2} = 1$, or $\dfrac{y^2}{25} - \dfrac{x^2}{49} = 1$.

33 The graph of $y^2 - 4x^2 = 16$, or, equivalently, $\dfrac{y^2}{16} - \dfrac{x^2}{4} = 1$,

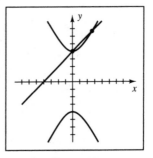

is that of a hyperbola with y-intercepts at $(0, \pm 4)$. The

graph of $y - x = 4$, or, equivalently, $y = x + 4$, is that of a

line with y-intercept 4 and slope 1. Substituting $y = x + 4$

into $y^2 - 4x^2 = 16$ yields $(x + 4)^2 - 4x^2 = 16 \Rightarrow$

$3x^2 - 8x = 0 \Rightarrow x(3x - 8) = 0 \Rightarrow x = 0, \frac{8}{3}; y = 4, \frac{20}{3}$.

The two points of intersection are $(0, 4)$ and $\left(\frac{8}{3}, \frac{20}{3}\right)$.

Figure 33

37 Refer to the definition of a hyperbola. As in the discussion on page 724,

we will let the positive constant, k, equal $2a$. $k = 2a = 16 \Rightarrow a = 8$.

$F(0, 10)$ and $F'(0, -10) \Rightarrow c = 10$. $b^2 = c^2 - a^2 = 100 - 64 = 36$.

An equation is $\dfrac{y^2}{64} - \dfrac{x^2}{36} = 1$.

39 $x = \frac{5}{4}\sqrt{y^2 + 16} \Rightarrow \frac{4}{5}x = \sqrt{y^2 + 16} \Rightarrow \frac{16}{25}x^2 = y^2 + 16 \Rightarrow \frac{16}{25}x^2 - y^2 = 16 \Rightarrow \dfrac{x^2}{25} - \dfrac{y^2}{16} = 1$.

Since $x > 0$ in the original equation, its graph is the right branch of the hyperbola.

41 $y = \frac{3}{7}\sqrt{x^2 + 49} \Rightarrow \frac{7}{3}y = \sqrt{x^2 + 49} \Rightarrow \frac{49}{9}y^2 = x^2 + 49 \Rightarrow \dfrac{y^2}{9} - \dfrac{x^2}{49} = 1$.

Since $y > 0$ in the original equation, its graph is the upper branch of the hyperbola.

43 $y = -\frac{9}{4}\sqrt{x^2 - 16} \Rightarrow -\frac{4}{9}y = \sqrt{x^2 - 16} \Rightarrow \frac{16}{81}y^2 = x^2 - 16 \Rightarrow \frac{x^2}{16} - \frac{y^2}{81} = 1.$

Since $y \le 0$ in the original equation,

its graph is the lower halves of the branches of the hyperbola.

45 $x = -\frac{2}{3}\sqrt{y^2 - 36} \Rightarrow -\frac{3}{2}x = \sqrt{y^2 - 36} \Rightarrow \frac{9}{4}x^2 = y^2 - 36 \Rightarrow \frac{y^2}{36} - \frac{x^2}{16} = 1.$

Since $x \le 0$ in the original equation,

its graph is the left halves of the branches of the hyperbola.

49 The path is a hyperbola with $V(\pm 3,\, 0)$ and $W(0,\, \pm\frac{3}{2})$.

An equation is $\dfrac{x^2}{(3)^2} - \dfrac{y^2}{\left(\frac{3}{2}\right)^2} = 1$ or equivalently, $x^2 - 4y^2 = 9$.

If only the right branch is considered, then $x = \sqrt{9 + 4y^2}$ is an equation of the path.

51 Set up a coordinate system like the one in Example 6. Let the origin be located on the shoreline halfway between A and B and let $P(x,\, y)$ denote the coordinates of the ship. The coordinates of A and B (which can be thought of as the foci of the hyperbola) are $(-100,\, 0)$ and $(100,\, 0)$, respectively. Hence, $c = 100$. As in Exercise 37, the difference in distances is a constant—that is, $d(P,\, A) - d(P,\, B) = 2a = 160$ $\Rightarrow a = 80$. $b^2 = c^2 - a^2 = 100^2 - 80^2 \Rightarrow b = 60$. An equation of the hyperbola is

$\dfrac{x^2}{80^2} - \dfrac{y^2}{60^2} = 1$. Now, $y = 100 \Rightarrow \dfrac{x^2}{80^2} = 1 + \dfrac{100^2}{60^2} \Rightarrow x^2 = 80^2 \cdot \dfrac{13{,}600}{60^2} \Rightarrow$

$x = 80 \cdot \frac{10}{60}\sqrt{136} = \frac{80}{3}\sqrt{34}$. The ship's coordinates are $\left(\frac{80}{3}\sqrt{34},\, 100\right) \approx (155.5,\, 100)$.

11.4 Exercises

Note: Let D denote the value of the discriminant $B^2 - 4AC$. The following is a general outline of the solutions for Exercises 1–13 in this section and (33)–(35) in the Chapter Review Exercises.

(a) Discriminant value and conic type are given.

(b) The 5 steps in part (b) are:

1) $\cot 2\phi = \dfrac{A-C}{B} \Rightarrow \phi = \frac{1}{2}\cot^{-1}\!\left(\dfrac{A-C}{B}\right)$, where the range of $\cot^{-1}$ is

$0°$ to $180°$. Note that the range of ϕ is $0° < \phi < 90°$.

2) $\cos 2\phi = \dfrac{\pm(A-C)}{\sqrt{(A-C)^2 + B^2}}$, $\sin\phi = \sqrt{\dfrac{1-\cos 2\phi}{2}}$, and $\cos\phi = \sqrt{\dfrac{1+\cos 2\phi}{2}}$;

Note that $\cos 2\phi$ will have the same sign as $\cot 2\phi$ since $\cot 2\phi = \dfrac{\cos 2\phi}{\sin 2\phi}$ and

$\sin 2\phi$ is positive. Since ϕ is acute, $\sin\phi$ and $\cos\phi$ are positive.

3) The rotation of axes formulas are given. $\begin{cases} x = x'\cos\phi - y'\sin\phi \\ y = x'\sin\phi + y'\cos\phi \end{cases}$

4) The rotation of axes formulas are substituted into the original equation to obtain an equation in x' and y'. This equation is then simplified into a standard form.

5) The vertices (V') of the graph on the $x'y'$-plane are listed along with the corresponding vertices (V) of the graph of the original equation on the xy-plane.

1 (a) $D = (-2)^2 - 4(1)(1) = 0$, parabola

(b) 1) $\cot 2\phi = 0$; $\phi = 45°$

2) $\cos 2\phi = 0$; $\sin\phi = \frac{1}{2}\sqrt{2}$; $\cos\phi = \frac{1}{2}\sqrt{2}$

3) $\begin{cases} x = \frac{1}{2}\sqrt{2}\,x' - \frac{1}{2}\sqrt{2}\,y' = \frac{1}{2}\sqrt{2}\,(x' - y') \\ y = \frac{1}{2}\sqrt{2}\,x' + \frac{1}{2}\sqrt{2}\,y' = \frac{1}{2}\sqrt{2}\,(x' + y') \end{cases}$

4) $2(y')^2 - 4x' = 0 \Rightarrow (y')^2 = 2(x')$

5) $V'(0,\,0) \rightarrow V(0,\,0)$ $\{x$ & y int. @ $2\sqrt{2}\}$

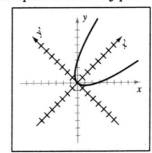

Figure 1

The following provides more detail on step 4 for Exercise 1. It is helpful to arrange your work in a table format as follows. In the leftmost column we list each term of the original equation. We then make 5 columns—one for each of the terms $(x')^2$, $x'y'$, $(y')^2$, x', and y'. The column sums are the equation in the $x'y'$ system. The 0 in the leftmost column total comes from the original equation.

		$(x')^2$	$x'y'$	$(y')^2$	x'	y'
x^2	$=$	$\frac{1}{2}(x')^2$	$-x'y'$	$+\frac{1}{2}(y')^2$		
$-2xy$	$=$	$-(x')^2$		$+(y')^2$		
y^2	$=$	$\frac{1}{2}(x')^2$	$+x'y'$	$+\frac{1}{2}(y')^2$		
$-2\sqrt{2}x$	$=$				$-2x'$	$+2y'$
$-2\sqrt{2}y$	$=$				$-2x'$	$-2y'$
0	$=$			$2(y')^2$	$-4(x')$	

This result, $0 = 2(y')^2 - 4(x')$, is equivalent to $(y')^2 = 2(x')$.

Using this format, it is easy to check that the $x'y'$ term is 0.

3 (a) $D = (-8)^2 - 4(5)(5) = -36 < 0$, ellipse

(b) 1) $\cot 2\phi = 0$; $\phi = 45°$

2) $\cos 2\phi = 0$; $\sin \phi = \frac{1}{2}\sqrt{2}$; $\cos \phi = \frac{1}{2}\sqrt{2}$

3) $\begin{cases} x = \frac{1}{2}\sqrt{2}\,x' - \frac{1}{2}\sqrt{2}\,y' = \frac{1}{2}\sqrt{2}\,(x' - y') \\ y = \frac{1}{2}\sqrt{2}\,x' + \frac{1}{2}\sqrt{2}\,y' = \frac{1}{2}\sqrt{2}\,(x' + y') \end{cases}$

4) $1(x')^2 + 9(y')^2 = 9 \Rightarrow \dfrac{(x')^2}{9} + \dfrac{(y')^2}{1} = 1$

5) $V'(\pm 3, 0) \rightarrow V(\pm\frac{3}{2}\sqrt{2}, \pm\frac{3}{2}\sqrt{2})$

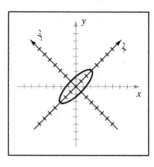

Figure 3

5 (a) $D = (10\sqrt{3})^2 - 4(11)(1) = 256 > 0$, hyperbola

(b) 1) $\cot 2\phi = \frac{1}{3}\sqrt{3}$; $\phi = 30°$

2) $\cos 2\phi = \frac{1}{2}$; $\sin \phi = \frac{1}{2}$; $\cos \phi = \frac{1}{2}\sqrt{3}$

3) $\begin{cases} x = \frac{1}{2}\sqrt{3}\,x' - \frac{1}{2}y' = \frac{1}{2}(\sqrt{3}\,x' - y') \\ y = \frac{1}{2}x' + \frac{1}{2}\sqrt{3}\,y' = \frac{1}{2}(x' + \sqrt{3}\,y') \end{cases}$

4) $\frac{64}{4}(x')^2 - \frac{16}{4}(y')^2 = 4 \Rightarrow \dfrac{(x')^2}{\frac{1}{4}} - \dfrac{(y')^2}{1} = 1$

5) $V'(\pm\frac{1}{2}, 0) \rightarrow V(\pm\frac{1}{4}\sqrt{3}, \pm\frac{1}{4})$

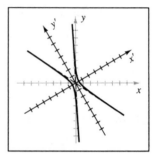

Figure 5

7 (a) $D = (-24)^2 - 4(16)(9) = 0$, parabola

(b) 1) $\cot 2\phi = -\frac{7}{24}$; $\phi \approx 53.13°$

2) $\cos 2\phi = -\frac{7}{25}$; $\sin \phi = \frac{4}{5}$; $\cos \phi = \frac{3}{5}$

3) $\begin{cases} x = \frac{3}{5}x' - \frac{4}{5}y' = \frac{1}{5}(3x' - 4y') \\ y = \frac{4}{5}x' + \frac{3}{5}y' = \frac{1}{5}(4x' + 3y') \end{cases}$

4) $\frac{625}{25}(y')^2 - 100\,x' + 100 = 0 \Rightarrow$
$$(y')^2 = 4(x' - 1)$$

5) $V'(1, 0) \rightarrow V(\frac{3}{5}, \frac{4}{5})$

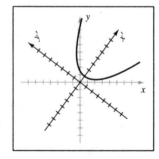

Figure 7

9 (a) $D = (-36)^2 - 4(40)(25) = -2704 < 0$, ellipse

(b) 1) $\cot 2\phi = -\frac{5}{12}$; $\phi \approx 56.31°$

2) $\cos 2\phi = -\frac{5}{13}$; $\sin \phi = \frac{3}{13}\sqrt{13}$; $\cos \phi = \frac{2}{13}\sqrt{13}$

3) $\begin{cases} x = \frac{2}{13}\sqrt{13}\,x' - \frac{3}{13}\sqrt{13}\,y' = \frac{1}{13}\sqrt{13}\,(2x' - 3y') \\ y = \frac{3}{13}\sqrt{13}\,x' + \frac{2}{13}\sqrt{13}\,y' = \frac{1}{13}\sqrt{13}\,(3x' + 2y') \end{cases}$

4) $\frac{169}{13}(x')^2 + \frac{676}{13}(y')^2 - 52\,x' = 0 \Rightarrow$
$$\dfrac{(x' - 2)^2}{4} + \dfrac{(y')^2}{1} = 1$$

5) $V'(2 \pm 2, 0) \rightarrow V([\frac{4}{13} \pm \frac{4}{13}]\sqrt{13}, [\frac{6}{13} \pm \frac{6}{13}]\sqrt{13})$

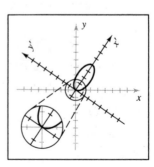

Figure 9

11. (a) $D = (6\sqrt{3})^2 - 4(5)(-1) = 128 > 0$, hyperbola

(b) 1) $\cot 2\phi = \frac{1}{3}\sqrt{3}$; $\phi = 30°$

2) $\cos 2\phi = \frac{1}{2}$; $\sin \phi = \frac{1}{2}$; $\cos \phi = \frac{1}{2}\sqrt{3}$

3) $\begin{cases} x = \frac{1}{2}\sqrt{3}\,x' - \frac{1}{2}y' = \frac{1}{2}(\sqrt{3}\,x' - y') \\ y = \frac{1}{2}x' + \frac{1}{2}\sqrt{3}\,y' = \frac{1}{2}(x' + \sqrt{3}\,y') \end{cases}$

4) $\frac{32}{4}(x')^2 - \frac{16}{4}(y')^2 - 16\,y' - 12 = 0 \Rightarrow$

$$\frac{(y'+2)^2}{1} - \frac{(x')^2}{\frac{1}{2}} = 1$$

5) $V'(0, -2 \pm 1) \rightarrow V(1 \mp \frac{1}{2}, -\sqrt{3} \pm \frac{1}{2}\sqrt{3})$

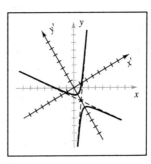

Figure 11

13. (a) $D = (-72)^2 - 4(32)(53) = -1600 < 0$, ellipse

(b) 1) $\cot 2\phi = \frac{7}{24}$; $\phi \approx 36.87°$

2) $\cos 2\phi = \frac{7}{25}$; $\sin \phi = \frac{3}{5}$; $\cos \phi = \frac{4}{5}$

3) $\begin{cases} x = \frac{4}{5}x' - \frac{3}{5}y' = \frac{1}{5}(4x' - 3y') \\ y = \frac{3}{5}x' + \frac{4}{5}y' = \frac{1}{5}(3x' + 4y') \end{cases}$

4) $\frac{125}{25}(x')^2 + \frac{2000}{25}(y')^2 = 80 \Rightarrow \frac{(x')^2}{16} + \frac{(y')^2}{1} = 1$

5) $V'(\pm 4, 0) \rightarrow V(\pm \frac{16}{5}, \pm \frac{12}{5})$

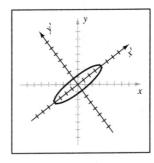

Figure 13

15. *Note:* Depending on the type of software used, you may need to solve the given equation for y. $2.1x^2 - 4xy + 1.5y^2 - 4x + y - 1 = 0 \Rightarrow$

$1.5y^2 - (4x - 1)y + (2.1x^2 - 4x - 1) = 0 \Rightarrow$

$$y = \frac{(4x - 1) \pm \sqrt{(4x - 1)^2 - 6(2.1x^2 - 4x - 1)}}{3}$$

$[-18, 18]$ by $[-12, 12]$

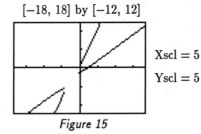

Xscl $= 5$

Yscl $= 5$

Figure 15

Note: For the following exercises, the substitutions $y = r \sin \theta$, $x = r \cos \theta$, $r^2 = x^2 + y^2$, and $\tan \theta = \frac{y}{x}$ are used without mention. We have found it helpful to find the "pole" values {when the graph intersects the pole} to determine which values of θ should be used in the construction of an r-θ chart. The numbers listed on each line of the r-θ chart correspond to the numbers labeled on the figures.

$\boxed{1}$ $r = 5 \Rightarrow r^2 = 25 \Rightarrow x^2 + y^2 = 25$, a circle centered at the origin with radius 5.

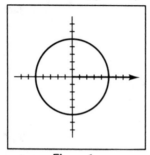

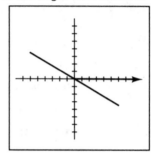

Figure 1   Figure 3

$\boxed{3}$ $\theta = -\frac{\pi}{6} \Rightarrow \tan \theta = \tan\left(-\frac{\pi}{6}\right) \Rightarrow \frac{y}{x} = -\frac{1}{\sqrt{3}} \Rightarrow y = -\frac{1}{3}\sqrt{3}\,x$.

This is a line through the origin with slope $-\frac{1}{3}\sqrt{3}$.

$\boxed{5}$ $r = 3\cos\theta \Rightarrow \{$ multiply by r to obtain $r^2 \}$

$r^2 = 3r\cos\theta \Rightarrow x^2 + y^2 = 3x \Rightarrow$

{recognize this as an equation of a circle and complete the square}

$\left(x^2 - 3x + \frac{9}{4}\right) + y^2 = \frac{9}{4} \Rightarrow \left(x - \frac{3}{2}\right)^2 + y^2 = \frac{9}{4}$.

This is a circle with center $\left(\frac{3}{2}, 0\right)$ and radius $\sqrt{\frac{9}{4}} = \frac{3}{2}$. From the table, we see that as θ varies from 0 to $\frac{\pi}{2}$, r will vary from 3 to 0. This corresponds to the portion of the circle in the first quadrant. As θ varies from $\frac{\pi}{2}$ to π, r varies from 0 to -3. Remember that -3 in the π direction is the same as 3 in the 0 direction. This corresponds to the portion of the circle in the fourth quadrant.

Variation of θ	Variation of r
1) $0 \rightarrow \frac{\pi}{2}$	$3 \rightarrow 0$
2) $\frac{\pi}{2} \rightarrow \pi$	$0 \rightarrow -3$

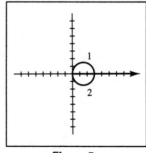

Figure 5

7 $r = 4\cos\theta + 2\sin\theta \Rightarrow r^2 = 4r\cos\theta + 2r\sin\theta \Rightarrow$

$x^2 + y^2 = 4x + 2y \Rightarrow$

$x^2 - 4x + \underline{4} + y^2 - 2y + \underline{1} = \underline{4} + \underline{1} \Rightarrow$

$(x-2)^2 + (y-1)^2 = 5.$

Variation of θ	Variation of r
1) $\quad 0 \;\to\; \frac{\pi}{2}$	$4 \to \quad 2$
2) $\quad \frac{\pi}{2} \;\to\; \pi$	$2 \to \quad -4$

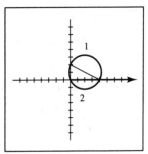

Figure 7

9 $r = 4(1 - \sin\theta)$ is a cardioid since the coefficient of $\sin\theta$ has the same magnitude as the constant term.

$0 = 4(1 - \sin\theta) \Rightarrow \sin\theta = 1 \Rightarrow \theta = \frac{\pi}{2} + 2\pi n$. The "v" in the heart-shaped curve corresponds to the pole value $\frac{\pi}{2}$.

Variation of θ	Variation of r
1) $\quad 0 \;\to\; \frac{\pi}{2}$	$4 \to \quad 0$
2) $\quad \frac{\pi}{2} \;\to\; \pi$	$0 \to \quad 4$
3) $\quad \pi \;\to\; \frac{3\pi}{2}$	$4 \to \quad 8$
4) $\quad \frac{3\pi}{2} \;\to\; 2\pi$	$8 \to \quad 4$

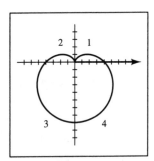

Figure 9

11 $r = -6(1 + \cos\theta)$ is a cardioid.

$0 = -6(1 + \cos\theta) \Rightarrow \cos\theta = -1 \Rightarrow \theta = \pi + 2\pi n$. The "v" in the heart-shaped curve corresponds to the pole value π.

Variation of θ	Variation of r
1) $\quad 0 \;\to\; \frac{\pi}{2}$	$-12 \to \quad -6$
2) $\quad \frac{\pi}{2} \;\to\; \pi$	$-6 \to \quad 0$
3) $\quad \pi \;\to\; \frac{3\pi}{2}$	$0 \to \quad -6$
4) $\quad \frac{3\pi}{2} \;\to\; 2\pi$	$-6 \to \quad -12$

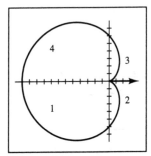

Figure 11

13 $r = 2 + 4\sin\theta$ is a limaçon with a loop since the constant term has a smaller magnitude than the coefficient of $\sin\theta$. $0 = 2 + 4\sin\theta \Rightarrow \sin\theta = -\frac{1}{2} \Rightarrow \theta = \frac{7\pi}{6} + 2\pi n$, $\frac{11\pi}{6} + 2\pi n$. Trace through the table and the figure to make sure you understand what values of θ form the loop.

	Variation of θ			Variation of r	
1)	0	$\rightarrow$	$\frac{\pi}{2}$	$2 \rightarrow$	6
2)	$\frac{\pi}{2}$	$\rightarrow$	π	$6 \rightarrow$	2
3)	π	$\rightarrow$	$\frac{7\pi}{6}$	$2 \rightarrow$	0
4)	$\frac{7\pi}{6}$	$\rightarrow$	$\frac{3\pi}{2}$	$0 \rightarrow$	-2
5)	$\frac{3\pi}{2}$	$\rightarrow$	$\frac{11\pi}{6}$	$-2 \rightarrow$	0
6)	$\frac{11\pi}{6}$	$\rightarrow$	2π	$0 \rightarrow$	2

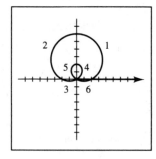

Figure 13

15 $r = \sqrt{3} - 2\sin\theta$ is a limaçon with a loop. $0 = \sqrt{3} - 2\sin\theta \Rightarrow \sin\theta = \sqrt{3}/2 \Rightarrow$
$\theta = \frac{\pi}{3} + 2\pi n, \frac{2\pi}{3} + 2\pi n$. Let $a = \sqrt{3} - 2 \approx -0.27$ and $b = \sqrt{3} + 2 \approx 3.73$.

	Variation of θ			Variation of r	
1)	0	$\rightarrow$	$\frac{\pi}{3}$	$\sqrt{3} \rightarrow$	0
2)	$\frac{\pi}{3}$	$\rightarrow$	$\frac{\pi}{2}$	$0 \rightarrow$	a
3)	$\frac{\pi}{2}$	$\rightarrow$	$\frac{2\pi}{3}$	$a \rightarrow$	0
4)	$\frac{2\pi}{3}$	$\rightarrow$	π	$0 \rightarrow$	$\sqrt{3}$
5)	π	$\rightarrow$	$\frac{3\pi}{2}$	$\sqrt{3} \rightarrow$	b
6)	$\frac{3\pi}{2}$	$\rightarrow$	2π	$b \rightarrow$	$\sqrt{3}$

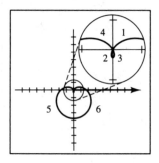

Figure 15

17 $r = 2 - \cos\theta$ •

$0 = 2 - \cos\theta \Rightarrow \cos\theta = 2 \Rightarrow$ no pole values.

	Variation of θ			Variation of r	
1)	0	$\rightarrow$	$\frac{\pi}{2}$	$1 \rightarrow$	2
2)	$\frac{\pi}{2}$	$\rightarrow$	π	$2 \rightarrow$	3
3)	π	$\rightarrow$	$\frac{3\pi}{2}$	$3 \rightarrow$	2
4)	$\frac{3\pi}{2}$	$\rightarrow$	2π	$2 \rightarrow$	1

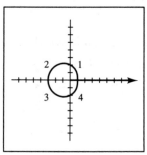

Figure 17

19 $r = 4\csc\theta \Rightarrow r\sin\theta = 4 \Rightarrow y = 4$.

r is undefined at $\theta = \pi n$.

This is a horizontal line with y-intercept $(0, 4)$.

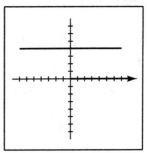

Figure 19

21 $r = 8\cos 3\theta$ is a 3-leafed rose since 3 is odd.

$0 = 8\cos 3\theta \Rightarrow \cos 3\theta = 0 \Rightarrow 3\theta = \frac{\pi}{2} + \pi n \Rightarrow \theta = \frac{\pi}{6} + \frac{\pi}{3}n$.

	Variation of θ		Variation of r	
1)	0	$\rightarrow \frac{\pi}{6}$	$8 \rightarrow$	0
2)	$\frac{\pi}{6}$	$\rightarrow \frac{\pi}{3}$	$0 \rightarrow$	-8
3)	$\frac{\pi}{3}$	$\rightarrow \frac{\pi}{2}$	$-8 \rightarrow$	0
4)	$\frac{\pi}{2}$	$\rightarrow \frac{2\pi}{3}$	$0 \rightarrow$	8
5)	$\frac{2\pi}{3}$	$\rightarrow \frac{5\pi}{6}$	$8 \rightarrow$	0
6)	$\frac{5\pi}{6}$	$\rightarrow \pi$	$0 \rightarrow$	-8

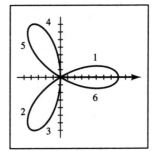

Figure 21

23 $r = 3\sin 2\theta$ is a 4-leafed rose. $0 = 3\sin 2\theta \Rightarrow \sin 2\theta = 0 \Rightarrow 2\theta = \pi n \Rightarrow \theta = \frac{\pi}{2}n$.

	Variation of θ		Variation of r	
1)	0	$\rightarrow \frac{\pi}{4}$	$0 \rightarrow$	3
2)	$\frac{\pi}{4}$	$\rightarrow \frac{\pi}{2}$	$3 \rightarrow$	0
3)	$\frac{\pi}{2}$	$\rightarrow \frac{3\pi}{4}$	$0 \rightarrow$	-3
4)	$\frac{3\pi}{4}$	$\rightarrow \pi$	$-3 \rightarrow$	0
5)	π	$\rightarrow \frac{5\pi}{4}$	$0 \rightarrow$	3
6)	$\frac{5\pi}{4}$	$\rightarrow \frac{3\pi}{2}$	$3 \rightarrow$	0
7)	$\frac{3\pi}{2}$	$\rightarrow \frac{7\pi}{4}$	$0 \rightarrow$	-3
8)	$\frac{7\pi}{4}$	$\rightarrow 2\pi$	$-3 \rightarrow$	0

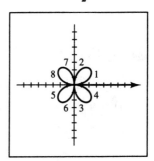

Figure 23

25 $r^2 = 4\cos 2\theta$ (lemniscate) • $0 = 4\cos 2\theta \Rightarrow \cos 2\theta = 0 \Rightarrow 2\theta = \frac{\pi}{2} + \pi n \Rightarrow \theta = \frac{\pi}{4} + \frac{\pi}{2}n$.

Note that as θ varies from 0 to $\frac{\pi}{4}$, we have $r^2 = 4$ or $r = \pm 2$ to $r^2 = 0$—both parts labeled "1" are traced with this range of θ. When θ varies from $\frac{\pi}{4}$ to $\frac{3\pi}{4}$, 2θ varies from $\frac{\pi}{2}$ to $\frac{3\pi}{2}$, and $\cos 2\theta$ is negative. Since r^2 can't equal a negative value, no portion of the graph is traced for these values of θ.

Variation of θ			Variation of r		
1)	0	$\rightarrow \frac{\pi}{4}$	$\pm 2 \rightarrow$		0
2)	$\frac{\pi}{4}$	$\rightarrow \frac{\pi}{2}$	undefined		
3)	$\frac{\pi}{2}$	$\rightarrow \frac{3\pi}{4}$	undefined		
4)	$\frac{3\pi}{4}$	$\rightarrow \pi$	$0 \rightarrow$		± 2

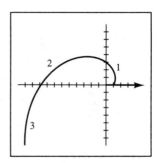

Figure 25

27 $r = 2^\theta$, $\theta \geq 0$ (spiral) •

Variation of θ			Variation of r		
1)	0	$\rightarrow \frac{\pi}{2}$	1	$\rightarrow$	2.97
2)	$\frac{\pi}{2}$	$\rightarrow \pi$	2.97	$\rightarrow$	8.82
3)	π	$\rightarrow \frac{3\pi}{2}$	8.82	$\rightarrow$	26.22
4)	$\frac{3\pi}{2}$	$\rightarrow 2\pi$	22.62	$\rightarrow$	77.88

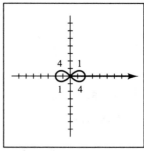

Figure 27

29 $r = 2\theta$, $\theta \geq 0$ •

Variation of θ			Variation of r		
1)	0	$\rightarrow \frac{\pi}{2}$	0	$\rightarrow$	π
2)	$\frac{\pi}{2}$	$\rightarrow \pi$	π	$\rightarrow$	2π
3)	π	$\rightarrow \frac{3\pi}{2}$	2π	$\rightarrow$	3π

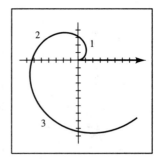

Figure 29

31 To simplify $\sin^2\left(\frac{\theta}{2}\right)$, recall the half-angle identity for the

sine. $r = 6\sin^2\left(\frac{\theta}{2}\right) = 6\left(\frac{1-\cos\theta}{2}\right) = 3(1-\cos\theta)$ is a

cardioid. $0 = 3(1-\cos\theta) \Rightarrow \cos\theta = 1 \Rightarrow \theta = 2\pi n$.

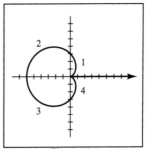

Variation of θ		Variation of r	
1) 0	$\rightarrow \frac{\pi}{2}$	$0 \rightarrow$	3
2) $\frac{\pi}{2}$	$\rightarrow \pi$	$3 \rightarrow$	6
3) π	$\rightarrow \frac{3\pi}{2}$	$6 \rightarrow$	3
4) $\frac{3\pi}{2}$	$\rightarrow 2\pi$	$3 \rightarrow$	0

Figure 31

33 Note that $r = 2\sec\theta$ is equivalent to $x = 2$. If $0 < \theta < \frac{\pi}{2}$ or $\frac{3\pi}{2} < \theta < 2\pi$, then $\sec\theta > 0$ and the graph of $r = 2 + 2\sec\theta$ is to the right of $x = 2$. If $\frac{\pi}{2} < \theta < \frac{3\pi}{2}$, $\sec\theta < 0$ and $r = 2 + 2\sec\theta$ is to the left of $x = 2$. r is undefined at $\theta = \frac{\pi}{2} + \pi n$.

$0 = 2 + 2\sec\theta \Rightarrow \sec\theta = -1 \Rightarrow \theta = \pi + 2\pi n$.

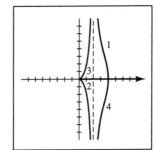

Variation of θ		Variation of r	
1) 0	$\rightarrow \frac{\pi}{2}$	$4 \rightarrow$	∞
2) $\frac{\pi}{2}$	$\rightarrow \pi$	$-\infty \rightarrow$	0
3) π	$\rightarrow \frac{3\pi}{2}$	$0 \rightarrow$	$-\infty$
4) $\frac{3\pi}{2}$	$\rightarrow 2\pi$	$\infty \rightarrow$	4

Figure 33

35 Use (1) in the "relationship between rectangular and polar coordinates."

(a) $x = r\cos\theta = 3\cos\frac{\pi}{4} = 3\left(\frac{\sqrt{2}}{2}\right) = \frac{3}{2}\sqrt{2}.$ $y = r\sin\theta = 3\sin\frac{\pi}{4} = 3\left(\frac{\sqrt{2}}{2}\right) = \frac{3}{2}\sqrt{2}.$

Hence, rectangular coordinates for $\left(3, \frac{\pi}{4}\right)$ are $\left(\frac{3}{2}\sqrt{2}, \frac{3}{2}\sqrt{2}\right)$.

(b) $x = -1\cos\frac{2\pi}{3} = -1\left(-\frac{1}{2}\right) = \frac{1}{2}.$ $y = -1\sin\frac{2\pi}{3} = -1\left(\frac{\sqrt{3}}{2}\right) = -\frac{1}{2}\sqrt{3}.$

37 (a) $x = 8\cos\left(-\frac{2\pi}{3}\right) = 8\left(-\frac{1}{2}\right) = -4.$ $y = 8\sin\left(-\frac{2\pi}{3}\right) = 8\left(-\frac{\sqrt{3}}{2}\right) = -4\sqrt{3}.$

(b) $x = -3\cos\frac{5\pi}{3} = -3\left(\frac{1}{2}\right) = -\frac{3}{2}.$ $y = -3\sin\frac{5\pi}{3} = -3\left(-\frac{\sqrt{3}}{2}\right) = \frac{3}{2}\sqrt{3}.$

39 Let $\theta = \arctan\frac{3}{4}$. Then we have $\cos\theta = \frac{4}{5}$ and $\sin\theta = \frac{3}{5}$.

$$x = 6\cos\theta = 6\left(\frac{4}{5}\right) = \frac{24}{5}. \quad y = 6\sin\theta = 6\left(\frac{3}{5}\right) = \frac{18}{5}.$$

41 Use (2) in the "relationship between rectangular and polar coordinates."

(a) $r^2 = x^2 + y^2 = (-1)^2 + (1)^2 = 2 \Rightarrow r = \sqrt{2}$ {since we want $r > 0$}.

$$\tan\theta = \frac{y}{x} = \frac{1}{-1} = -1 \Rightarrow \theta = \frac{3\pi}{4} \text{ {θ in QII}}.$$

(b) $r^2 = (-2\sqrt{3})^2 + (-2)^2 = 16 \Rightarrow r = \sqrt{16} = 4.$

$$\tan\theta = \frac{-2}{-2\sqrt{3}} = \frac{1}{\sqrt{3}} \Rightarrow \theta = \frac{7\pi}{6} \,\{\theta \text{ in QIII}\}.$$

43 (a) $r^2 = 7^2 + (-7\sqrt{3})^2 = 196 \Rightarrow r = \sqrt{196} = 14.$

$$\tan\theta = \frac{-7\sqrt{3}}{7} = -\sqrt{3} \Rightarrow \theta = \frac{5\pi}{3} \,\{\theta \text{ in QIV}\}.$$

(b) $r^2 = 5^2 + 5^2 = 50 \Rightarrow r = \sqrt{50} = 5\sqrt{2}.$ $\tan\theta = \frac{5}{5} = 1 \Rightarrow \theta = \frac{\pi}{4} \,\{\theta \text{ in QI}\}.$

45 (a) Since $\frac{7\pi}{3}$ is coterminal with $\frac{\pi}{3}$, $(3, \frac{7\pi}{3})$ represents the same point as $(3, \frac{\pi}{3})$.

(b) The point $(3, -\frac{\pi}{3})$ is in QIV, not QI, as is $(3, \frac{\pi}{3})$.

(c) The angle $\frac{4\pi}{3}$ is π radians larger than $\frac{\pi}{3}$, so its terminal side is on the same line as the terminal side of the angle $\frac{\pi}{3}$. $r = -3$ in the $\frac{4\pi}{3}$ direction is equivalent to $r = 3$ in the $\frac{\pi}{3}$ direction, so $(-3, \frac{4\pi}{3})$ represents the same point as $(3, \frac{\pi}{3})$.

(d) The point $(3, -\frac{2\pi}{3})$ is diametrically opposite the point $(3, \frac{\pi}{3})$.

(e) From part (d), we deduce that the point $(-3, -\frac{2\pi}{3})$ would represent the same point as $(3, \frac{\pi}{3})$.

(f) The point $(-3, -\frac{\pi}{3})$ is in QII, not QI.

Thus, choices (a), (c), and (e) represent the same point as $(3, \pi/3)$.

47 $x = -3 \Rightarrow r\cos\theta = -3 \Rightarrow r = \frac{-3}{\cos\theta} \Rightarrow r = -3\sec\theta$

49 $x^2 + y^2 = 16 \Rightarrow r^2 = 16 \Rightarrow r = \pm 4 \,\{\text{both are circles with radius 4}\}.$

51 $2y = -x \Rightarrow \frac{y}{x} = -\frac{1}{2} \Rightarrow \tan\theta = -\frac{1}{2} \Rightarrow \theta = \tan^{-1}(-\frac{1}{2})$

53 $y^2 - x^2 = 4 \Rightarrow r^2\sin^2\theta - r^2\cos^2\theta = 4 \Rightarrow -r^2(\cos^2\theta - \sin^2\theta) = 4 \Rightarrow$

$$-r^2\cos 2\theta = 4 \Rightarrow r^2 = \frac{-4}{\cos 2\theta} \Rightarrow r^2 = -4\sec 2\theta$$

55 $(x-1)^2 + y^2 = 1 \Rightarrow x^2 - 2x + 1 + y^2 = 1 \Rightarrow x^2 + y^2 = 2x \Rightarrow$

$$r^2 = 2r\cos\theta \Rightarrow r = 2\cos\theta$$

57 $r\cos\theta = 5 \Rightarrow x = 5.$ This is a vertical line with x-intercept $(5, 0)$.

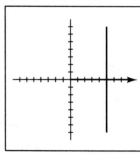

Figure 57

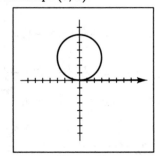

Figure 59

59 $r - 6\sin\theta = 0 \Rightarrow r^2 = 6r\sin\theta \Rightarrow x^2 + y^2 = 6y \Rightarrow x^2 + y^2 - 6y + \underline{9} = \underline{9} \Rightarrow$

$$x^2 + (y-3)^2 = 9.$$

61 $\theta = \frac{\pi}{4} \Rightarrow \tan\theta = \tan\frac{\pi}{4} \Rightarrow \frac{y}{x} = 1 \Rightarrow y = x.$

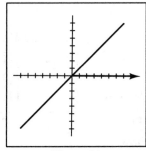

Figure 61

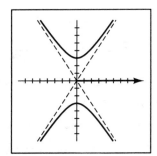

Figure 63

63 $r^2(4\sin^2\theta - 9\cos^2\theta) = 36 \Rightarrow 4r^2\sin^2\theta - 9r^2\cos^2\theta = 36 \Rightarrow$

$$4y^2 - 9x^2 = 36 \Rightarrow \frac{y^2}{9} - \frac{x^2}{4} = 1.$$

65 $r^2\cos 2\theta = 1 \Rightarrow r^2(\cos^2\theta - \sin^2\theta) = 1 \Rightarrow r^2\cos^2\theta - r^2\sin^2\theta = 1 \Rightarrow x^2 - y^2 = 1.$

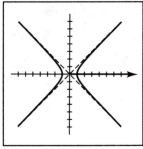

Figure 65

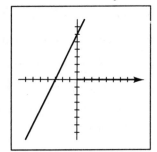

Figure 67

67 $r(\sin\theta - 2\cos\theta) = 6 \Rightarrow r\sin\theta - 2r\cos\theta = 6 \Rightarrow y - 2x = 6.$

69 $r(\sin\theta + r\cos^2\theta) = 1 \Rightarrow r\sin\theta + r^2\cos^2\theta = 1 \Rightarrow y + x^2 = 1,$ or $y = -x^2 + 1.$

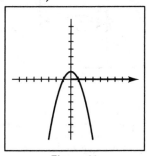

Figure 69

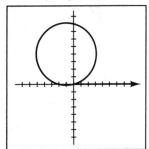

Figure 71

71 $r = 8\sin\theta - 2\cos\theta \Rightarrow r^2 = 8r\sin\theta - 2r\cos\theta \Rightarrow x^2 + y^2 = 8y - 2x \Rightarrow$

$$x^2 + 2x + \underline{1} + y^2 - 8y + \underline{16} = \underline{1} + \underline{16} \Rightarrow (x+1)^2 + (y-4)^2 = 17.$$

$\boxed{73}$ $r = \tan\theta \Rightarrow r^2 = \tan^2\theta \Rightarrow x^2 + y^2 = \dfrac{y^2}{x^2} \Rightarrow$

$x^4 + x^2y^2 = y^2 \Rightarrow y^2 - x^2y^2 = x^4 \Rightarrow$

$y^2(1 - x^2) = x^4 \Rightarrow y^2 = \dfrac{x^4}{1 - x^2}.$

This is a tough one even after getting the equation in x

and y. Since we have solved for y^2, the right side must

be positive. Since x^4 is always nonnegative, we must

have $1 - x^2 > 0$, or equivalently, $|x| < 1$. We have

vertical asymptotes at $x = \pm 1$, i.e., when the denominator is 0.

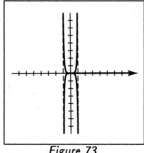

Figure 73

$\boxed{77}$ For TI-81 users, to sketch $r = f(t)$ $\{r = f(\theta)\}$, we make use of the equations

$$x = f(t)\cos t \quad \text{and} \quad y = f(t)\sin t.$$

First switch to "Param" from "Function" under the $\boxed{\text{MODE}}$ menu. Next, go to the

$\boxed{\text{Y} =}$ menu. To avoid repetitive typing, assign $X_{2T}\cos T$ to X_{1T} and $X_{2T}\sin T$ to

Y_{1T}. Now we need only assign the function, $f(t)$, to X_{2T}. For this exercise, assign

$2(\sin T)^2(\tan T)^2$ to X_{2T}, -1.05 to Tmin, 1.05 to Tmax, and 0.105 to Tstep $\{$ Tmin,

Tmax, and Tstep are found under $\boxed{\text{RANGE}}\}$. The graph is symmetric with respect

to the polar axis.

$[-9, 9]$ by $[-6, 6]$

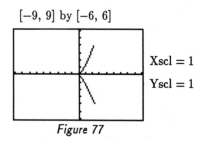

Xscl $= 1$

Yscl $= 1$

Figure 77

$[-12, 12]$ by $[-9, 9]$

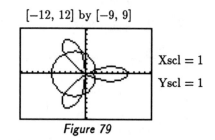

Xscl $= 1$

Yscl $= 1$

Figure 79

$\boxed{79}$ Assign $8\cos(3T)$ to X_{2T}, $Y_{2T}\cos T$ to X_{3T}, $Y_{2T}\sin T$ to Y_{3T}, $4 - 2.5\cos T$ to Y_{2T}

$\{$ turn *off* X_{2T} and Y_{2T}, leaving X_{1T}, Y_{1T}, X_{3T}, and Y_{3T} *on*$\}$, 0 to Tmin, 6.28 to

Tmax, and 0.105 to Tstep. From the graph, there are six points of intersection. The

approximate polar coordinates are $(1.75, \pm 0.45)$, $(4.49, \pm 1.77)$, and $(5.76, \pm 2.35)$.

11.6 Exercises

Note: For the ellipse, the major axis is vertical if the denominator contains $\sin\theta$, horizontal if the denominator contains $\cos\theta$. For the hyperbola, the transverse axis is vertical if the denominator contains $\sin\theta$, horizontal if the denominator contains $\cos\theta$. The focus at the pole is called F and V is the vertex associated with (or closest to) F. $d(V, F)$ denotes the distance from the vertex to the focus. The foci are not asked for in the directions, but are listed. For the parabola, the directrix is on the right, on the left, above, or below the focus depending on the term "+cos", "−cos", "+sin", or "−sin", respectively, appearing in the denominator.

[1] Divide the numerator and denominator by the constant term in the denominator, i.e., 6. $r = \dfrac{12}{6 + 2\sin\theta} = \dfrac{2}{1 + \frac{1}{3}\sin\theta} \Rightarrow e = \frac{1}{3} < 1$, ellipse. From the previous note, we see that the denominator has "$+\sin\theta$" and we have vertices when $\theta = \frac{\pi}{2}$ and $\frac{3\pi}{2}$. They are $V(\frac{3}{2}, \frac{\pi}{2})$ and $V'(3, \frac{3\pi}{2})$. The distance from the focus at the pole to the vertex V is $\frac{3}{2}$. The distance from V' to F' must also be $\frac{3}{2}$ and we see that $F' = (\frac{3}{2}, \frac{3\pi}{2})$. We will use the following notation to summarize this in future problems:

$$d(V, F) = \tfrac{3}{2} \Rightarrow F' = (\tfrac{3}{2}, \tfrac{3\pi}{2}).$$

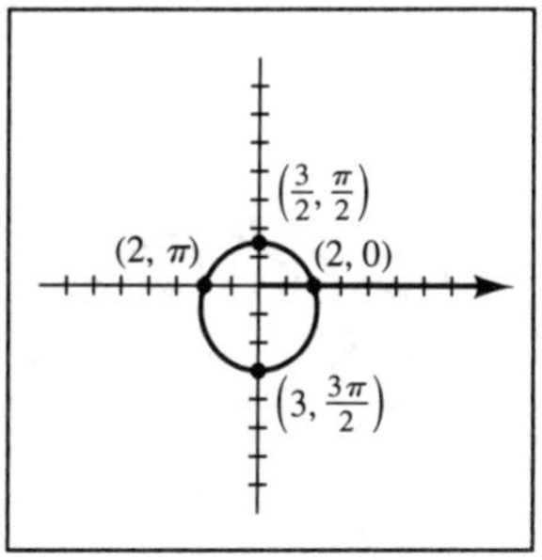

Figure 1

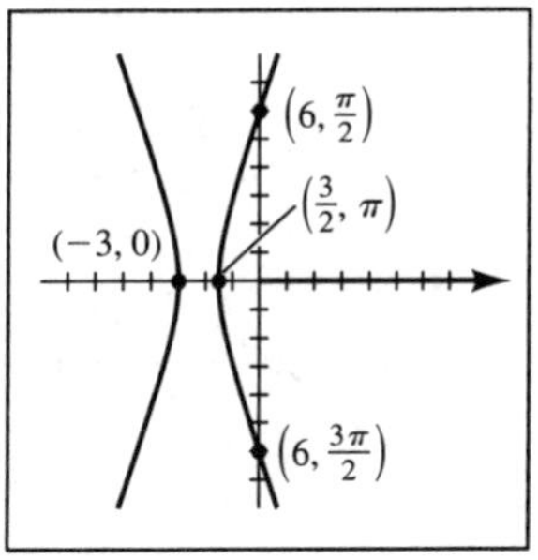

Figure 3

[3] Divide both the numerator and the denominator by 2 to obtain the "1" in the standard form. $r = \dfrac{12}{2 - 6\cos\theta} = \dfrac{6}{1 - 3\cos\theta} \Rightarrow e = 3 > 1$, hyperbola.

$V(\frac{3}{2}, \pi)$ and $V'(-3, 0)$. $d(V, F) = \frac{3}{2} \Rightarrow F' = (-\frac{9}{2}, 0)$.

⑤ $r = \dfrac{3}{2 + 2\cos\theta} = \dfrac{\frac{3}{2}}{1 + 1\cos\theta} \Rightarrow e = 1$, parabola. Note that the expression is

undefined in the $\theta = \pi$ direction. The vertex is in the $\theta = 0$ direction, $V(\frac{3}{4},\, 0)$.

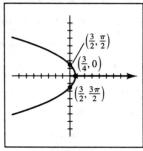

Figure 5

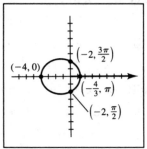

Figure 7

⑦ $r = \dfrac{4}{\cos\theta - 2} = \dfrac{-2}{1 - \frac{1}{2}\cos\theta} \Rightarrow e = \frac{1}{2} < 1$, ellipse.

$$V(-\tfrac{4}{3},\, \pi) \text{ and } V'(-4,\, 0). \quad d(V,\, F) = \tfrac{4}{3} \Rightarrow F' = (-\tfrac{8}{3},\, 0)$$

⑨ We multiply by $\dfrac{\sin\theta}{\sin\theta}$ to obtain the standard form of a conic.

$$r = \dfrac{6\csc\theta}{2\csc\theta + 3}\cdot\dfrac{\sin\theta}{\sin\theta} = \dfrac{6}{2 + 3\sin\theta} = \dfrac{3}{1 + \frac{3}{2}\sin\theta} \Rightarrow e = \tfrac{3}{2} > 1, \text{ hyperbola.}$$

$V(\frac{6}{5},\, \frac{\pi}{2})$ and $V'(-6,\, \frac{3\pi}{2})$. $d(V,\, F) = \frac{6}{5} \Rightarrow F' = (-\frac{36}{5},\, \frac{3\pi}{2})$.

Since the original equation is undefined when $\csc\theta$ is undefined { which is

when $\theta = \pi n$ }, the points $(3,\, 0)$ and $(3,\, \pi)$ are excluded from the graph.

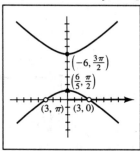

Figure 9

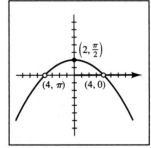

Figure 11

⑪ $r = \dfrac{4\csc\theta}{1 + \csc\theta}\cdot\dfrac{\sin\theta}{\sin\theta} = \dfrac{4}{1 + 1\sin\theta} \Rightarrow e = 1$, parabola. The vertex is in the $\theta = \frac{\pi}{2}$

direction, $V(2,\, \frac{\pi}{2})$. Since the original equation is undefined when $\csc\theta$ is undefined,

the points $(4,\, 0)$ and $(4,\, \pi)$ are excluded from the graph.

$\boxed{13}$ $r = \dfrac{12}{6 + 2 \sin\theta} \Rightarrow 6r + 2r \sin\theta = 12 \Rightarrow 6r + 2y = 12 \Rightarrow$

{ Isolate the term with the "r" so that by squaring both sides of the equation,

we can make the substitution $r^2 = x^2 + y^2$. }

$3r = 6 - y \Rightarrow 9r^2 = 36 - 12y + y^2 \Rightarrow 9(x^2 + y^2) = 36 - 12y + y^2 \Rightarrow$

$$9x^2 + 8y^2 + 12y - 36 = 0$$

Note: For the following exercises, the substitutions

$x = r \cos\theta$, $y = r \sin\theta$, and $r^2 = x^2 + y^2$ are made without mention.

$\boxed{15}$ $r = \dfrac{12}{2 - 6 \cos\theta} \Rightarrow 2r - 6x = 12 \Rightarrow r = 3x + 6 \Rightarrow r^2 = 9x^2 + 36x + 36 \Rightarrow$

$$8x^2 - y^2 + 36x + 36 = 0$$

$\boxed{17}$ $r = \dfrac{3}{2 + 2 \cos\theta} \Rightarrow 2r + 2x = 3 \Rightarrow 2r = 3 - 2x \Rightarrow 4r^2 = 4x^2 - 12x + 9 \Rightarrow$

$$4y^2 + 12x - 9 = 0$$

$\boxed{19}$ $r = \dfrac{4}{\cos\theta - 2} \Rightarrow x - 2r = 4 \Rightarrow x - 4 = 2r \Rightarrow x^2 - 8x + 16 = 4r^2 \Rightarrow$

$$3x^2 + 4y^2 + 8x - 16 = 0$$

$\boxed{21}$ $r = \dfrac{6 \csc\theta}{2 \csc\theta + 3} \cdot \dfrac{\sin\theta}{\sin\theta} = \dfrac{6}{2 + 3 \sin\theta} \Rightarrow 2r + 3y = 6 \Rightarrow 2r = 6 - 3y \Rightarrow$

$4r^2 = 36 - 36y + 9y^2 \Rightarrow 4x^2 - 5y^2 + 36y - 36 = 0$.

r is undefined when $\theta = 0$ or π. For the rectangular equation, these points

correspond to $y = 0$ (or $r \sin\theta = 0$). Substituting $y = 0$ into the above rectangular

equation yields $4x^2 = 36$, or $x = \pm 3$. $\therefore$ exclude $(\pm 3, 0)$

$\boxed{23}$ $r = \dfrac{4 \csc\theta}{1 + \csc\theta} \cdot \dfrac{\sin\theta}{\sin\theta} = \dfrac{4}{1 + 1 \sin\theta} \Rightarrow r + y = 4 \Rightarrow r = 4 - y \Rightarrow$

$r^2 = y^2 - 8y + 16 \Rightarrow x^2 + 8y - 16 = 0$. r is undefined when $\theta = 0$ or π.

For the rectangular equation, these points correspond to $y = 0$ (or $r \sin\theta = 0$).

Substituting $y = 0$ into the above rectangular equation yields $x^2 = 16$, or $x = \pm 4$.

$\therefore$ exclude $(\pm 4, 0)$

$\boxed{25}$ $r = 2 \sec\theta \Rightarrow r \cos\theta = 2 \Rightarrow x = 2$. Remember that d is the distance from the focus at

the pole to the directrix. Thus, $d = 2$ and since the directrix is on the

right of the focus at the pole, we use "$+\cos\theta$". $r = \dfrac{2(\frac{1}{3})}{1 + \frac{1}{3} \cos\theta} \cdot \dfrac{3}{3} = \dfrac{2}{3 + \cos\theta}$.

$\boxed{27}$ $r \cos\theta = -3 \Rightarrow x = -3$. Thus, $d = 3$ and since the directrix is on the left of the

focus at the pole, we use "$-\cos\theta$". $r = \dfrac{3(\frac{4}{3})}{1 - \frac{4}{3} \cos\theta} \cdot \dfrac{3}{3} = \dfrac{12}{3 - 4 \cos\theta}$.

$\boxed{29}$ $r \sin\theta = -2 \Rightarrow y = -2$. Thus, $d = 2$ and since the directrix is below the focus at

the pole, we use "$-\sin\theta$". $r = \dfrac{2(1)}{1 - 1 \sin\theta} = \dfrac{2}{1 - \sin\theta}$.

31 $r = 4\csc\theta \Rightarrow r\sin\theta = 4 \Rightarrow y = 4$. Thus, $d = 4$ and since the directrix is above

the focus at the pole, we use "$+\sin\theta$". $r = \dfrac{4(\frac{2}{5})}{1 + \frac{2}{5}\sin\theta} \cdot \dfrac{5}{5} = \dfrac{8}{5 + 2\sin\theta}$.

33 For a parabola, $e = 1$. The vertex is 4 units above the focus at the pole,

so $d = 2(4)$ and we should use "$+\sin\theta$" in the denominator. $r = \dfrac{8}{1 + \sin\theta}$

35 (a) See *Figure 35*. $e = \frac{c}{a} = \dfrac{d(C, F)}{d(C, V)} = \dfrac{3}{4}$.

Figure 35

(b) Since the vertex is below the focus at the pole,

use "$-\sin\theta$". $r = \dfrac{d(\frac{3}{4})}{1 - \frac{3}{4}\sin\theta}$ and $r = 1$ when

$\theta = \frac{3\pi}{2} \Rightarrow 1 = \dfrac{d(\frac{3}{4})}{1 - \frac{3}{4}(-1)} \Rightarrow 1 = \dfrac{\frac{3}{4}d}{\frac{7}{4}} \Rightarrow$

$d = \frac{7}{3}$. Thus, $r = \dfrac{(\frac{7}{3})(\frac{3}{4})}{1 - \frac{3}{4}\sin\theta} \cdot \dfrac{4}{4} = \dfrac{7}{4 - 3\sin\theta}$.

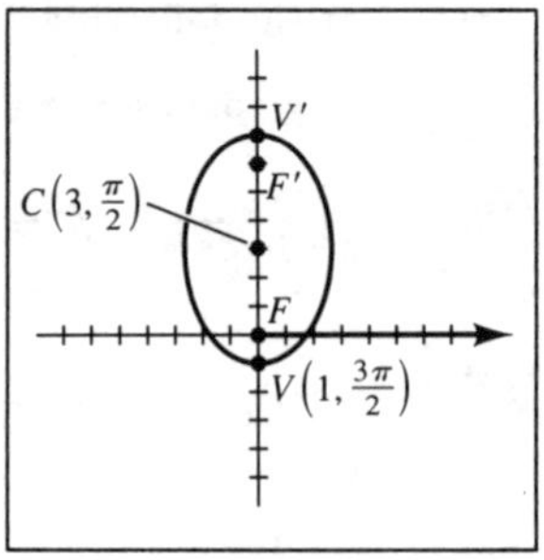

An equivalent rectangular equation is $\dfrac{x^2}{7} + \dfrac{(y - 3)^2}{16} = 1$.

37 (a) Let V and C denote the vertex closest to the sun and the center of the ellipse,

respectively. Let s denote the distance from V to the directrix to the left of V.

$d(O, V) = d(C, V) - d(C, O) = a - c = a - ea = a(1 - e)$.

Also, by the first theorem in §11.6, $\dfrac{d(O, V)}{s} = e \Rightarrow s = \dfrac{d(O, V)}{e} = \dfrac{a(1 - e)}{e}$.

Now, $d = s + d(O, V) = \dfrac{a(1 - e)}{e} + a(1 - e) = \dfrac{a(1 - e^2)}{e}$ and $de = a(1 - e^2)$.

Thus, the equation of the orbit is $r = \dfrac{(1 - e^2)a}{1 - e\cos\theta}$.

(b) The minimum distance occurs when $\theta = \pi$. $r_{\text{per}} = \dfrac{(1 - e^2)a}{1 - e(-1)} = a(1 - e)$.

The maximum distance occurs when $\theta = 0$. $r_{\text{aph}} = \dfrac{(1 - e^2)a}{1 - e(1)} = a(1 + e)$.

11.7 Exercises

1 For this exercise (and others), we solve for t in terms of x, and then substitute that expression for t in the equation that relates y and t.

$x = t - 2 \Rightarrow t = x + 2.$ $y = 2t + 3 = 2(x + 2) + 3 = 2x + 7.$

As t varies from 0 to 5, (x, y) varies from $(-2, 3)$ to $(3, 13)$.

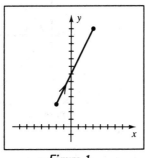

Figure 1

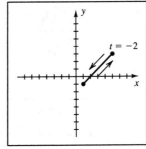

Figure 3

3 $x = t^2 + 1 \Rightarrow t^2 = x - 1.$ $y = t^2 - 1 = x - 2.$ As t varies from -2 to 2,

(x, y) varies from $(5, 3)$ to $(1, -1)$ { when $t = 0$ } and back to $(5, 3)$.

5 Since y is linear in t, it is easier to solve the second equation for t than it is to solve the first equation for t. Hence, we solve for t in terms of y, and then substitute that expression for t in the equation that relates x and t. $y = 2t + 3 \Rightarrow t = \frac{1}{2}(y - 3).$ $x = 4\left[\frac{1}{2}(y - 3)\right]^2 - 5 \Rightarrow (y - 3)^2 = x + 5.$ This is a parabola with vertex at $(-5, 3)$. Since t takes on all real values, so does y, and the curve C is the entire parabola.

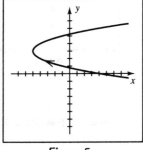

Figure 5

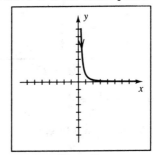

Figure 7

7 $y = e^{-2t} = (e^t)^{-2} = x^{-2} = 1/x^2.$ This is a rational function. Only the first quadrant portion is used and as t varies from $-\infty$ to ∞, x varies from 0 to ∞, excluding 0.

9 $x = 2\sin t$ and $y = 3\cos t \Rightarrow \frac{x}{2} = \sin t$ and $\frac{y}{3} = \cos t \Rightarrow$

$\frac{x^2}{4} + \frac{y^2}{9} = \sin^2 t + \cos^2 t = 1.$ As t varies from 0 to 2π,

(x, y) traces the ellipse from $(0, 3)$ in a clockwise direction back to $(0, 3)$.

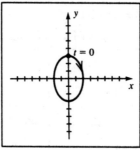

Figure 9

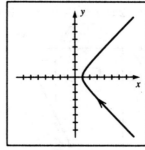

Figure 11

11 $x = \sec t$ and $y = \tan t \Rightarrow x^2 - y^2 = \sec^2 t - \tan^2 t = 1.$

As t varies from $-\frac{\pi}{2}$ to $\frac{\pi}{2}$, (x, y) traces the right branch of the hyperbola along the

asymptote $y = -x$ to $(1, 0)$ and then along the asymptote $y = x$.

13 $y = 2\ln t = \ln t^2$ {since $t > 0$} $= \ln x$.

As t varies from 0 to ∞, so does x, and y varies from $-\infty$ to ∞.

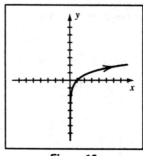

Figure 13

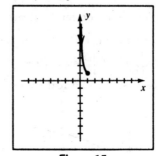

Figure 15

15 $y = \csc t = \frac{1}{\sin t} = \frac{1}{x}$.

As t varies from 0 to $\frac{\pi}{2}$, (x, y) varies asymptotically from the positive y-axis to $(1, 1)$.

17 $x = t$ and $y = \sqrt{t^2 - 1} \Rightarrow y = \sqrt{x^2 - 1} \Rightarrow x^2 - y^2 = 1$.

Since y is nonnegative, the graph is the top half of both branches of the hyperbola.

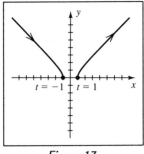

Figure 17

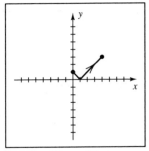

Figure 19

19 $x = t$ and $y = \sqrt{t^2 - 2t + 1} \Rightarrow y = \sqrt{x^2 - 2x + 1} = \sqrt{(x-1)^2} = |x - 1|$.

As t varies from 0 to 4, (x, y) traces $y = |x - 1|$ from $(0, 1)$ to $(4, 3)$.

21 $x = (t + 1)^3 \Rightarrow t = x^{1/3} - 1$. $y = (t + 2)^2 = (x^{1/3} + 1)^2$.

This is probably an unfamiliar graph. The graph of $y = x^{1/3}$ is similar to the graph of $y = x^{1/2}$ $\{y = \sqrt{x}\}$ but is symmetric with respect to the origin. The "+1" shifts $y = x^{1/3}$ up 1 unit and the "squaring" makes all y values nonnegative. Since we have restrictions on the variable t, we only have a portion of this graph. As t varies from 0 to 2, (x, y) varies from $(1, 4)$ to $(27, 16)$.

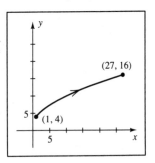

Figure 21

23 All of the curves are a portion of the parabola $x = y^2$.

C_1: $x = t^2 = y^2$. y takes on all real values and we have the entire parabola.

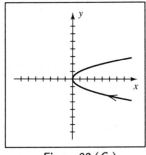

Figure 23 (C_1)

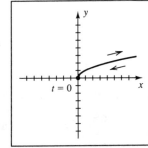

Figure 23 (C_2)

C_2: $x = t^4 = (t^2)^2 = y^2$. C_2 is only the top half since $y = t^2$ is nonnegative.

As t varies from $-\infty$ to ∞, the top portion is traced twice.

C_3: $x = \sin^2 t = (\sin t)^2 = y^2$. C_3 is the portion of the curve from $(1, -1)$ to $(1, 1)$.

     The point $(1, 1)$ is reached at $t = \frac{\pi}{2} + 2\pi n$ and the point $(1, -1)$ when

$$t = \frac{3\pi}{2} + 2\pi n.$$

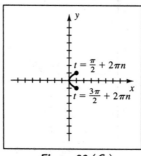

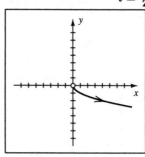

Figure 23 (C_3)   Figure 23 (C_4)

C_4: $x = e^{2t} = (e^t)^2 = (-e^t)^2 = y^2$. C_4 is the bottom half of the parabola since y

                is negative. As t approaches $-\infty$, the parabola approaches the origin.

25 In each part, the motion is on the unit circle since $x^2 + y^2 = 1$.

    (a) For $0 \le t \le \pi$, x $\{\cos t\}$ varies from 1 to -1 and y $\{\sin t\}$ is nonnegative.

                            $P(x, y)$ moves from $(1, 0)$ counterclockwise to $(-1, 0)$.

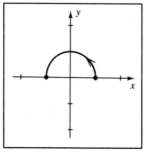 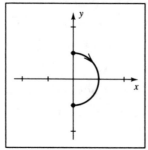 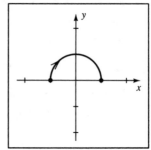

Figure 25(a)   Figure 25(b)   Figure 25(c)

    (b) For $0 \le t \le \pi$, y $\{\cos t\}$ varies from 1 to -1 and x $\{\sin t\}$ is nonnegative.

                          $P(x, y)$ moves from $(0, 1)$ clockwise to $(0, -1)$.

    (c) For $-1 \le t \le 1$, x $\{t\}$ varies from -1 to 1 and y $\{\sqrt{1-t^2}\}$ is nonnegative.

                          $P(x, y)$ moves from $(-1, 0)$ clockwise to $(1, 0)$.

27 $x = a \cos t + h$ and $y = b \sin t + k \Rightarrow \frac{x-h}{a} = \cos t$ and $\frac{y-k}{b} = \sin t \Rightarrow$

$\frac{(x-h)^2}{a^2} + \frac{(y-k)^2}{b^2} = \cos^2 t + \sin^2 t = 1$.   This is the equation of an ellipse with

        center (h, k) and semiaxes of lengths a and b (axes of lengths $2a$ and $2b$).

29 Let $\theta = \angle FDP$ and $\alpha = \angle GDP = \angle EDP$. Then $\angle ODG = (\frac{\pi}{2} - t)$ and

$\alpha = \theta - (\frac{\pi}{2} - t) = \theta + t - \frac{\pi}{2}$. Arcs AF and PF are equal in length since each is

the distance rolled. Thus, $at = b\theta$, or $\theta = (\frac{a}{b})t$ and $\alpha = \frac{a+b}{b}t - \frac{\pi}{2}$.

Note that $\cos\alpha = \sin\left(\frac{a+b}{b}t\right)$ and $\sin\alpha = -\cos\left(\frac{a+b}{b}t\right)$.

For the location of the points as illustrated, the coordinates of P are:

$$x = d(O,\, G) + d(G,\, B) = d(O,\, G) + d(E,\, P) = (a+b)\cos t + b\sin\alpha$$

$$= (a+b)\cos t - b\cos\left(\frac{a+b}{b}t\right).$$

$$y = d(B,\, P) = d(G,\, D) - d(D,\, E) = (a+b)\sin t - b\cos\alpha$$

$$= (a+b)\sin t - b\sin\left(\frac{a+b}{b}t\right).$$

It can be verified that these equations are valid for all locations of the points.

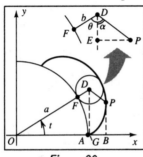

Figure 29

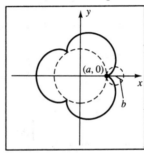

Figure 31

31 $b = \frac{1}{3}a \Rightarrow a = 3b$. Substituting into the equations from Exercise 29 yields:

$$x = (3b + b)\cos t - b\cos\left(\frac{3b+b}{b}t\right) = 4b\cos t - b\cos 4t$$

$$y = (3b + b)\sin t - b\sin\left(\frac{3b+b}{b}t\right) = 4b\sin t - b\sin 4t$$

As an aid in graphing, to determine where the path of the smaller circle will intersect
the path of the larger circle (for the original starting point of intersection at $A(a,\, 0)$),
we can solve $x^2 + y^2 = a^2$ for t.

$$x^2 + y^2 = 16b^2\cos^2 t - 8b^2\cos t\cos 4t + b^2\cos^2 4t +$$
$$16b^2\sin^2 t - 8b^2\sin t\sin 4t + b^2\sin^2 4t$$

$$= 17b^2 - 8b^2(\cos t\cos 4t + \sin t\sin 4t)$$

$$= 17b^2 - 8b^2[\cos(t - 4t)] = 17b^2 - 8b^2\cos 3t.$$

Thus, $x^2 + y^2 = a^2 \Rightarrow 17b^2 - 8b^2\cos 3t = a^2 = 9b^2 \Rightarrow 8b^2 = 8b^2\cos 3t \Rightarrow$

$1 = \cos 3t \Rightarrow 3t = 2\pi n \Rightarrow t = \frac{2\pi}{3}n$.

It follows that the intersection points are at $t = \frac{2\pi}{3}, \frac{4\pi}{3}$, and 2π.

33 (a) $x = a \sin \omega t$ and $y = b \cos \omega t \Rightarrow \frac{x}{a} = \sin \omega t$ and $\frac{y}{b} = \cos \omega t \Rightarrow \frac{x^2}{a^2} + \frac{y^2}{b^2} = 1$.

The figure is an ellipse with center $(0, 0)$ and axes of lengths $2a$ and $2b$.

(b) $f(t + p) = a \sin[\omega_1(t + p)] = a \sin[\omega_1 t + \omega_1 p] = a \sin[\omega_1 t + 2\pi n] =$

$$a \sin \omega_1 t = f(t).$$

$g(t + p) = b \cos[\omega_2(t + p)] = b \cos\left[\omega_2 t + \frac{\omega_2}{\omega_1} 2\pi n\right] = b \cos\left[\omega_2 t + \frac{m}{n} 2\pi n\right] =$

$$b \cos[\omega_2 t + 2\pi m] = b \cos \omega_2 t = g(t).$$

Since f and g are periodic with period p,

the curve retraces itself every p units of time.

35 Change "Param" from "Function" under $\boxed{\text{MODE}}$. Make the assignments

$3(\sin T)\hat{\ }5$ to X_{1T}, $3(\cos T)\hat{\ }5$ to Y_{1T}, 0 to Tmin, 6.28 to Tmax, and 0.105 to Tstep.

Algebraically, we have $x = 3\sin^5 t$ and $y = 3\cos^5 t \Rightarrow \frac{x}{3} = \sin^5 t$ and $\frac{y}{3} = \cos^5 t \Rightarrow$

$\left(\frac{x}{3}\right)^{2/5} + \left(\frac{y}{3}\right)^{2/5} = \sin^2 t + \cos^2 t = 1$. The graph traces an astroid.

$[-6, 6]$ by $[-4, 4]$

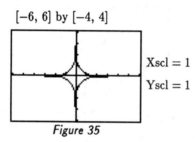

Xscl $= 1$
Yscl $= 1$

$[-30, 30]$ by $[-20, 20]$

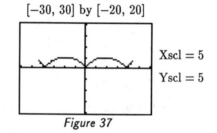

Xscl $= 5$
Yscl $= 5$

Figure 35   *Figure 37*

37 The graph traces a curtate cycloid.

39 The figure is a mask with a mouth, nose, and eyes. This graph may be obtained with a graphing utility that has the capability to graph 5 sets of parametric equations.

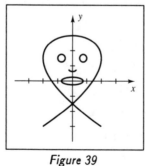

Figure 39

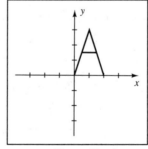

Figure 41

41 C_1 is the line $y = 3x$ from $(0, 0)$ to $(1, 3)$. For C_2, $x - 1 = \tan t$ and $1 - \frac{1}{3}y = \tan t \Rightarrow x - 1 = 1 - \frac{1}{3}y \Rightarrow y = -3x + 6$. This line is sketched from $(1, 3)$ to $(2, 0)$. C_3 is the horizontal line $y = \frac{3}{2}$ from $\left(\frac{1}{2}, \frac{3}{2}\right)$ to $\left(\frac{3}{2}, \frac{3}{2}\right)$. The figure is the letter A.

Chapter 11 Review Exercises

Note: Let the notation be the same as in §11.1–11.7.

boxed{1} $y^2 = 64x \Rightarrow x = \frac{1}{64}y^2 \Rightarrow a = \frac{1}{64}.$ $p = \dfrac{1}{4(\frac{1}{64})} = 16.$ $V(0, 0);$ $F(16, 0);$ $l: x = -16$

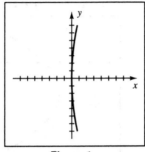

Figure 1

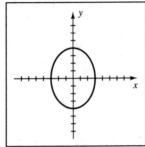

Figure 3

boxed{3} $9y^2 = 144 - 16x^2 \Rightarrow \dfrac{x^2}{9} + \dfrac{y^2}{16} = 1.$ $c^2 = 16 - 9 \Rightarrow c = \pm\sqrt{7}.$

$$V(0, \pm 4); \; F(0, \pm\sqrt{7}); \; M(\pm 3, 0)$$

boxed{5} $x^2 - y^2 - 4 = 0 \Rightarrow \dfrac{x^2}{4} - \dfrac{y^2}{4} = 1.$ $c^2 = 4 + 4 \Rightarrow c = \pm 2\sqrt{2}.$

$$V(\pm 2, 0); \; F(\pm 2\sqrt{2}, 0); \; W(0, \pm 2); \; y = \pm x$$

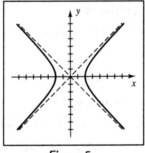

Figure 5

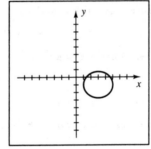

Figure 8

boxed{8} $3x^2 + 4y^2 - 18x + 8y + 19 = 0 \Rightarrow$

$3(x^2 - 6x + \underline{\;9\;}) + 4(y^2 + 2y + \underline{\;1\;}) = -19 + \underline{\;27\;} + \underline{\;4\;} \Rightarrow$

$3(x - 3)^2 + 4(y + 1)^2 = 12 \Rightarrow \dfrac{(x - 3)^2}{4} + \dfrac{(y + 1)^2}{3} = 1.$ $c^2 = 4 - 3 \Rightarrow c = \pm 1.$

$$C(3, -1); \; V(3 \pm 2, -1); \; F(3 \pm 1, -1); \; M(3, -1 \pm \sqrt{3})$$

12 $4x^2 - y^2 - 40x - 8y + 88 = 0 \Rightarrow$

$\quad 4(x^2 - 10x + \underline{\;25\;}) - (y^2 + 8y + \underline{\;16\;}) = -88 + \underline{\;100\;} - \underline{\;16\;} \Rightarrow$

$\quad 4(x-5)^2 - (y+4)^2 = -4 \Rightarrow \dfrac{(y+4)^2}{4} - \dfrac{(x-5)^2}{1} = 1. \;\; c^2 = 4 + 1 \Rightarrow c = \pm\sqrt{5}.$

$\qquad C(5, -4); \; V(5, -4 \pm 2); \; F(5, -4 \pm \sqrt{5}); \; W(5 \pm 1, -4); \; (y+4) = \pm 2(x-5)$

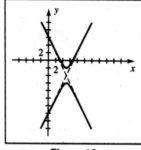 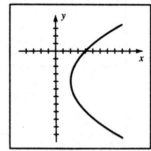

| Figure 12 | Figure 13 |

13 $y^2 - 8x + 8y + 32 = 0 \Rightarrow x = \frac{1}{8}y^2 + y + 4 \Rightarrow a = \frac{1}{8}. \;\; p = \dfrac{1}{4(\frac{1}{8})} = 2.$

$\qquad\qquad\qquad\qquad\qquad\qquad\qquad\qquad V(2, -4); \; F(4, -4); \; l : x = 0$

17 The vertex is halfway between the x-intercepts, so it is of the form $V(-7, k)$.

$\quad y = a(x + 10)(x + 4)$ and $x = 0$, $y = 80 \Rightarrow 80 = a(10)(4) \Rightarrow a = 2.$

$\qquad\qquad x = -7 \Rightarrow y = 2(3)(-3) = -18. \;\;$ Hence, $y = 2(x+7)^2 - 18.$

20 $F(-4, 0)$ and $l : x = 4 \Rightarrow p = -4$ and $V(0, 0).$

$\qquad\qquad\qquad$ An equation is $(y - 0)^2 = \left[4(-4)\right](x - 0),$ or $y^2 = -16x.$

22 The general equation of a parabola that is symmetric to the x-axis and has its vertex

$\quad$ at the origin is $x = ay^2.$ Substituting $x = 5$ and $y = -1$ into that equation yields

$\qquad\qquad\qquad\qquad\qquad a = 5. \;\;$ An equation is $x = 5y^2.$

24 $F(\pm 10, 0)$ and $V(\pm 5, 0) \Rightarrow b^2 = 10^2 - 5^2 = 75.$

$\qquad\qquad\qquad$ An equation is $\dfrac{x^2}{5^2} - \dfrac{y^2}{75} = 1$ or $\dfrac{x^2}{25} - \dfrac{y^2}{75} = 1.$

26 $F(\pm 2, 0) \Rightarrow c^2 = 4.$ Now $\dfrac{x^2}{a^2} + \dfrac{y^2}{b^2} = 1$ can be written as $\dfrac{x^2}{a^2} + \dfrac{y^2}{a^2 - 4} = 1$ since

$\quad b^2 = a^2 - c^2.$ Substituting $x = 2$ and $y = \sqrt{2}$ into that equation yields

$\quad \dfrac{4}{a^2} + \dfrac{2}{a^2 - 4} = 1 \Rightarrow 4a^2 - 16 + 2a^2 = a^4 - 4a^2 \Rightarrow a^4 - 10a^2 + 16 = 0 \Rightarrow$

$\quad (a^2 - 2)(a^2 - 8) = 0 \Rightarrow a^2 = 2, 8. \;\;$ Since $a > c$, a^2 must be 8 and b^2 is equal to 4.

$\qquad\qquad\qquad\qquad\qquad$ An equation is $\dfrac{x^2}{8} + \dfrac{y^2}{4} = 1.$

27 $M(\pm 5, 0) \Rightarrow b = 5. \;\; e = \dfrac{c}{a} = \dfrac{\sqrt{a^2 - b^2}}{a} = \dfrac{\sqrt{a^2 - 25}}{a} = \dfrac{2}{3} \Rightarrow \dfrac{2}{3}a = \sqrt{a^2 - 25} \Rightarrow$

$\quad \dfrac{4}{9}a^2 = a^2 - 25 \Rightarrow \dfrac{5}{9}a^2 = 25 \Rightarrow a^2 = 45. \;\;$ An equation is $\dfrac{x^2}{25} + \dfrac{y^2}{45} = 1.$

30 The vertex of the square in the first quadrant has coordinates (x, x).

Since it is on the ellipse, $\dfrac{x^2}{a^2} + \dfrac{y^2}{b^2} = 1 \Rightarrow b^2x^2 + a^2x^2 = a^2b^2 \Rightarrow x^2 = \dfrac{a^2b^2}{a^2 + b^2}$.

x^2 is $\frac{1}{4}$ of the area of the square, hence $A = \dfrac{4a^2b^2}{a^2 + b^2}$.

31 The focus is a distance of $p = 1/(4a) = 1/(4 \cdot \frac{1}{8}) = 2$ units from the origin.

An equation of the circle is $x^2 + (y - 2)^2 = 2^2 = 4$.

32 $P(x, y)$ is a distance of $(2 + d)$ from $(0, 0)$ and a distance of d from $(4, 0)$. The difference of these distances is $(2 + d) - d = 2$, a *positive constant*. By the definition of a hyperbola, $P(x, y)$ lies on the right branch of the hyperbola with foci $(0, 0)$ and $(4, 0)$. The center of the hyperbola is halfway between the foci, i.e., $(2, 0)$. The vertex is halfway from $(2, 0)$ to $(4, 0)$ since the distance from the circle to P equals the distance from P to $(4, 0)$. Thus, the vertex is $(3, 0)$ and $a = 1$. $b^2 = c^2 - a^2 = 2^2 - 1^2 = 3$ and an equation of the right branch of the hyperbola is

$$\frac{(x-2)^2}{1} - \frac{y^2}{3} = 1, \; x \geq 3 \quad \text{or} \quad x = 2 + \sqrt{1 + \frac{y^2}{3}}.$$

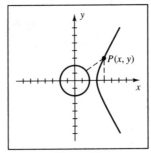

Figure 32

Note: See §11.4 for a discussion of the general solution outline for Exercises 33–35.

33 $D = (-8)^2 - 4(1)(16) = 0$, parabola

1) $\cot 2\varphi = \frac{15}{8}$; $\varphi \approx 14.04°$

2) $\cos 2\varphi = \frac{15}{17}$; $\sin \varphi = \frac{1}{17}\sqrt{17}$; $\cos \varphi = \frac{4}{17}\sqrt{17}$

3) $\begin{cases} x = \frac{4}{17}\sqrt{17}\,x' - \frac{1}{17}\sqrt{17}\,y' = \frac{1}{17}\sqrt{17}\,(4x' - y') \\ y = \frac{1}{17}\sqrt{17}\,x' + \frac{4}{17}\sqrt{17}\,y' = \frac{1}{17}\sqrt{17}\,(x' + 4y') \end{cases}$

4) $\frac{289}{17}(y')^2 - 51(x') = 0 \Rightarrow (y')^2 = 3(x')$

5) $V'(0, 0) \rightarrow V(0, 0)$

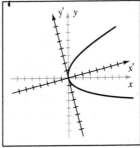

Figure 33

$\boxed{35}$ $D = (-10\sqrt{3})^2 - 4(11)(1) = 256 > 0$, hyperbola

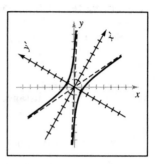

1) $\cot 2\varphi = -\frac{1}{3}\sqrt{3}$; $\varphi = 60°$

2) $\cos 2\varphi = -\frac{1}{2}$; $\sin \varphi = \frac{1}{2}\sqrt{3}$; $\cos \varphi = \frac{1}{2}$

3) $\begin{cases} x = \frac{1}{2}x' - \frac{1}{2}\sqrt{3}\,y' = \frac{1}{2}(x' - \sqrt{3}\,y') \\ y = \frac{1}{2}\sqrt{3}\,x' + \frac{1}{2}y' = \frac{1}{2}(\sqrt{3}\,x' + y') \end{cases}$

4) $-\frac{16}{4}(x')^2 + \frac{64}{4}(y')^2 = 20 \Rightarrow \dfrac{(y')^2}{\frac{5}{4}} - \dfrac{(x')^2}{5} = 1$

5) $V'(0, \pm\frac{1}{2}\sqrt{5}) \rightarrow V(\mp\frac{1}{4}\sqrt{15}, \pm\frac{1}{4}\sqrt{5})$

Figure 35

$\boxed{36}$ (a) $D = B^2 - 4AC = (-3)^2 - 4(2)(4) = -23 < 0$, ellipse

$\boxed{37}$ $x = r\cos\theta = 5\cos\frac{7\pi}{4} = 5\left(\frac{\sqrt{2}}{2}\right) = \frac{5}{2}\sqrt{2}$. $\quad y = r\sin\theta = 5\sin\frac{7\pi}{4} = 5\left(-\frac{\sqrt{2}}{2}\right) = -\frac{5}{2}\sqrt{2}$.

$\boxed{38}$ $r^2 = x^2 + y^2 = (2\sqrt{3})^2 + (-2)^2 = 16 \Rightarrow r = 4$.

$$\tan\theta = \frac{y}{x} = \frac{-2}{2\sqrt{3}} = -\frac{1}{\sqrt{3}} \Rightarrow \theta = \frac{11\pi}{6} \ \{\theta \text{ in QIV}\}.$$

$\boxed{39}$ $r = -4\sin\theta \Rightarrow r^2 = -4r\sin\theta \Rightarrow x^2 + y^2 = -4y \Rightarrow$

$$x^2 + y^2 + 4y + \underline{4} = \underline{4} \Rightarrow x^2 + (y+2)^2 = 4.$$

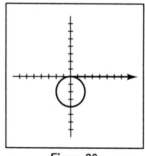

Figure 39

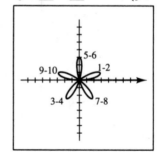

Figure 41

$\boxed{41}$ $r = 3\sin 5\theta$ is a 5-leafed rose. $\quad 0 = 3\sin 5\theta \Rightarrow \sin 5\theta = 0 \Rightarrow 5\theta = \pi n \Rightarrow \theta = \frac{\pi}{5}n$. The numbers 1–10 correspond to θ ranging from 0 to π in $\frac{\pi}{10}$ increments. One leaf is centered on the line $\theta = \frac{\pi}{2}$ and the others are equally spaced apart $\left(\frac{360°}{5} = 72°\right)$.

$\boxed{43}$ $r = 3 - 3\sin\theta$ is a cardioid since the coefficient of $\sin\theta$ has the same magnitude as the constant term.

$0 = 3 - 3\sin\theta \Rightarrow \sin\theta = 1 \Rightarrow \theta = \frac{\pi}{2} + 2\pi n$.

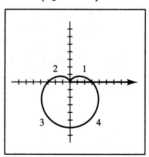

Variation of θ			Variation of r	
1)	0	$\rightarrow$ $\frac{\pi}{2}$	3 $\rightarrow$	0
2)	$\frac{\pi}{2}$	$\rightarrow$ π	0 $\rightarrow$	3
3)	π	$\rightarrow$ $\frac{3\pi}{2}$	3 $\rightarrow$	6
4)	$\frac{3\pi}{2}$	$\rightarrow$ 2π	6 $\rightarrow$	3

Figure 43

45 $r^2 = 9 \sin 2\theta$ •

$0 = 9 \sin 2\theta \Rightarrow \sin 2\theta = 0 \Rightarrow 2\theta = \pi n \Rightarrow \theta = \frac{\pi}{2}n.$

Variation of θ	Variation of r
1) $0 \rightarrow \frac{\pi}{4}$	$0 \rightarrow \pm 3$
2) $\frac{\pi}{4} \rightarrow \frac{\pi}{2}$	$\pm 3 \rightarrow 0$
3) $\frac{\pi}{2} \rightarrow \frac{3\pi}{4}$	undefined
4) $\frac{3\pi}{4} \rightarrow \pi$	undefined

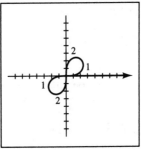

Figure 45

46 $2r = \theta \Rightarrow r = \frac{1}{2}\theta$. Positive values of θ yield the "counterclockwise spiral" while the "clockwise spiral" is obtained from the negative values of θ.

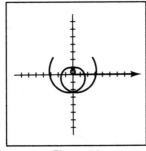

Figure 46

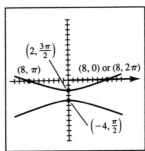

Figure 47

47 $r = \dfrac{8}{1 - 3 \sin \theta} \Rightarrow e = 3 > 1$, hyperbola. See §11.6 for more details on this problem.

$$V\left(2, \tfrac{3\pi}{2}\right) \text{ and } V'\left(-4, \tfrac{\pi}{2}\right). \quad d(V, F) = 2 \Rightarrow F'\left(-6, \tfrac{\pi}{2}\right).$$

48 $r = 6 - r \cos \theta \Rightarrow r + r \cos \theta = 6 \Rightarrow r(1 + \cos \theta) = 6 \Rightarrow r = \dfrac{6}{1 + 1 \cos \theta} \Rightarrow$

$e = 1$, parabola. The vertex is in the $\theta = 0$ direction, $V(3, 0)$.

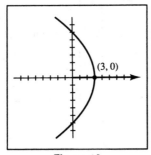

Figure 48

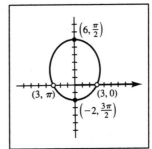

Figure 50

50 $r = \dfrac{-6 \csc \theta}{1 - 2 \csc \theta} \cdot \dfrac{-\sin \theta}{-\sin \theta} = \dfrac{6}{2 - \sin \theta} = \dfrac{3}{1 - \frac{1}{2}\sin \theta} \Rightarrow e = \frac{1}{2} < 1$, ellipse.

$V\left(2, \tfrac{3\pi}{2}\right)$ and $V'\left(6, \tfrac{\pi}{2}\right).$ $d(V, F) = 2 \Rightarrow F' = \left(4, \tfrac{\pi}{2}\right).$

Since the original equation is undefined when $\csc \theta$ is undefined,

the points $(3, 0)$ and $(3, \pi)$ are excluded from the graph.

$\boxed{51}$ $y^2 = 4x \Rightarrow r^2 \sin^2\theta = 4r \cos\theta \Rightarrow r = \dfrac{4r \cos\theta}{r \sin^2\theta} = 4 \cdot \dfrac{\cos\theta}{\sin\theta} \cdot \dfrac{1}{\sin\theta} \Rightarrow r = 4\cot\theta \csc\theta.$

$\boxed{54}$ $x^2 + y^2 = 2xy \Rightarrow r^2 = 2r^2 \cos\theta \sin\theta \Rightarrow 1 = 2\sin\theta \cos\theta \Rightarrow \sin 2\theta = 1 \Rightarrow$

$\qquad 2\theta = \frac{\pi}{2} + 2\pi n \Rightarrow \theta = \frac{\pi}{4}, \frac{5\pi}{4}$ on $[0,\, 2\pi)$, which are the same lines.

In rectangular coordinates: $x^2 + y^2 = 2xy \Rightarrow x^2 - 2xy + y^2 = 0 \Rightarrow$

$\qquad\qquad\qquad\qquad\qquad (x-y)^2 = 0 \Rightarrow x - y = 0$, or $y = x$.

$\boxed{55}$ $r^2 = \tan\theta \Rightarrow x^2 + y^2 = \frac{y}{x} \Rightarrow x^3 + xy^2 = y.$

$\boxed{57}$ $r^2 = 4\sin 2\theta \Rightarrow r^2 = 4(2\sin\theta \cos\theta) \Rightarrow r^2 = 8\sin\theta \cos\theta \Rightarrow$

$\qquad\qquad\qquad r^2 \cdot r^2 = 8(r \sin\theta)(r \cos\theta) \Rightarrow (x^2 + y^2)^2 = 8xy.$

$\boxed{58}$ $\theta = \sqrt{3} \Rightarrow \tan^{-1}\left(\frac{y}{x}\right) = \sqrt{3} \Rightarrow \frac{y}{x} = \tan\sqrt{3} \Rightarrow y = (\tan\sqrt{3})\,x.$

Note that $\tan\sqrt{3} \approx -6.15$. This is a line through the origin making an angle of

$\qquad$ approximately $99.24°$ with the positive x-axis. The line is *not* $y = \frac{\pi}{3}x$.

$\boxed{60}$ $r^2 \sin\theta = 6\csc\theta + r\cot\theta \Rightarrow$

$\qquad\qquad r^2 \sin^2\theta = 6 + r\cos\theta$ { multiply by $\sin\theta$ to get $r^2 \sin^2\theta$ } $\Rightarrow y^2 = 6 + x$

$\boxed{61}$ $y = t - 1 \Rightarrow t = y + 1.$ $x = 3 + 4t = 3 + 4(y+1) = 4y + 7.$

$\qquad\qquad$ As t varies from -2 to 2, $(x,\, y)$ varies from $(-5,\, -3)$ to $(11,\, 1)$.

$\boxed{63}$ $x = \cos^2 t - 2 \Rightarrow x + 2 = \cos^2 t$; $y = \sin t + 1 \Rightarrow (y-1)^2 = \sin^2 t.$

$\sin^2 t + \cos^2 t = 1 = x + 2 + (y-1)^2 \Rightarrow (y-1)^2 = -(x+1).$ This is a parabola with

vertex at $(-1,\, 1)$ and opening to the left. $t = 0$ corresponds to the vertex and as t

varies from 0 to 2π, the point $(x,\, y)$ moves to $(-2,\, 2)$ at $t = \frac{\pi}{2}$, back to the vertex at

$t = \pi$, down to $(-2,\, 0)$ at $t = \frac{3\pi}{2}$, and finishes at the vertex at $t = 2\pi$.

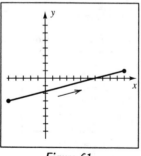

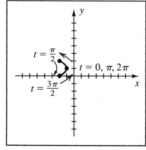

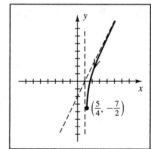

| Figure 61 | Figure 63 | Figure 65 |

$\boxed{65}$ $x = \frac{1}{t} + 1 \Rightarrow x - 1 = \frac{1}{t} \Rightarrow t = \dfrac{1}{x-1}$ and

$\qquad\qquad y = \frac{2}{t} - t = 2(x-1) - \left(\dfrac{1}{x-1}\right) = \dfrac{2(x^2 - 2x + 1) - 1}{x-1} = \dfrac{2x^2 - 4x + 1}{x-1}.$

This is a rational function with a vertical asymptote at $x = 1$ and an oblique

asymptote of $y = 2x - 2$. The graph has a minimum point at $\left(\frac{5}{4},\, -\frac{7}{2}\right)$ when $t = 4$ and

then approaches the oblique asymptote as t approaches 0.

66 All of the curves are a portion of the circle $x^2 + y^2 = 16$.

C_1: $y = \sqrt{16 - t^2} = \sqrt{16 - x^2}$.

Since $y = \sqrt{16 - t^2}$, y must be nonnegative and we have the top half of the circle.

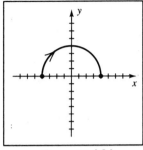

Figure 66 (C_1)

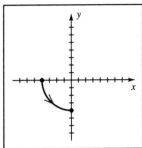

Figure 66 (C_2)

C_2: $x = -\sqrt{16 - t} = -\sqrt{16 - (-\sqrt{t})^2} = -\sqrt{16 - y^2}$.

This is the left half of the circle. Since $y = -\sqrt{t}$, y can only be nonpositive.

Hence we have only the third quadrant portion of the circle.

C_3: $x = 4\cos t$, $y = 4\sin t \Rightarrow \frac{x}{4} = \cos t$, $\frac{y}{4} = \sin t \Rightarrow \frac{x^2}{16} = \cos^2 t$, $\frac{y^2}{16} = \sin^2 t \Rightarrow$

$\frac{x^2}{16} + \frac{y^2}{16} = \cos^2 t + \sin^2 t = 1 \Rightarrow x^2 + y^2 = 16$. This is the entire circle.

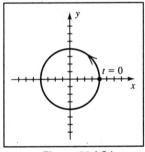

Figure 66 (C_3)

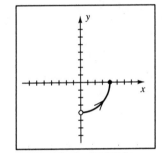

Figure 66 (C_4)

C_4: $y = -\sqrt{16 - e^{2t}} = -\sqrt{16 - (e^t)^2} = -\sqrt{16 - x^2}$.

This is the bottom half of the circle. Since e^t is positive, x takes on all positive

real values. Note that $(0, -4)$ is *not* included on the graph since $x \neq 0$.

Appendix I: Using Logarithmic and Trigonometric Tables

Note: For the solutions in this appendix, it is assumed that the reader is familiar with all the properties of the logarithm. The = sign is used for table values even though the case could be made that $\approx$ should be used. The values are not those that you would get using a calculator, and sufficient work is shown as to how the answers were arrived at.

1. $\log 3.47 \times 10^k = 0.5403 + k$; $k = 2, -3, 0 \Rightarrow 2.5403, 7.5403 - 10, 0.5403$

3. $\log 5.40 \times 10^k = 0.7324 + k$; $k = -1, 2, 5 \Rightarrow 9.7324 - 10, 2.7324, 5.7324$

5. $\log 6.02 \times 10^k = 0.7796 + k$; $k = 1, -5, 2 \Rightarrow 1.7796, 5.7796 - 10, 2.7796$

7. $\log (44.9)^n = n \log 44.9 = n(1.6522)$; $n = 2, \frac{1}{2}, -2 \Rightarrow 3.3044, 0.8261, 6.6956 - 10$

9. $\log (0.943)^n = n \log 0.943 = n(9.9745 - 10)$; $n = 3, -3, \frac{1}{3} \Rightarrow 9.9235 - 10, 0.0765,$
$$9.9915 - 10$$

11. $\log (638)(17.3) = \log 638 + \log 17.3 = 2.8048 + 1.2380 = 4.0428$

13. $\log \dfrac{(47.4)^3}{(29.5)^2} = 3 \log 47.4 - 2 \log 29.5 = 3(1.6758) - 2(1.4698) = 2.0878$

15. $\log \sqrt[3]{20.6} \, (371)^3 = \frac{1}{3} \log 20.6 + 3 \log 371 = \frac{1}{3}(1.3139) + 3(2.5694) = 8.1462$

17. $\log x = 3.6274 \Rightarrow x = 4.24 \times 10^3 = 4240$

19. $\log x = 0.9469 \Rightarrow x = 8.85 \times 10^0 = 8.85$

21. $\log x = 5.2095 \Rightarrow x = 1.62 \times 10^5 = 162{,}000$

23. $\log x = 9.7348 - 10 \Rightarrow x = 5.43 \times 10^{-1} = 0.543$

25. $\log x = 8.8306 - 10 \Rightarrow x = 6.77 \times 10^{-2} = 0.0677$

27. $\log x = 2.2765 \Rightarrow x = 1.89 \times 10^2 = 189$

29. $\log x = -1.6253 = 8.3747 - 10 \Rightarrow x = 2.37 \times 10^{-2} = 0.0237$

31. Log $2.54 = 0.4048$ and $\log 2.55 = 0.4065$. The difference between 2.54 and 2.55 is 0.01. It is helpful to think of this difference as 10 units. We want eight-tenths of this difference. The difference between 0.4048 and 0.4065 is 0.0017. The log of 25.48 will then be 1.4048 plus $\frac{8}{10}$ of 0.0017. We will use the following shorter version to compute logarithms involving interpolation.
$$\log 25.48 = 1.4048 + (0.8)(0.0017) = 1.4062$$

33. $\log 5363 = 3.7292 + (0.3)(0.0008) = 3.7294$

35. $\log 0.001259 = 7.0969 - 10 + (0.9)(0.0035) = 7.1001 - 10$

37. $\log 123{,}400 = 5.0899 + (0.4)(0.0035) = 5.0913$

39. $\log 0.7786 = 9.8910 - 10 + (0.6)(0.0005) = 9.8913 - 10$

$\boxed{41}$ $\log 384.7 = 2.5843 + (0.7)(0.0012) = 2.5851$

$\boxed{43}$ $\log 0.9462 = 9.9759 - 10 + (0.2)(0.0004) = 9.9760 - 10$

$\boxed{45}$ $\log 66{,}590 = 4.8228 + (0.9)(0.0007) = 4.8234$

$\boxed{47}$ $\log 0.04321 = 8.6355 - 10 + (0.1)(0.0010) = 8.6356 - 10$

$\boxed{49}$ $\log 3.003 = 0.4771 + (0.3)(0.0015) = 0.4776$

$\boxed{51}$ $\log x = 1.4437 \Rightarrow x = 27.7 + \frac{12}{15}(0.1) = 27.78$

$\boxed{53}$ $\log x = 4.6931 \Rightarrow x = 49{,}300 + \frac{3}{9}(100) = 49{,}330$

$\boxed{55}$ $\log x = 9.1664 - 10 \Rightarrow x = 0.146 + \frac{20}{29}(0.001) = 0.1467$

$\boxed{57}$ $\log x = 3.8153 - 6 \Rightarrow x = 0.00653 + \frac{4}{7}(0.00001) = 0.006536$

$\boxed{59}$ $\log x = 2.3705 \Rightarrow x = 234 + \frac{13}{19}(1) = 234.7$

$\boxed{61}$ $\log x = 0.1358 \Rightarrow x = 1.36 + \frac{23}{32}(0.01) = 1.367$

$\boxed{63}$ $\log x = 8.9752 - 10 \Rightarrow x = 0.0944 + \frac{2}{4}(0.0001) = 0.09445$

$\boxed{65}$ $\log x = 5.0409 \Rightarrow x = 109{,}000 + \frac{35}{40}(1000) = 109{,}900$

$\boxed{67}$ $\log x = -2.8712 = 7.1288 - 10 \Rightarrow x = 0.00134 + \frac{17}{32}(0.00001) = 0.001345$

$\boxed{69}$ $\log x = -0.6123 = 9.3877 - 10 \Rightarrow x = 0.244 + \frac{3}{18}(0.001) = 0.2442$

$\boxed{71}$ $\sin 0.46 = 0.4436 + \frac{4}{29}(0.0026) = 0.4440$

$\boxed{73}$ $\tan 3 = -\tan(\pi - 3) = -\tan 0.1416 = -\left[0.1405 + \frac{20}{29}(0.0030)\right] = -0.1426$

$\boxed{75}$ $\sec \frac{1}{4} = 1.031 + \frac{27}{29}(0.001) = 1.032$

$\boxed{77}$ $\cos 37°43' = 0.7916 - \frac{3}{10}(0.0018) = 0.7911$

$\boxed{79}$ $\cot 62°27' = 0.5243 - \frac{7}{10}(0.0037) = 0.5217$

$\boxed{81}$ $\csc 16°55' = 3.453 - \frac{5}{10}(0.033) = 3.437$

$\boxed{83}$ $\cos t = 0.8620 \Rightarrow t = 0.5294 + \frac{11}{15}(0.0029) = 0.5315$

$\boxed{85}$ $\tan t = 4.501 \Rightarrow t = 1.3497 + \frac{52}{62}(0.0029) = 1.3521$

$\boxed{87}$ $\csc t = 1.436 \Rightarrow t = 0.7679 + \frac{4}{5}(0.0030) = 0.7703$

$\boxed{89}$ $\sin \theta = 0.3672 \Rightarrow \theta = 21°30' + \frac{7}{27}(10') = 21°33'; \quad 180° - 21°33' = 158°27'$

$\boxed{91}$ $\tan \theta = 0.5042 \Rightarrow \theta = 26°40' + \frac{20}{37}(10') = 26°45'; \quad 180° + 26°45' = 206°45'$

$\boxed{93}$ $\cos \theta = 0.3465 \Rightarrow \theta = 69°40' + \frac{10}{27}(10') = 69°44'; \quad 360° - 69°44' = 290°16'$

$\boxed{95}$ $\sec \theta = 1.385 \Rightarrow \theta = 43°40' + \frac{3}{4}(10') = 43°48'; \quad 360° - 43°48' = 316°12'$